Nonlinear Microwave Signal Processing: Towards a New Range of Devices

NATO ASI Series

Advanced Science Institutes Series

A Series presenting the results of activities sponsored by the NATO Science Committee, which aims at the dissemination of advanced scientific and technological knowledge, with a view to strengthening links between scientific communities.

The Series is published by an international board of publishers in conjunction with the NATO Scientific Affairs Division

A	**Life Sciences**	Plenum Publishing Corporation
B	**Physics**	London and New York
C	**Mathematical and Physical Sciences**	Kluwer Academic Publishers
D	**Behavioural and Social Sciences**	Dordrecht, Boston and London
E	**Applied Sciences**	
F	**Computer and Systems Sciences**	Springer-Verlag
G	**Ecological Sciences**	Berlin, Heidelberg, New York, London,
H	**Cell Biology**	Paris and Tokyo
I	**Global Environmental Change**	

PARTNERSHIP SUB-SERIES

1.	**Disarmament Technologies**	Kluwer Academic Publishers
2.	**Environment**	Springer-Verlag / Kluwer Academic Publishers
3.	**High Technology**	Kluwer Academic Publishers
4.	**Science and Technology Policy**	Kluwer Academic Publishers
5.	**Computer Networking**	Kluwer Academic Publishers

The Partnership Sub-Series incorporates activities undertaken in collaboration with NATO's Cooperation Partners, the countries of the CIS and Central and Eastern Europe, in Priority Areas of concern to those countries.

NATO-PCO-DATA BASE

The electronic index to the NATO ASI Series provides full bibliographical references (with keywords and/or abstracts) to more than 50000 contributions from international scientists published in all sections of the NATO ASI Series.
Access to the NATO-PCO-DATA BASE is possible in two ways:

– via online FILE 128 (NATO-PCO-DATA BASE) hosted by ESRIN,
Via Galileo Galilei, I-00044 Frascati, Italy.

– via CD-ROM "NATO-PCO-DATA BASE" with user-friendly retrieval software in English, French and German (© WTV GmbH and DATAWARE Technologies Inc. 1989).

The CD-ROM can be ordered through any member of the Board of Publishers or through NATO-PCO, Overijse, Belgium.

3. High Technology – Vol. 20

Nonlinear Microwave Signal Processing: Towards a New Range of Devices

Proceedings of the III International Workshop Nonlinear Microwave Magnetic and Magnetooptic Information Processing

edited by

Romolo Marcelli
National Research Council of Italy,
Rome, Italy

and

Sergei A. Nikitov
Russian Academy of Sciences,
Moscow, Russia

Springer-Science+Business Media, B.V.

Proceedings of the NATO Advanced Research Workshop on
Nonlinear Microwave Signal Processing: Towards a New Range of Devices
Rome, Italy
3–6 October, 1996

A C.I.P. Catalogue record for this book is available from the Library of Congress

ISBN 978-94-010-6407-1   ISBN 978-94-011-5708-7 (eBook)
DOI 10.1007/978-94-011-5708-7

Printed on acid-free paper

To Our Parents

Romolo Marcelli and Sergei A. Nikitov

CONTENTS

Preface.. xi
Introduction.. xiii

Chapter I. Classical and Novel Nonlinear Effects, Applications in Different Physical Systems.

1. Notes on the Problem of Magnetization Reversal........................... 3
Rodrigo Arias, H. Neil Bertram, and Harry Suhl
Center for Magnetic Recording Research and Physics Department
University of California, San Diego, USA

2. An Overview of Nonlinear Microwave and Millimeter Wave Generation in Magnetic, Acoustic and Electromagnetic Distributed Nonlinear Physical Systems.. 13
Mircea Dragoman
Research Institute of Electron Devices, Bucarest, ROMANIA
Daniela Dragoman
University of Bucharest, Physics Department, ROMANIA

3. Inhomogeneous Internal Field Distribution in Planar Microwave Ferrite Devices... 45
Martha Pardavi-Horvath and Guobao Zheng
Institute of Magnetics Research, George Washington University, USA

4. Design of Nonlinear Transmission Lines: GaAs and Magnetic Film Devices... 71
Romolo Marcelli and Paolo De Gasperis
Istituto di Elettronica dello Stato Solido del CNR, Roma, ITALY
Giancarlo Bartolucci and Fabrizio Pini
Università di Roma "Tor Vergata", Dipartimento di Ingegneria Elettronica, ITALY
Mircea Dragoman
Research Institute of Electron Devices, Bucarest, ROMANIA

5. Nonlinear Dynamics of Optical Solitons....................................... 101
Stefan Wabnitz
Fondazione Ugo Bordoni, Roma, ITALY

Chapter II. Spin Wave Instabilities.

6. Theory of Spin-Wave Interactions in Heisenberg Ferromagnetic Thin
Films... 121
M.G. Cottam and N.J. Zhu
University of Western Ontario, Physics Department, London, CANADA

7. Kinetic Instability and Bose Condensation of Magnons - The Sources of
Controlled Microwave Emission from Magnetic Crystals............................ 139
G.A. Melkov and A.Yu. Taranenko
Faculty of Radiophysics, Taras Shevchenko Kiev University, UKRAINE

8. Exchange Spin Waves in Nonuniform Films of Yttrium Iron
Garnet.. 165
A.G. Temiryazev, M.P. Tikhomirova, and P.E. Zilberman
Institute of Radioengineering & Electronics
Russian Academy of Sciences, Moscow, RUSSIA

9. Parametric Instability of Spin Waves in Ferromagnets under a Spatially
Localized Longitudinal Magnetic Pump Field... 213
Yu.V. Gulyaev, A.V. Lugovskoi and P.E. Zilberman
Institute of Radioengineering & Electronics
Russian Academy of Sciences, Moscow, RUSSIA

10. Spatial Nonuniformity and Spin Wave Turbulence in
Antiferromagnet... 253
A.I. Smirnov
P.L. Kapitza Institute for Physical Problems, Moscow, RUSSIA

Chapter III. Solitons and Chaos.

11. Microwave Solitons in Magnetic Media: a Review of Fundamental Properties... 277

Allan D. Boardman, R.C.J. Putman, K. Xie, and H.M. Metha
Department of Physics, University of Salford, UNITED KINGDOM
Sergei A. Nikitov
Institute of Radioengineering & Electronics
Russian Academy of Sciences, Moscow, RUSSIA

12. Soliton-like Packets of Parametrically Coupled Spin-Waves.................. 305

A.F. Popkov
Zelenograd Research Institute of Physical Problems, Moscow, RUSSIA
N.V. Ostrovskaya
Moscow Institute of Electronic Technology, RUSSIA
L.L. Savchenko
Moscow State University, RUSSIA

13. Macroscopic Quantum Tunneling of Solitons in Ultrathin Films.. 325

A.K. Zvezdin
Institute of General Physics, Moscow, RUSSIA
V.V. Dobrovitski
Moscow State University, Physics Department, RUSSIA

14. Controllong Chaos in Thin YIG Films at Microwave Frequencies.......... 355

Derrick W. Peterman and Philip E. Wigen
The Ohio State University, Columbus, Ohio, USA

15. Suppressing and controlling Chaos in Spin-Wave Instabilities................ 381

T. Bernard, R. Henn, W. Just, E. Reibold, F. Rödelsperger and H. Benner
Institüt für Festkörperphysik, Techn. Hochschule Darmstadt, GERMANY

Chapter IV. Magneto-Optic Interaction: Non-linear Effects and Devices.

16. Applications of Magnetic Garnet Films in Optical Communication........   **411**
H. Dötsch, A. Erdmann, M. Fehndrich, R. Gerhardt, P. Hertel, B. Lührmann,
M. Shamonin, H.P. Winkler, and M. Wallenhorst
University of Osnabrück, GERMANY

**17. Interactions Between Optical Guided Modes and Nonlinear
Magnetostatic Waves..**   **467**
Daniel D. Stancil
*Dept. of Electrical and Computer Engineering, Carnagie Mellon University,
Pittsburgh, USA*
Anil Prabhakar
Department of Physics, Carnagie Mellon University, Pittsburgh, USA

**18. Integrated Magnetooptic Devices with Applications to RF Signal
Processing and Communications..**   **487**
Chen S. Tsai
*Dept. of Electrical and Computer Engineering and Institute for Surface and
Interface Sciences, University of California, Irvine, USA*

Subject Index..   **509**

Preface

This book is the result of the contributions coming from the more than thirty key speakers of the 3rd international Workshop on Nonlinear Microwave Magnetics held in Roma, Italy from the 3rd to the 6th of October 1995.

Since the 1990, in Ulyanovsk, when the Russian Academy of Sciences promoted the first Workshop of the series, the basic idea was to have a sort of Institutional Meeting collecting Scientists of the Magnetics Community devoted to Spin Wave Electronics at Microwave Frequencies. It was a succesful organization, and the birth of an effective interaction between eastern and western researchers overcame the meaning of the Workshop itself.

Three years later, in Irvine, California, 1993, the Spin Wave Community was joined again. It was clear that the growing interest on hot topics of Nonlinear Microwave Magnetics involving both, applicative and fundamental aspects of microwave magnetic media, favoured the organization of further meetings on the same subject. So far, during the social dinner, in the middle between a serious proposal and the joke encouraged by the Californian Wine, Roma was proposed as the third place for the Workshop. Day after day, the joke became serious, and it was possible to solve the financial and logistic problems in time for the predicted deadline.

There was, in fact, a sentimental reason for having the meeting on 1995, because fourty years ago before the Workshop it has began the seminal work of Professor Harry Suhl, University of California, concerning the nonlinear ferromagnetic resonance in magnetic media, which became a milestone for all the subsequent investigations. For above reason, Professor H. Suhl was the Honorary Chairman of the Third International Workshop on Nonlinear Microwave Magnetics and Magnetooptical Signal Processing.

As a remarkable and well accepted idea, the Workshop participation was extended to scientists working on semiconductors and optical media and also a poster session was established to encourage further exchanges. It was an effort to unify the language of people having different background, formal approaches, measurement verifications and applications, but working on quite similar phenomena.

Now, as a final act of that memorable event, this book comes as the first available published contribution of this kind of Workshop. In the spirit of the books resulting from NATO organizations, it represents not only a collection of papers, merely a sort of proceedings, but it is in principle a reference book of up-dated topics in the framework of Non-linear Microwave Magnetic and Magneto-optic Materials and Devices.

Our wish in presenting this collection of papers is in having the first, but we hope not the last, written contribution involving the major part of Scientists working in this area xii

We are grateful to:

- The NATO Scientific Affairs Division for funding this initiative;
- The National Research Council of Italy (CNR) for providing the place of the Workshop;
- Dr. Paolo De Gasperis, Director of the Istituto di Elettronica dello Stato Solido del CNR for his encouragement and helpful advices in the Workshop Organization, and for partially covering the Workshop management expenses;
- Mr. Luigi Maita, responsible of the Accounts Department of the Istituto di Elettronica dello Stato Solido del CNR for his availability and experience in solving financial matters;
- Mrs. Claudia Fraiegari for her kind participation as the Workshop Secretary;
- the Workshop Scientific Community, for Her prompt cooperation, which has allowed us to respect the tight time schedule for the organization.

Roma, 2ⁿᵈ of June 1996

Romolo Marcelli and Sergei A. Nikitov

Introduction

Objective of the Workshop was to provide a comprehensive picture of the state of the art and future applications in microwave and millimeter wave signal processing by means of magnetic, magnetooptic and dielectric media. The main focus in this book is on the use of Nonlinear Microwave Magnetics and exploitation of novel magnetooptic phenomena for the information processing.

Presently, the field of Microwave Magnetics is part of the larger technological area of Magnetics, and regular international conferences support many areas of investigations on magnetic media. During the last few years, Microwave Magnetics and Magneto-Optic Information Processing become extremely important and essential topics of different International Meetings and Workshops of the Magnetics Community have been dedicated to the wide possibilities to use the results of investigations in this area for different applications.

Microwave Magnetics is, in fact, a growing and strategically important area with the drive toward higher frequencies, and there exists the need for non-linear, non-reciprocal components for the microwave market applications. From a broad perspective, Microwave Magnetics includes: ferromagnetic resonance, magneto-optics, magnetic substrate combinations in strip line type waveguides, hexagonal ferrites, solitons, ferrite-ferroelectric and ferrite-high T_C-superconductor sandwich structures. Over the next 10 years these areas are expected to evolve into useful technologies.

Accompanying the drive toward higher frequencies, new phenomena and mechanisms become attractive for device integration. One such mechanism involves generation and control of cyclotron resonances in high electron mobility device structures. This raises the possibility of a new class of non-reciprocal components authomatically integrated into semiconductor, monolithic circuits. Solitons represent another such mechanism. They involve non-linear response in the form of propagating non-perturbed short pulses. Current vortices in high T_C superconductors can also have millimeter wave properties. Optical techniques are being increasingly utilized to meet the ever-growing data rate requirements of signal processing and communication applications.

Furthermore, Microwave Magnetics can be developed into many analog devices. Analog signal processing performs an essential function in many current radar and EW systems. The main performance advantage that analog devices have over their digital competitors is the ability to operate with wide instantaneous bandwidths and moderately high dynamic range at microwave frequencies.

All these and related topics are now in progress in scientific laboratories of different countries. The Workshop Organization was an occasion to exchange the scientific results in this fastly growing area of interest.

This book, which collects a selection of the topics discussed during the Workshop, is organized in chapters having the same names used for the sessions of the Workshop itself. Classical and novel non-linear effects, applications to microwave and magneto-optic devices have been included. Attention has been paid to both, theoretical and experimental aspects, and also to the design of prototype configurations useful for microwave signal processing.

Romolo Marcelli and Sergei A. Nikitov

Chapter I
Classical and Novel Nonlinear Effects, Applications in Different Physical Systems

NOTES ON THE PROBLEM OF MAGNETIZATION REVERSAL

RODRIGO ARIAS, H. NEIL BERTRAM, and HARRY SUHL
Center for Magnetic Recording Research and Physics Department
University of California San Diego, La Jolla, Ca, USA

1. Introduction

A perennial non-linear problem in magnetic storage technology is the problem of magnetization reversal. Its complete analytic solution has never been accomplished, and heavy reliance has had to be placed on computer simulations. Although these really are a type of experiment, they furnish some insight into the switching process, simply because they allow one to look at the motion of computer 'spins' as a function of time and of position, whereas the motion of the actual physical spin system can at best be studied by inference from data of limited resolution. Nevertheless, the computer simulations do not tell us why the spins move the way they do. They only help us to dismiss model theories that do not agree with them. The need for a good theory is becoming more and more apparent as magnetic recording pushes towards higher and higher data rates (a nanosecond time scale is the immediate goal), and ever higher recording densities.

Since an all-encompassing, 'heroic' solution of this problem is still out of the question, we must content ourselves with attacking the problem piece by piece. For example, before attacking the dynamics, we may try to determine the static energy landscape of the spin system. For general values of the switching field below a critical value, reversal is presumably initiated by thermally activated escape over one or more energy barriers.

2. The Energy Landscape

We therefore begin by determining these barriers for a particularly simple case: a linear chain of spins, interacting via dipole forces only (exchange could be added without any significant modification of the results). To simplify the problem further, we restrict the interaction to nearest neighbors; however, we retain the full angular symmetry of the dipolar coupling. For simplicity, we restrict rotation of the spins to a plane. (This corresponds to thin films, or to very flat coplanar particles). The energy of such a chain is evidently

R. Marcelli and S.A. Nikitov (eds.), Nonlinear Microwave Signal Processing: Towards a New Range of Devices, 3–11.
© 1996 *Kluwer Academic Publishers.*

4

$$E = \sum_i \sum_\delta [\cos(\theta_i - \theta_{i+\delta}) - 3\cos\theta_i \cos\theta_{i+\delta}] - H\cos\theta_i - \tfrac{1}{2}K\cos^2\theta_i \quad (1)$$

where the δ-summation is over the two nearest neighbors of the ith spin. Also, H, K are respectively the applied field and the anisotropy field (shape or crystalline). These fields are measured in units of the dipole field, and distances are measured in units of spacing between the spins. For positive K, the easy axis is along the chain. The stationary states are obtained by equating all derivatives of E with respect to each θ_i to zero. This gives a set of nonlinear recursion relations:

$$2\sin\theta_i(\cos\theta_{i+1} + \cos\theta_{i-1}) + \cos\theta_i(\sin\theta_{i+1} + \sin\theta_{i-1}) + (H + K\cos\theta_i)\sin\theta_i = 0$$

Their solution requires the use of the computer, but only in a trivial way. Assuming the chain to be semi-infinite, the results are as follows:

2.1 SOLUTIONS THAT PERVADE THE ENTIRE CHAIN

All these are periodic (we have not found any chaotic solutions). There are l-cycle, 2-cycle, 3-cycle etc. solutions. By l-cycle is meant the solution with all $\cos\theta$ equal (other than the solution $\theta=0$, which is, of course, the ground state). There are two such solutions; one with all angles equal and in the same quadrant, the other with successive angles still having the same cosine, but with $\sin\theta$ alternating in sign, so that successive angles are reflections of each other in the x-axis. For the first of these, we have is $\theta=\theta_a$, where

$$\cos\theta_a = -\frac{H}{6+K} \qquad (3)$$

For the second one we have $\theta=\theta_b$, where:

$$\cos\theta_b = -\frac{H}{2+K} \qquad (4)$$

For an isolated spin, these cosine would equal $-H/K$, and would give the angular position of the barrier height due to the anisotropy. Evidently, according to formula 3, the dipole interaction has increased the anisotropy from K to $K + 6$ and to $K + 2$ respectively. Substituting the result 3 in the formula for the energy gives the energy as a function of the fields (fig. 1 for formula 3, and fig.2 for formula 4). Evidently, the second case has lower energy.

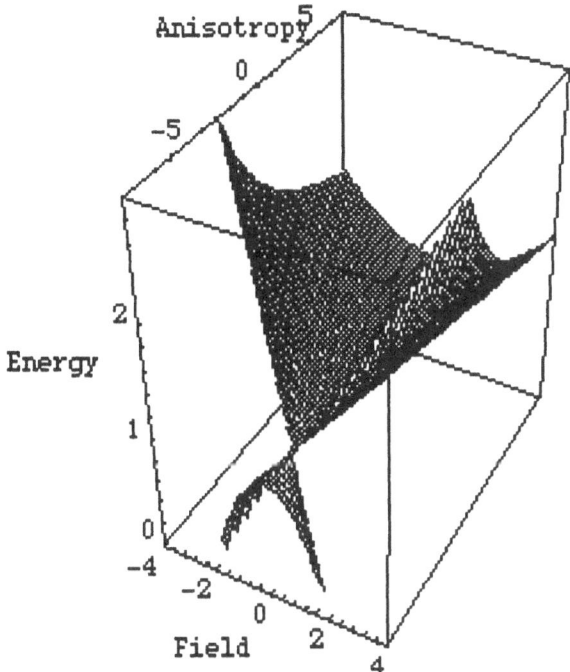

Figure 1

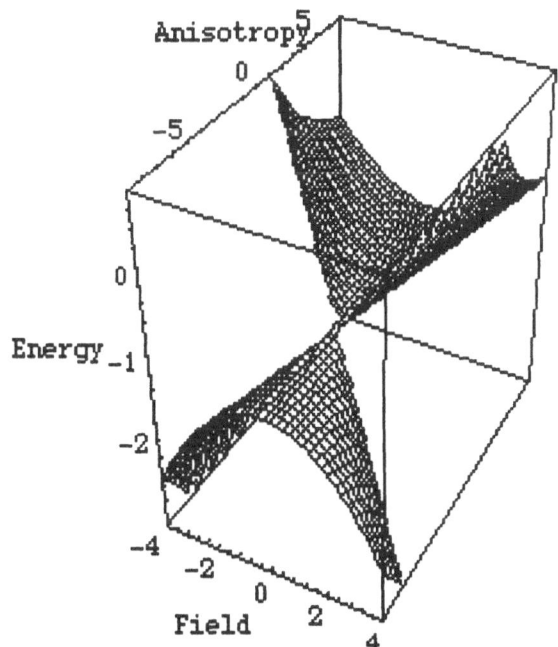

Figure2

6

For 2-cycle solutions, the value of the angle, as well as its cosine, alternates between two different values θ_1, θ_2. Writing $\cos\theta_1 = x$, $\cos\theta_2 = y$, we can represent the fields parametrically as:

$$H = -4(x + y) + \frac{2xy}{y - x}\sigma_1\sigma_2\left(\sqrt{\frac{1-x^2}{1-y^2}} - \sqrt{\frac{1-y^2}{1-x^2}}\right) = (x + y)\left(-4 + \frac{2xy\sigma_1\sigma_2}{\sqrt{(1-x^2)(1-y^2)}}\right)$$

$$K = 4 + \frac{2}{y - x}\sigma_1\sigma_2\left(x\sqrt{\frac{1-y^2}{1-x^2}} - y\sqrt{\frac{1-x^2}{1-y^2}}\right) = 4 - 2\frac{(1+xy)\sigma_1\sigma_2}{\sqrt{(1-x^2)(1-y^2)}}$$

(5)

Here σ's is either +1 or -1 according to whether the sign of $\sin\theta$ is the same or opposite to that of the cosine. Substitution in expression 1 then gives the energy surface in parametric form, shown for the case $\sigma_1\sigma_2 = 1$ in figure 3.

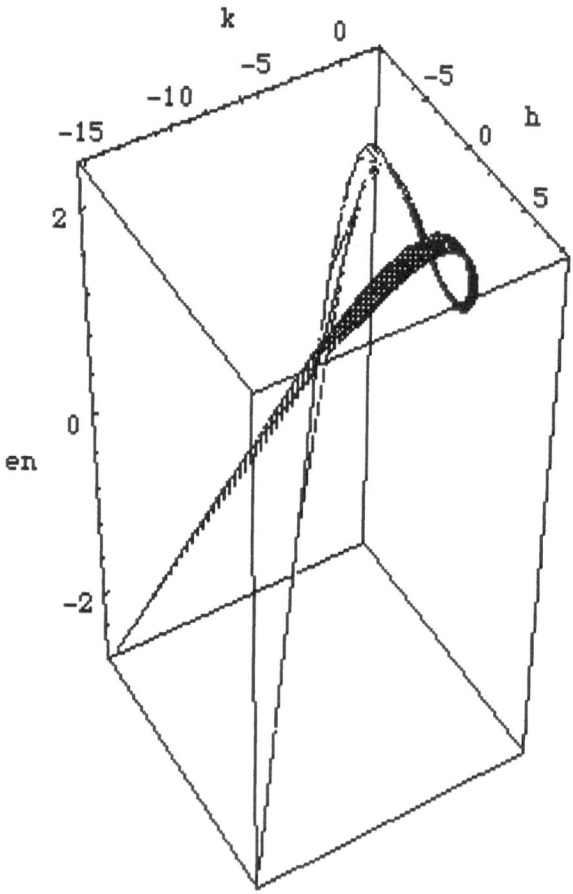

Figure 3

The case $\sigma_1\sigma_2=-1$ is shown in figure 4.

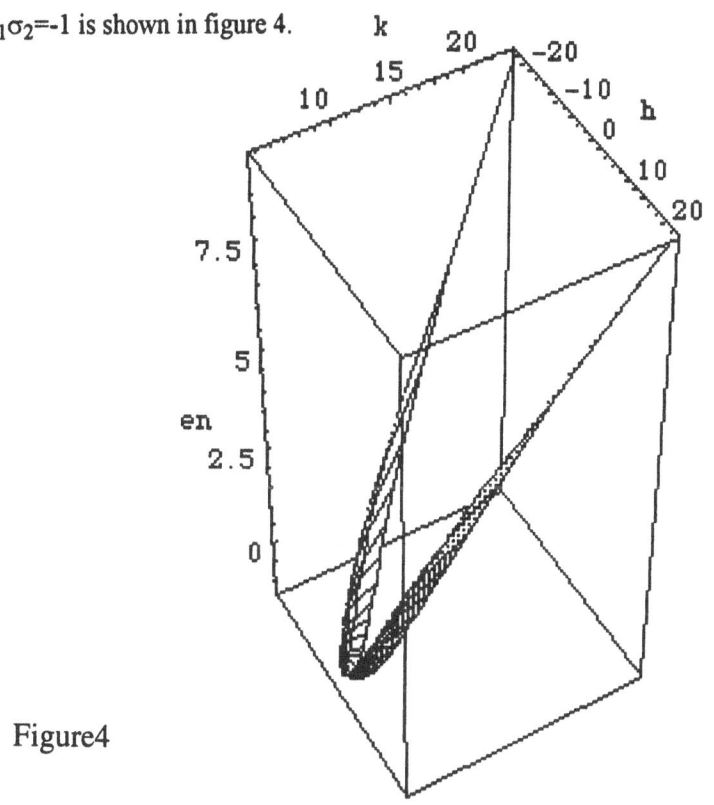

Figure4

3-cycles and higher cycles are much harder to evaluate. However, they all have more or less tha same ribbon like appearence in E, H, K space.

A calculation of the second derivatives of the energy, and solution of the secular equation shows that all the stationary states discussed here are saddle points of various orders and are therefore dynamically unstable. We have not been able to Prove this in the general case of the n-cycle, but feel that it is in fact true. This is important; it enables one to dismiss in this particular case the popular notion that a complex system has a very large number of metastable states, in which the system can get trapped and cannot reach the ground state within a reasonable time. However, if imperfections are admitted into the chain, metastable states might arise.

2.2 SOLUTIONS CONFINED TO THE BEGINNING OF THE CHAIN

In the infinite, uniform chain, only the stationary states enumerated above exist .

For a semi-infinite chain, however, there are also some edge-states at the beginning of the chain. These extend over only a few spins (typically two to ten for the cases we have studied numerically. They also differ from the extended states in that the angular excursions are generally smaller. For modest deviations of two initial θ's from zero (two values are needed, since the recursion relation 2 is second order), the solutions for θ_i become complex for quite small values of the subscript i. Naturally the energy of these states, while still greater than the ground state energy, is smaller by a factor (length of chain)$^{-1}$ than the energy of the extended states. It is not easy to map out the basin of attraction within which the initial values θ_1, θ_2 must lie if these local states, rather than the extended states, are to be found. A crude idea of that basin can be formed by considering just the first non-trivial recursion relation and to determine in what region of the θ_1, θ_2 plane, θ_3 becomes complex (see fig 5). (The actual basin, of course, will be a bit bigger; in particular it will engulf the origin completely). The contours in figure 5 are labeled with the value of $\cos\theta_3$, the axes are θ_1, and θ_2,. Values of these variables within the black area belong only to localized states.

These results have a counterpart in the linearized approximation to equation 2. That linearized version is

$$\delta\theta_{n+2} + h\delta\theta_{n+1} + \delta\theta_n = 0$$

where

$$h = H + K + 4$$

The solution has the form

$$\delta\theta_n = A\gamma_+^{\,n} + B\gamma_-^{\,n}$$

where

$$\gamma_\pm = -\tfrac{1}{2}h \pm \sqrt{\tfrac{1}{4}h^2 - 1}$$

If $|h| < 2$, the γ's are unimodular complex numbers, and the stationary state oscillates and is extended. For $|h| > 2$, that is for $H > -K - 2$, or else for $H < -K - 6$, the disturbance decays along the chain (the growing root must be rejected for the usual reasons). In the full nonlinear theory this corresponds to the state localized at the beginning of the chain.

3. Relation to More Realistic Structures

One may hope that these conclusions have some limited relevance to switching of an actual magnetic sample. Thus, the analog of switching via the 1-cycle barrier state would be Stoner-Wohlfarth uniform rotation of the entire magnetization. The higher cycle states would correspond to switching via various buckling modes. In the present model, there is no analog of the curling modes, because spin rotation has been confined to a plane.

The analog of the localized states would be surface states in an actual sample. There, too, these states would present a much lower barrier to switching than the extended states, the 'rest' of the magnetization will then switch by front propagation (i.e. domain wall motion). These matters will be the subject of a forthcoming publicationl. Note that imperfections, just like the surface, can result in localized stationary states, so that a multi-barrier treatment is still indicated for samples with ample defects.

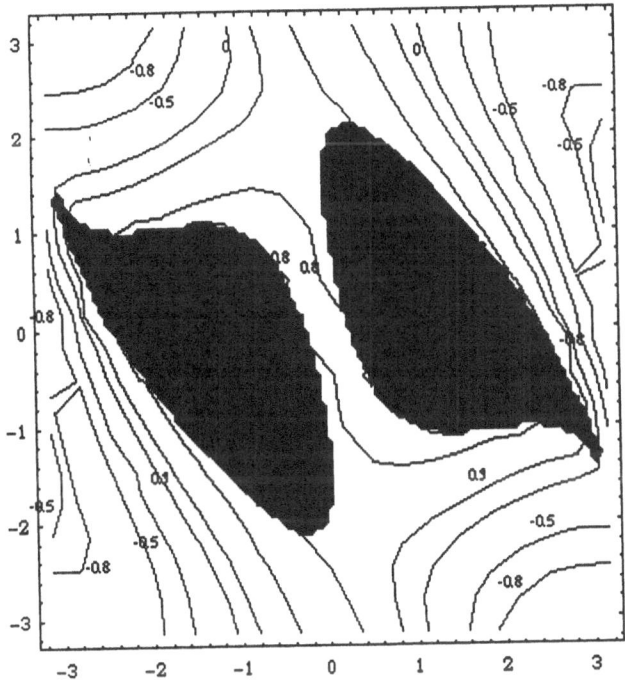

Figure 5

4. Switching in the Presence of Many Barriers.

Front propagation is probably not the rate-limiting step in the magnetization reversal of a realistic imperfect sample. Therefore it is reasonable to separately examine the question of switching impeded by many barriers.

For a single barrier of height E, the escape rate v should obey an Arrhenius formula:

$$v = \omega e^{-E_B/kT} \tag{6}$$

where ω is some attempt frequency into whose nature we do not inquire at this point.

If n_0, n_1 are the probabilities of finding the system in a metastable state (O) to the ground state or another, lower, metastable state (1), the two master equations governing these two probabilities give for n_0, i.e. for the probability of not having switched after time t

$$n_0(t) = A\{e^{-(E_0-E_1)/kT} + \exp-(vt/A)\}, \quad A = (1+e^{-(E_0-E_1)/kT})^{-1} \tag{7}$$

This is also essentially the magnetization remaining in the initial direction after time t.

If $E_1 \ll E_0$, so that the reverse barrier is much higher than the forward barrier, this reduces to e^{-vt}. The observations seem to give a linear decay of the curve of n_0, versus $\log t$ over several decades. Clearly a single exponential cannot deliver this result. To approximately fit the data, it has been suggested that several decay processes must be going on in parallel, so that one should write (in the case of high reverse barriers)

$$n_0 = \sum_k W_k e^{-v_k t} \tag{8}$$

all the weights W_k, being positive, and $\sum_k W_k = 1$. Accordingly,

$$\frac{dn_0}{d\log t} = -\ln(10)\sum_k W_k v_k t e^{-v_k t} \tag{9}$$

Unfortunately the absolute value of the right hand side has a maximum, which cannot exceed $e^{-1}\log(10) \cong 0.85$, because xe^{-x} is maximal at $x=1$, where its value is e^{-1}.

By contrast, the measurements reported in literature give a slope of about 1.4 near the coercive field, i. e. almost twice as high as the simple formula 8.

Two of the authors (R.A. and H.N.B.) have proposed that this difficulty may be overcome by a more detailed solution of the master equation that allows for the Dossibility that some of the barriers may have to be surmounted in series rather than in parallel. With barriers in series, the probability of not switching obviously becomes enhanced . The enhancement becomes relatively easy to calculate when the barriers are all equal, but are much higher for reverse than for forward motion (respectively away from, and towards, the global minimum). Further, detailed analysis shows that not only the probability of not switching, but also its maximum time rate of change is enhanced. This appears in the experiments as a steepening of the dM/$d$log$t$ plot.

5. Relation of High Data Rate Recording to Front Propagation and Soliton-like Motion

Whereas in ordinary materials, front propagation may not be the rate limiting step, it is most certainly an upper limit on the rate for perfect magnetic systems. In view of current pressures to maximize recording rates it is therefore worth while to extend the above analysis. Extending it to higher dimensional arrays will involve greater computational effort, but will probably not produce major qualitative changes. Conceivably, some surprises will result from inclusion of precessional effects in the analysis. Perhaps the most important problem to be addressed will be the interaction of fronts under actual recording conditions, such as during inscription of di-bits, etc. Hopefully we will be able to profit from current knowledge of such nonlinear interactions in other areas of communications technology.

AN OVERVIEW OF NONLINEAR MICROWAVE AND MILLIMETER WAVE GENERATION IN MAGNETIC, ACOUSTIC AND ELECTROMAGNETIC DISTRIBUTED NONLINEAR PHYSICAL SYSTEMS

M. DRAGOMAN
Res. Inst. Electron Devices
Str. Erou Iancu Nicolae 32B
Bucharest 72996, Romania

D. DRAGOMAN
Univ. of Bucharest, Physics Dept.
P.O.Box MG-63, Bucharest, Romania

1. Introduction

It was generally believed by the scientific community that the nonlinear behaviour of the electromagnetic field in different media can be used only in the optical range of frequencies for device implementations and other specific applications. Nonlinear optics is nowadays a mature branch of applied optics and its principles and applications are found in any book of optics or in books dedicated only for nonlinear optics [1], [2].

Up to recently, the nonlinear properties of the electromagnetic field have been poorly studied or applied for device realizations at the frequency range lower than optical frequencies. The active and nonlinear wave propagation devices at microwave and millimeter frequencies have been reviewed in 1970 by Scott [3], in 1985 by Jäger [4] and in 1994 by Rodwell [5]. Despite the huge progress realized in electronics due to the utilization of the nonlinear electromagnetic wave propagation devices, i.e. the fastest electronic devices ever known [5], a branch of microwave science called *Nonlinear Microwaves* does not exists as in the case of the optical science

The aim of this review paper is to demonstrate that Nonlinear Microwaves, the analogue of the Nonlinear Optics, is an active and developing area of the applied sciences and which has shown impressive results in the last years. As a result important advances of the microwave and millimeter wave technology have been evidenced in the area of device applications and basic research. Another reason for such a paper is that while in the area of nonlinear waves and solitons at optical frequencies there are excellent books [2], [6], [7] written on this subject, in that of microwave and millimeter frequencies there is no book and no review paper having a broad view on the nonlinear microwave and millimeter wave devices and applications.

R. Marcelli and S.A. Nikitov (eds.), Nonlinear Microwave Signal Processing: Towards a New Range of Devices, 13–43.
© 1996 *Kluwer Academic Publishers.*

In the last mentioned area there are some excellent review papers, some of them just mentioned, but treating only specific domains of interest on this subject. The role of this paper is to review the results on nonlinear devices and applications in the area of microwave acoustics, microwave magnetics, microwave nonlinear transmission line circuits and high temperature superconductor devices, partially separately reviewed up to now. Based on all these important but disparate results the paper shows that their unification in a single and very promising branch of applied science, i.e. *Nonlinear Microwaves*, is possible and desirable for further progresses in the area of microwave and millimeter wave technologies.

One can argue that although important realizations were made in the area of nonlinear microwave and millimeter wave propagation devices as in the case of nonlinear transmission lines NLTL MMICs [5], other topics such as nonlinear acoustics or nonlinear superconductor microwave circuits has only a limited academic interest and therefore *Nonlinear Microwaves* is still immature to be considered as a distinct branch of the applied sciences. Such arguments are not valid if we think that only some years ago the NLTLs were also subjects of academic interest, while nonlinear microwave magnetics was an almost unknown scientific subject. Much more, the role of a new branch in the applied science is to offer a common conceptual basis for disparable scientific efforts in related areas and to focus them towards new applications and performances never obtained using other means. This has happened in the last years in the area of nonlinear wave propagation devices at microwave and millimeter frequencies. The basic concepts and experiments regarding nonlinear wave propagation in different physical systems such as NLTL, hydrodynamics, mechanics or optics were recently excellent reviewed [8]. This reference is very useful for the concepts that will be utilized here such as: nonlinear and/or dispersive electromagnetic wave propagation. Here we point out that all physical systems that will be described further are both dispersive and nonlinear. In a dispersive physical system a pulse is spreaded due to the fact that different parts of the wave propagate at different velocities. In a nonlinear physical system the top (the large amplitude part) of a pulse propagates faster than the rest. The dispersivity produces a pulse broadening effect, while the nonlinearity produces a narrowing effect. When the nonlinearity dominates the dispersion, shock waves are formed, while when these two effects are balanced soliton waves are produced, both types of waves being a compressed version of the input excitation and which propagate undeformed further. The Nonlinear Microwave is based on the properties of these waves determined by the interaction of two basic wave characteristics, i.e. nonlinearity and dispersivity. In the next paragraphs we will see different (electromagnetic, magnetic, acoustics, superconductor) nonlinear and dispersive wave systems acting at microwave or millimeter frequencies, how they must be designed in order to produce such nonlinear waves and what kind of devices can be made using such waves and what are their performances.

2. Nonlinear Electromagnetic Wave Propagation Devices at Microwave and Millimeter Wave Frequencies

In the area of microwave and millimeter wave frequencies the nonlinear effect is involved in two ways:

a) The devices (diodes, bipolar or field transistors) have nonlinear characteristics which produce harmonics of the input signals used for example in the case of mixers or multipliers. The nonlinearity is induced *locally* by the d.c. or a.c. signal characteristics of the semiconductor devices while the transmission lines which connect these devices between them and the input and the output propagate and process *linearly* the signals of the nonlinear semiconductor devices arranged in different configurations depending of their applications. These circuits are called *nonlinear microwave and millimeterwave circuits.*

b) The nonlinear characteristics of dielectrics or semiconductor devices are used to produce changes in electromagnetic propagation parameters such as phase velocity of the transmission lines which connect the nonlinear devices between them and the input and the output. Therefore, these type of distributed devices are called *nonlinear electromagnetic wave propagation devices* or *nonlinear transmission lines* (NLTL). The behaviour of this type of devices is quite different compared with the nonlinear microwave or millimeter circuits since the nonlinearity is acting in the *entire* device having a *global* effect. Much more, while in the case of the NLTLs the dispersion plays an *important role* depending on the device applications (for example, to balance the nonlinearity, to determine the time delay and the Bragg frequency of these devices, to change the frequency gaps in the case of periodic configurations), in the case of the nonlinear microwave circuits the dispersion has *no role* or is an undesired feature.

The *Nonlinear Microwave* should contain both categories of devices. The nonlinear microwave and millimeter wave circuits is an active area of research and is periodically reviewed [9]. A new review of this area is beyond the aim of this paper. Much more, if we ant to preserve the analogy with the Nonlinear Optics, we have to consider the devices with features based on the nonlinear and dispersive propagation characteristics of the electromagnetic field, i.e. a nonlinear and dispersive propagating media.

There are some different ways to realize such nonlinear and dispersive propagating media.

2.1. TRANSMISSION LINES FILLED WITH NONLINEAR DIELECTRICS

A rectangular waveguide filled with a nonlinear dielectric can produce microwave and millimeterwave solitons.

A waveguide structure is always dispersive while the nonlinearity is introduced by the variation of the dielectric permitivity as a function of the applied field:

$$\varepsilon = \varepsilon_0 \left(\varepsilon_r + \Delta\varepsilon \left| \vec{E} \right|^2 \right) \tag{1}$$

The relation (1) shows that the dielectric is a Kerr media. Some types of these dielectrics called artificial Kerr media have been recently studied [10], [11] for different purposes, but all having a common point: to realize at microwave and millimeterwave frequencies specific functions attributed only to nonlinear optics such as bistability or phase conjugation. If we consider the propagation in such a structure where the xy plane is the cross-section plane of the waveguide which has a width a along the Ox axis and a height b along Oy and where Oz is the propagation direction, from Maxwell's equations we obtain the following relation:

$$\nabla^2 \vec{E} - \frac{1}{c^2} \frac{\partial^2}{\partial t^2} \left[\left(\varepsilon_r + \Delta\varepsilon |\vec{E}|^2 \right) \vec{E} \right] = \nabla \left(\nabla \vec{E} \right) \tag{2}$$

For TE modes $E_z = 0$, $E_x = 0$,

$$E_y(x, z, t) = \sum_{m=1}^{\infty} \Psi_m \sin(m\pi x / a) \exp[-j(\omega t + \beta_m z)] \tag{3}$$

and one can obtain the Nonlinear Schrödinger equation (NLS) if (3) is introduced in (2) and then integrated from 0 to a, and if the slow varying approximation is used. Finally, we have [12]:

$$j\frac{\partial \Psi}{\partial t} + P \frac{\partial^2 \Psi}{\partial u^2} - Q |\Psi|^2 \Psi = 0 \tag{4}$$

where $P = -\omega_0/2\varepsilon_r\Delta\varepsilon^2$, $Q = \omega_0/\Delta\varepsilon^2$, $u = z - v_g t$, $v_g = \beta c/\varepsilon_r\Delta\varepsilon$ is the group velocity and ω_0 is the carrier frequency. Equation (4) is the same as the NLS equation which describes the optical pulse propagation in nonlinear optical fibers [2], [6]. Since large nonlinearities can be attained in microwave artificial Kerr media, pulse compression, bistability or phase conjugation could be evidenced by using these nonlinear dielectric waveguides.

2.2. DISTRIBUTED NONLINEAR TRANSMISSION LINES

The NLTL are distributed devices which consist of a transmission line periodically loaded with nonlinear elements such as diodes or transistors. The nonlinearity is due to the nonlinear dependence of the capacitance (C_{gs} for transistors) as a function of the applied voltage.

In Fig. 1a a schematic diagram of the NLTL is represented while in Fig. 1b its monolithic realization is shown in a coplanar waveguide (CPW) technique having the equivalent circuit presented in Fig. 1c.

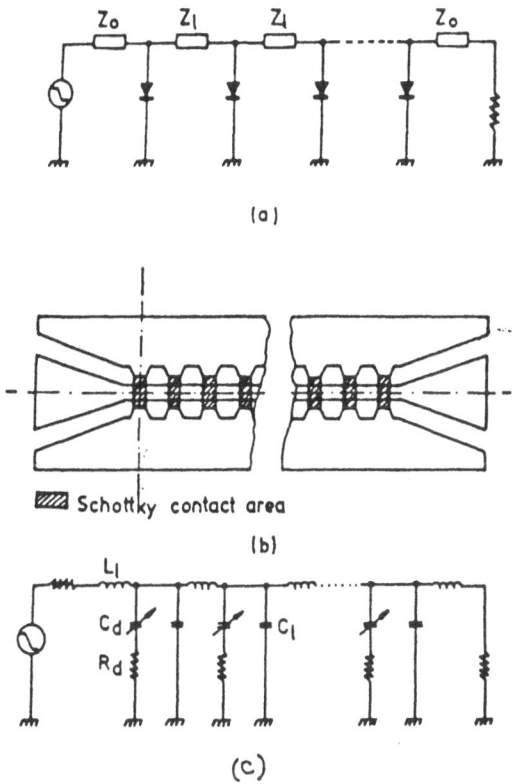

Fig.1 NLTL MMIC a) NLTL concept b) NLTL MMIC in a CPW technique c) the equivalent circuit of the NLTL MMIC in the CPW technique

There are a variety of NLTLs depending on their implementation. For example, lumped varactors or step recovery diodes can periodically load an air-transmission line, a microstrip or a CPW realized as any hybrid integrated microwave circuit, i.e. having a ceramic or a low-loss dielectric substrate [13], [14]. These hybrid NLTLs can be used as microwave multipliers [12], [13] or like a scale model of a NLTL MMIC (as presented in Fig.1c) having the ability for an easy measurement and characterization of its monolithic version [15]. However, this type of NLTLs has a quite restricted area of utilization due to the problems generated by the hybrid technology, i.e. parasitics, losses, interferences, all producing a drastically reduction of the NLTL amplitude and frequency performances. In order to increase these performances a monolithic implementation is desired. A monolithic NLTL, i.e. a NLTL MMIC incorporates multiple advantages compared to its discrete version:
 (a) Schottky diodes as nonlinear elements have a cutoff frequency $f_d = (2\pi R_d C_d)^{-1}$ of some THz, (b) CPW stubs have low loss radiation, relatively low attenuation and skin losses. In this manner, a NLTL MMIC is able to operate at frequencies of the same order of magnitude as f_d, being in this manner the electronic device operating at the highest frequency ever .

In order to analyze the NLTL MMIC nonlinear propagation properties let us consider R_d small (Fig. 1c) and the transfer matrix of one arbitrary cell:

$$\begin{pmatrix} A & B \\ C & D \end{pmatrix} = \begin{pmatrix} 1 & j\omega L \\ 0 & 1 \end{pmatrix}\begin{pmatrix} 1 & 0 \\ j\omega C & 1 \end{pmatrix}\begin{pmatrix} 1 & j\omega L \\ 0 & 1 \end{pmatrix} \qquad (5)$$

The propagation constant is given by $\cos(kd) = (A+D)/2$ and the characteristic impedance is $Z_c = \sqrt{B/C}$, where d is the length of the CPW. We obtain the dispersion relation:

$$\omega^2 = \frac{1}{LC}(1 - \cos(kd)) \qquad (6)$$

and the impedance:

$$Z_c = \sqrt{\frac{L}{C} - \frac{\omega^2 L^2}{4}} \qquad (7)$$

The dispersion relation (6) shows a lowpass characteristic, therefore at frequencies above $\omega_B = 2/\sqrt{LC}$ determined from (6) for $kd = \pi$. This frequency $f_B = \omega_B/2\pi$ is called Bragg frequency.

If $f_d \leq f_B$ and the dispersion is small shock waves are generated [16], due to the strong nonlinearity introduced by $C(V)$. Depending on the technology which has been used different Schottky diodes can be obtained having higher and higher cutoff frequencies. Using the MBE technology some remarkable results have been obtained: (a) a 0.68 ps 10% - 90% fall time of a step shock wave output of a NLTL and having 3.5 V in amplitude [5], (b) a 0.48 fs 3.5 V step shock wave output of a NLTL having delta-doped Schottky diodes [17].

At lower millimeter frequencies extremely good results in their generation and processing can be obtained using NLTL MMICs having Schottky diodes made without the MBE process. The result is a dramatically decrease of their cost [18], [19], [20]. In this respect, we have designed a NLTL MMIC based on the shock wave effect. The distance between two adjacent diodes is $d = 90$ μm, the delay time between one cell of the periodic structure being $\tau = d/V_{CPW} = 0.8$ ps, where $V_{CPW} = 1.13 \times 10^8$ m/s for GaAs. The characteristic impedance for the CPW is 70 Ω. Shock waves are expected in the range of ps if $R_d = 5$ - 10 Ω and $f_b \sim 250$ GHz the number of diodes being $N = 45$. In the simulation presented in Fig. 2 the devices has been excited with a sinusoidal wave at 18 GHz and having amplitudes of 3V, 3.5 V and 4 V. It can be observed that the fall time of the output voltage is 2.2 ps for $V_{in} = 4$ V, with an almost flat spectrum up to 200 GHz with a loss of 15 dB.

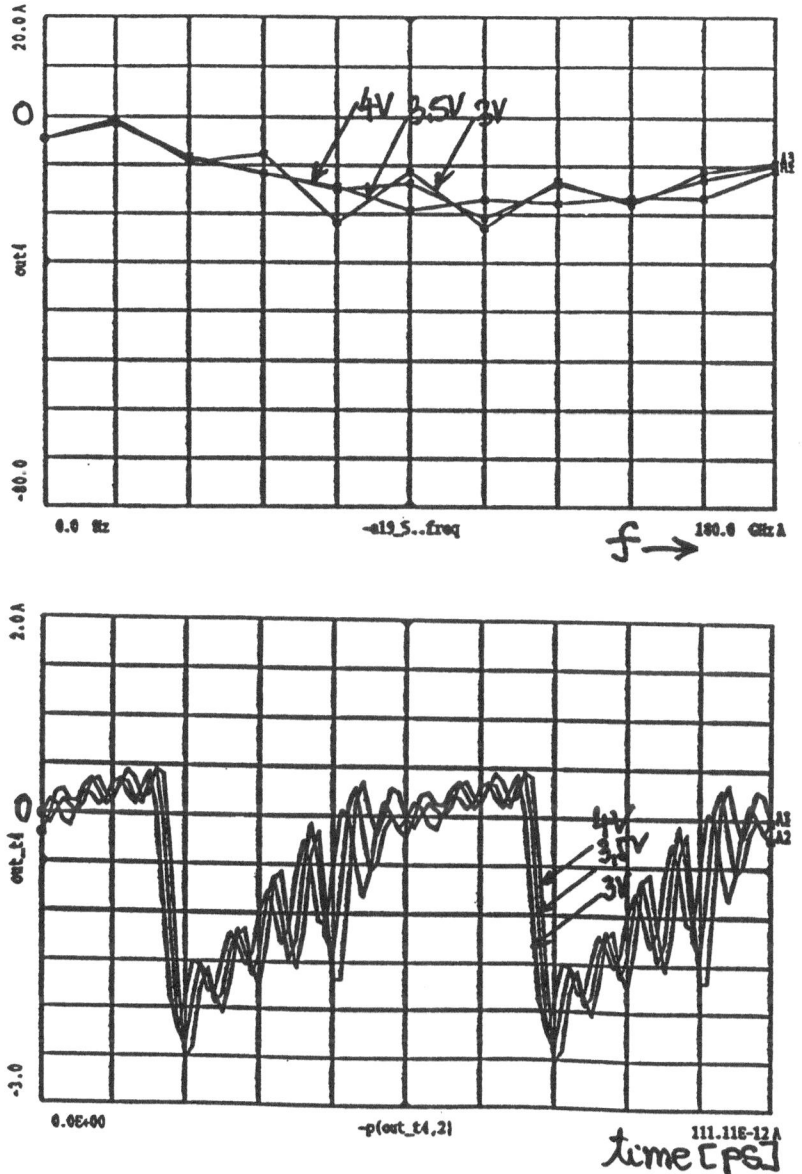

Fig.2 Simulation of the shock wave generation in the NLTL MMICs at different amplitudes and frequencies of the input sinusoidal excitation. The simulation is performed in the frequency and in the time domain

Taking into account that the performance limitations are due to the diode losses it is expected that the fall time decreases with a factor 2.5 at 77 K [21]. The simulation was performed for a Schottky varactor diode with a voltage varying capacitance C(V)

$= C_0/(1-V/V_0)^p$, where $p = 0.87$, $C_0 = 40$ fF, $V_0 = 0.8$ V. The losses due to CPW and GaAs have been included in the simulation.

The device is prepared on GaAs epitaxial structures grown by chloride type VPE technology [19-21]. A four mask technological process was developed. The first mask is the mesa mask which defines the area where the Schottky diodes array will be located. The mesa etch reaches the Si substrate. A CVD SiO_2 0.5 - 1 μm layer covers the wafer and, with the second mask, the windows for the ohmic contact are opened. The ohmic contact AuGeNi is deposited directly on the n^- layer before, or for a smaller series resistance on the n^+ layer. A lift-off technique is used in the metallization process. The next step consists in the opening of Schottky contacts in the SiO_2 layer. After the Schottky metal deposition (Al, TiAu or CrAu), the coplanar transmission line with a 12 μm width central conductor is realized using a lift-off process. A cross section of one cell of the device is presented in Fig.3

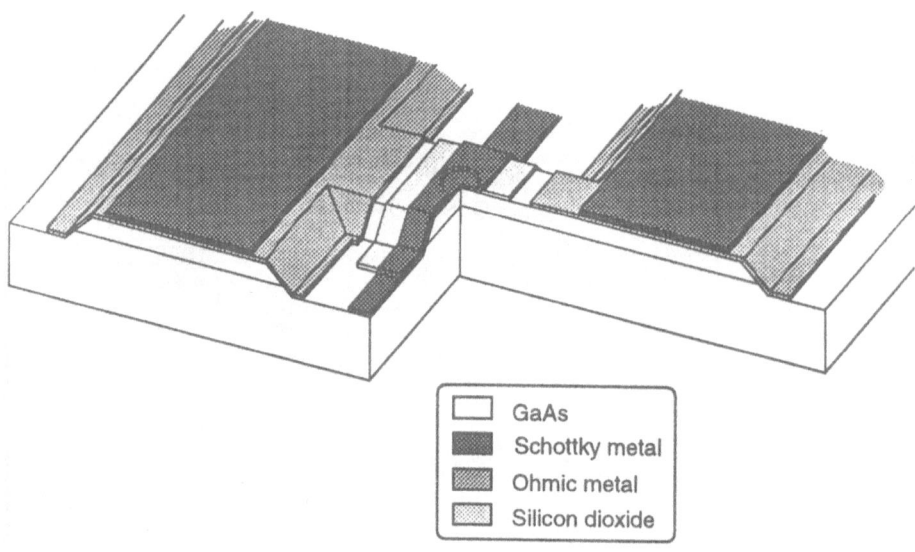

Fig.3 The cross-section of one cell of the NLTL.

In Fig.4 a scanning electron microscope photo of the device is presented, while in Fig.5 a preliminary measurement of the device is displayed for the second harmonic generation.

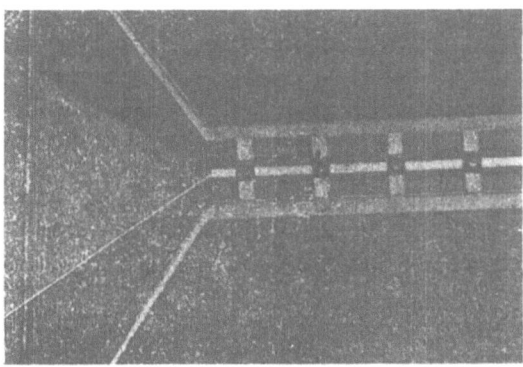

Fig.4 A scanning electron microscope photo of the NLTL MMIC

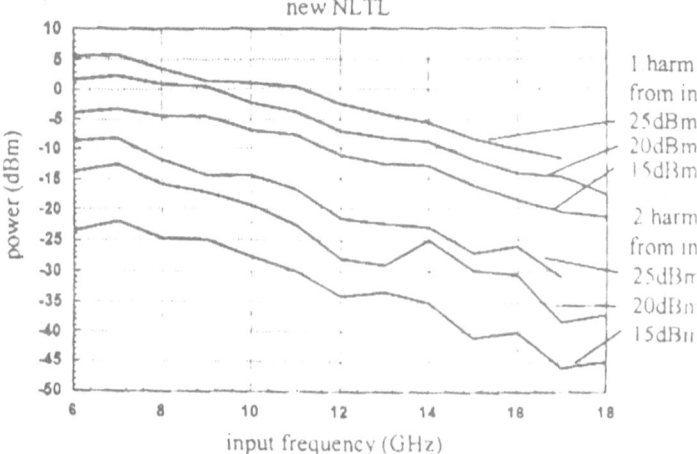

Fig.5 Second harmonic generation of the NLTL MMIC

Recent advances in the area of NLTL MMICs have demonstrated the concept of the quantum NLTL MMIC (QNLTL MMIC). These devices are similar to the NLTL mentioned above, but have as nonlinear elements quantum devices such as quantum varactors or resonant tunneling diodes. In Fig.6 a QNLTL is presented and consists of a transmission line periodically loaded with bbBNN (back to back connected barrier-n-layer-n^+) quantum varactor diodes [22]. A cross-section of the device and its equivalent circuit are shown.

Due to this back to back connection the transmission line will be not a CPW, but a slot line [23]. The structure of the bbBNN is a GaAs n^+ layer followed by a n GaAs which serves as a depletion layer and a δ doped layer (used to improve the nonlinear characteristic of the diode). On the top of the barrier a Ti/Pt/Au slot line is realized using a lift-off technique. Such a diode has a symmetric C-V characteristic and exhibits a very steep dependence of the capacitance on the applied voltage. We consider such a diode with $C(V) = \Delta C/(1+\exp[(V_s+|V|)/V_t])$, where $\Delta C = C_{max} - C_{min}$, $C_{max} = 40$ fF, $C_{min} = 10$ fF, $V_s = 2$ V, $V_t = 0.25$ V and the breakdown voltage $V_b = 14$ V. The series resistance is 4 Ω, the diode area being 35 μm^2.

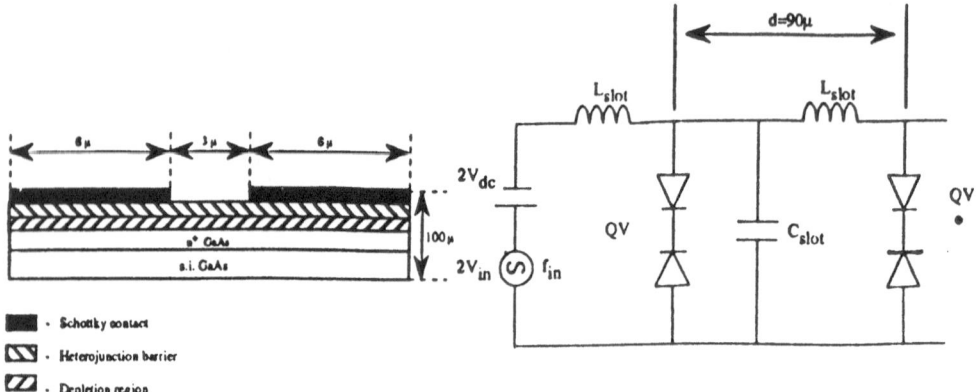

Fig.6 The QNLTL MMIC cross-section and its equivalent circuit

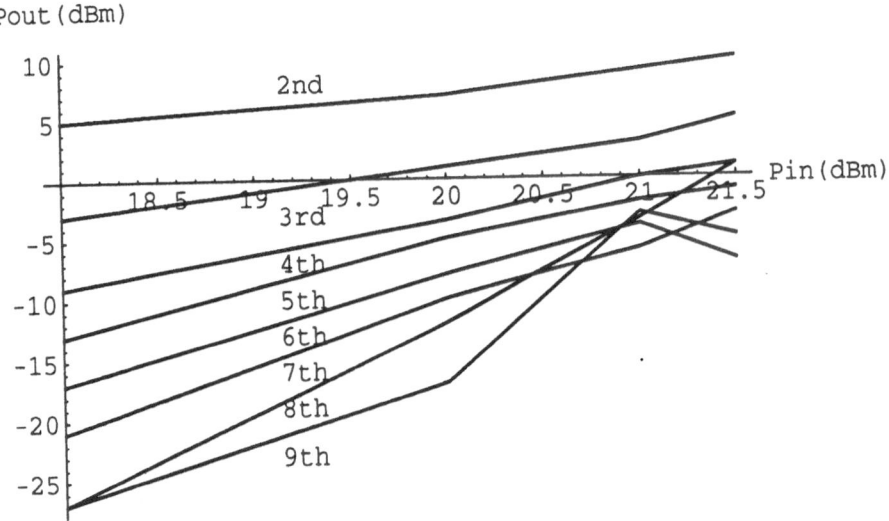

Fig.7 The output power as a function of the input power of the first 9th harmonics of QNLTL MMIC.

In Fig.7 the input-output characteristics are shown for a QNLTL with 49 diodes and a distance of 90 μm between them; the input frequency is 10 GHz. the results presented in Fig.7 for the first 9 harmonics show that up to an input of 21 dBm all the 9th harmonics are generated with a quite good efficiency.

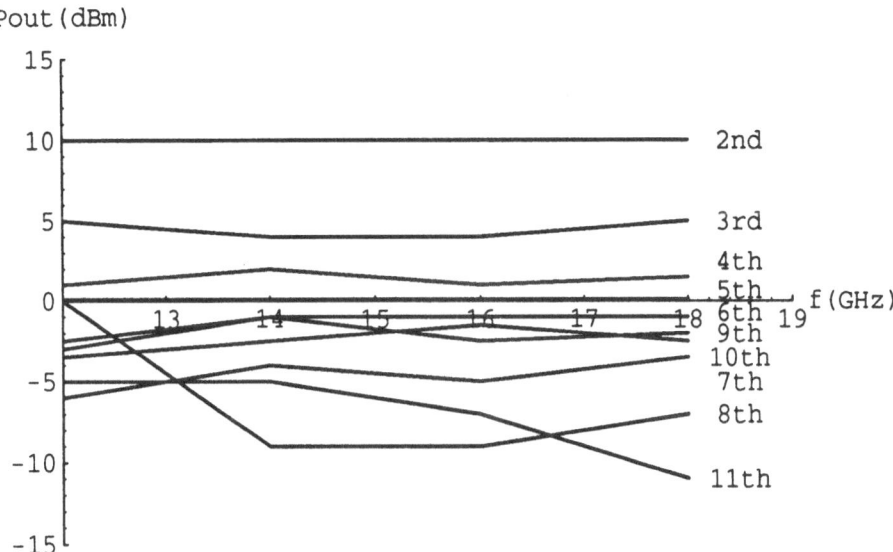

Fig.8 The output power of the QNLTL MMIC as a function of the frequency up to the 11th harmonics.
P_{in} = 21 dBm

Fig.8 shows the simulation performed for the same device at different input frequencies. The first 11 harmonics are shown. It is obvious from this plot that especially the 10th harmonic is generated with a high efficiency. Consequently an output power of -2.5 dBm can be obtained at 200 GHz, which gives a conversion loss of -2 dBm

In order to enhance the efficiency of the harmonics as far as possible by decreasing the fall time of the shock waves a QNLTL ring configuration is used as it is displayed in Fig.9. As can be seen from Figs.10, 11, 12 there are optimum values of the input power P_{in} (P_{in} = 25 dBm), input frequency (f_{in} = 36 GHz) and delay (τ = 37 ps) where all harmonics are generated with a quite good efficiency. A decrease of 30 % of the fall time of the QNLTL can be obtained working at the specified optimum input parameters.

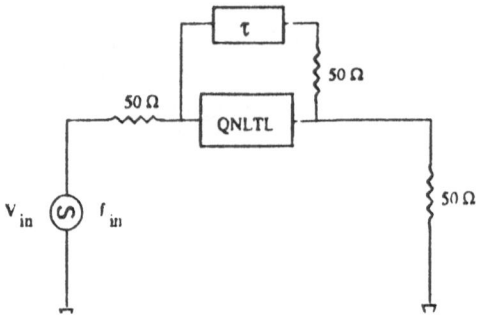

Fig.9 The QNLTL MMIC ring configuration

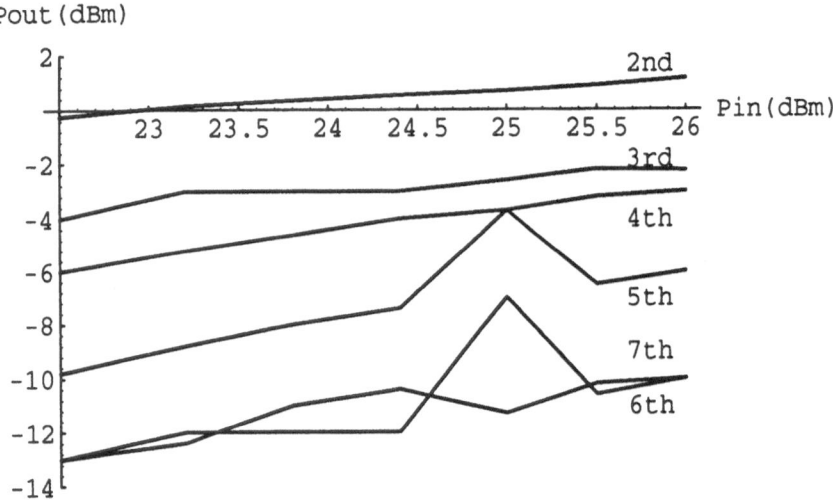

Fig. 10 The output power as a function of the input power of the ring QNLTL MMIC for the first 6th harmonics
(f_{in} = 36 GHz, τ = 37 ps)

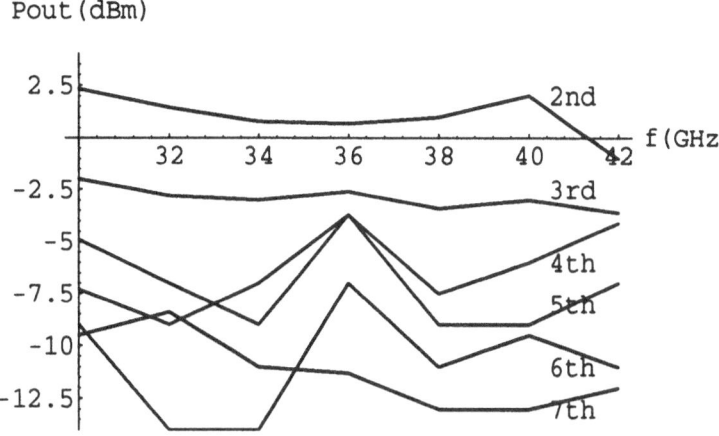

Fig.11 The output power as a function of the frequency of the ring QNLTL MMICs for the first 7th harmonics
(P_{in} = 25 dBm, τ = 37 ps)

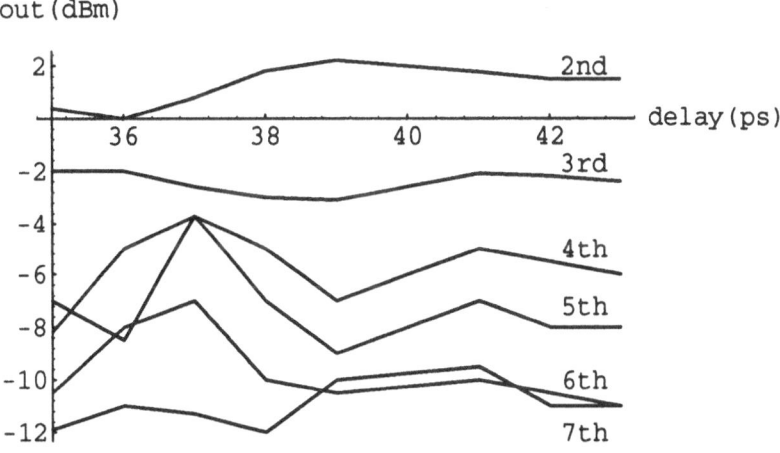

Fig.12 The output power as a function of the delay of the ring QNLTL MMICs for the first 7th harmonics
(P_{in} = 25 dBm, f_{in} = 36 GHz)

Another QNLTL can be realized by loading periodically a transmission line (CPW) with resonant tunneling diodes (RTD) [24-25] acting as a nonlinear element. In the following the RTD is that used in [24] with an area of 1 μm^2 and R = 5.1 Ω, C = 2.8 pF and an I-V characteristic with an N shape and PVR of 2. The experimental I-V curve reported in [24] was fitted using polynomial expressions [25] for the modeling of harmonic generation of this kind of QNLTL. It was considered a generator impedance of R_g = 50 Ω, a load Z_L = 50 Ω and an input frequency of 40 GHz. In order to investigate the performances of this multiplier for the second and the third harmonic generation, a systematic study was made [25] by changing the bias voltage V_{DC}, the number of the diodes of the QNLTL (N) and the separation distances between them (S).

The result of the optimized values for a maximum output of the second and the third harmonic is presented in Table 1.

TABLE 1.

	Bias(V)	Pin(dBm)	Pout(2nd)	Pout(3rd)
2nd harm.	1.46	1.72	-4.68	-17
3rd harm.	1.51	1.95	-26	-9

In the case of the 2nd harmonic generation the value of -4.68 dBm at 80 GHz is obtained when N = 40, S = 11.8 μm. In the case of the 3rd harmonic the value of -9 dBm at 120 GHz is obtained when N = 50, S = 20.8 μm. If the dispersion is not weak and if R_d is small (see eq.6) at any node n (see Fig.1a) the following equation is valid:

$$I_n - I_{n-1} = -j\omega C_{dn} V_n \qquad (8)$$
$$V_{n+1} - V_n = -j\omega L_1 I_n \qquad (9)$$

If we suppose that $C_d(V_n)$ is given by:

$$C_d(V_n) = C_0 / (1 - V_n / V_0) \qquad (10)$$

or

$$Q_n = \int C_{dn}(V_n) dV_n = C_0 V_0 \ln(1 + V_n / V_0) \qquad (11)$$

then from (8), (9) and (11) we have:

$$LC_0 \frac{\partial^2}{\partial t^2} \ln(1 + V_n / V_0) = (V_{n+1} / V_0 + V_{n-1} / V_0 - 2V_n / V_0) \qquad (12)$$

This is Hirota equation [26] and if n is large it transforms to the Korteweg de Vries equation (KdV) which has soliton solutions. The soliton solution of the Hirota equation is:

$$V_n(t) = V_M \sec h^2\left[1.76(t - nT_D)/T_{FWHM}\right] \qquad (13)$$

where T_D is the delay time through each section and T_{FWHM} is the soliton full width at half maximum duration.

$$T_D = \frac{1}{\pi f_B \sqrt{\ln(1 + V_M/V_0)}} \sinh^{-1}\left(\sqrt{V_M/V_0}\right) \qquad (14)$$

$$T_{FWHM} = \frac{1.76}{\pi f_B \sqrt{\ln(1 + V_M/V_0)}} \qquad (15)$$

If $T_{FWHM} > 1.76\sqrt{L_1 C_0 V_0/V_M}$ then an input pulse excitation is splitted during the propagation in a train of two or more solitons. Each negative part of a sine wave input is splitted in a train of solitons during the propagation along a NLTL, the result being a waveform with multiple pulse/cycle and with a rich harmonic content. The compression ratio of the input excitation is limited to 2:1 and in order to increase it the structure is tapered by varying the diode spacing as $\tau_n = k^n \tau_0$ with $C_n = k^{n-1} C_0$ and $L_n = k^{n-1} L_1$ [5].

We have simulated the soliton generation on a NLTL with $Z_c = 75\ \Omega$, $R_{d1} = 3\ \Omega$, $R_{dn} = R_{d1}/k^{n-1}$, $C_{nd} = C_{01}/[1-(V_n/V_0)^p]$ where $p = 0.87$, $V_0 = 0.8$ V. In addition to the diode model consisting of a series connection between R_d and C_{nd}, a parallel source current with C_d and a series inductance with R_d are introduced. We have considered $L_{dn} = L_d = 20$ pH, $I_{Sn} = I_{0n}[\exp(qV_n/k_BT)-1]$, $I_{0n} = k^{n-1}I_{01}$ and $I_{01} = 1$ nA.

In the Figs.13-15 it is shown the soliton formation and its splitting in a train of solitons for the NLTL MMICs considered above [27]. The device is excited by a sinusoid having a 20 GHz frequency and different voltage amplitudes and bias voltages. In this respect we have considered in Fig.13: $V_{in} = 4$ V, $f_{in} = 20$ GHz, $V_{dc} = -2$ V, in Fig.14: $V_{in} = 6$ V, $f_{in} = 20$ GHz, $V_{dc} = -3$ V, in Fig.15: $V_{in} = 8$ V, $f_{in} = 20$GHz, $V_{dc} = -4$ V.

The soliton formation (Fig.13) and its splitting in two solitons with different amplitudes can be easily observed from these simulations.

Experiments concerning the harmonic generation by using the soliton splitting have demonstrated the generation of millimeter waves up to 108 GHz with an efficiency greater than 45% for the second and third harmonics [28], [29].

28

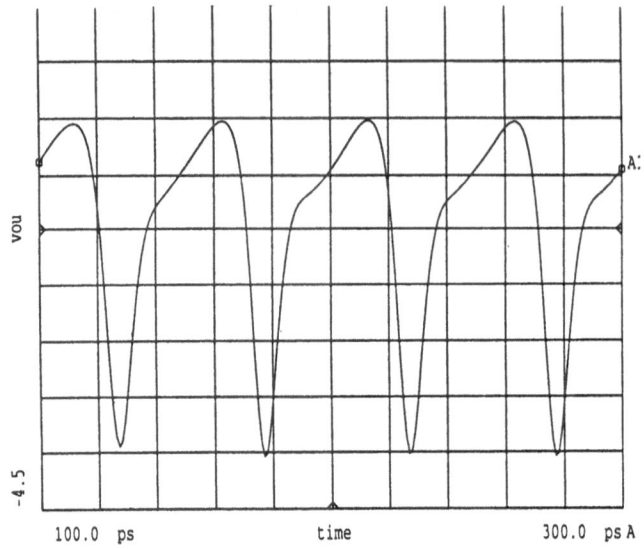

Fig. 13 Soliton generation in a nonuniform NLTL MMIC (V_{in} = 4 V, V_{dc} = -2 V)

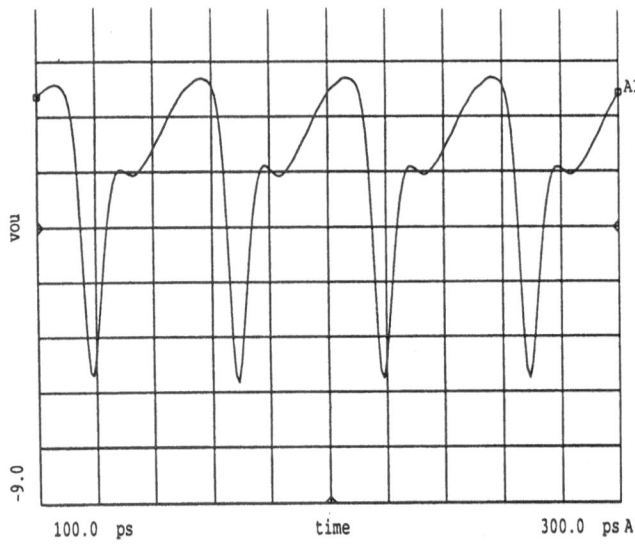

Fig. 14 Soliton splitting in a nonuniform NLTL MMIC (V_{in} = 6 V, V_{dc} = -3 V)

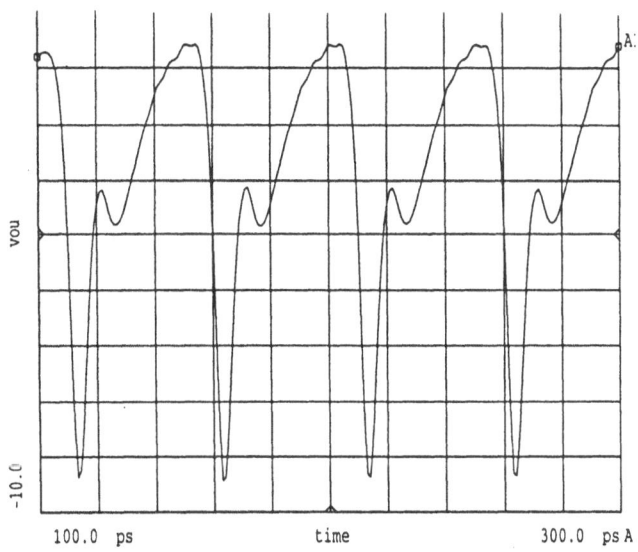

Fig.15 Two soliton generation in a nonuniform NLTL MMIC (V_{in} = 8 V, V_{dc} = -4 V)

The applications of the NLTL MMICs are:

(a) *Sampling circuits* for signal measurements, instrumentation and frequency conversion yielding a bandwidth of 500 GHz [30], and higher than 725 GHz [5] has been obtained using NLTL MMICs. 1THz bandwidth sampling circuits should be realized in the near future.

(b) *On-wafer Millimeter Network Analysis*. The HEMT devices and the RTDs oscillators are only two examples from a large category of millimeterwave solid state devices which must be characterized at several hundreds of GHz. The commercial network analysis has a bandwidth of 0-65 GHz which is completely unsatisfactory for the devices mentioned above.

In the last years, based on NLTL MMICs, on-wafer network analyzers have been developed with a bandwidth of 200 GHz [5]. The NLTL is used as a signal generator and as a strobe signal for a directional sampler.

Time domain analyzers based on NLTL MMICs are able to measure sub-picosecond pulses [5], [30]

(c) *Photodetectors*. High speed photodetectors can be integrated on the same chip with NLTL MMICs sampling devices for the detection of optical signals. These MOEMICs (Millimeterwaves Optoelectronic Monolithic Integrated Circuits) are able to detect optical signals of ps duration [31].

(d) *Free-Space Analysis*. The NLTL MMICs are connected with bow-tie antennas and serves as transmitter and receiver in a free-space analyzer which measures gain/attenuation as a function of frequency. The transmitted and received signals are collimated by a Si lens mounted on the back side of the bow-tie-NLTL MMICs. Different materials or passive devices can be characterized in the range of 0-300 GHz [32].

(e) *Intelligent sensors* for traffic application. A single NLTL MMIC generates, amplifies and down converts millimeter signals for 40 GHz or 60 GHz traffic applications. These functions are based on the nonlinear properties of NLTLs. Quite good performances are estimated for these smart sensors [32].

2.3. FULLY DISTRIBUTED NONLINEAR TRANSMISSION LINES

The devices described in 2.2. are distributed circuits, where the nonlinearity is introduced periodically by a number of semiconductor devices. An alternatively way is to consider a hole nonlinear and distributed device such as a Schottky CPW traveling wave device with different geometries [4], [34]. Some examples are presented in Figs.16a and 16b. In such structures shock waves, KdV solitons as well as NLS

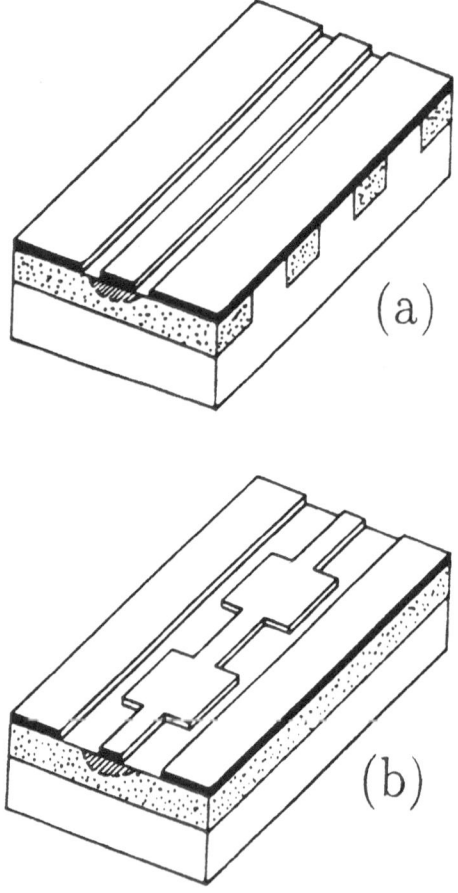

Fig.16 Fully distributed NLTLs a) a periodically doped CPW Schottky structure b) a periodically nonuniform CPW Schottky structure. The substrate can be s.i. GaAs or InP

equation solitons can be generated. It was predicted that solitons with less than 1 ps duration can propagate in these structures [33]. There are two problems: (a) high microwave losses due to doped semiconductor layer, (b) large disparity in the dimensions for the semiconductor devices versus CPW in order to minimize skin-effect losses and radiation losses [5]. It seems that the design of the semiconductor devices and the transmission lines (CPW) should be separated as well as their role in a NLTL in order to achieve high performances in time and frequency. However, fully distributed NLTLs have demonstrated experimentally good performances as pulse compressors, phase shifters or multipliers in X and Ku band.

3. Magnetostatic Nonlinear Wave Propagation in Thin Magnetic Films

Magnetostatic waves (MSW) are slow, magnetically dominated electromagnetic waves in the microwave range of frequencies which can be excited in magnetic materials. The ferrimagnetic yttrium iron garnet (YIG) is the key magnetic material for MSW propagation.

There are two types of magnetostatic waves that can be launched in a low loss YIG film biased in a d.c. external magnetic field: volume waves (forward and backward) and surface waves. All types of magnetostatic waves are dispersive and their propagation becomes nonlinear at moderate power levels of excitation. The NLS equation (4) is used to describe the onset and the propagation of magnetostatic wave envelope solitons in magnetic garnet films *for all types* of magnetostatic waves:

$$j\frac{\partial \Phi}{\partial t} + P\frac{\partial^2 \Phi}{\partial u^2} - Q|\Phi|^2 \Phi = 0 \qquad (16)$$

where $\Phi = m/2M_s$ is the normalized magnetization and t and $u = x-v_g t$ are the time and space variables, respectively (u is along the propagation direction x). The NLS equation is able to describe the soliton propagation for all types of MSW despite their different physical mechanism , due to the fact that for all types of MSW the NLS equation is deduced from the second order Taylor expansion of the dispersion relation. We have:

$$\omega\left(k,|\Phi|^2\right) \cong \omega_0 + \frac{\partial \omega}{\partial k}(k - k_0) + \frac{1}{2}\frac{\partial^2 \omega}{\partial k^2}(k - k_0)^2 + \frac{\partial \omega}{\partial |\Phi|^2}|\Phi|^2 \qquad (17)$$

If we consider $\partial/\partial t \leftrightarrow \omega-\omega_0$, $\partial/\partial x \leftrightarrow k-k_0$, $v_g = \partial\omega/\partial k$, $P = \partial^2\omega/\partial k^2$ and $Q = \partial\omega/\partial|\Phi|^2$ then from (17) the NLS equation (16) can be deduced. For each type of MSW, P and Q differ as well as their sign.

In the last period important review papers were published [35-40] where the theoretical and experimental evidence for microwave envelope solitons were revealed for all MSW types. The condition for the existence of bright solitons is the fulfillment of the Lighthill criterion: PQ < 0. Otherwise, dark solitons can be generated but with complete different initial conditions compared to bright solitons.

In some of these review papers it was discussed that it is not possible to excite purely dipolar magnetostatic surface wave (MSSW) bright solitons because the Lighthill criterion is not satisfied, the product PQ being always positive. The introduction of ground planes affects strongly the dispersion properties of the MSW,

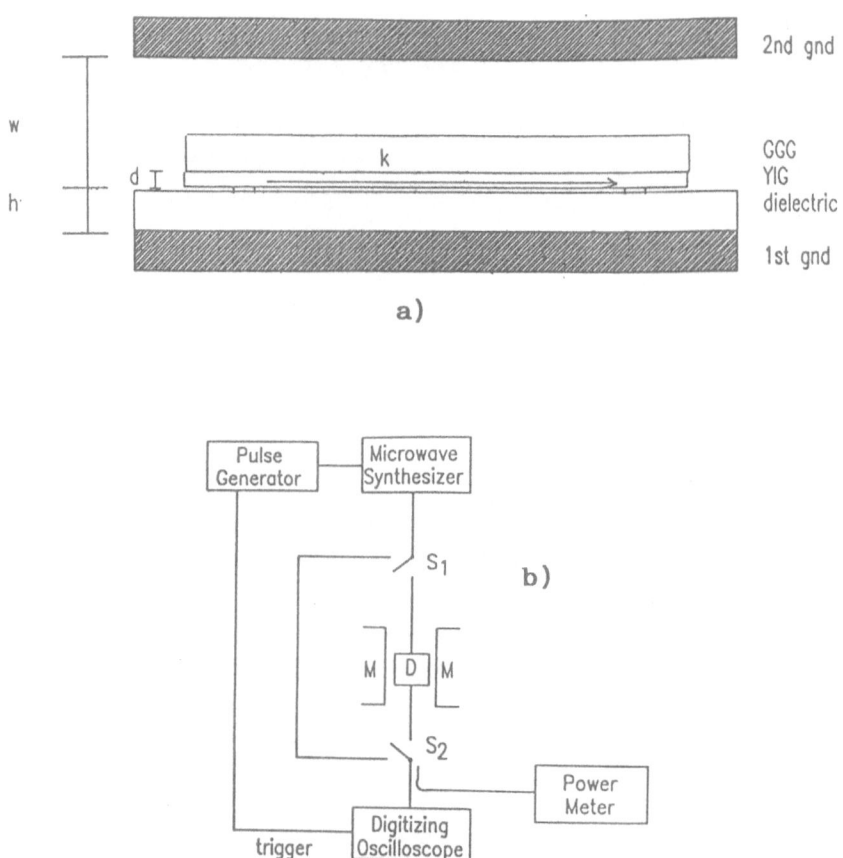

Fig. 17 a) Typical delay line arrangement. d is the film thickness, h and w the separation between the film surface and the device ground planes; a YIG film, grown onto a commercial GGG substrate is excited by two microstrip transducers and biased by a d.c. field orthogonal to the device cross-section and parallel to the film plane. k is the excited wavevector. b) Experimental setup used for the soliton detection experiment. A microwave synthesizer generates short pulses by means of pulse modulation. Both the input and output pulses are visualized on a digitizing oscilloscope screen and the carried power is measured with a powermeter. S_1 and S_2 are switches, M are the polar expansions of an electromagnet and D is the device test.

inducing important changes in the frequency dependence of the group velocity (from up chirp to down chirp or vice-versa) [39],[41-42]. These results have allowed the experimental observation of purely dipolar magnetostatic surface wave bright solitons, due to the fulfillment of the Lighthill criterion in grounded YIG delay lines.

A 28 μm thick pure yttrium iron garnet film, 0.4 cm and 1.5 cm long has been used in the experiment. Microwave pulses of tens of ns have been launched in the YIG film by means of microstrip transducers evaporated on a 254 μm alumina substrate. The microstrips were 50 μm wide in the coupling region, the distance between them was 1.3 cm and both were grounded. The device under test is presented in Fig.17a while in Fig.17b the experimental setup is shown.

The measurements were made at H = 592 Oe and f = 3580 MHz. The dispersion curve for the device in Fig.17a is shown in Fig.18 where also the predicted insertion loss is displayed. In Fig.19, the dispersion relation for the ungrounded configuration (curve a) and for the grounded configuration (curve b) of the device are presented and can be easy compared. One can observe that the Lighthill criterion is satisfied in a bandwidth Δf = 100 MHz. This frequency range is easily excitable since it corresponds to the minimum insertion loss of the delay line.

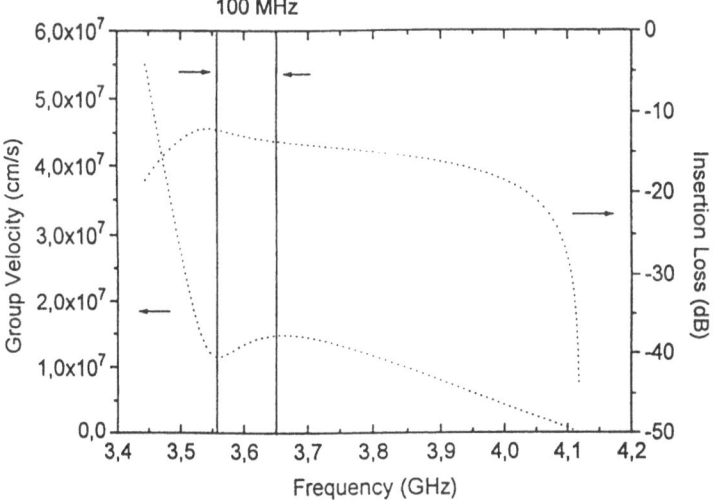

Fig.18 Predicted insertion loss and group velocity versus frequency for a MSSW delay line in a one ground plane configuration. The YIG film is 28 μm thick, the bias field is H = 592 Oe and the saturation magnetization of the film is $4\pi M_S$ = 1760 Oe. The ground plane is 254 μm far from the film surface and the microstrips are 50 μm wide. The positive dispersion trend is within 100 MHz.

34

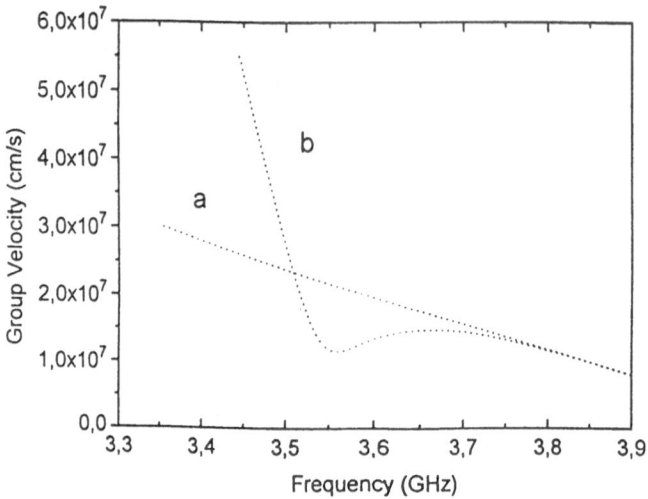

Fig. 19 Group velocity versus frequency for an ungrounded (curve a) and a grounded (curve b) MSSW delay line with the parameters given in Fig. 18

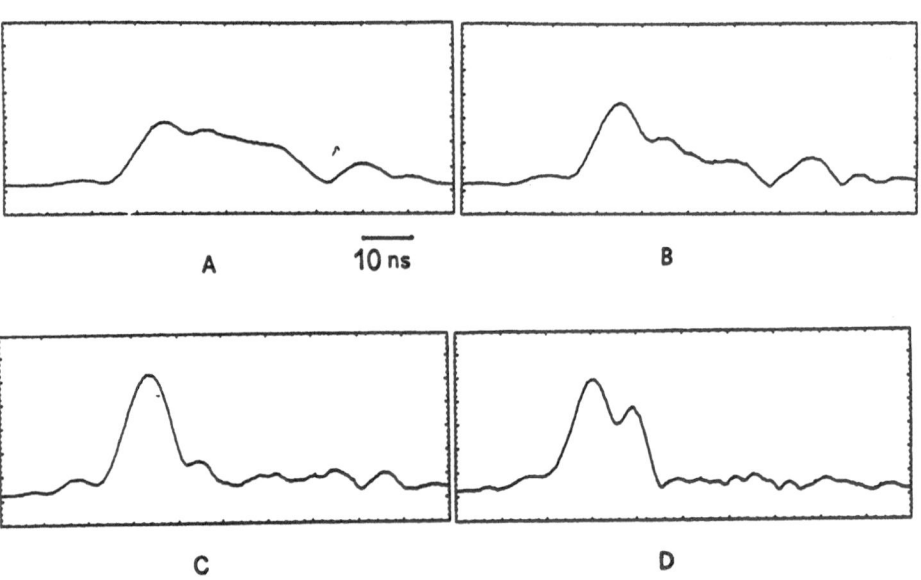

Fig. 20 Measured of soliton formation at f = 3580 MHz. The input pulse is 40 ns wide and the pulse train has a period T = 200 ns. A is the output pulse for an input peak power P_{in} = 136μW, B is for P_{in} = 816μW, C is for P_{in} = 3.85 mW and D is for P_{in} = 18.26 mW.

In Fig.20 the microwave envelope soliton for pure dipolar MSSW is shown as well as its splitting in two solitons. The input excitation was $\tau = 40$ ns, the carrier frequency $f = 3580$ MHz and a duty cycle was imposed. $D = T/\tau = 5$, where $T = 200$ ns is the pulse train, so that we have a single pulse boundary condition for the delay line.

The soliton in Fig.20c is represented in Fig.21 fitted by a hyperbolic secant function -the shape of the envelope soliton solution of the NLS equation.

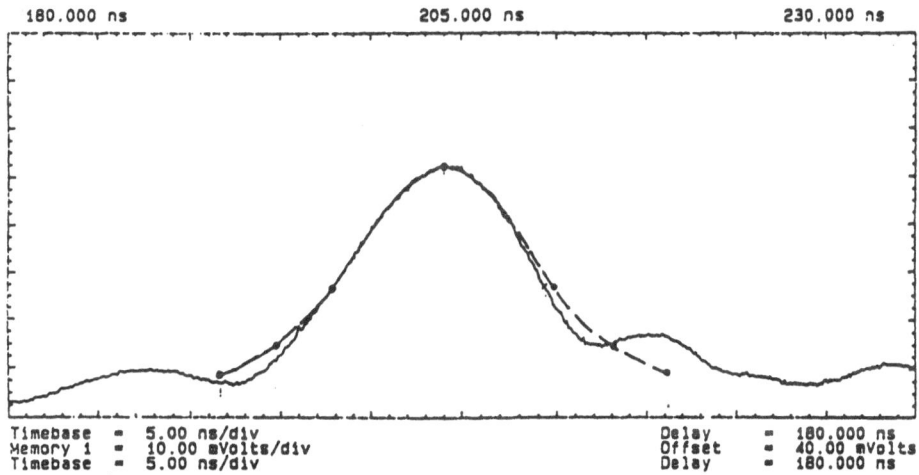

```
180.000 ns                205.000 ns                230.000 ns
```

```
Timebase  =  5.00 ns/div                    Delay   =  180.000 ns
Memory 1  =  10.00 mVolts/div               Offset  =  40.00 mVolts
Timebase  =  5.00 ns/div                    Delay   =  180.000 ns
```

Fig.21 Hyperbolic secant reshaping of the pulse in Fig.20c. The continuous curve represents the experimental data and the dashed line is the fitting result which is in very good agreement with the experimental data.

The behaviour of the output peak power versus the input power is shown in Fig.22 and exhibits two threshold levels at the input peak power levels of 1.3 mW and 12 mW, respectively. This shape is due to the threshold activation of the soliton mechanism which is expected to change the output power trend compared to its linear behaviour. The ratio between the two threshold powers, where one and two solitons are generated, is close to 9, as predicted by the theory [41-42]. A 2 dB enhancement of the output power with respect to the input power is observed in Fig.22b due to the soliton effect.

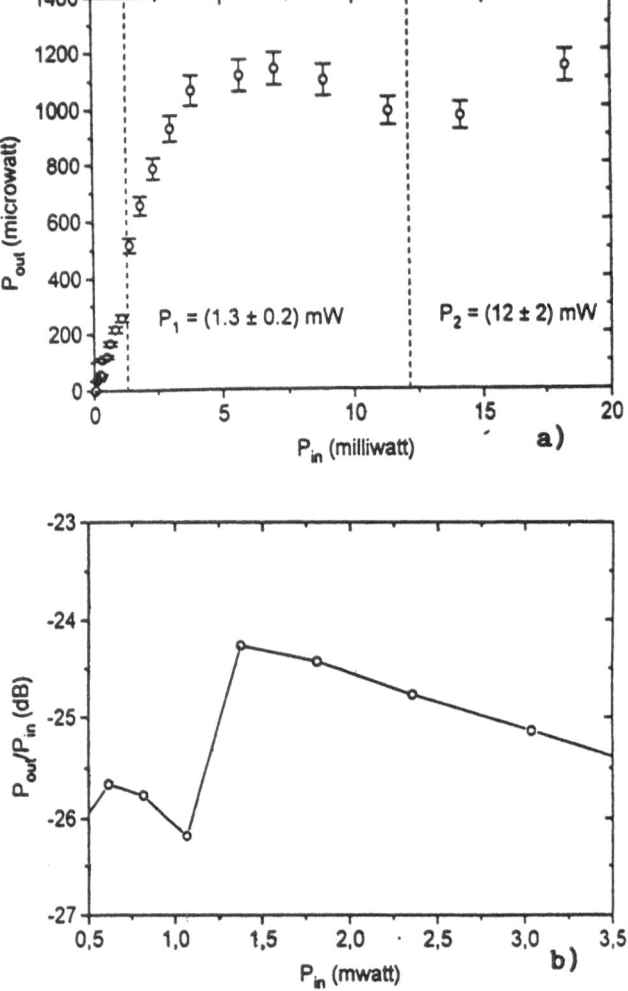

Fig.22 a) Measured output peak power P_{out} versus input peak power P_{in} measured at f = 3580 MHz for a pulsewidth of τ = 40 ns. Two thresholds are obtained, at P_1 = 1.3 mW and P_2 = 12 mW, with a ratio R = P_2/P_1 = 9.3 close to the expected theoretical value R_{th} = 9 between the threshold power of the second and the first soliton of the NLS equation. b) Normalized power response P_{out}/P_{in} in dB as a function of P_{in} in the region of the fist threshold. A 2dB enhancement of the output peak power with respect to the input power has been measured due to the pulse narrowing.

These results demonstrate that the MSSW soliton devices are promising for pulse compression and signal to noise ratio enhancement applications at low power levels in the 2-5 GHz range.

Other types of microwave MSW envelope solitons have been predicted and/or experimentally evidenced such as dark solitons [43] and gap solitons [44].

These briefly reported results demonstrate how attractive are the MSW devices for microwave applications as well as for the fundamental research.

4. Nonlinear Wave Propagation in High Temperature Superconductor Transmission Lines at Microwave Frequencies

The existence of shock waves [45] and solitons in several configurations which contain Josephson junction elements is well established [46]. In the following we consider a high temperature superconductor (HTS) stripline with a length L approximately equal to λ_d and which has capacitive coupling gaps at the input and the output. The HTS stripline together with the gaps represent a HTS resonator for which extensive experimental results were obtained [47]. A top view of this resonator is represented in Fig.23. The HTS resonator works at 77 K and has a central frequency $f_0 = 3.0778$ GHz, a length L = 1.95 cm and a total area of 1 cm^2. The width of the YBCO microstrip is 150 μm. The HTS resonator has $Z_c = 34$ Ω, Q = 7800 and an insertion loss of -43 dBm.

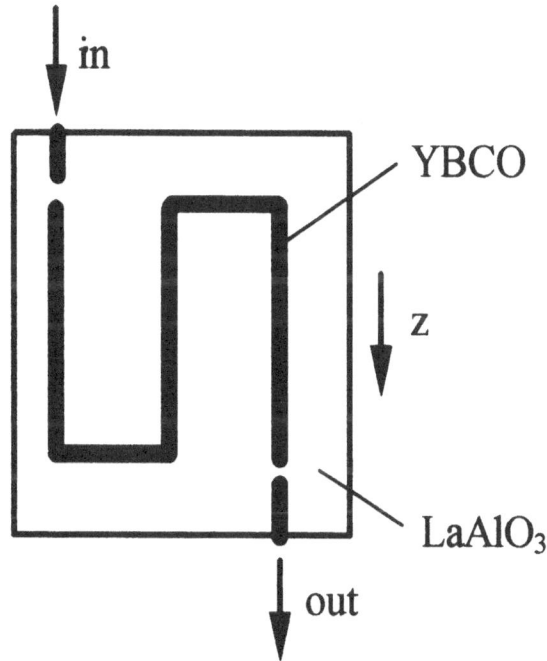

Fig.23 A top view of a HTS resonator

Due to the high Q of the HTS resonator, large currents are produced which circulate in the HTS resonator and lead to a nonlinear behaviour of the device above a threshold of the input power. A model of a NLTL can be used to describe the behaviour of this HTS resonator. The normalized NLTL equations are:

$$\begin{cases} \dfrac{\partial v}{\partial y} = -\dfrac{\partial}{\partial \tau}(li) - ri \\[3mm] \dfrac{\partial i}{\partial y} = -\dfrac{\partial}{\partial \tau}(cv) \end{cases} \tag{18}$$

where $\tau = \omega_0 t$, $y = \beta_0 z$, $v = V$, $i = IR_0$, $c = C/C_0$, $l = L/L_0$, $r = R/(\beta_0 R_0)$, $R_0 = \sqrt{L_0/C_0}$, $\beta_0 = \omega_0 \sqrt{L_0 C_0}$. v and i are, respectively, the normalized values of the voltage and the current waves and L, C, R denote the elements of the HTS NLTL which have respectively values equal to L_0, C_0, R_0 at the resonant frequency f_0.

The resistance and the inductance of the HTS NLTL have the following nonlinear variation with the current:

$$\begin{cases} R = R_1 + R_2 \exp(R_3 I^2) \quad [\Omega/m] \\[2mm] L = L_1 + L_2 I^2 \quad [nH/m] \end{cases} \tag{19}$$

where $R_1 = 3.91$, $R_2 = 5.27$, $R_3 = 18$, $L_1 = 550$, $L_2 = 11.98$, values which have been determined from the experimental results [47].

By using the reductive perturbation method [48] a NLS equation for the envelope of the current $\Phi(x,p)$ can be found:

$$2j\,\partial\Phi/\partial x + P\,\partial^2\Phi/\partial p^2 + Q|\Phi|^2\Phi = 0 \tag{20}$$

where $x = \varepsilon^2 y$, $p = \varepsilon(\tau - v_g t)$. The coefficients P and Q in (20) are $P = -jc^2(r_1 + r_2)^2/4k^3$, $Q = 3\Omega c(r_2 r_3 - j\Omega l_2)/kl_2 r_2$, and the dispersion relation is $k^2 = \Omega^2 l_1 c + j\Omega c(r_1 + r_2)$ with $k = \beta/\beta_0$ and $\Omega = \omega/\omega_0$.

By denoting $u = ax$ and $s = bp$ with $a = j\,\mathrm{Im}Q/2$ and $b = \sqrt{-\mathrm{Im}Q/2jP}$ we obtain:

$$j\frac{\partial\Phi}{\partial u} + \frac{1}{2}\frac{\partial^2\Phi}{\partial s^2} + (1 + j\Gamma)|\Phi|^2\Phi = 0 \tag{21}$$

where $\Gamma = -\mathrm{Re}Q/\mathrm{Im}Q = r_2 r_3/\Omega l_2$. In the lossless case $r_3 = 0$, $\Gamma = 0$; otherwise these quantities are different from zero.

The experimental dependence of P_{out} on P_{in} is shown in Fig.24. As for the MSW [49], the same power threshold effect can be observed: beyond an input power threshold the excitation is compressed, solitons are formed and the $P_{out} = f(P_{in})$ curve deviates from its linear behaviour. In the case presented in Fig.24 two situations are considered: (a) the excitation frequency deviates with $\Delta f = \pm 1$ MHz from f_0 and (b) the excitation frequency deviates with $\Delta f = \pm 2$ MHz from f_0. Envelope solitons can be obtained if the length of the NLTL is relatively long. This condition is obtained by slightly detuning the resonator such that the microwave excitation is internally reflected. Thus, the length of the HTS NLTL is artificially increased by internal reflections of the microwave signal at the expense of an increase in the input power. A similar effect was also studied in YIG resonators [50].

The threshold values determined from the experiment in Fig.24 are (a) P_{thr} = 14.53 dBm and (b) P_{thr} = 19.8 dBm and correspond fairly well to the threshold input powers determined theoretically by using (19)-(21).

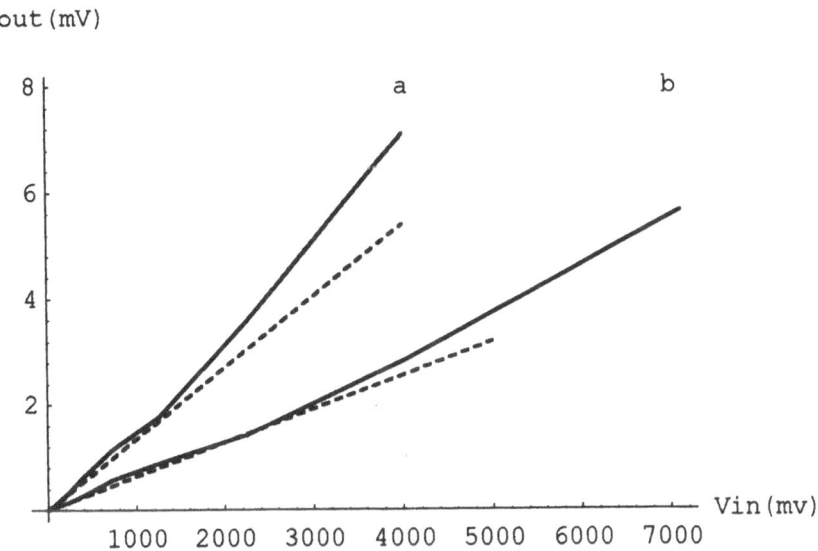

Fig.24 The output voltage as a function of the input voltage of the HTS resonator. The dashed lines indicate the linear trend

5. Solitons in Surface Acoustic Wave Devices

The evidence of solitons in surface acoustic wave (SAW) devices was an active area of research for more than a decade [51-52]. Envelope solitons described by the NLS equation and baseband solitons described by the KdV equation have been theoretically predicted and experimentally measured in SAW devices [53-57].

Recently, the KdV solitons were experimentally evidenced in a metallic grating waveguide. This SAW propagating structure is presented in Fig.25. The substrate is a 128° Y-X LiNbO$_3$ and the Al metallic grating has a width of w = 650 μm. The hole structure has the dimensions of 300x1x1 mm^3.

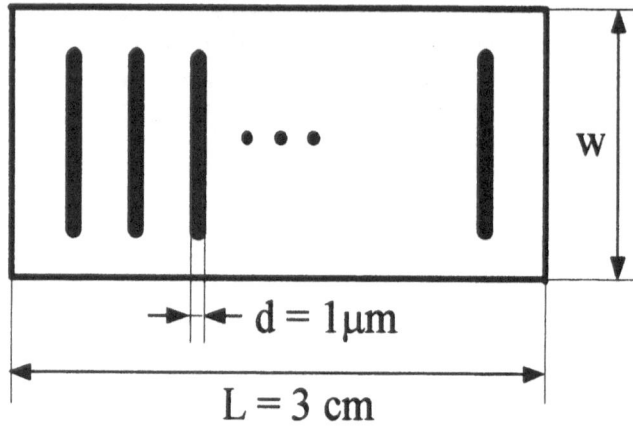

Fig.25 The SAW metallic grating waveguide

The equation which describes the dynamics of the strain is:

$$\frac{\partial S}{\partial \tau} - \beta S \frac{\partial S}{\partial y} + \frac{d^2}{24} \frac{\partial^3 S}{\partial y^3} = 0 \tag{22}$$

where $y = x - v_g t$, $\tau = v_g t$, $S = \partial \Phi / \partial x$ with Φ the scalar potential and v_g is the group velocity. The one soliton solution of (22) is:

$$S = -A \sec h^2 \frac{\sqrt{2\beta A}}{d} \left(y - \beta A \tau / 3 - x_0\right) \tag{23}$$

The solitons of (23) are the same as those described in the first paragraph. In this manner, second and third harmonics can be generated, phase conjugation can be expected and other nonlinear signal processes in the rf or microwave range are possible.

6. Conclusions

The paper presents an overview of microwave and millimeter nonlinear wave propagation in magnetic, acoustic and electromagnetic distributed nonlinear physical systems. It contains a large number of theoretical predictions and experimental results which demonstrate that Nonlinear Microwaves is a new area of applied science and that it is well related to Nonlinear Optics. Nonlinear Microwaves is also an active area of research, a fact that is best demonstrated by the report of new topics such as microwave spatial solitons [58] during the preparation of this paper. The applications of Nonlinear Microwaves are very promising; the paper has reviewed only a part of them since a lot of proposals still wait their experimental implementation.

Acknowledgment

One author (MD) would like to thank to Dr. R. Marcelli and Dr. P. de Gasperis from IESS-CNR, Rome, Italy, to Prof. G. Bartolucci from Univ. "Tor Vergata", Rome, Italy, to Prof. D. Jäger from the Univ. of Duisburg, Germany, to Dr. B. Szentpali from MFKI, Budapest, Hungary and to Dr. A. Müller and S. Iordanescu from ICCE, Bucharest, Romania for the cooperation during the author's research in the area of Nonlinear Microwaves. The author has co-authored with them a large number of papers which have been used here to demonstrate the basis of the Nonlinear Microwave Science. MD would like also to thank the Alexander von Humboldt Foundation, NATO and the National Research Council of Italy for the financial support.

REFERENCES
1. Mills, D.L. (1991) *Nonlinear Optics*, Springer Verlag, Berlin.
2. Agrawal, G.P. (1989) *Nonlinear Fiber Optics*, Academic Press, New York.
3. Scott, A. (1970) *Active and Nonlinear Waves Propagation in Electronics*, Wiley Interscience, New York.
4. Jäger, D. (1985) Characteristics of traveling waves along the non-linear transmission lines for monolithic integrated circuits: a review, *Int. J. Electronics* **58**, 649-669.
5. Rodwell, M.J.W., Allen, S.T., Yu, R.Y., Case, M.G., Bhattacharya, V., Reddy, M., Kamegawa, M., Konishi, Y., Pusi, J. and Pullela, R. (1994) Active and nonlinear wave propagation devices in ultrafast electronics and optoelectronics, *Proc. IEEE* **82**, 1037-1059.
6. Hasegawa, A. (1989) *Optical Solitons in Fibers*, Springer Verlag, Berlin.
7. Newell, A.C., and Moloney, J.V. (1991) *Nonlinear Optics*, Addison-Wesley, Reading.
8. Remoissent, M. (1994) *Waves Called Solitons. Concepts and Experiments*, Springer Verlag, Berlin.
9. Maas, S.A. (1992) *Nonlinear Microwave Circuits*, Artech House, London.
10. McGraw, R., Rogovin, D., Ho, W., Bobbs, B., Shih, R. and Feterman, H. (1988) Nonlinear response of a suspension medium to millimeterwavelength radiation, *Phys. Rev. Lett.* **61**, 943-945.
11. Rogovin, D. and Shen, T.P. (1991) Active optomechanical media for nonlinear microwave processes, *IEEE Microwave and Guided Lett.* **1**, 388-390.
12. Ma, J.G. (1990) Propagation properties of the TE-mode soliton in rectangular waveguides with nonlinear dielectrics, *Int. J. Infrared and Millimeter Waves* **11**, 1033-1045.
13. Weide, D.W. (1993) A YIG tunned nonlinear transmission line multiplier, *IEEE MTT Symp. Dig.*, 557-560.
14. Whiteley, W.C., Kunz, W. and Anklam, W.J. (1991) Utilizing a small shockline and an internal SRD, *IEEE MTT Symp. Dig.*, 895-898.
15. Jäger, D. and Tegude, F.J. (1978) Nonlinear wave propagation along periodic loaded transmission lines, *Appl. Phys.* **15**, 393-397.
16. Rodwell, M.J.W., Kamegawa, M., Yu, R., Case, M., Carman, E. and Giboney, K.S. (1991) GaAs nonlinear transmission lines for picosecond pulse generation and millimeter wave sampling, *IEEE MTT* **39**, 1194-1204.
17. Weide, D.W. (1994) Delta-doped Schottky diode nonlinear transmission lines for 480 fs 3.5 V transients, *Appl. Phys. Lett.* **65**, 881-883.
18. Dragoman, M., Szentpali, B., Müller, A., Somogyi, K., Craciunoiu, F., Riesz, F., Iordanescu, S., Varga, S. and Simion, S. (1994) Nonlinear transmission line CPW MMIC, *IEEE MELECON Conf.*, Antalia, Turkey, 617-620.
19. Dragoman, M., Szentpali, B., Müller, A., Somogyi, K., Craciunoiu, F., Riesz, F., Iordanescu, S., Varga, S. and Simion, S. (1994) Millimeter frequency generation of a traveling wave MMIC Schottky diode array and application in an automotive sensor, *European GaAs Applic. Symp.*, Torino, Italy, 293-296.

42

20. Dragoman, M., Szentpali, B., Müller, A., Somogyi, K., Craciunoiu, F., Riesz, F., Iordanescu, S., Várga, S., Simion, S. and Rizescu, R. (1994) Quasi-optical THz radar and spectroscopy instrumentation based on NLTL-MMIC, *3rd Int. Workshop on Integrated Nonlinear Microwave and Millimeterwave Conf.*, Duisburg, Germany, 259-264.

21. Weide, D.W. (1993) All electronic generation of 880 fs, 3.5 V shock waves and their applications to a 3 THz free-space, *Appl. Phys. Lett.* **62**, 22-24.

22. Frerking, M.A., and East, J.R. (1992) Novel heterojunction varactors, *Proc. IEEE* **80**, 1853-1860.

23. Dragoman, M., Bartolucci, G., Pini, F., Marcelli, R. and Jäger, D. (1995) Very high efficient frequency generation beyond 200 GHz using quantum nonlinear transmission lines, *MIOP Conf.*, Sindelfingen, Germany, 300-305.

24. Yu, R.Y., Konishi, K., Allen, S.T., Reddy, M. and Rodwell, M.J.W. (1994) A travelling-wave resonant tunneling diode pulse generator, *IEEE Microwave Guided Wave Lett.* **4**, 220-222.

25. Bartolucci, G., Dragoman, M., Marcelli, R. and Pini, F. (1995) Design considerations for tunnel diode nonlinear transmission lines, *Int. J. Infrared and Millimeter Waves*, to be published, Nov. 1995.

26. Hirota, R. and Suzuki, K. (1973) Theoretical and experimental studies of lattice solitons in nonlinear lumped networks, *Proc. IEEE* **61**, 1483-1491.

27. Bartolucci, G., Dragoman, M., Marcelli, R., Pini, F. and Simion, S. (1996) Optimization of non-uniform transmission lines for harmonic generation, to be published.

28. Carman, E., Giboney, K., Case, M., Kamegawa, M., Yu, R., Abe, K., Rodwell, M. and Franklin, J. (1991) 28-29 GHz distributed harmonic generation on a soliton nonlinear transmission line, *IEEE Microwave and Guided Lett.* **1**, 28-31.

29. Carman, E., Case, M., Kamegawa, M., Yu, R., Giboney, K. and Rodwell, M.J.W. (1992) V-band and W-band broadband monolithic distributed frequency multipliers, *IEEE Microwave and Guided Lett.* **2**, 253-256.

30. Shakouri, M.S., Black, A., Auld, B.A., and Bloom, D.M. (1993) 500 GHz GaAs MMIC sampling wafer probes, *Electr. Lett.* **29**, 557-558.

31. Ozbay, E., Li, K.D., and Bloom, D.M. (1991) 2 ps, 150 Ghz monolithic photodiode and all-electronic sampler, *IEEE Phot. Tech. Lett.* **3**, 570-572.

32. Konishi, V., Kamegawa, M., Case, M., Yu, R., Allen, S.T., and Rodwell, M.J.W. (1994) A broadband free space millimeter wave vector transmission measurement system, *IEEE MTT* **42**, 1131-1139.

33. Simion, S., Iordanescu, S. and Dragoman, M. (1994) Distributed mixer for Doppler applications, *INMMC '94*, Duisburg, Germany, 253-258.

34. Dragoman, M., Kremer, R. and Jäger, D. (1993) Pulse generation and compression on traveling-wave MMIC Schottky diode array, in H. Bertoni (eds.) *Ultra-Wideband, Short-Pulse Electromagnetics*, Plenum Press, pp.67-74.

35. Kalinikos, B.A., Kovshikov, N.G., and Slavin, A.N. (1990) Spin-wave envelope solitons in thin ferromagnetic films, *J. Appl. Phys.* **67**, 5633-5638.

36. Kalinikos, B.A., Kovshikov, N.G., and Slavin, A.N. (1991) Envelope solitons of highly dispersive and low dispersive spin waves in magnetic films, *J. Appl. Phys.* **69**, 5712-5717.

37. Chen, M., Tsankov, M.A., Nash, J.M., and Patton, C.E. (1994) Backward-vlume wave microwave envelope solitons in YIG films, *Phys. Rev. B* **49**, 12773-12790.

38. Chen, M., Tsankov, M.A., Nash, J.M., and Patton, C.E. (1994) Forward volume wave microwave envelope solitons in YIG films: propagation, decay and collision, *J. Appl. Phys.* **76**, 4274-4289.

39. Marcelli, R. and De Gasperis, P. (1994) Non-linear propagation of short microwave pulses in magnetostatic volume wave delay lines, *IEEE Trans. Magn.* **30**, 26-36.

40. Boardman, A.D., Nikitov, S.A., Xie, K. and Mehta, H. (1995) Bright magnetostatic spin-wave envelope solitons in ferromagnetic films, *J. Magnetism Magnetic Mat.* **145**, 357-378.

41. Dragoman, M., Marcelli, R. and De Gasperis, P. (1994) Experimental observation of microwave solitons of dipolar magnetostatic waves, *Appl. Phys. Lett.* **65**, 249-250.

42. Marcelli, R., De Gasperis, P., Dragoman, M. and Jun, S. (1995) Microwave pulse compression and reshaping at low levels input powers using bright surface wave solitons in YIG film delay lines, *25th European Microwave Conf.*, Bologna, Italy, 599-604.

43. Dragoman, M. and Jäger, D. (1993) Microwave gap solitons and bistability in magnetostatic periodic structures, *Appl. Phys. Lett.* **62**, 110-112.

44. Chen, M., Tsankov, M.A., Nash, J.M., and Patton, C.E. (1993) Microwave magnetic envelope dark solitons in yttrium iron garnet films, *Phys. Rev. Lett.* **70**, 1707-1710.

45. Chen, G.J., and Beasley, M.R. (1991) Shock wave generation and pulse sharpening on a series array Josephson junction transmission line, *IEEE Trans. Appl. Superconductivity* 1, 140-144.

46. Barone, A. and Paterno, G. (1982) *Physics and Applications of the Josephson Effect*, John Wiley and Sons, New York.

47. Oates, J.H., Shin, R.T., Oates, D.E., Tsuk, M.J., and Nguyen, P.P. (1993) A nonlinear transmission line model for superconducting stripline resonators, *IEEE Trans. Appl. Superconductivity* 3, 17-22.

48. Taniuti, T. and Yajima, N. (1969) Perturbation method for a nonlinear wave modulator, *J. Math. Phys.* **10**, 1369-1372.

49. Dragoman, M. and Georgescu, D. (1991) Experimental evidence of magnetostatic soliton propagation at microwave frequencies, *Appl. Phys. Lett.* **59**, 1788-1789.

50. Dragoman, M. and Catoiu, M. (1989) Observation of envelope soliton propagation in a new nonlinear transmission line, *Appl. Phys. Lett.* **54**, 1472-1473.

51. Ewen, J.F., Gunshor, R.L., and Weston, W.H. (1982) An analysis of solitons in surface acoustic waves devices, *J. Appl. Phys.* **53**, 5682-5688.

52. Anderson, D.R., Datta, S. and Gunshor, R.L. (1983) A coupled mode approach to modulation instability and envelope solitons, *J. Appl. Phys.* **54**, 5608-5612.

53. Sakuma, T. and Kawanami, Y. (1984) Theory of the surface acoustic soliton, *Phys. Rev. B* **29**, 869-879.

54. Sakuma, T. and Saito, O. (1987) Theory of surface acoustic solitons. Superlattices, *Phys. Rev. B* **35**, 1294-1299.

55. Sakuma, T. and Nishiguchi, N. (1990) Theory of surface acoustic solitons. Approximate soliton solution of the KdV type, *Phys. Rev. B* **41**, 12117-12121.

56. Planat, M. and Hoummady, M. (1989) Observation of soliton like envelope modulations in an anizotropic quartz plate by metallic integrated transducers, *Appl. Phys. Lett.* **55**, 103-105.

57. Cho, Y., Wakita, J. and Miyagawa, N. (1993) Nonlinear equivalent circuit model analysis of acoustic devices and propagation of surface acoustic waves, *Jap. J. Appl. Phys.* **32**, 2261-2264.

58. Hayata, K. and Koshiba, M. (1995) Chirosolitons: unique spatial solitons in chiral media, *IEEE MTT* **43**, 1814-1818.

INHOMOGENEOUS INTERNAL FIELD DISTRIBUTION IN PLANAR MICROWAVE FERRITE DEVICES

M. PARDAVI-HORVATH and GUOBAO ZHENG
Inst. for Magnetics Research, The George Washington University,
Washington, DC 20052, U.S.A.

Abstract

Ferrite elements are widely used in microwave devices, isolators, circulators, phase shifters. The traditional elements have a spherical shape and uniform magnetization. The magnetization distribution in modern planar ferrite elements in monolithic/integrated microwave devices, due to the non-ellipsoidal shape of the ferrite, is no more uniform. This gives rise to non-uniform internal fields, affecting the operation of the device.

For a non-ellipsoidal magnetic sample, the internal magnetic field is inhomogeneous even when a uniform external magnetic field is applied to it. In such a case the elements of the demagnetizing tensor become position-dependent. The knowledge of the *local* demagnetization tensor is important for the analysis and design of devices using finite size and shape magnetic elements. The demagnetizing tensor elements have been calculated numerically. However, often a quick check of the extent of the inhomogeneity of the magnetization distribution of a sample, or a single approximate value of the demagnetizing factor for the given non-ellipsoidal geometry, would be satisfactory. Therefore a method to define and measure an effective demagnetizing factor for rectangular and circular shapes has been developed. Experiments have been performed on YIG samples up to 220 μm thickness, on 140 μm thick square YIG films with different areas, and on MgMn-ferrite discs up to 0.02" thickness. The experimental results show that the numerical results and analytical approximation can be used to define an N_{eff}, even when the aspect ratio of the ferrite is finite.

1. Introduction

Ferrite elements are widely used in microwave devices, as isolators, circulators, and phase shifters. The traditional elements, have an ellipsoidal or spherical shape and uniform magnetization.

As the industry turns to monolithic integrated/hybrid non-reciprocal microwave devices, planar geometries have to be used. This requires the development of planar magnetic elements,

45

R. Marcelli and S.A. Nikitov (eds.), Nonlinear Microwave Signal Processing: Towards a New Range of Devices, 45–69.
© 1996 *Kluwer Academic Publishers.*

compatible with strip-line systems. As high frequency systems become fabricated using monolithic microwave integrated circuit (MMIC) designs, the size of the ferrite circulator must be compatible with the MMIC chip technology. At microwave frequencies, the size of a distributed micro-strip circulator is related to the wavelength in the ferrite ($\sim r/2$).

The magnetization distribution in modern planar ferrite elements, due to the nonellipsoidal shape of the ferrite, is no more uniform. This gives rise to non-uniform internal fields, affecting the operation of the device. For a really thin film, having thickness on the order of tens of nanometers, the aspect ratio of thickness to diameter can be regarded as zero, and the demagnetizing factors can be approximated by the corresponding ellipsoid values. In planar microwave devices the shape of the magnetic (ferrite) elements is not an infinite plane. The thickness of these layers is in the range of 100 μm, and the lateral dimensions are comparable to the thickness. As a result, demagnetizing effects can not be neglected. The demagnetizing factors, and/or the local demagnetizing tensor elements have been calculated analytically in [1], [2], [3], and [4] using various approximations. Due to inhomogeneous demagnetizing fields, the value of the external bias field in general has to be increased as compared to the case of the spherical geometry. Another problem, arising from the fact that the ferrite element is not a thin film, is that one has to solve the problem in three dimensions. The distribution of the internal magnetization of the ferrite and the electromagnetic field around the edges has a great importance for device performance [5].

The inhomogeneity of magnetization should be taken into account in device design, i.e. the knowledge of *local* distribution of demagnetization factors $N_{ij}(x, y, z)$ is necessary. If the *local* demagnetizing tensor is known, the demagnetizing factor distribution can be obtained, then the equilibrium distribution of the magnetization can be calculated by micromagnetic methods. Then it can be used to optimize and correct the analysis of microwave devices. The goal of this paper is to provide a method to obtain the nonuniform distribution of demagnetizing "factors", and the nonuniform distribution of magnetization from the distribution of demagnetizing factors. The methods include numerical, analytical and experimental techniques.

The demagnetizing correction is non-trivial for samples in open magnetic circuits. An exact correction can be obtained only for ellipsoids, where both the magnetization M and the demagnetizing field H_d are uniform under a uniform applied field H_{app} [6], [7], [8]. If the three principal ellipsoid axes coincide with the x, y, and z axes, then the internal field is:

$$H_i = H_{app} + H_d = H_{app} - \overline{N}M \tag{1}$$

$$\overline{N} = \begin{bmatrix} N_{xx} & 0 & 0 \\ 0 & N_{yy} & 0 \\ 0 & 0 & N_{zz} \end{bmatrix} \tag{2}$$

where H_d is demagnetizing field, or interaction field, N is a diagonal demagnetizing tensor, and

$$N_{xx} + N_{yy} + N_{zz} = 1 \tag{3}$$

If the applied field is along one of the principal axes, then

$$H_i = H_{app} + H_d = H_{app} - NM \tag{4}$$

where N is called the demagnetizing factor. $0 \leq N \leq 1$ in SI units, and $0 \leq N \leq 4\pi$ in CGS units. For example, for a thin film circular disk having an aspect ratio $h/d << 0.01$, magnetized along its surface normal, $N_{zz} = 1$, and for in-plane magnetic fields, $N_{xx} = N_{yy} = 0$.

For nonellipsoidal samples, if the sample is placed in a uniform applied field H_{app} along its axis, a demagnetization factor N can be defined as the ratio of the average demagnetizing field to the average magnetization of the entire sample [9]:

$$\int_V H_d(r)\, dS = -N \int_V M(r)\, dS \tag{5}$$

N is the function of the aspect ratio. So, the ellipsoidal approximation is no longer valid and a relationship between the shape and size of the ferrite element and the demagnetizing factors should be developed.

The problem of the calculation of demagnetizing factors or demagnetizing tensor elements is not new. Joseph and Schloemann [1] developed an analytical approximation for calculating the nonuniform demagnetizing factor for ferromagnetic bodies of arbitrary shape. Kraus [10] determined the complete local demagnetizing tensor for uniformly magnetized cylinders. Brug and Wolf [11] calculated the magnetization distribution in discs and obtained the local demagnetizing factors for materials that undergo phase transitions. Brown [4] has shown how demagnetizing factor could be determined using self-inductance calculations where a uniformly magnetized cylinder was modeled as a solenoid. Chen [3] evaluated the fluxmetric and magneto-metric demagnetizing factors for cylinders as a function of susceptibility and aspect ratio. Most simple, people often look thin film as the oblate ellipsoid, long cylinder as prolate cylinder. In general, thin samples are treated as a special case of an oblate ellipsoid, and long cylinders as prolate ellipsoids. The formulas are given for these two cases in [2].

All the above methods are based on simplifying assumptions. For example, Ref. [1] is based upon the assumption that the magnitude of the magnetization vector is constant throughout the sample and that its direction coincides with the direction of the local magnetic field at any point within the sample. This assumption may cause a 20% error. It is desirable to know the difference between the real demagnetizing factors and that calculated by these methods. With the advance of numerical computing methods, these approximations can be relaxed and the local demagnetizing tensor elements can be calculated numerically. However, often a quick check of the extent of the inhomogeneity of the magnetization distribution of a sample, or a single approximate value of the demagnetizing factor for the given non-ellipsoidal geometry, would be satisfactory. Therefore a method to define and measure an effective demagnetizing factor for rectangular and circular shapes has been developed by us [12].

We have developed a quick experimental method to determine an effective demagnetizing factor N_{eff} for thick, small rectangular and circular ferrite samples, as those applied in microwave devices. Finite Difference Method (FDM) was used to determine the local values of the demagnetizing tensor elements for arbitrary shaped samples. All results are compared to the results derived from [1], [2], [3], [4]. The comparisons between the demagnetizing factor N from experimental and numerical/analytical methods are based on statistical averaging all the local demagnetizing tensor elements over the volume.

2. Inhomogeneous distribution of internal field

The distribution of magnetization is uniform if an uniform external field is applied to an ellipsoidal body. However, it is not true for nonellipsoidal shapes, as a consequence, the internal field, acting on the local magnetization, will be non-uniform. If the distribution of the internal field is known, the problem of the magnetization distribution can be solved numerically by micromagnetic methods.

The demagnetizing tensor elements were calculated for an arbitrary rectangular body having dimensions $2a \times 2b \times 2h$ $(a > b > h)$, as shown in Fig. 1. The origin of the coordinate system is located at the center of the sample. For this configuration, according to the results of the first order approximation of [1], the demagnetizing factor along the z direction, $N_z^{(1)}(r)$ can be obtained by

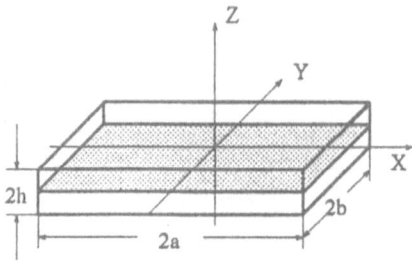

Figure 1: Rectangular sample, $a > b > h$

$$N_{zz}^{(1)}(\vec{r}) = -\frac{1}{4\pi} \log\{\frac{G(\vec{r}|a,b,h)G(\vec{r}|-a,-b,h)G(\vec{r}|-a,b,-h)G(\vec{r}|a,-b,-h)}{G(\vec{r}|-a,b,h)G(\vec{r}|a,-b,h)G(\vec{r}|a,b,-h)G(\vec{r}|-a,-b,-h)}\} \qquad (6)$$

$$G(\vec{r}|a,b,h) = (b-y) + [(a-x)^2 + (b-y)^2 + (h-z)^2]^{\frac{1}{2}} \qquad (7)$$

$N_{zz}^{(1)}(r)$ can be used to calculate the distribution of internal field in this direction.

Figure 2(a) shows the distribution of demagnetizing factor N_{zz} on the top or bottom plane ($z = \pm h$) of the rectangular slab. The distribution of demagnetizing factors is clearly not uniform, it is position-dependent, like the demagnetizing fields. Therefore the distribution of internal magnetic field will follow the distribution of H_d, and the local magnetization will be aligned with the internal field in the case of zero anisotropy. For the external field applied along the z axis of a sample having saturation magnetization M_r, Eq. (4) gives

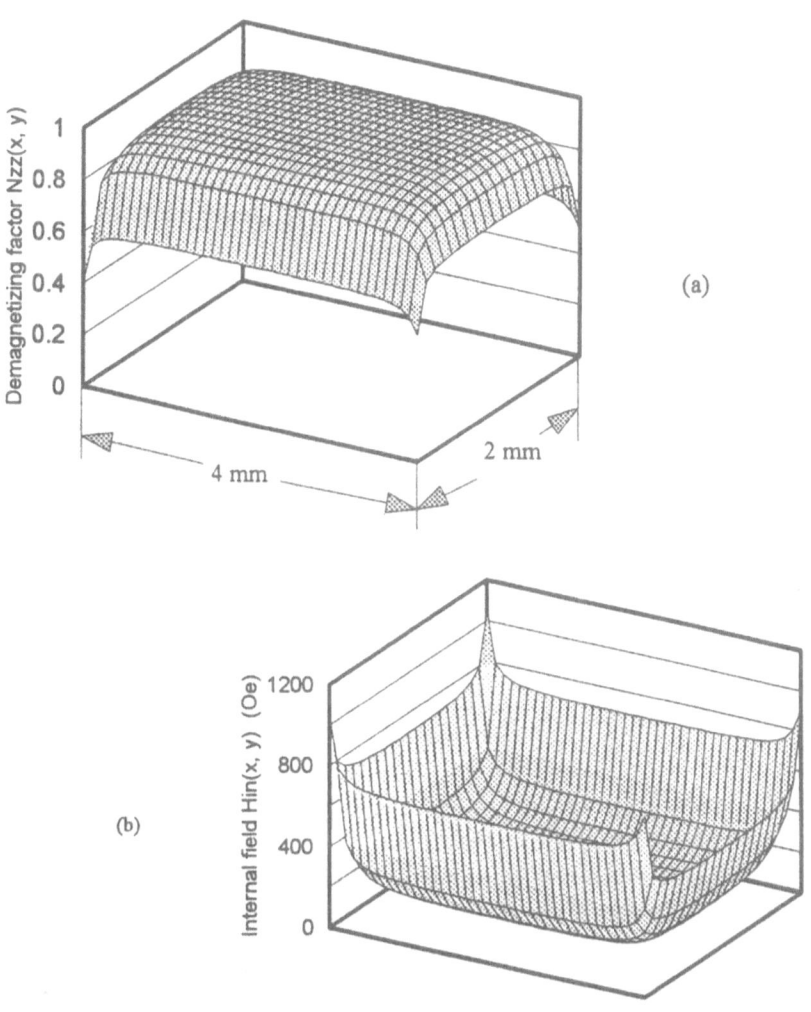

Figure 2. Distributions of (a) 3-D demagnetizing factor N_{zz}, (b) internal field H_{in}, © 2-D demagnetizing factor N_{zz}. The sample is 4 mm× 2 mm × 115 μm YIG film.

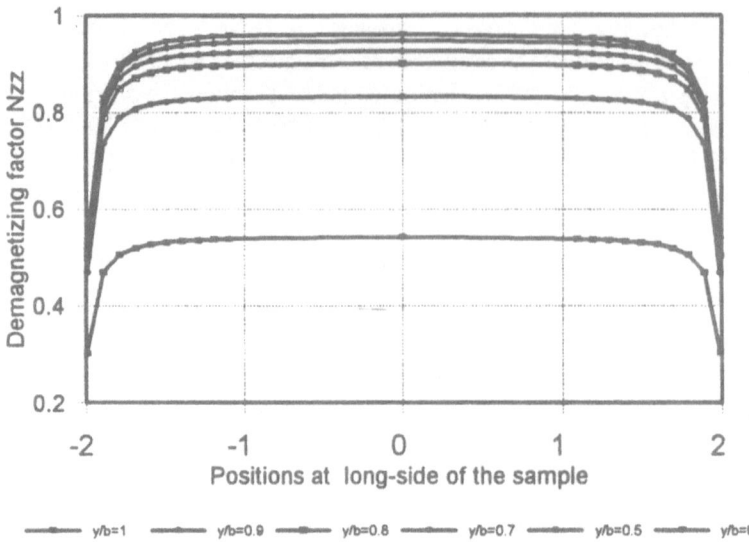

Figure 2(c). The distribution of 2-D demagnetizing factor N_{zz}

$$H_{in}=H_{app}-4\pi M_s \qquad\qquad\qquad (8)$$

Using (8), the distribution of internal fields in ($z=\pm h$) plane can be calculated, as shown in Figure 2(b). Further calculations show that the distribution of the internal field in this plane is mostly uniform for thin films (for 4x2 mm² YIG films if the thickness is less than 50 μm), inhomogeneous fields occur only around the edges of the sample. It implies that the formula could be a very good approximation for thin films. In order to have a more clear picture of the distribution of demagnetizing factors, in Figure 2© we show the changes along the long-side of the sample for different cross sections (y/b) in the top/bottom plane.

Figure 3(a) represents the distribution of demagnetizing factors along the edges of the sample. From the figures it is clear that the minimum value of N_{zz} is on the center of the short side edge, the maximum is on the long side edge. Figure 3(b) is an another example of the distributions of N_{zz} along the long side of the sample.

Calculating the demagnetizing factors all over the sample, the maximum N_{zz} will appear at the center on the top(bottom) surface. The smallest value will be at the corners of sample. If N_{zz} calculated over planes, corresponding to different z, and the average of all N_{zz} on that plane is taken, the maximum average N_{zz} will appear on the top(bottom) plane of sample, the minimum N_{zz} appears on the center plane of the sample. It will be shown that the average of N_{zz} all over the sample will correspond to the experimentally measured $N_{zz\,eff}$.

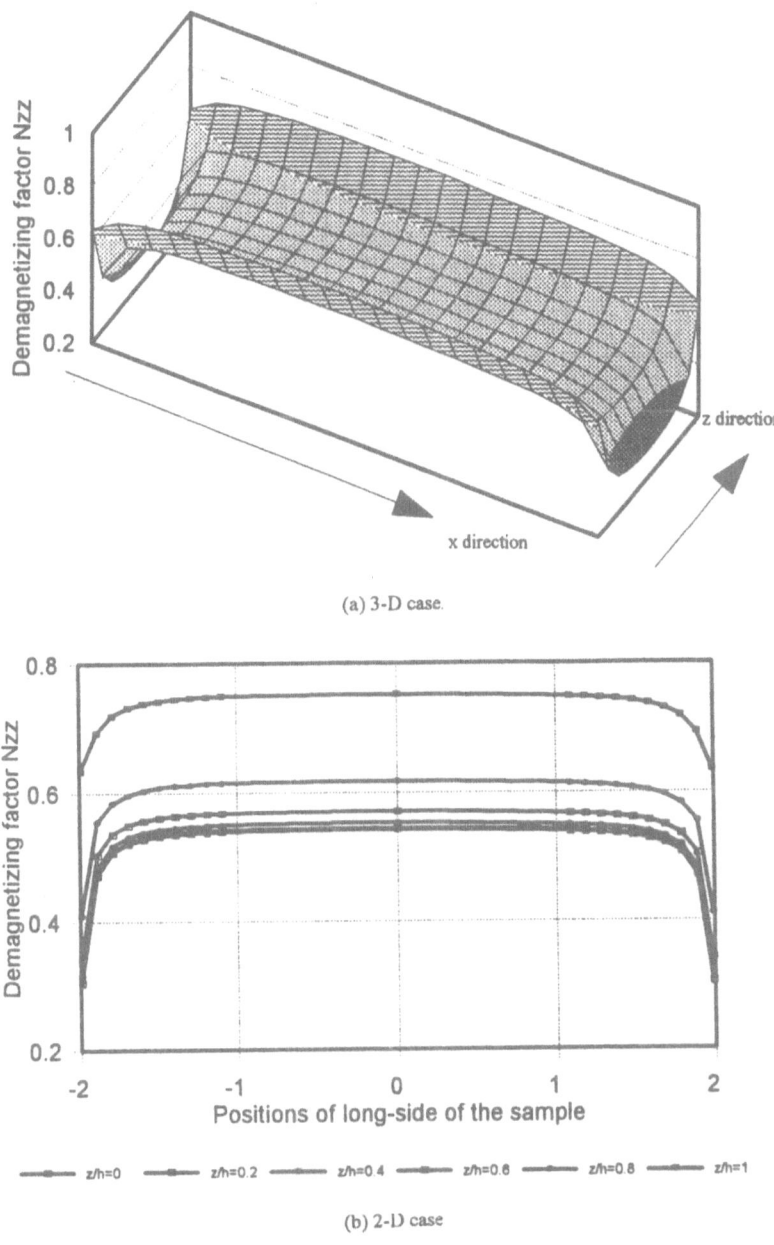

(a) 3-D case.

(b) 2-D case

Figure 3. Distributions of demagnetizing factor N_{zz} along the edge of the sample (a) 3-D case, (b) 2-D case. The sample is 4 mm × 2 mm × 115 μm YIG film.

3. Experimental method

The experimental method to determine the demagnetizing factor for a nonellipsoidal sample is based on the measurement of the in-plane and out-of-plane major hysteresis loops using a Vibrating Sample Magnetometer (VSM). Here, in-plane measurement means the applied magnetic field is parallel to the sample plane, out-of-plane means the applied field is perpendicular to the sample plane, as shown in Figure 1.

Thickness dependence measurements were done on 4×2 mm^2 YIG samples, in the thickness range from 5 to 220 μm, grown by a special Liquid Phase Epitaxial technique at the Institute for Materials Science, Budapest, Hungary. Samples were cut from the same 220 μm thick wafer and polished mechano-chemically to the final thickness. Size dependence was investigated on 140 μm thick square samples, 3, 4 and 5 mm on side, cut from the same wafer. Circular MgMn ferrite disks (EMS Technologies), diameter 0.15" and the thickness range from 0.005" to 0.020" with 0.005" step size, were used to measure $N_{z,\text{eff}}$ for cylindrical samples.

The in-plane and out-of-plane hysteresis loops for disk samples are shown in Figure 4(a). Due to the different demagnetizing factors and the inhomogeneity of the internal field in thick films or discs, the in-plane loop differs from the out-of-plane loop. The loops should be congruent after correction for the different demagnetizing factors, determined using Eq.(8).

The procedure of the correction is the following. First, it is assumed that the samples are thin enough, so that N_{xx} approaches zero, and there is no need to correct the in-plane hysteresis loop at this moment. Now, the out-of-plane loop is corrected with a proper N_z until it overlaps with the in-plane loop. This procedure gives the minimum N_{zz}, see Figure 4(b).

The next step is to consider that the inhomogeneous field exists not only in the out-of-plane direction, but also in-plane. So, we have to correct both hysteresis loops with N_z and N_x to arrive at the same shape. Usually, the exact shape of the loop is unknown. But, according to the characteristics of magnetic films, the final hysteresis loops, after corrected by demagnetizing factors, should not "overshoot", i.e. have a negative slope. This limiting N_z is the maximum N_{zz}. Thus, at the same time, N_{xx} is determined. The final loops are shown in Figure 4(c). The relationship between N_{xx}, $Max\, N_{zz}$ and $Min\, N_{zz}$ is given by:

$$Max\, N_{zz} = Min\, N_{zz} + N_{xx} \tag{9}$$

The values of $Max\, N_{zz}$ and $Min\, N_{zz}$ determined experimentally, are very useful to verify our numerical and analytical results[1], [2], [3], [4]. In designing magnetic devices, if the range of N_{zz} is known, one can visualize how strong is the inhomogeneity of the fields in the magnetic sample. Also, in most of the cases, the average demagnetizing factors, determined from experimental data, are the $Max\, N_{zz}$, in other words, one can measure the demagnetizing factors.

We have to point out that there is another important element we have to consider before correcting the loops, that is the paramagnetic susceptibility χ_{pm} of the substrate for epitaxial YIG films. For example, our YIG films are grown on the 500 μm GGG substrate, and its effect

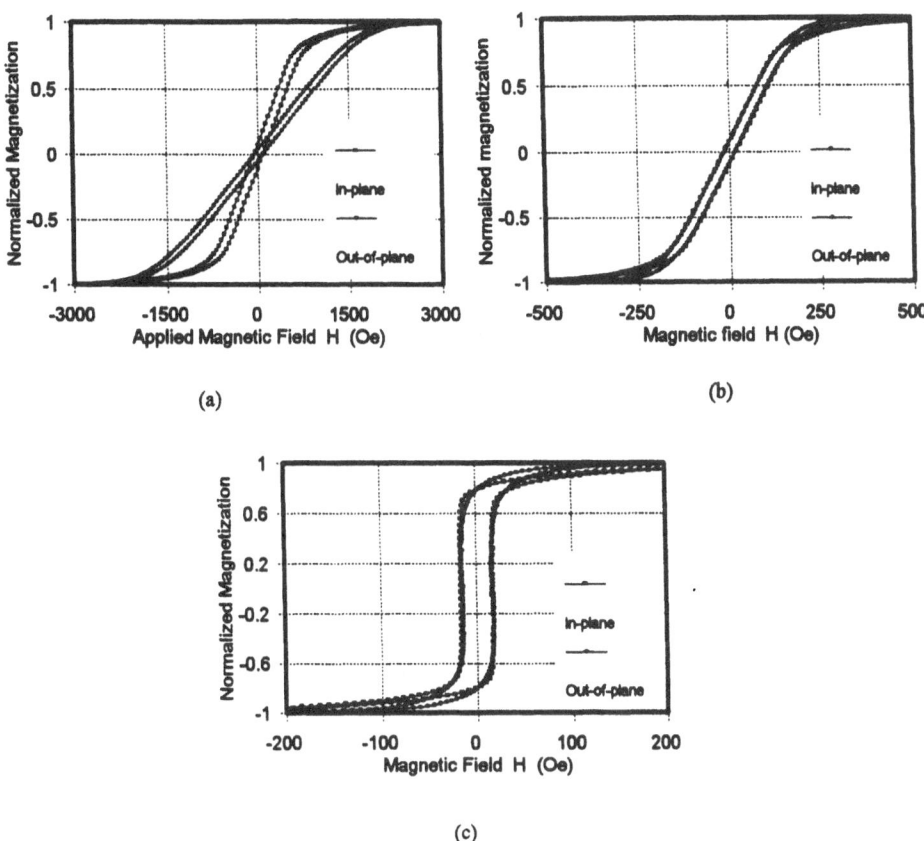

Figure 4. Hysteresis loops (a) oringinal measured by VSM, (b) only corrected by $Min\ N_{zz}$ (c) corrected by both $Max\ N_{zz}$ and N_{xx}. The sample is MgMn ferrite disc with 0.15" diameter and 0.015" thickness.

might overcome the magnetization of a thin film sample. The elimination of the substrate effects from the measured data is based on the linear regression analysis of the linear part of the hysteresis loops above saturation.

$$M(H) = M_{meas} - \chi_{pm}H \qquad (10)$$

For thin samples with a small thickness, the magnetization is small (assuming the size of sample is fixed). For example, for the typical YIG sample size of 4 mm x 2 mm and a thickness of 1 µm, the magnetization is about 1 memu. In this case, the sensitivity, the calibration accuracy, and the zero offset of the VSM become essential elements to the measurements. The integrating time constant and the number of points to be measured should be increased as the film thickness is reduced.

54

Because the susceptibility, $\chi_{GGG} = const$ at a given temperature, it can be obtained from the slope of the magnetization curve far above saturation. The susceptibility of the single crystal GGG is a tensor, it depends on the direction in the substrate plane. Because our samples were cut along the same crystallographic direction, the anisotropy of the susceptibility can be neglected. The measured value at room temperature is $\chi_{GGG} = 10^{-6}$ emu/G, i.e. the contribution to the magnetization is $M_{GGG} = 10^{-6} H_{app}$ emu. For the 4 mm x2 mm x 1μm sample, in a field of $H_{appl} = 4\pi M_s = 1780$ G, $M_{GGG} = 1.78$ memu, it is larger than the magnetization of the YIG sample itself. As a result, the shape of the hysteresis loop is highly distorted, especially for the out-of-plane loop. After correcting for the GGG paramagnetism, and applying our procedure to the corrected loops, the discrepancy of non-physical demagnetization factor values disappears.

Figure 5 shows the loops before and after eliminating the effects of the GGG substrate. The loops are measured on a 4 mm x 2 mm x 22 μm YIG sample.

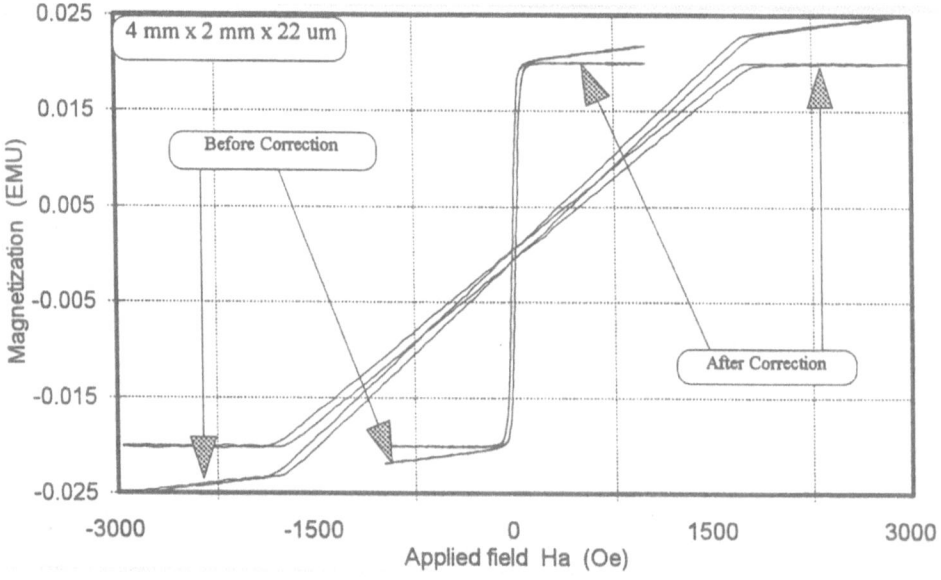

Figure 5. Hysteresis loops before and after eliminated the effect of the GGG substrate
The data was measured on 4 mm x 2 mm x 22 μm YIG film.

After the correction for GGG, numerical values for $4\pi M_s$ are obtained. For single crystalline YIG films, grown by Liquid Phase Epitaxy, $4\pi M_s$ is about 1730-1780 G, depending on the actual growth method and impurity content. Taking into account the inaccuracy of the thickness measurement (measured optically from interference pattern, what includes the index of refraction error) of about 5%, the VSM values for the magnetization indicate the very good accuracy of the method. The results will be given later.

4. Numerical method of calculation of the demagnetizing tensor elements

The internal field in a magnetized body is determined by the vector sum of the contributions from the exchange, anisotropy, Zeeman and demagnetizing fields. The equilibrium state of the magnetization corresponds to the minimum of the free energy [13]. In general for the ferrites applied in planar microwave devices, the demagnetizing fields are nonuniform, and in the

numerical micromagnetic problem the most time-consuming part of the calculation is the computation of the demagnetizing field inside the sample. The calculation can be accelerated by computing a demagnetizing tensor D between each pair of elements separately and then storing them, since the field is the product of this tensor and the corresponding magnetization. The field at element i due to element j, H_{ij}, is thus expressed as:

$$H_{ij} = D(r_{ij}) \cdot M_j \tag{11}$$

where $r_{ij} = r_i - r_j$ is the relative distance between the elements, and M_j is the magnetization at the element j. It should be noted that this tensor D has to be defined for each pair of elements, that is, it is a point function demagnetizing tensor, while for ideal ellipsoids, it only has to be defined at one point in order to define it for the entire sample. Once a discretization grid is given, then the elements of D can be obtained by calculating the field H_{ij} as

$$H_{ij} = -\nabla\phi \tag{12}$$

where ϕ is the scalar potential due to the magnetostatic charges, and

$$\phi = \oint_{S_j} (M \cdot dS)/(4\pi r_{ij}) \tag{13}$$

where S_j is the surface of the element. The integration is performed only over the surfaces since the magnetization is assumed to be uniform in each element. Here, a cubic discretization procedure is chosen which greatly simplifies the computation of the D tensor elements. A first order approximation is the dipole-dipole interaction between elements, which is independent of the discretization.

The integral expression (12) of the interaction field can be found by expanding $1/r_{ij}$ in a Taylor series. The magnetic dipole-dipole interaction is the first term in the series for the tensor elements of D:

$$D_{zz}(x, y, z) = d^3(3z^2/r^2 - 1)/(4\pi r^3) \tag{14}$$

and

$$D_{zx}(x, y, z) = 3zxd^3/(4\pi r^5) \tag{15}$$

where d is the size of the cubic elements, and $r^2 = x^2 + y^2 + z^2$.

One direct way to compute the tensor elements for the nearest neighbors is to integrate the pole densities over the six surfaces of each cube. The volume poles are zero due to the uniform magnetization in each element. This can be accomplished by numerical integration. In order to simplify the integration, the magnetization vector can be set in a certain direction. For example, in order to calculate D_{zz} and D_{zx}, one would let $M_x = M_y = 0$, and $M_z = 1$. Once H_z and H_x are integrated then

$$D_{zz} = H_z/M_z = H_z \tag{16}$$

and

$$D_{zx} = H_x/M_z = H_x \tag{17}$$

It is clearly seen from expressions for the dipole D tensor elements in (13), (14), that

$$D_{xx}(x, y, z) = D_{zz}(z, y, x), \ D_{yy}(x, y, z) = D_{zz}(x, z, y) \tag{18}$$

and

$$D_{xy}(x, y, z) = D_{zx}(x, z, y), \ D_{yz}(x, y, z) = D_{zx}(z, y, x), \ \text{ect.} \tag{19}$$

In other words, the diagonal element D_{xx} can be obtained from D_{zz} by permuting variable x with z, and D_{yy} is obtained by permuting variable y with z; the off-diagonal element D_{xy} can be obtained from D_{zx} by permuting variables y and z, D_{yz} by permuting the variables x and z. The property holds for other elements as well. Therefore when one diagonal element and one off-diagonal element are calculated, the other seven elements of the D tensor are obtained by permuting the variables in the corresponding way. This property is true in general for the elements derived from (12) and are currently used in the calculations. Thus only two sets of elements are computed and stored in which the dipole field approximation has been corrected by the numerical integration of the magnetostatic surface charges of each cubic element.

Table I shows the values of D_{zz} calculated by dipole approximation and surface integration for 2D case. The dipole approximation overestimates the exact surface integral calculation D_{zz} by 17.4 % for the first neighbor, and by 1.4 % for the 2nd neighbor; the deviation being negligible at larger distances.

Now, the demagnetizing factors can be calculated. Assuming that the applied field H_{app} is is along the z direction, then the demagnetizing field at a given cubic element is equal to the sum of demagnetizing fields from all neighbors

$$H_D(i, j, k) = \sum_i \sum_j \sum_k D_{zz}(i, j, k) \, M_z(i, j, k) = M_z \sum_i \sum_j \sum_k D_{zz}(i, j, k) \tag{20}$$

Therefore the demagnetizing factor can be obtained by

$$N_{zz}(i, j, k) = H_D(i, j, k) / M_z = \sum_i \sum_j \sum_k D_{zz}(i, j, k) \tag{21}$$

TABLE I. Comparison between the dipole approximation and surface integral for demagnetization tensor element $D_{zz}(x, y)$ (units:10^{-3})

Shell	1st	2nd	3rd	4th	5th
Distance	d	$\sqrt{2}$ d	d	$2\sqrt{d}$	$\sqrt{5}$d
Dipole	-79.58	-28.13	-9.947	-7.118	-3.517
Surf.Int	-67.79	-27.73	-9.848	-7.103	-3.513
Error, %	17.4	1.4	1.0	0.21	0.11

The FDM code has been written to calculate the demagnetizing tensor and demagnetizing factors for whole sample. Figure 6 shows the demagnetizing tensor of the corner cubic element in the 20 × 20 × 20 sample. In order to compare the numerical results with the results from [1], we choose two extreme cases. One is for a larger aspect ratio sample, 180×90×10, shown in Figure 7(a). Another one for a smaller aspect ratio sample, 180×90×2, shown in Figure 7(b).

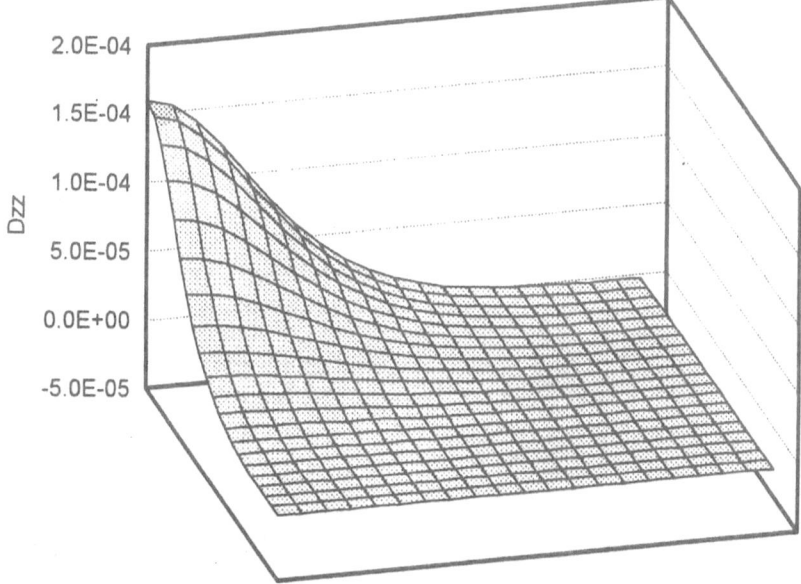

Figure 6. Demagnetizing tensor of cubic element at the conner from the elements on the center plane of 20×20×20 sample

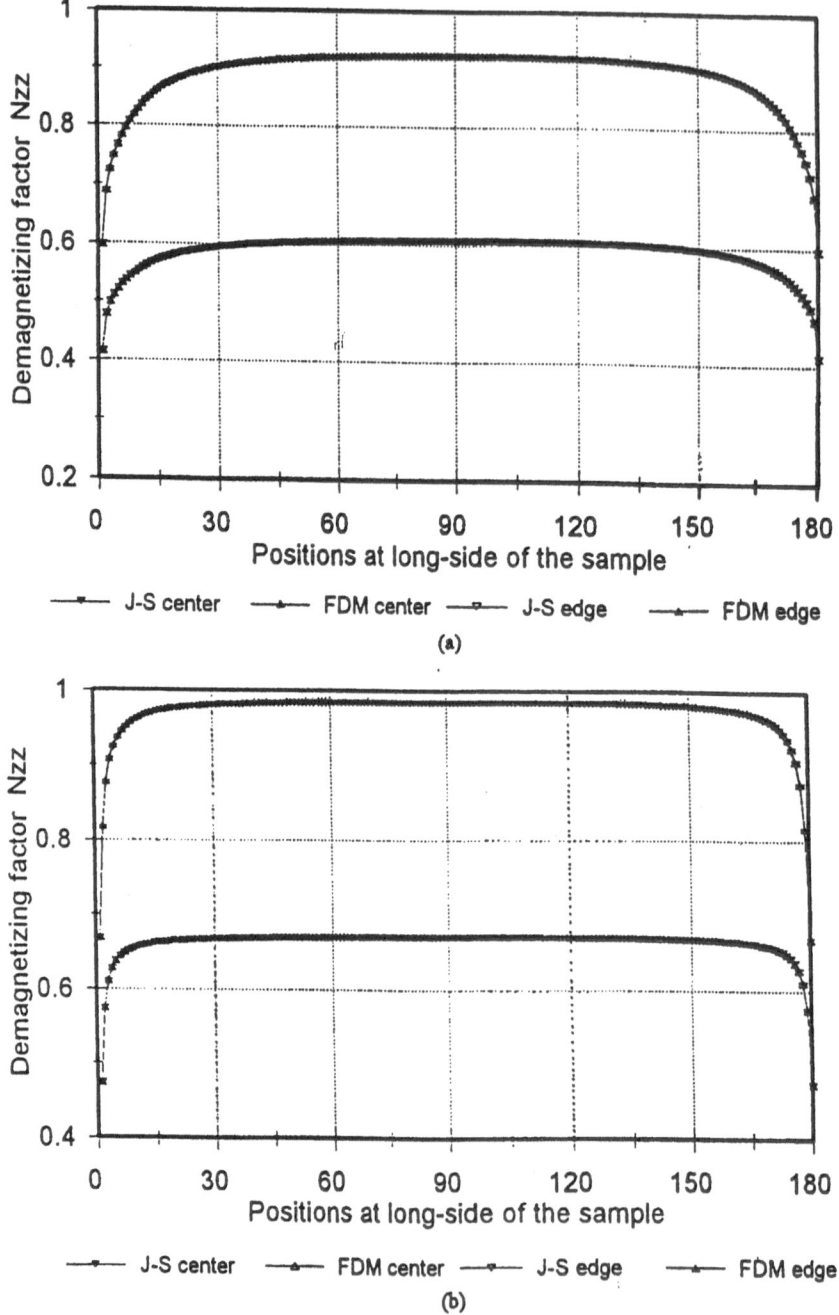

Figure 7. Comparison between analytic and numerical results of demagnetizing factors at the top-plane of the samples, (a) 180 x 90 x 10 sample, (b) 180 x 90 x 2 sample.

5. Applications and comparisons

In designing magnetic devices, the range of N_z is the measure of the inhomogeneity of the internal field in the sample. The correction procedure has been applied to a series of YIG samples of same size, but having different thickness, and same thickness but different size; and circular MgMn ferrites of different thickness.

According to the experimental method developed in Section 3., measurements have been made to find the demagnetizing factors for different geometries. The experimental results were compared to the results from the [1], [2], [3], and [4].

5.1. CIRCULAR FERRITE SAMPLES

Measurement were done on circular MgMn-ferrite samples. The diameter (D) is 0.15 inch, the thickness (L) is from 0.005 " to 0.025", the step size is 0.005" , $4\pi M_s$= 2350 G. The results are shown in Figure 8.

Only for infinitely large or infinitely thin circular ferrite samples can the edge fields be neglected and the ferrite can be treated as a homogeneously magnetized ellipsoidal body. Although this is not true in actual microwave ferrite devices, especially when we try to reduce the size of ferrite devices to be compatible with MMIC, still the ellipsoidal approximation for demagnetizing factors of cylinders are mostly used in the design and analysis of microwave ferrite devices.

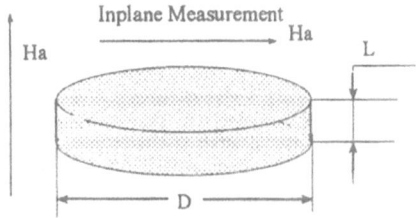

Figure 8. MgMn ferrite disc with D=0.15"

For a thin circular sample, N_{xx} is equal to N_{yy}. If the ratio $m=D/L$ is large, the sample can be referred to oblate ellipsoid approximately [2]. Then

$$N_{xx} = N_{yy} = 2\,\pi\times \{ \frac{m^2}{(m^2-1)^{3/2}} \sin^{-1}(\frac{\sqrt{m^2-1}}{m}) - \frac{1}{m^2-1} \} \qquad (22)$$

$$N_{zz} = 1 - N_{xx} - N_{yy} \qquad (23)$$

From (22), we know N_{xx} and N_y are very small for large m , approaching zero, and N approaches 1. Figure 9 illustrates the dependence of the demagnetizing factors N_{zz} on the aspect ratio m.

The second approximation of magnetizing factors for cylinder was developed in Refs. [3] and [4]. The demagnetizing factors can be determined using self-inductance calculations where

60

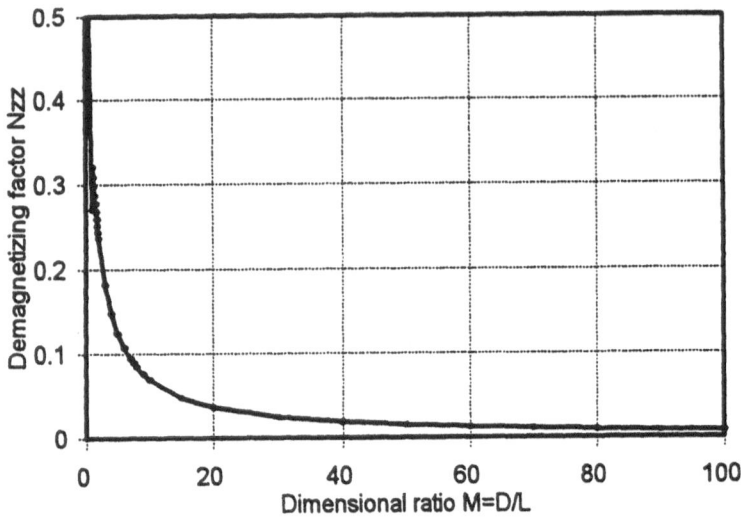

Figure 9. Demagnetizing factor N_{zz} vs aspect ratio m curve for circular sample

a uniformly magnetized cylinder was modeled as a solenoid. The main formulas for this approximation are

$$N_{zz}=1-4LL_s/(\mu_o\pi D^2) \qquad (24)$$

$$L_s=u_o/(3L^2)*\{(D^2+L^2)^{1/2}*[L^2F(K_s)+(D^2-L^2)E(K_s)]-D^3\} \qquad (25)$$

$$K_s^2=D^2/(D^2+L^2) \qquad (26)$$

where $F(K_s)$ and $E(K_s)$ are the complete elliptic integrals of first and second kind of modulus K_s, and μ_o is the permeability of vacuum.

Figures 10(a) and 10(b) show the measured hysteresis loops both in-plane and out-of-plane for m=10 amd m=15. For the in-plane case, there are obvious differences between the hysteresis loops, indicating that the difference in the distribution of internal fields; at a given applied field, the smaller the thickness, the larger is the magnetization. In other words, the thinnest samples has a smallest demagnetizing factor. For the out-of-plane case, again we can see the difference between the loops. Figure 4(a) shows the in-plane and out-of-plane loops for $m=D/L=10$. First, keeping the in-plane loop unchanged,

$$H=H_a-4\pi M_s N_{zz} \qquad (27)$$

to correct the out-of-plane loop. The result is shown in Figure 4(b). After this procedure, we got Min N_{zz}=0.647.

61

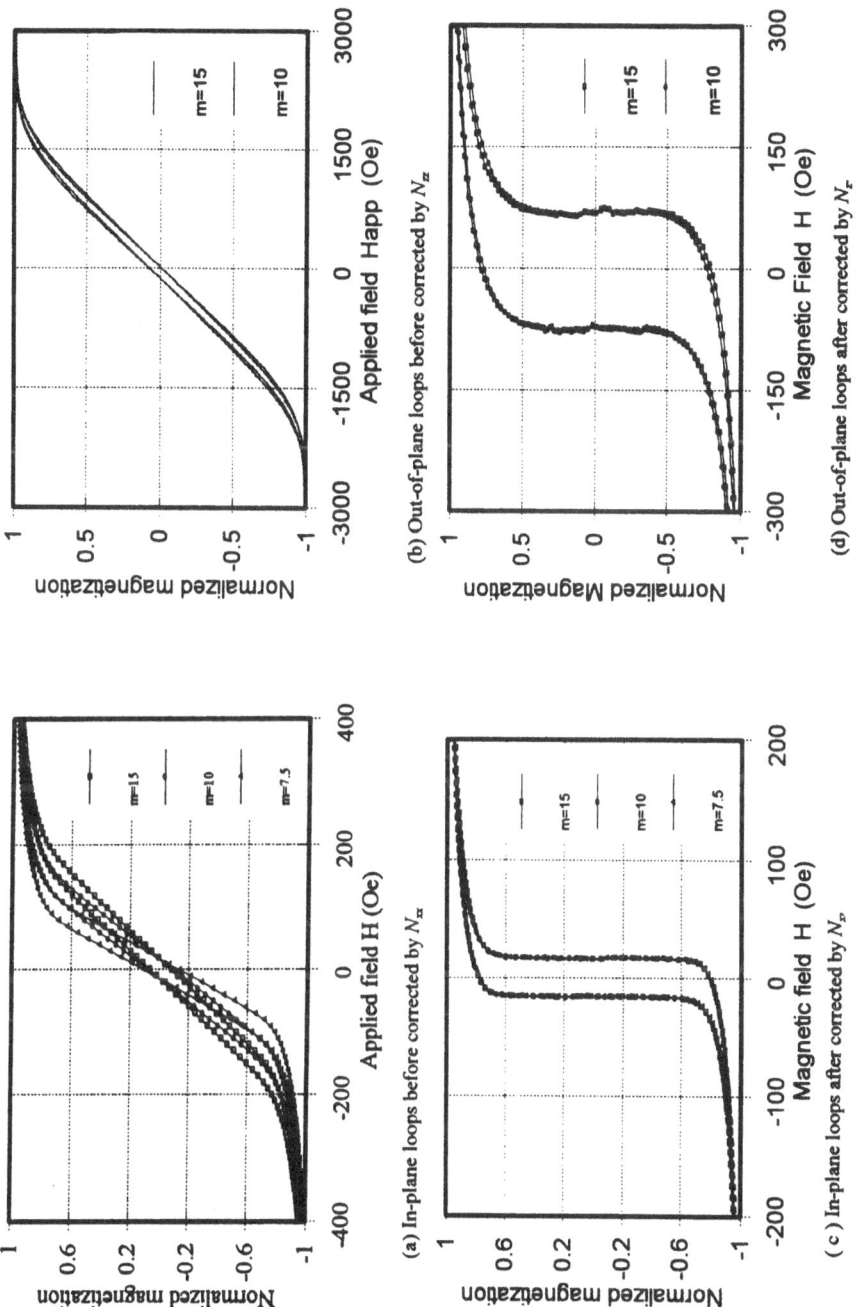

(a) In-plane loops before corrected by N_{xx}

(b) Out-of-plane loops before corrected by N_{zz}

(c) In-plane loops after corrected by N_{x}

(d) Out-of-plane loops after corrected by N_{z}

Figure 10. Hysteresis loops of MgMn ferrite samples for m=10 and m=15

Second, we apply Eq (27) to both the in-plane and out-of-plane loops, and the result is shown in Figure 4(c). At this time, we got $Max\ N_{zz}$=0.725 and $Max\ N_{xx}$=0.0785. It is clear that $MaxN_{zz}$=$Min\ N_{zz}$+ $Max\ N_{xx}$. As we expected, the loops corrected by $Max\ N_{zz}$ and $Max\ N_{xx}$ now overlap.

Repeating the same process to the measured hysteresis loops of all other samples, we get the values shown in Table II.

TABLE II: Thickness dependence of demagnetizing factor MgMn-ferrite with 0.15 " Diameter

N		D/L=15	D/L = 10	D/L = 7.5
N_{xx}	Theory[2]	0.0482	0.0696	0.0894
	Measured	0.0562	0.0785	0.0955
	Error	0.0081	0.0089	0.0061
N_{zz}	Theory[2]	0.904	0.861	0.821
	Theory[3]	0.848	0.797	0.753
	Measured	0.758	0.725	0.698

Figure 10(c) illustrates the in-plane hysteresis loops after the corrections are performed for circular ferrite samples of different aspect ratios, D=0.15". They are exactly same, as it is expected for the same material, after getting rid of shape effects. Figure 10(d) shows out-of-plane case. Theoretical values for N_{xx} in Table II are calculated by [2], and the corresponding N_{zz}=1-$2N_{xx}$ are given in Table II. The difference between the experimental data and the theoretical results of the ellipsoidal approximation [2] is much larger than expected. The solenoid approximation for demagnetizing factors given in [3] is much more close to the experimental results. These results are also given in Table II.

5.2. RECTANGULAR YIG SAMPLES

Measurements were done on the rectangular 4×2 mm^2 epitaxial YIG films with 4.7 μm, 12 μm, 21 μm, 22 μm, 35 μm, 55 μm, 70 μm, 115 μm, 145 μm, and 220 μm thickness. The theoretical results for the local demagnetizing tensor element $N_{zz}(x,y,z)$, are based on our numerical method and the first order approximation of [1]. In order to compare the experimentally measured values of the effective demagnetizing factor to the numerical data, the numerical values of the local demagentizing tensor elemnets have been statistically averaged over the sample volume.

Table **III** shows the results from both numerical and the 1st order approximation of [1]. The agreement between these values is very good, what is not surprising because both methods use the same approximation of the constancy of the magnetization vectors all over the sample.

TABLE III: Comparison between the first order J-S formula and our 3-D numerical method, the size of samples is 4 mm x 2 mm with different thicknesses.

Thickness (μm)	43	65	87	108	130	152	174	195	217
Ref. [1]	0.9500	0.9308	0.9134	0.8980	0.8830	0.8687	0.8552	0.8430	0.8307
3D FDM	0.9516	0.9320	0.9144	0.8983	0.8833	0.8692	0.8559	0.8430	0.8003
Error (%)	0.18	0.14	0.13	0.05	0.05	0.05	0.10	0.00	0.03

Measured hysteresis loops are shown in Figures 11(a) and (b). Demagnetizing factors are obtained from the in-plane and out-of-plane loops for each sample after correction for substrate paramagnetism. For thicknesses above 50 μm, the GGG substrate paramagnetism hardly affects the results. Figure 11(c) shows the loops after being corrected, meanwhile N_z and N_x are obtained from these loops.

The complete results are given in Table **IV** and Figure 12. In the Table IV, $MaxN_z$ is the average of demagnetization factors on the top (bottom) plane, $MinN_z$ is on the center plane; $Avg\ N_z$ is the average all over the sample. The agreement between the averaged values of the first order approximation of [1] and the measured demagnetization factors is very good when the thickness is less than 70 um. As it is expected, the difference between results becomes larger as the thickness increases. This indicates the increasing inhomogeneity of the internal field, and the corresponding non-collinearity of the magnetization.

5.3. SHAPE DEPENDENCE OF THE DEMAGNETIZING FACTOR

Measurements are performed on square YIG samples with aspect ratio of $20 \le L/h \le 200$. The geometry of samples is shown in Figure 13. The thickness of samples is 140 μm, the area are 3x3 mm², 4x4 mm², and 5x5 mm².

Figures 14(a) and (b) show the in-plane and out-of-plane hysteresis loops of these samples. At a given field, the smallest sample shows the largest demagnetization effects, because of the

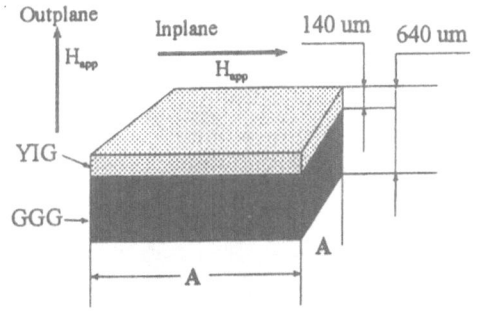

Figure 13. Square YIG Sample

64

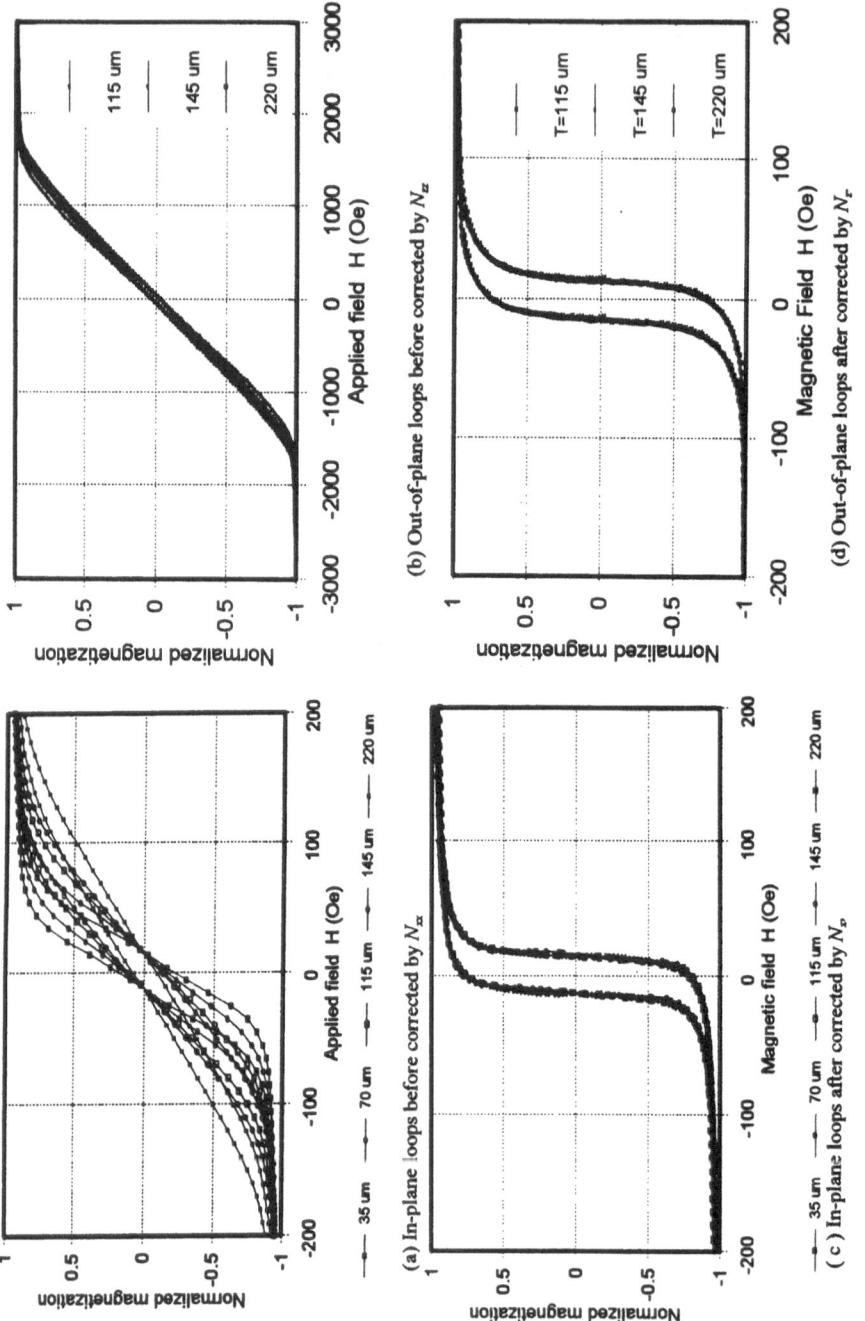

Figure 11. Hysteresis loops of 4 x 2 mm² rectangular YIG films with different thicknesses

TABLE IV: Thickness dependence of demagnetizing factors of rectangular (4x2 mm²) YIG samples

Thickness (μm)	Theoretical Estimation			Measurement			
	Max N_z	Min N_z	Average N_z	Max N_z	Min N_z	Max N_x	$4\pi Ms$ (Gauss)
12	0.984	0.979	0.982	0.985	0.967	0.018	1778
21	0.973	0.970	0.972	0.971	0.941	0.030	1769
35	0.960	0.957	0.958	0.955	0.922	0.033	1800
55	0.941	0.937	0.939	0.940	0.910	0.030	1778
70	0.929	0.925	0.927	0.915	0.872	0.043	1732
115	0.901	0.886	0.893	0.870	0.812	0.058	1786
145	0.885	0.8649	0.873	0.848	0.785	0.064	1790
220	0.849	0.818	0.829	0.790	0.700	0.09	1772

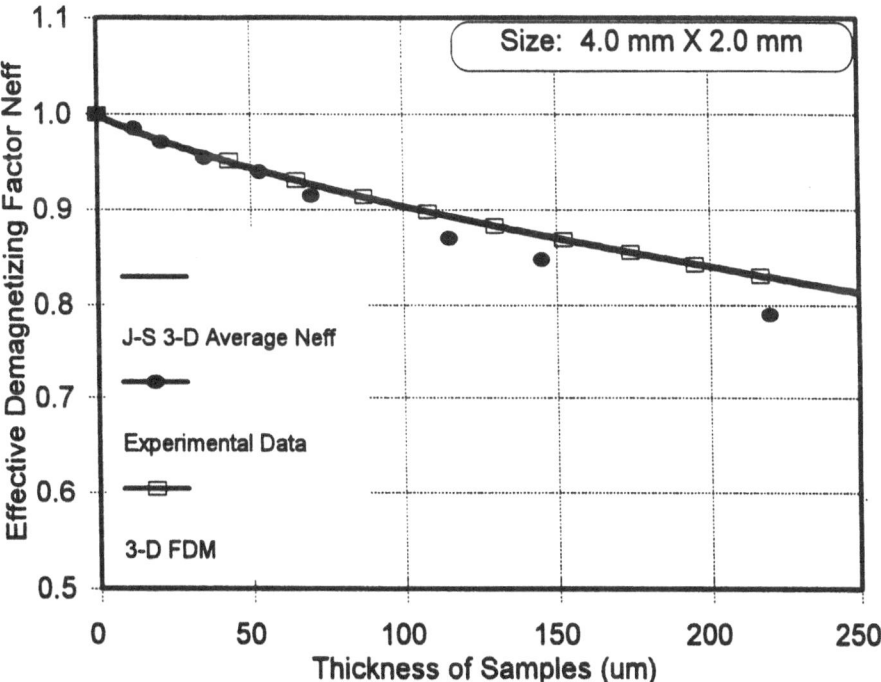

Figure 12. Comparisons among experimented, numerical, and analytic results for different thicknesses

66

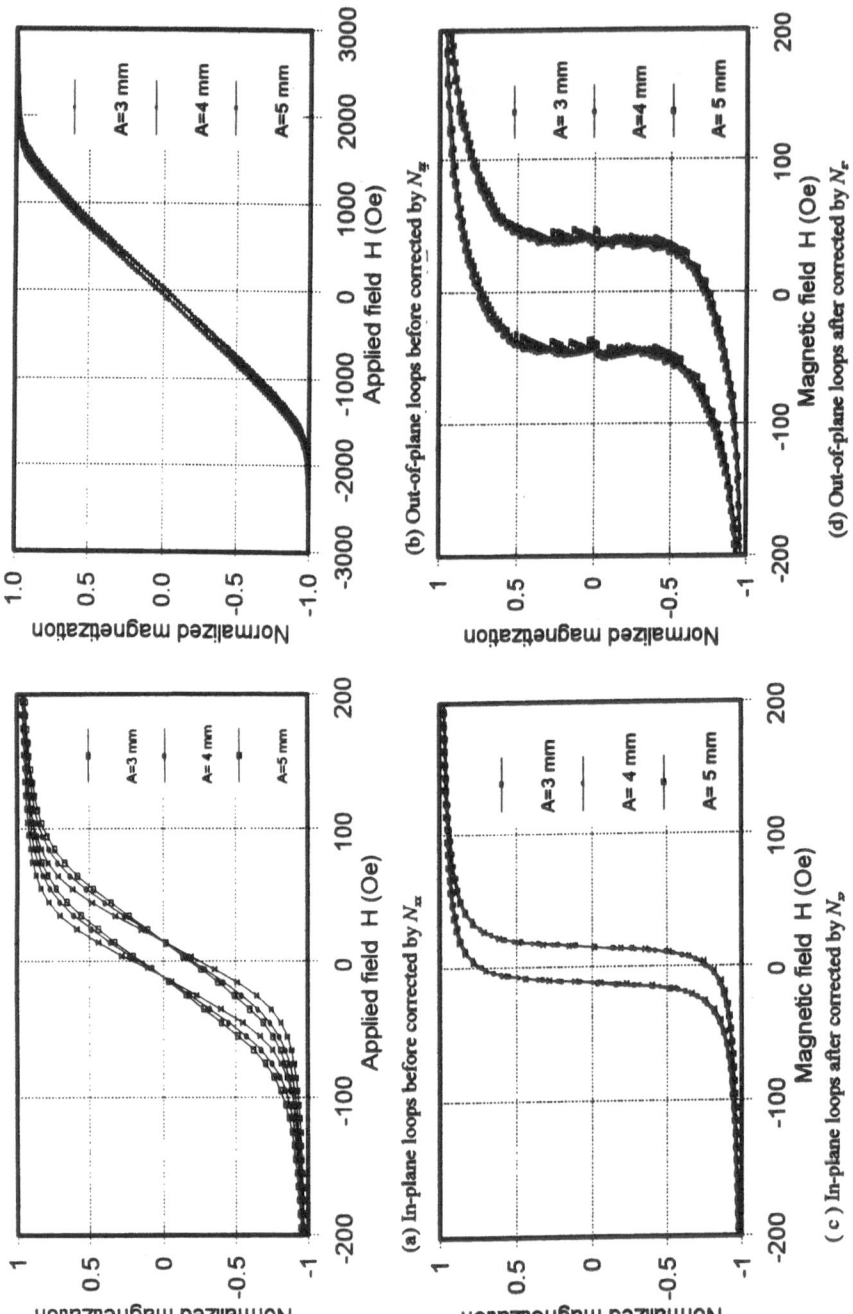

(a) In-plane loops before corrected by N_{xx}

(b) Out-of-plane loops before corrected by N_{zz}

(c) In-plane loops after corrected by N_{xx}

(d) Out-of-plane loops after corrected by N_{zz}

Figure 14. Hysteresis loops of 140 μm square YIG films with different sizes

relative large volume ration around the edges. For this configuration, the same method has been used to obtain the demagnetizing factors. The theoretical results for local demagnetizing tensor $N_{zz}(x, y, z)$ and the effective demagnetizing factors are also based on our numerical method and [1].

Figure 14(c) gives the in-plane hysteresis loops of these samples which were corrected by proper demagnetizing factors. Figure 14(d) is the out-of-plane case.

The calculated and measured values of N_{zz} are given in Table V. The difference between the measured $MaxN_{zz}$ and the calculated value is 4.7% for the smallest sample, and it is less than 4.3% for the 5 mm sample. However, the calculated $MaxN_{zz}$ is only 0.964 for L/h=214, i.e. the validity of the thin film approximation is still worse than 2%.

TABLE V: Shape dependence of demagnetizing factor for square YIG samples with 140 μm thickness

Demag. Factor N			$3\times3mm^2$	$4\times4mm^2$	$5\times5mm^2$	$30\times30mm^2$ (theor.)
N_{xx}	Measured		0.042	0.0368	0.030	
N_{zz}	Theory	Max	0.893	0.913	0.927	0.986
		Min	0.882	0.901	0.918	0.976
		Ave	0.887	0.909	0.923	0.982
	Measured	Max	0.840	0.867	0.880	
		Min	0.798	0.830	0.850	

6. Conclusions

The internal field of a finite size, finite thickness magnetic material is non-uniform, which affects the operation of microwave ferrite devices. The knowledge of the *local* demagnetizing tensor elements $N_{ij}(x,y,z)$ is necessary to analyze and design such devices. Analytical calculations of the demagnetizing tensor elements are based on approximations. To investigate the applicability of these theories, a method, based on in-plane and out-of-plane hysteresis loop measurements, has been developed to define and measure an effective demagnetizing factor for non-ellipsoidal samples. The measurements dependence of the effective N_{zz}, N_{xx} on thickness, sample size and shape have been performed on rectangular YIG and circular MgMn ferrite samples. The measured data have been compared to the properly averaged, calculated results, and it is concluded, that the agreement between the calculated and measured values is very good.

Acknowledgment

The authors would like to thank Dr. E. Della Torre and their colleagues at the Ferrite Development Consortium for cooperations and stimulating discussions.

References

This work was partially supported by ARPA Ferrite Development Consortium

1. R. I. Josephs and E. Schloemann (1965) Demagnetizing field in nonellipsoidal bodies, *J. Appl. Phys.*, 36, 1579.
2. Richard M. Bozorth (1993) *Ferromagnetism*, IEEE Press, New York.
3. Du-Xing Chen, James A. Brug, and Ronald B. Goldfarb (1991) Demagnetizing factors for cylinders, *IEEE Trans on Mag.* 27, 3601.
4. W.F. Brown, Jr. (1960) Single-domain particles: New uses of old theorems, *Am. J. Phys.* Vol. 28, 542-55.
5. G. Vertesy, M. Pardavi-Horvath, L. Bodis and I. Pinter (1988), Dependence of coercivity on the measurement method in epitaxial magnetic garnet films, *J. Magn. Magn. Mater.*, 75, 389.
6. J. Clerk Maxwell (1892) *A Treatise on Electricity and Magnetism*, 3rd ed., vol. 2. Oxford: Clarendon, Reprinted New York: Dover, 1954.
7. J. A. Osborn (1945) Demagnetizing factors of the general ellipsoids, *Phys. Rev.*, 67, 35.
8. E. C. Stoner (1945) The demagnetizing factors for ellipsoids, *Phil Mag.*, Ser. 7, 36, 803.
9. H. Zijlstra (1967) *Experimental Methods in Magnetism*, vol. 2. North-Holland, Amsterdam.
10. L. Kraus (1973) The demagnetization tensor of a cylinder, *Czech. J. Phys. B.* Vol. 23, 512-519
11. J.A. Brug and W.P. Wolf (1985) Demagnetizing fields in magnetic measurements, I. Thin discs, *J. Appl. Phys.* Vol. 57, 4685-4694.
12. Guobao Zheng, M. Pardavi-Horvath, X. Huang, B. Keszei and J. Vandlik (1996) Experimental determination of an effective demagnetization factor for non-ellipsoidal geometries, *J. Appl. Phys.*, in press.
13. E. Della Torre (1986) Fine particle micromagnetics, *IEEE Trans. Magn.*, Vol. MAG-22, 484-489.
14. M. Pardavi-Horvath (1996) Switching properties of a regular two-dimensional array of small uniaxial particles, INTERMAG 96, Seattle, Digest EA-06, to be publ. in *IEEE Trans. Mag.*, September 1996

DESIGN OF NONLINEAR TRANSMISSION LINES: GaAs AND MAGNETIC FILM DEVICES

R. MARCELLI and P. DE GASPERIS
Istituto di Elettronica dello Stato Solido del CNR
via Cineto Romano 42, 00156 Roma, Italy

G. BARTOLUCCI and F. PINI
Dip. Ingegneria Elettronica, Università di Tor Vergata
via della Ricerca Scientifica s/n, 00133 Roma, Italy

M. DRAGOMAN
Institutul de Cercetari Componente Electronice
Str. Erou Iancu Nicolae 32 B, 72996 Bucharest, Romania

1. Introduction

Computer Aided Engineering (CAE) of microwave devices is an essential tool to study the feasibility of new device configurations. Actually, the modeling and the prediction of the device performances allows lower costs in terms of human capital and materials by reducing the "cut and try" procedures before the realization of a test prototype. Simulation of devices and systems for both, mass production and feasibility of new configurations can be usually performed. So far, prototype devices useful for confirming theoretical expectations can be designed, as well as interpretations of novel effects can be, at least, attempted.

Linear passive devices (filters, resonators and linear transmission lines) can be predicted and boundary effects interpreted in terms of continuity conditions by using commerical software based on finite elements analysis or ladder network simulations. Nonlinear devices like oscillators and harmonic generators can be also modeled, but novel, unusual non-linear effects related to modulation of *cw* and pulsed signals are more difficult to be efficiently simulated, and also the modeling of real, 3D geometry devices accounting for the proper boundary effects is not a straightforward application.

In the case of microwave propagation in a non-linear transmission line (NLTL), traditional CAE packages based on the Harmonic Balance Method and on time domain analysis are able to simulate harmonic generation and frequency modulation of the input signal in the frequency and in the time domain. The Fast Fourier Transform (FFT) technique usually provides the transformation between the two domains. In this case, lumped elements have to be defined in the network, and the equivalent electrical

R. Marcelli and S.A. Nikitov (eds.), Nonlinear Microwave Signal Processing: Towards a New Range of Devices, 71–99.
© *1996 Kluwer Academic Publishers.*

circuit and the basic characteristics of each individual element are data of the problem. Presently, the finite elements method is in progress to be implemented for non-linear simulation, and the final goal of Companies producing Software Simulators is to create a link between linear and non-linear simulators and electromagnetic modules to realize a complete package which properly accounts for lumped elements and for electromagnetic boundary conditions and interference phenomena.

Monolithic Microwave Integrated Circuits (MMIC) NLTLs based on GaAs technology are modeled by LC networks, where a non-linear capacitance C=C(V) dependent on the applied voltage V is responsible for harmonic generation and solitons. Until now, the KdV equation is the best theoretical picture to describe semiconductor NLTLs: it allows for the continuum approximation of a discrete series of components, in the limit of weakly dispersive and non-linear regime. Dissipation is usually neglected, but the effect of low losses level on the device performances can be included as a perturbative term by defining a resistor in the equivalent circuit of the NLTL.

The individual LC stages of the non-linear transmission line can be coplanar waveguides (CPWs) loaded by Schottky varactor diodes or any other element having a C(V) non-linear characteristic. The onset of shock waves or solitons will be caused by the predominance of non-linear terms or by the balance between non-linearity and dispersion, respectively. The number of cells is important to activate the non-linear regime, favoured by the cooperative response of a number of individual strongly non-linear stages. The distance between the cells is also important, and it is a parameter to be accounted for an effective integrability of the exploited structure.

The continuum approximation is only an optional tool for the modeling of GaAs NLTLs, but it is the natural environment of non-linear waves propagating in a continuum medium. This is the case of integrated optical devices, and of dielectric and magnetic waveguides. In particular, magnetic films like the ferrimagnet yttrium iron garnet (YIG), characterized by low propagation losses, support the propagation of magnetically dominated waves, whose behaviour has been analytically described by means of the linear magnetostatic wave (MSW) approximation and by non-linear wave equations, as, for instance, the non-linear Schrödinger equation (NLSE). On the other hand, an electrical modeling in terms of a transmission line does not result from the chanonical definitions based on the Maxwell equations, and a YIG film delay line can not be easily modeled as a two-wire transmission line or as a parallel plate waveguide. Further to this, owing to the peculiarity of the YIG film configuration to be a waveguide with tensorial elements of the magnetic permeability, the equations for a pure TE or TM mode can be not straightforwardly written by defining electrical elements, as it is the case of free space waves.

The aim of this chapter is to discuss lumped element equivalent circuits for GaAs and YIG film non-linear transmission lines to get a physical picture of the onset of non-linear waves propagation in this kind of structures. Actually, the definition of an equivalent circuit model in terms of a ladder network allows the design and the simulation of the NLTL by using a commercial software package.

2. Non-Linear Transmission Lines and Wave Packet Propagation

In this section, the excitation of non-linear waves in ladder networks and the non-linear propagation of wave packets are considered. In particular, the Koerteweg and de Vries (KdV) equation is discussed in the first case, while in the second one the non-linear Schrödinger equation (NLSE) is obtained starting from the frequency modulation of the carrier signal.

2.1. NONLINEAR TRANSMISSION LINES AND KdV EQUATION

In the lossless approximation, a linear dispersive transmission line is a distributed structure whose equivalent circuit in an infinitesimal section dx (being x the propagation direction) is composed of a series connected inductance Ldx and a shunt connected capacitance Cdx, being L and C quantities per unitary length. A resistive contribution R can be defined to account for the losses along the transmission line and a conductance G for the non-linear device losses. In Figure 1 is shown the equivalent circuit of a non-linear transmission line. In this structure, the dispersion is naturally provided passing from discrete cells to the continuum approximation.

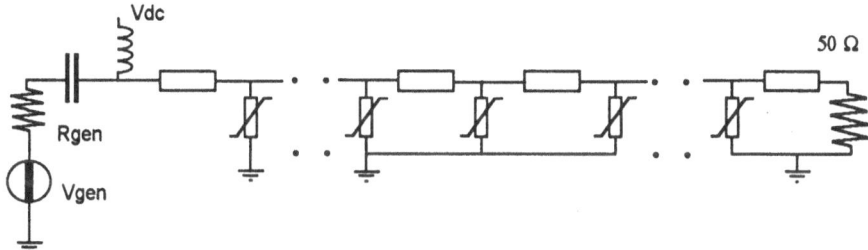

Figure 1. Equivalent circuit of a non-linear transmission line composed by uniform sections periodically loaded by non-linear impedances. R_{gen} is the generator resistance, V_{gen} the generator voltage and V_{dc} the bias. The NLTL is loaded by a 50 Ω resistor.

The behavior of a non-linear transmission line (NLTL) is accounted by introducing a voltage dependence of the capacitance C as [1]:

$$C = C_0(1 + a_1 V + a_2 V^2 + ...) \qquad (1)$$

where C_0 is the value of C in the linear approximation and the quantity in parenthesis is the result of a Taylor expansion in the variable V, which is the voltage applied to the non-linear capacitor. If V is small, Eq.(1) can be re-written accounting for the first order term only, thus obtaining:

$$C \approx C_0(1 - 2bV) \qquad (2)$$

where b = -a_1/2 is small, and it is defined as the non-linear coefficient of C. The approximation in Eq.(2) is used to obtain an equation for V in its space and time derivatives, which is non-linear and dispersive. Actually, passing from the discrete structure to the continuous one, a weakly dispersive non-linear equation can be written as [1]:

$$\frac{\partial^2 V}{\partial t^2} - \left(\frac{\delta^2}{LC_0}\right)\frac{\partial^2 V}{\partial x^2} = \left(\frac{\delta^4}{12LC_0}\right)\frac{\partial^4 V}{\partial x^4} + b\frac{\partial^2 V^2}{\partial t^2}$$ (3)

where x is the propagation direction, t is the time and δ is the cell length. The dispersion is naturally provided by the continuum approximation of the cell-based structure. If the non-linear element is the inductance L, the same equation is obtained for the current I. By properly defining reduced variables in Eq.(3), the Koerteveg and de Vries (KdV) equation is obtained as:

$$\frac{\partial V}{\partial T} + PV\frac{\partial V}{\partial s} + Q\frac{\partial^3 V}{\partial s^3} = 0$$ (4)

where P and Q are the non-linear and dispersive coefficients, respectively, and the slow variables T and s have substituted the natural ones t and x.

2.2 WAVE PACKET MODULATION AND NON-LINEAR SCHRÖDINGER EQUATION.

Non-linear propagation of wave packets can be described by means of the frequency and phase modulation of the carrier. Formally, it results in a Taylor expansion of the radian frequency ω up to the lowest possible order in dispersive and non-linear terms, and, after the substitution of the small variations of the wavevector k and of ω with the corresponding operators in the space and time derivatives, the Non-linear Schrödinger Equation (NLSE) is obtained [2]. Of course, this is a lossless case, and the approximation holds when the medium is responsible for very low propagation losses, which can be included in the NLSE as small perturbations. Also the Taylor expansion is an approximation itself, and, rigorously, higher order terms in dispersion and non-linearity should be included.

The dispersion relation for magnetostatic waves is a rather complicate implicit function describing the wavevector dependence of the radian frequency ω, and it can be written by properly accounting for the magnetic microwave fields boundary conditions [3], as it is the case for any dielectric slab excited by an electromagnetic field. In the linear regime, such a dispersion function is:

$$D(\omega, k) = D(\omega(k), k) = 0$$ (5)

Nonlinearity and dispersion induce a frequency and wavevector modulation, due to the not longer negligible amplitude of the microwave magnetization field and; consequently, Eq.(5) must be re-written as:

$$D'(\omega,k) = D'(\omega(k,|\varphi|\,),k,|\varphi|^2\,) = 0 \tag{6}$$

where $\omega = \omega(k, |\varphi|^2)$, and $|\varphi|^2 = (1/2)\cdot(m/M_s)^2$ is the correction factor for the magnetization component along the direction of the bias field H_0. By assuming it as the z-component, the linear regime approximation $M_z \approx M_s$ is replaced by the non-linear one where $M_z = M_s\sqrt{1-(m/M_s)^2} \approx M_s\cdot(1-1/2\cdot(m/M_s)^2) = M_s\cdot(1-|\varphi|^2)$, where m is the microwave magnetization in the plane (x,y) and the series expansion is valid in the approximation that m $\ll M_s$.

The second order correction term is responsible for the frequency and the wavevector modulation of the propagating wave, and the ω-variation will be related to the k- and $|\varphi|^2$-modulations in the following way [4]:

$$d\omega = \frac{\partial\omega}{\partial k}dk + \frac{1}{2}\frac{\partial^2\omega}{\partial k^2}dk^2 + \frac{\partial\omega}{\partial|\varphi|^2} + O(dk^3,|\varphi|^4) \tag{7}$$

Then, by substituting $d\omega$ and dk by means of the corresponding operators $i\partial/\partial t$ and $-i\partial/\partial x$, respectively, and applying both the operatorial members to φ, a second order nonlinear differential equation is derived as:

$$i\left(\frac{\partial\varphi}{\partial t} + v_s\frac{\partial\varphi}{\partial x}\right) + \frac{1}{2}\frac{\partial^2\omega}{\partial k^2}\frac{\partial^2\varphi}{\partial x^2} - \frac{\partial\omega}{\partial|\varphi|^2}|\varphi|^2\,\varphi = 0 \tag{8}$$

Eq.(8) describes the time and space evolution of a modulated wave and it can be easily re-written in the canonical form for NLSE by means of variables substitution. Eq.(8) can be inferred directly from the variation of the magnetostatic dispersion relation D. In this case, an explicit form of the group velocity, the dispersion and the nonlinearity by means of the D derivatives is obtained. A solution of a propagating wave, modulated in frequency and amplitude is given by $\varphi = \varphi_0 e^{i(kx-xt)}$, where φ_0 = constant in linear regime, and $\varphi_0 = \varphi_0(x, t)$ in the nonlinear one. The dispersion $\partial^2\omega/\partial k^2$ and the nonlinearity $\partial\omega/\partial|\varphi|^2$ are critical quantities, because the value and the sign of their ratio are related to the threshold power value for the onset of a solitonic behaviour and for the stability of the created nonlinear wave, respectively (Lighthill criterion).

In the case of MSW volume waves both, the forward waves (FVW's) and the backward waves (BVW's) can be described by the dispersion relation:

$$\tan\left(\frac{kd}{\beta^p}\right)\big(\tanh(kh)\tanh(kw) + \mu\big) - p\beta\big(\tanh(kh) + \tanh(kw)\big) = 0 \tag{9}$$

where k is the in-plane excited wavevector, d is the film thickness, h and w are the two ground planes distances from the magnetic film surface, $\beta = \sqrt{-\mu}$, p = -1 for FVW's and p = +1 for BVW's.

A typical arrangement of a MSW delay line is shown in Figure 2.

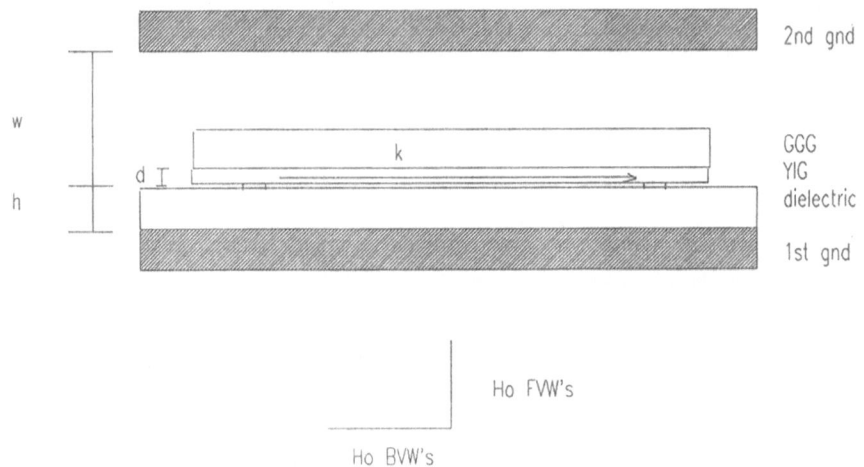

Figure 2. MSW delay line. A magnetic film (YIG) having thickness d, epitaxially grown onto a commercial gadolinium gallium garnet (GGG) substrate, is top coupled to two microstrip transducers (input and output of the device) evaporated onto a dielectric (alumina) substrate. Two grounds properly separated by the film provide the device enclosure and affect its dispersive behavior.

The onset of a non-linear behavior for the device in Figure 2 depends on the activation of multi-magnon mechanisms caused by increasing power levels, and it is also frequency dependent. In particular, the possibility to generate solitons in pulsed regime within the volume wave configuration is tightly related to the fulfilment of the so-called Lighthill criterion, which states that the non-linear term and the dispersive one of the NLSE in Eq. (8) must always have opposite sign. The Lighthill criterion is written as:

$$\frac{\partial^2 \omega / \partial k^2}{\partial \omega / \partial |\varphi|^2} < 0 \qquad (10)$$

In Figure 3 and in Figure 4, the trend of the Dispersion $\partial^2 \omega / \partial k^2$ and of the Nonlinearity $\partial \omega / \partial |\varphi|^2$ are shown vs the excited wavevector k for the configuration of Figure 2, where the bias magnetc field is H_0=4000 Oe and the film thickness is changed.

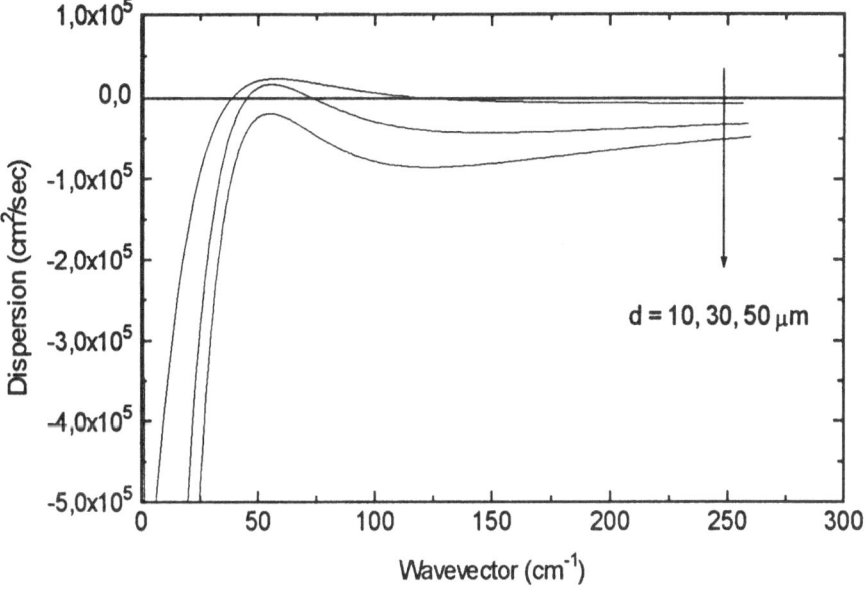

Figure 3a

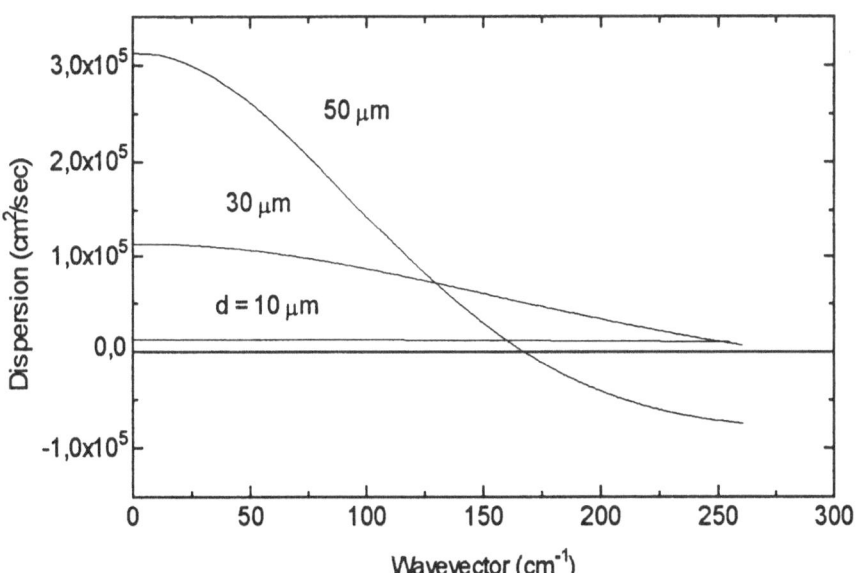

Figure 3b

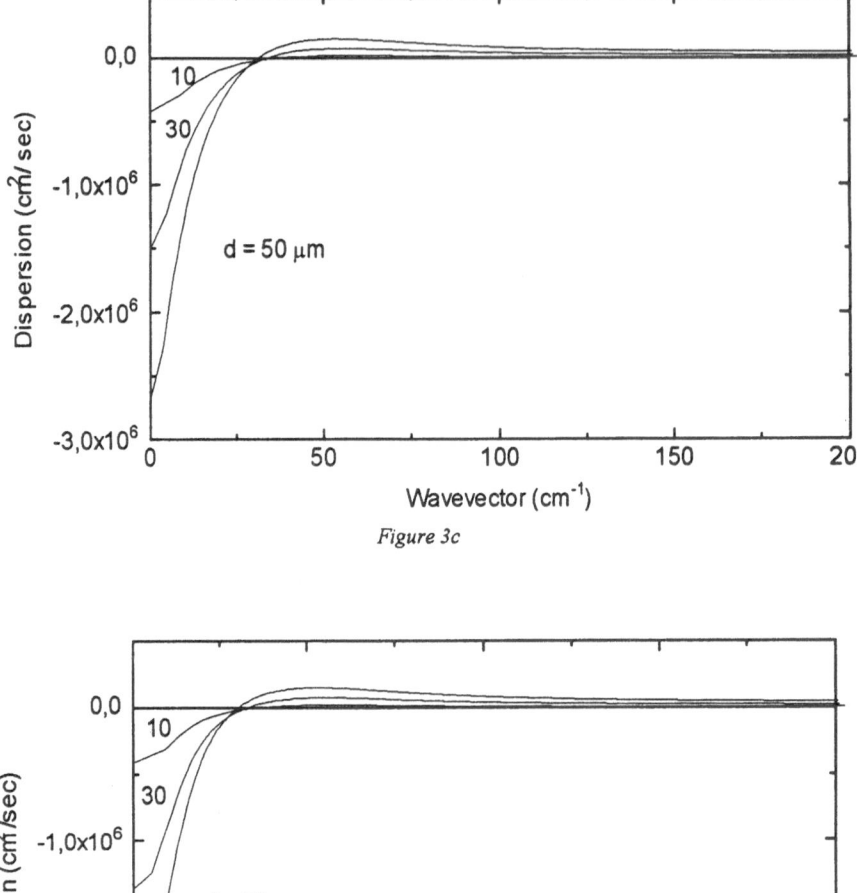

Figure 3c

Figure 3d

Figure 3. Dispersion of a MSFVW and of a MSBVW delay line. The effect of one and two ground planes has been accounted in the simulation. In Figure 3a) and b) is shown the dispersion of a MSFVW biased by a dc field H_0=4000 Oe, in a configuration with h=254 μm for a microstrip 240 μm wide, while w=∞ (a) and w=1 mm (b), respectively (see Figure 2). In Figure 3c) and d) the dispersion response of a MSBVW with H_0=2240 Oe is simulated for the same w-values (c, d). In both configurations, the sweeped parameter is the film thickness, from d = 10 μm to 50 μm.

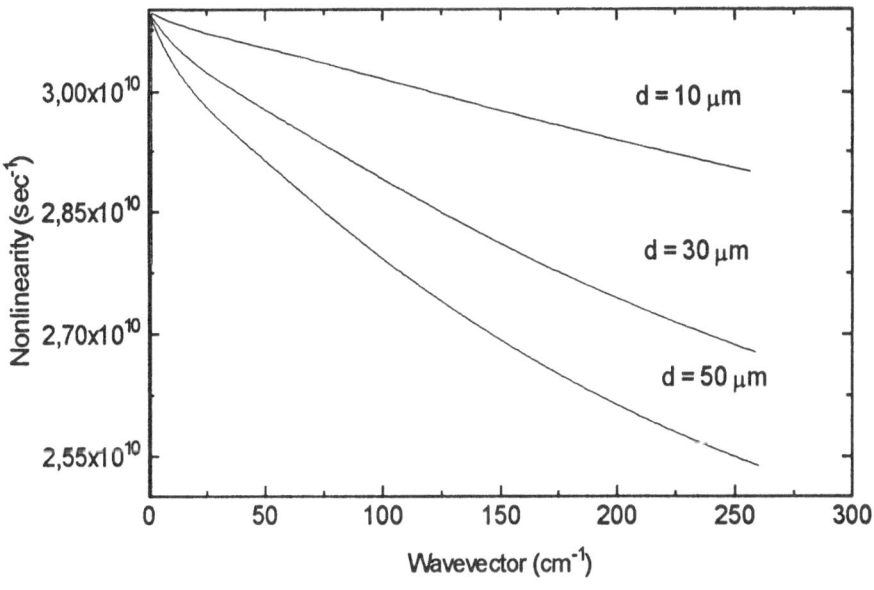

Figure 4a

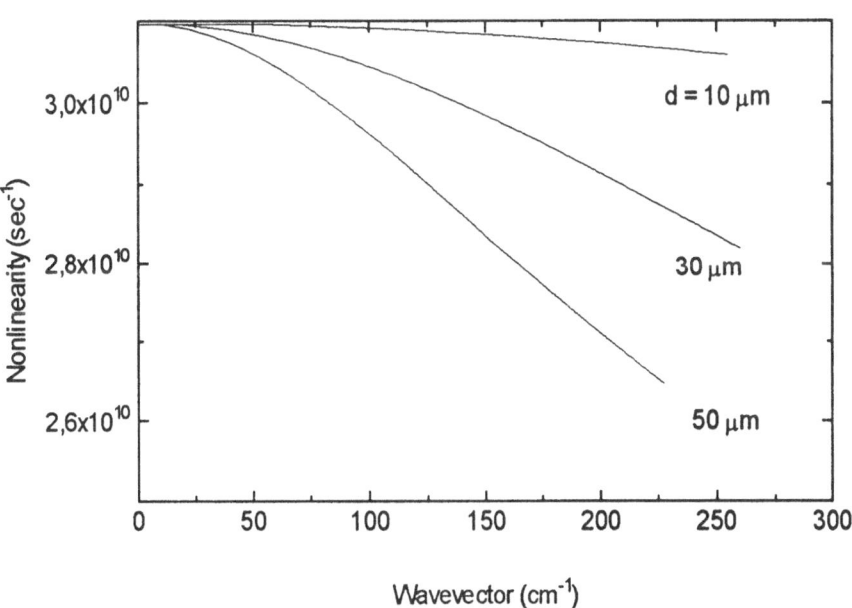

Figure 4b

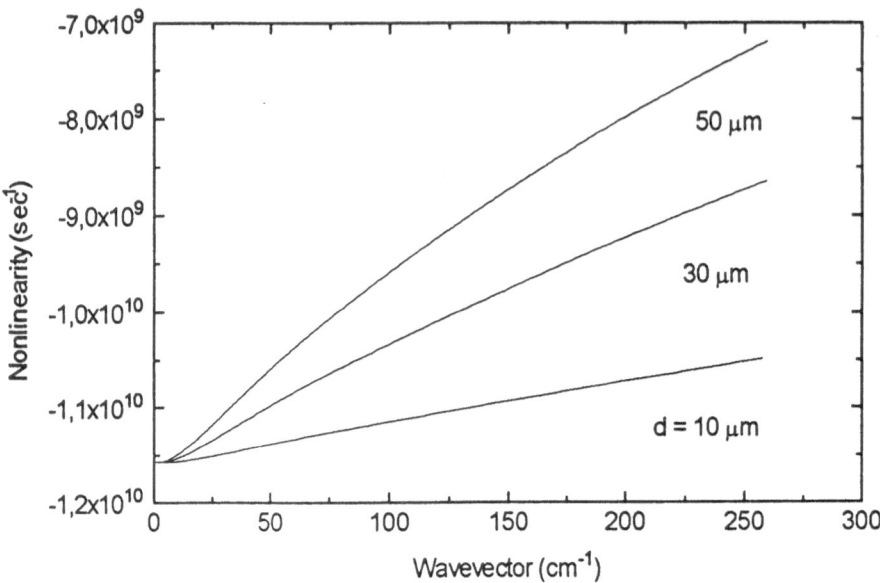

Figure 4c

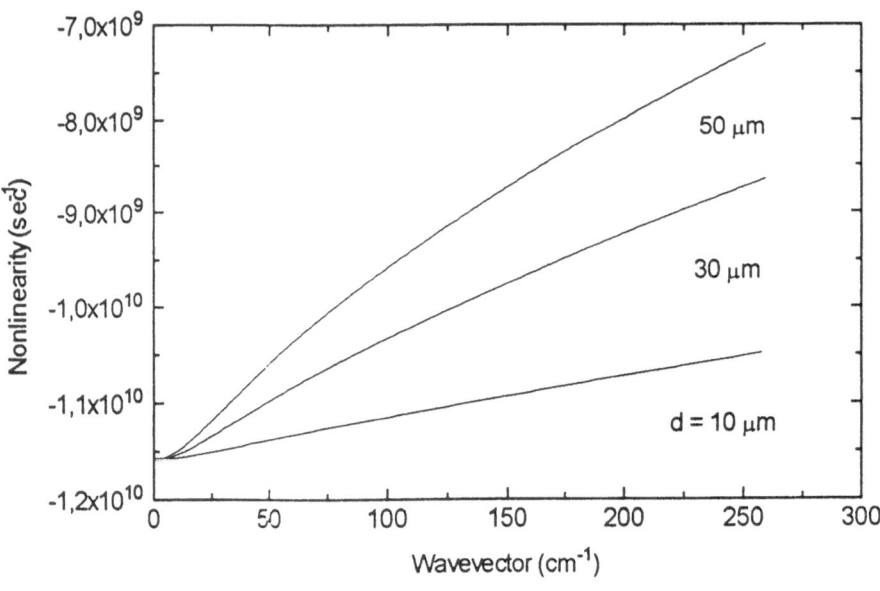

Figure 4d

Figure 4. Nonlinearity response of a MSFVW and of a MSBVW by using the same simulation parameters as in Figure 3.

As known, the dispersion is strongly influenced by the film thickness d and by the ground planes configuration. On the other hand, the nonlinearity is slightly or not at all influenced by the structure and by the ground planes position (see Figure 4). The ground planes presence is very important because of the change in sign of the dispersion passing from one ground to two grounds. In fact, the dispersion allows the Lighthill condition satisfaction for all the wavevector values inside the band only for the FVW's in the one ground configuration when the film is thick enough to get always a negative dispersion. In the two ground device, the dispersion could not have the same sign for all the values of k. The BVW's can satisfy the Lighthill criterion, but only when k is greater than a certain value, because nonlinearity and dispersion have the same sign for both ground arrangements for the low wavevector values. In fact, BVW's envelope solitons can exist for all the k-values when the ground planes are not present, as clearly demonstrated in [4] where a dispersion relation without ground planes has been used, but the change in sign of the dispersion due to the device structure introduces limitations in the excitation of BVW's solitons. Moreover, the BVW's are much less influenced than the FVW's by the presence of the second ground plane.

3. GaAs Non-Linear Trasmission Lines

In recent years electronic devices able to generate millimeter and sub-millimeter waves are often required in many electronic systems.
In order to produce so high frequency signals, non-linear transmission lines (NLTL) have been proposed [5 - 7]. The GaAs NLTL is a MMIC device composed by high impedance (tipically about 75 Ω) coplanar waveguide sections, periodically loaded by Schottky varactor diodes. The equivalent circuit of the NLTL is shown in Figure 5.

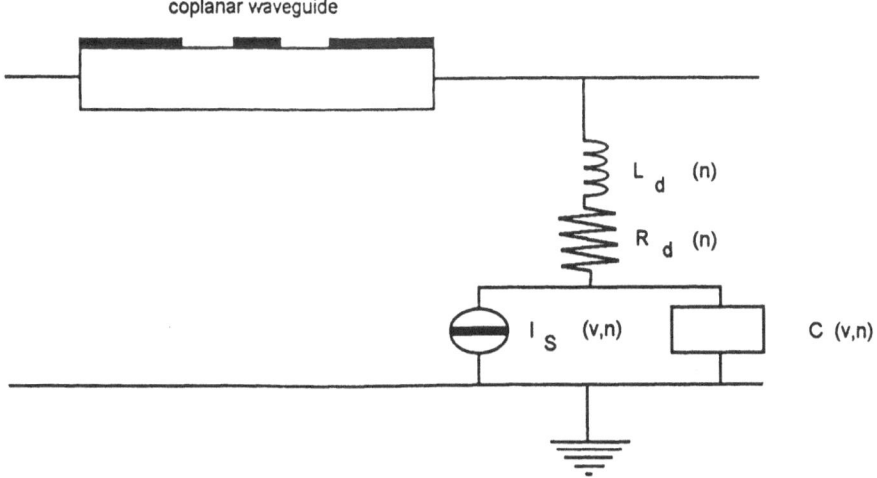

Figure 5. Equivalent circuit of a NLTL section with a coplanar waveguide loaded by a Schottky diode

82

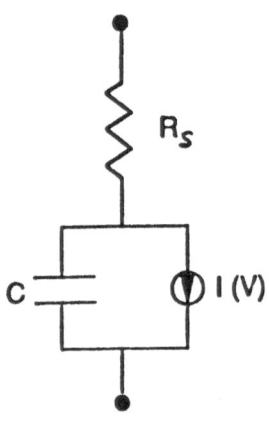

The number of diodes **N** as well as their separation distance **S** strongly affect the nonlinear behavior of the whole structure.

Recently, Yu et al. have described a NLTL where tunnel diodes are the non-linear elements [8]. This new device configuration is shown in the following Figure 6.

In this section, the possibility to use this tunnel diode non linear transmission line (TDNLTL) for harmonic generation is investigated. The analysis is carried out by using the time domain analysis implemented in the commercial software HP-MDS-IMPULSE package. In particular, the diode I-V characteristic utilized for the simulation has been obtained by fitting previously reported experimental data [9]. The effect of the diodes number, positioning and bias voltage V_{dc} on the nonlinear response of the TDNLTL is discussed, to develop design criteria for harmonic generation.

Figure 6. Equivalent circuit of the resonating tunnel diode, composed by a capacitor C shunt connected to a non-linear current generator I(V), and a parasitic series resistance R_s.

3.1. NON LINEAR TRANSMISSION LINE SIMULATION

In order to evaluate the performances of the TDNLTL as harmonic generator, the time domain method implemented in the HP-MDS-IMPULSE commercial software package is used. For analysis purposes, each tunnel diode must be replaced by its equivalent circuit.

In the following, the diode equivalent circuit elements will be assumed to be those used in [9], with a diode active area of 1 μm^2, and R_s=5.1 W, C=2.8 fF, both values including parasitic contributions. The I-V characteristic used in our paper to describe the response of each diode is that shown in Figure 4 of [9].

Since the theoretical treatment of the single diode behavior is rather complex and an analytical solution is not available, approximated solutions for the I-V characteristic have been used in the simulation of TDNLTLs. A suggested method is described in [10], where the I-V curve is approximated by a piecewise linear model. In that paper, the analysis of networks including also tunnel diodes, is performed approximating the non linear "N" I-V behavior by means of three linear segments: one having negative slope corresponding to the negative resistance region of the characteristic curve, the other ones having positive slope. However, this model is not very accurate and it critically dependes on the I-V curve shape. A more rigorous approach is to develop an approximation of the I-V behavior using other kind of functions. A possible solution is given by fitting the experimental data with polynomial expressions. Nevertheless, this method requires a high degree of the polynomial function to fullfil the agreement between the fitted curve and the experimental one over the entire range of the bias voltage. In order to avoid this kind of problem, the interesting voltage range can be divided into a number of sub-intervals, and the data in each of them can be fitted by a lower order polynomial function. On the other hand, the crossing points of adjacent voltage sub-intervals are the connection between two different fitting curves, and the derivative of the function I(V) performed at these voltage values could be not continuous (wedge points). To overcome the above introduced difficulty, the fit can be performed on the values of the derivative of the I-V characteristic, and the I-V characteristic is obtained by integrating the fitted curves. The choice of the arbitrary constants for the integration depends on the continuity conditions passing from one sub-interval to the next one. By using this procedure, we have fitted the dI/dV curve (derivative of the current with respect to the voltage) experimentally determined in [9] by dividing the voltage range in four intervals, and then the fitting curves have been integrated, thus resulting in the following I(V) equations:

$$I_1(V) = -53.63 + 188.48 \cdot V - 245.737 \cdot V^2 + 140.1256 \cdot V^3 - 29.24081 \cdot V^4$$

$$I_2(V) = (1.48 - 8.19 \cdot V + 19.859 \cdot V^2 - 27.4714 \cdot V^3 + 23.72091 \cdot V^4 - 13.091976 \cdot V^5$$
$$+ 4.5102305 \cdot V^6 - 0.88672522 \cdot V^7 + 0.076170382 \cdot V^8) \times 10^6$$

$$I_3(V) = 518.79 - 1048.68 \cdot V + 798.571 \cdot V^2 - 271.1316 \cdot V^3 + 34.62978 \cdot V^4$$

$$I_4(V) = 26.10 - 25.17 \cdot V + 6.25 \cdot V^2$$

(11)

$$I(V) = \begin{cases} I_1(V) & V \in [0,1.3] \\ I_2(V) & V \in [1.3,1.7] \\ I_3(V) & V \in [1.7,2.1] \\ I_4(V) & V \in [2.1,2.3] \end{cases}$$

where **V** is in volt and **I** is in milliampere.

3.2 DESIGN CONSIDERATIONS

In order to investigate the performances of the TDNLTL as a frequency multiplier for second and third harmonic generation, a systematic study has been developed by changing the bias voltage V_{dc}, the number of diodes and the separation distance between them. Preliminary considerations about the space occupancy of the exploited configurations will be also done. Following the discussion on the fitting results obtained in the previous section, the derivative method has been used to approximate each diode behavior in the simulation of the whole structure. All the simulation outputs have been obtained by imposing a generator impedance R_g=50 Ω, a Z_L=50 Ω load impedance and an input frequency f=40 GHz. The generator voltage is V_g=0.8 volt.

84

The first results are shown in Figure 7 for a TDNLTL with N=30 diodes as a function of the V_{dc} amplitude. In particular, in Figure 7a) is plotted the real part of the input impedance Z_{in}, in Figure 7b) is shown the trend of the imaginary part of Z_{in}, and in Figure 7c) the second and third harmonic output power behaviors are plotted. It is worth noting the good electrical matching predicted in Figure 7a) and in Figure 7b). Moreover, the existence of peak values in both the curves of Figure 7c) suggests that an harmonic can be generated by modulating the V_{dc} value, and the second or the third harmonic enhanced also depending on the V_{dc} choice.

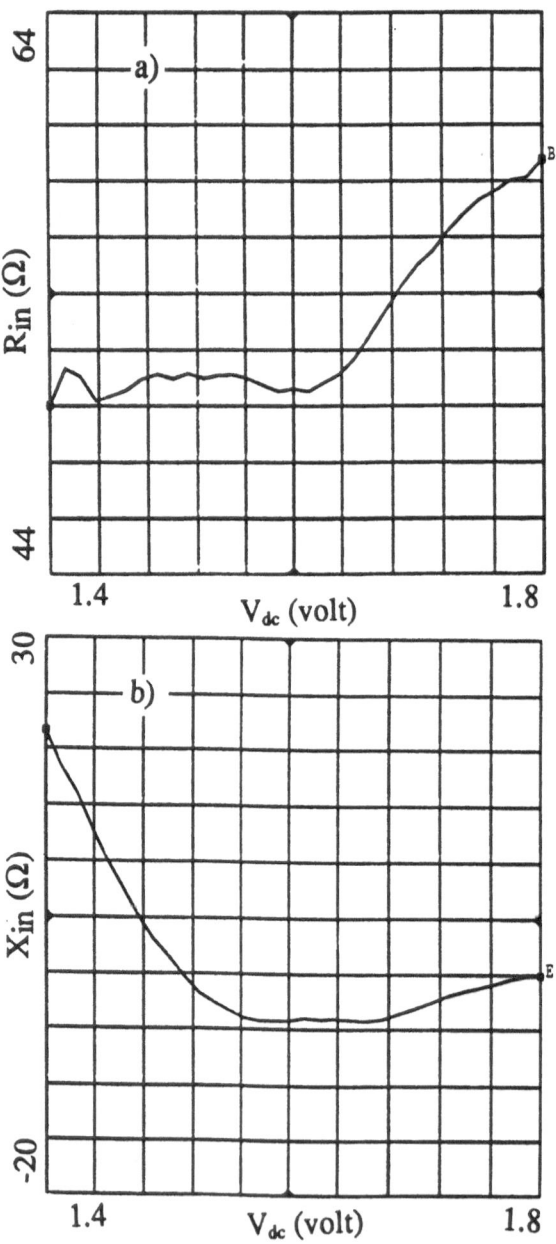

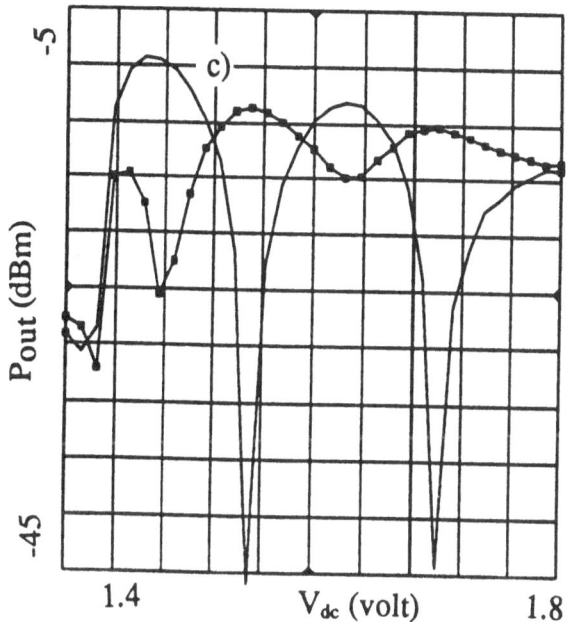

Figure 7. Input impedance Z_{in} and output power P_{out} for second and third harmonic generation of a 30 diode structure as a function of the bias voltage V_{dc}. In a) R_{in}, real part of Z_{in}, and in b) X_{in}, the imaginary part of Z_{in}, are shown respectiely, while in c) P_{out} is plotted for second (continuous curve) and third (dotted curve) harmonic.

In Figure 8 and in Figure 9 the output power vs V_{dc} of the second and of the third harmonic generated at the output of the TDNLTLs composed by N=40 and N=50 diodes respectively are shown. From the analysis of Figure 7c), Figure 8 and Figure 9 it turns out that, generally speaking, the maximum of the output power in generating second and third harmonic signals can be obtained at well defined V_{dc} values.

86

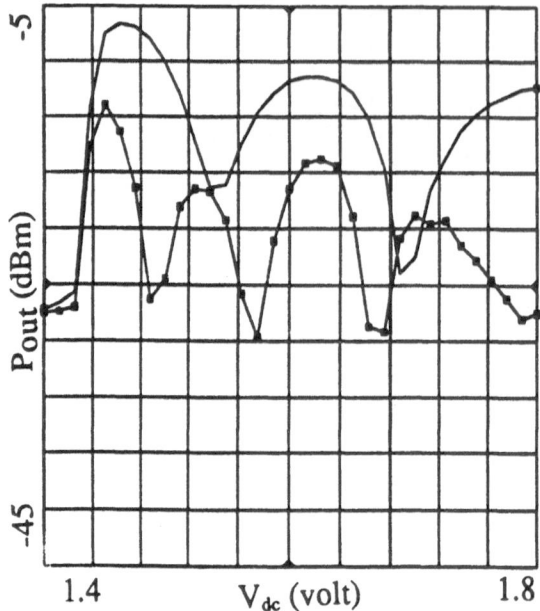

Figure 8. Output power Pout for second (continuous curve) and third (dotted curve) harmonic vs the bias voltage Vdc for the 40 diode structure.

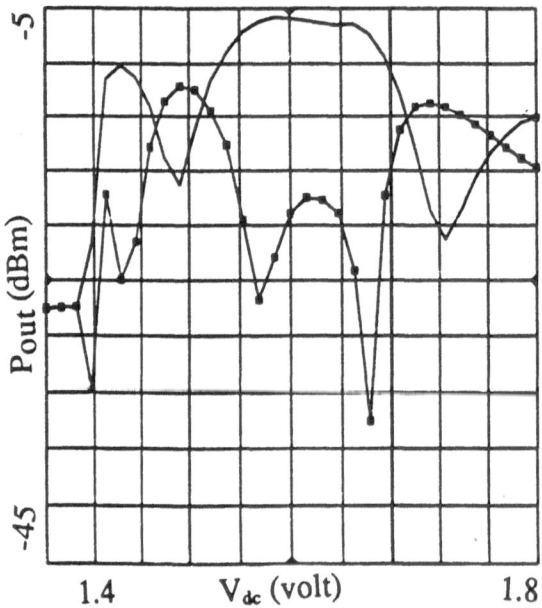

Figure 9. Output power Pout for second (continuous curve) and third (dotted curve) harmonic vs the bias voltage Vdc for the 50 diode structure.

By using the above presented simulation results, the voltage bias values corresponding to these maxima ($V_{dc,max}$) have been calculated and they are shown in the following TABLE 1.

TABLE 1. V_{dc} values corresponding to the maxima of the second harmonic ($V_{dc,max}$ (2nd H.)) and of the third harmonic ($V_{dc,max}$ (3rd H.)), derived by the simulation results shown in Figure 7, Figure 8 and Figure 9.

	30 diodes	40 diodes	50 diodes
Vdc,max (2nd H.)	1.462	1.462	1.587
Vdc,max (3rd H.)	1.55	1.45	1.512

These $V_{dc,max}$ values have been used as a parameter for further simulations on the N=30, 40, 50 diode structures to get the output power response of second and third harmonic signals as a function of the distance between the diodes. The simulations have been done by imposing a maximum value of the distance between the diodes L_{max}=100 μm, on the basis of practical considerations about the maximum space occupancy for a MMIC device. It means that a maximum dimension D_{max}=3, 4, 5 mm is allowed for the N=30, 40, and 50 diode structures, respectively. Actually, in many practical situations, a maximum value for the diodes distance between 50 μm and 70 μm is reasonably expected. In Figure 10a), 10b), and 10c) the output power for the three structures is shown, and the second harmonic response is optimized by using the values in the first line of TABLE 1. In this case, the output power P_{out} is a decreasing function of the diodes distance.

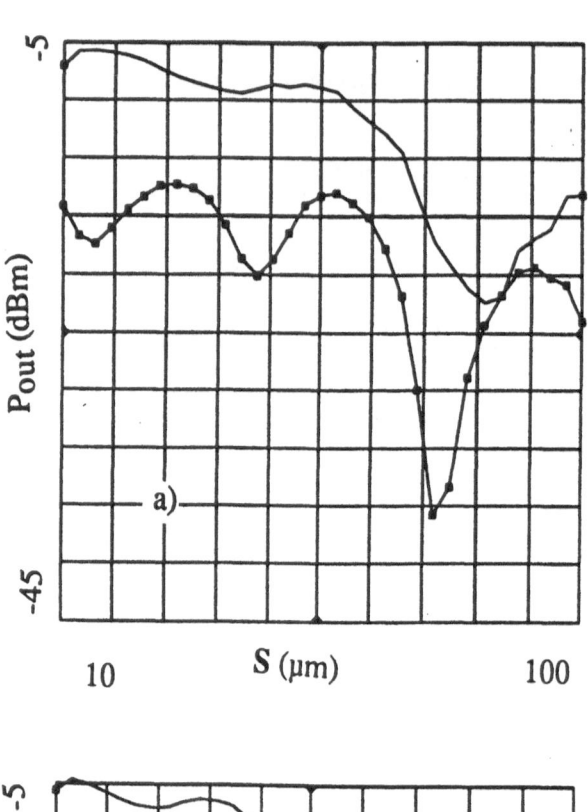

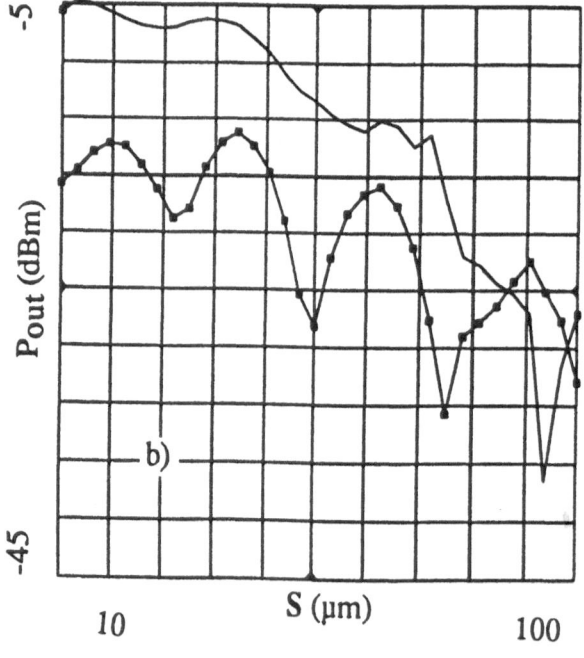

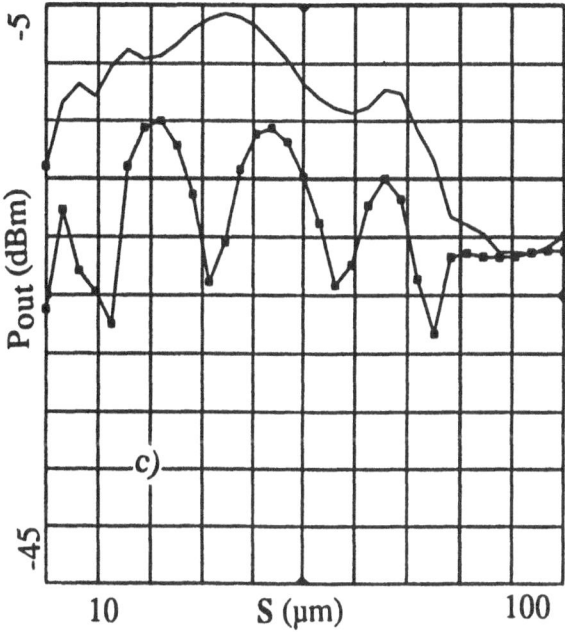

Figure 10. Output power P$_{out}$ for second (continuous curve) and third (dotted curve) harmonic generation as a function of the distance S between the tunnel diodes. In this case, the second harmonic output has been optimized by choosing a V$_{dc}$ value which maximizes the second harmonic output: a) is for N=30 diodes, b) is for N=40, and c) is for N=50. The maxima of second harmonic generation coincide with minima of third harmonic, thus allowing the selective harmonic generation.

In Figure 11a), 11b), and 11c), the same procedure has been followed to optimize the third harmonic generation (second line of V$_{dc}$ values in TABLE 1).

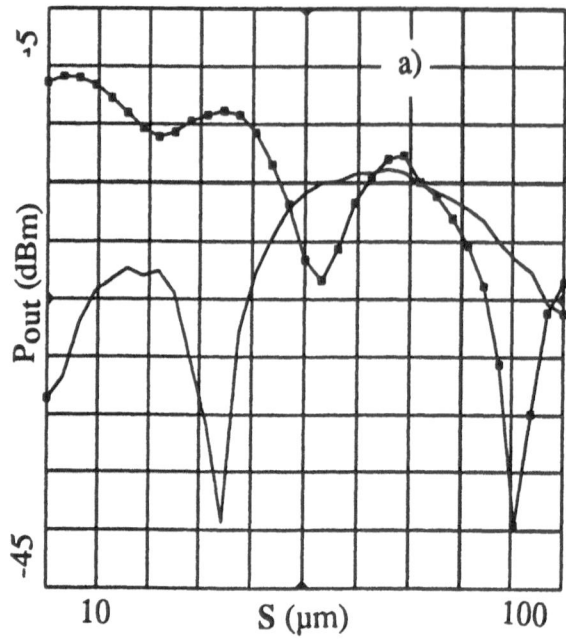

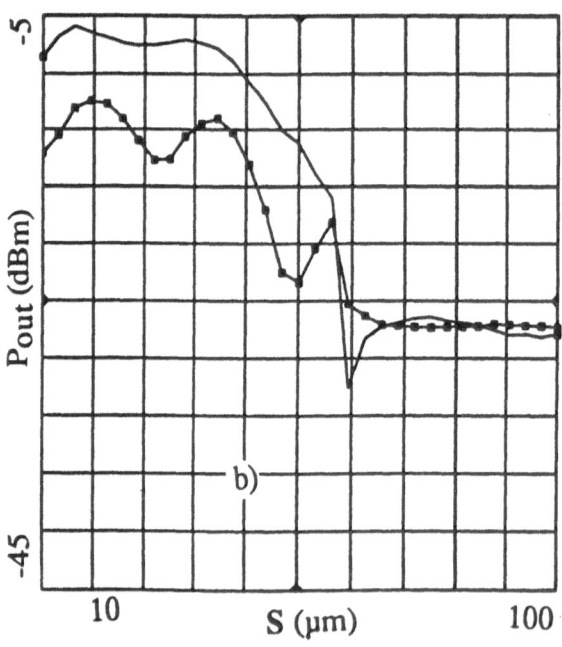

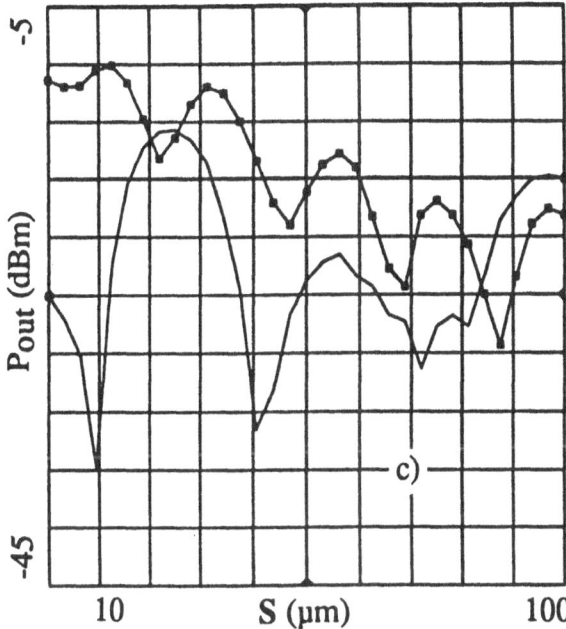

Figure 11. Output power P_{out} for second (continuous curve) and third (dotted curve) harmonic generation as a function of the distance S between the tunnel diodes. In this case, the third harmonic output has been optimized by choosing a V_{dc} value which maximizes the third harmonic output: a) is for N=30 diodes, b) is for N=40, and c) is for N=50. In the case N=40 the maxima of third and second harmonic coincide, thus forbidding the enhancement of the third harmonic with respect to the second one.

In Figure 10 as well as in Figure 11, both, second and third harmonic outputs have been produced. From the analysis of Figure 10, it results that when the second harmonic generation of the TDNLTL is optimized, the third one is excited with output power levels about 10 dB lower with respect to the second one for many values of the distances between the individual diodes. For the results presented in Figure 11, where the third harmonic is optimized, the same considerations can be done; that is, a lot of values for the distance between the diodes exist for which the third harmonic output power is considerably higher with respect to the second harmonic power. From above discussion, it turns out that a selective response of the TDNLTL can be obtained, depending on the harmonic generation required. As a summary of the predicted optimization for the exploited configurations, we present in TABLE 2 the best results for the second and third harmonic generation, respectively.

TABLE 2. Optimized values inferred for the 2nd and 3rd harmonic generation of the TDNLTL with 40 diodes, separated by 11.8 μm (first line, 2nd H.) and with 50 diodes separated by 20.8 μm (second line, 3rd H.). The bias voltage V_{dc} is in volt, the input resistence R_{in} is in ohm, the input reactance X_{in} is in ohm, the input power P_{in} is in dBm, as well as the output power of the second harmonic $P_{out}(2)$ and of the third harmonic $P_{out}(3)$.

	V_{dc}	R_{in}	X_{in}	P_{in}	$P_{out}(2)$	$P_{out}(3)$
2nd H.	1.462	30	3.2	1.72	-4.68	-17
3rd H.	1.512	42.5	3.2	1.95	-26	-9

4. Magnetic Media Non-Linear Trasmission Lines

As discussed in the introduction, transmission lines can be defined only in the case of two-wire or parallel plate configurations, to obtain optical as well as microwave waveguides in coaxial, circular and rectangular shapes. A plane wave propagating in free space evolves in a Transverse Electromagnetic (TEM) mode when it is confined between parallel plates, with a propagation constant $\beta = 2\pi/\lambda$, being λ the free space wavelength. Other fundamental modes can be introduced by assuming reflections on the plates: Transverse Electric (TE) modes, characterized by electric fields always perpendicular with respect to the propagation direction y, and Transverse Magnetic (TM) modes, with magnetic fields always transverse to y. By using the Maxwell equations, time periodic solutions can be found, with a $\exp(-j\omega t)$ dependence, assuming that partial derivatives of the fields vanish in the x direction ($\partial/\partial x = 0$). Because of the fields confinement, it is not possible to use the same condition for the z direction, which is along the thickness of the waveguide, then it will be $\partial/\partial z \neq 0$.

As well known, the time and space dependence of the electromagnetic fields are described by the Maxwell equations:

$$\begin{cases} \nabla \times e = -\dfrac{\partial b}{\partial t} \\ \nabla \times h = \dfrac{\partial d}{\partial t} \\ \nabla \cdot b = \nabla \cdot d = 0 \end{cases} \tag{12}$$

where e is the electric field, b the magnetic induction, h the magnetic field and d the electric induction. $d = \varepsilon e$ and $b = \mu h$ are the constitutive relations between the fields and the inductions by means of the permittivity ε and the permeability μ, which, in general are tensorial quantities. In the case of plane waves having a time dependence as $\exp(-j\omega t)$, the time derivatives are substituted by a $-j\omega$ term. When ε and μ of the medium are not tensorial quantities, simple equations can be found for the TE modes, which relate between them the derivatives of the fields e and h. In particular, the derivatives of the field components along the propagation direction will be:

$$\begin{cases} \dfrac{\partial e_z}{\partial y} = -j\omega\mu h_x \\ \dfrac{\partial h_x}{\partial y} = -j\omega\varepsilon e_z \end{cases} \tag{13}$$

From the above equations, by defining $e_z(y) = V(y)$ and $h_x(y) = I(y)$ as a voltage and as a current respectively, the telegraphist equations can be written as:

$$\begin{cases} \dfrac{\partial V}{\partial y} = -j\omega\mu I \\[3mm] \dfrac{\partial I}{\partial y} = -j\omega\varepsilon V \end{cases} \tag{14}$$

where the permeability μ plays the role of the inductance, and the permittivity ε is the capacitance of the propagating wave, both defined per unitary length. In this way, an equivalent circuit of the parallel plate configuration can be done in terms of an inductor Ldy and a capacitor Cdy, being dy the infinitesimal longitudinal section along the y-direction, with a characteristic impedance $Z_0 = \sqrt{(L/C)} = \sqrt{(\mu/\varepsilon)}$.

The time-dependent response of the magnetization of a ferro-magnetic medium biased by a dc field $\mathbf{H_0}$ and excited by a time-dependent field $\mathbf{h}$ is described by the precession of the magnetization $\mathbf{M}$ around an effective static internal field $\mathbf{H_i}$, which results from the contributions of $\mathbf{H_0}$, the anisotropy $\mathbf{H_A}$ and the demagnetization $\mathbf{H_{dem}}$, plus the time-dependent $\mathbf{h}$. The Landau-Lifschits (LL) equation formally summarizes such a situation by:

$$\frac{\partial M}{\partial t} = -\gamma\left(M \times H_i\right) \tag{15}$$

where $\gamma = 1.76\times10^7$ $Oe^{-1}sec^{-1}$ is the gyromagnetic ratio, and:

$$\begin{cases} H_i = H_0 + H_A + H_{dem} + h(x,y,z,t) \\ M = M_0 + m(x,y,z,t) \end{cases} \tag{16}$$

where $\mathbf{M_0}$ is the saturation magnetization and $\mathbf{m}$ is the induced time-dependent magnetization. In the linear approximation, when $|\mathbf{m}| \ll \mathbf{M_0}$ and $\mathbf{M_0}$ is directed along the z-axis, a permeability tensor μ can be inferred as:

$$\mu = 1 + \chi = \begin{pmatrix} 1+\chi_{11} & -j\chi_{12} & 0 \\ j\chi_{21} & 1+\chi_{22} & 0 \\ 0 & 0 & 1 \end{pmatrix} \tag{17}$$

where $\chi_{11} = \chi_{22} = \kappa = M_0 H_i/[H_i^2-(\omega/\gamma)^2]$, and $\chi_{12} = \chi_{21} = \nu = M_0(\omega/\gamma)/[H_i^2-(\omega/\gamma)^2]$. In an infinite ferro-magnetic medium, by assuming $\partial/\partial x = \partial/\partial z = 0$ naturally turns out the condition $e_y = 0$, which corresponds to the TE modes, and the Maxwell equations become:

$$\begin{cases} \dfrac{\partial e_z}{\partial y} = j\omega(\mu h_x - j\nu h_y) \\[2mm] j\nu h_x + \mu h_y = 0 \\[2mm] \dfrac{\partial e_x}{\partial y} = -j\omega h_z \\[2mm] \dfrac{\partial h_z}{\partial y} = -j\omega\varepsilon e_x \\[2mm] e_y = 0 \\[2mm] \dfrac{\partial h_x}{\partial y} = j\omega e_z \end{cases} \qquad (18)$$

from where the ellipticity between the x and the y components of the magnetic field is derived as $h_x/h_y = j(\mu/\nu)$. By properly combining the equations (18), wave equations for all the transverse fields can be obtained as:

$$\begin{cases} \dfrac{\partial^2 h_x}{\partial y^2} + \varepsilon\omega^2 \dfrac{\mu^2 - \nu^2}{\mu} h_x = 0 \\[3mm] \dfrac{\partial^2 h_z}{\partial y^2} + \varepsilon\omega^2 h_z = 0 \\[3mm] \dfrac{\partial^2 e_z}{\partial y^2} + \varepsilon\omega^2 \dfrac{\mu^2 - \nu^2}{\mu} e_z = 0 \\[3mm] \dfrac{\partial^2 e_x}{\partial y^2} + \varepsilon\omega^2 e_x = 0 \end{cases} \qquad (19)$$

Because of the linearization of the LL equation, by choosing the couple (e_x, h_z) it is difficult to define an inductance which is related to the permeability tensorial properties of the medium and, eventually, to its non-linear behaviour. Actually, in the linear approximation, μ has frequency-dependent components only in the plane orthogonal to the applied bias field, and its zz-term is unitary. In fact, matrix expressions different from those we have used in the above formulation of the problem have been derived only in the case of harmonic generation at high power levels, but not for the presence of small non-linearities. This is, actually, the bottleneck of the linearization of the LL-equation linked with the Maxwell equations: to try to find non-linear terms from the linearization of an intrinsic non-linear problem. On the other hand, the couple (h_x, e_z) is strictly dependent on the μ-components in the (x, y)-plane, and the telegraphist equations will be:

$$\begin{cases} \dfrac{\partial V}{\partial y} = j\omega \dfrac{\mu^2 - v^2}{\mu} I = -j\omega L I \\ \dfrac{\partial I}{\partial y} = -j\omega\varepsilon V = -j\omega C V \end{cases} \tag{20}$$

where $V = V(y) = e_z(y)$ and $I = I(y) = h_x(y)$ are the voltage and the current respectively. It is worthnoting that no magnetostatic approximation has been done until now. Only the Maxwell equations and the linear definition of the permeability tensor have been used. The previous result suggests that an equivalent inductor per unitary length can be defined as:

$$L = -\frac{\mu^2 - v^2}{\mu} \tag{21}$$

and from the knowledge of the tensor components a non-linear inductor can be derived by assuming a small non-linear displacement from the equilibrium value of the magnetization along the z-axis. Actually, since the non-linear contribution of the magnetization is related to the change of the z-component of the magnetization as $M_z \approx M_s(1-|\varphi|^2)$, it means that Eq.(21) can be considered as the Taylor series in $|\varphi|^2$-terms, up to the lowest possible order, as:

$$L = L_0\left(1 + A|\varphi|^2\right) \tag{22}$$

where L_0 is the linear term, and the value of A depends on all the magnetic quantities involved in the permeability tensor components definition and on the frequency.

By releasing the approximation $\partial/\partial z = 0$, which is valid for an unlimited medium but not in the real case of guided waves in dielectric media, the situation becomes more complicated, because of not vanishing terms in the Maxwell equations, and an inductor can be not easily defined.

In the configuration of a dielectric slab, guided waves are supported when total internal reflection of propagating rays is assumed, while an exponential decay of the electromagnetic fields happens outside of the medium. From the continuity conditions at the interfaces between medium and outer space, it turns out that both, TE and TM modes can propagate in the structure and the Maxwell equations for the TE mode become:

$$
\begin{cases}
\dfrac{\partial e_z}{\partial y} = j\omega\left(\mu h_x - jvh_y\right) \\[2mm]
\dfrac{\partial e_x}{\partial z} - \dfrac{\partial e_z}{\partial x} = j\omega\left(jvh_x + \mu h_y\right) \\[2mm]
-\dfrac{\partial e_x}{\partial y} = j\omega h_z \\[2mm]
\dfrac{\partial h_z}{\partial y} - \dfrac{\partial h_y}{\partial z} = -j\omega\varepsilon e_x \\[2mm]
\dfrac{\partial h_x}{\partial z} - \dfrac{\partial h_z}{\partial x} = 0 \\[2mm]
\dfrac{\partial h_y}{\partial x} - \dfrac{\partial h_x}{\partial y} = -j\omega\varepsilon e_z
\end{cases}
\tag{23}
$$

As it is usual, the approximation $\partial/\partial x=0$ can be mantained (along the direction transverse with respect to propagation), but no other term can be simplified. From the condition $\partial/\partial x=0$, the two couples of equations which can be used for the possible definition of an inductor and of a capacitor are:

$$
\begin{cases}
\dfrac{\partial e_z}{\partial y} = j\omega\mu h_x + \omega vh_y \\[2mm]
\dfrac{\partial h_x}{\partial y} = j\omega\varepsilon e_z
\end{cases}
\qquad
\begin{cases}
\dfrac{\partial e_x}{\partial y} = -j\omega h_z \\[2mm]
\dfrac{\partial h_z}{\partial y} = -j\omega\varepsilon e_x + \dfrac{\partial h_y}{\partial z}
\end{cases}
\tag{24}
$$

but the presence of mixed terms depending on h_y and on the μ-tensor, does not allow a straightforward definiton of L and C.

Until now, we have treated electromagnetic waves with no magnetostatic approximation, and we have found that no simple equivalent circuit models can be obtained for the tensorial magnetic medium. We will perform in what follows the same discussion for the MSW problem by using the scalar potential which is solution of the magnetostatic wave equation.

The MSW approximation is given by:

$$
\nabla \times h = 0
\tag{25}
$$

$h=\nabla\varphi$, where φ is a scalar potential, can be defined to solve the MSW problem. From above equation, a solution of the wave propagationg along the y-axis in a medium limited along the z-axis and the x-axis is given by:

$$
\varphi = \varphi_0 sin(k_z z + \theta_z) sin(k_x x + \theta_x) e^{jk_y y}
\tag{26}
$$

where φ_0 is the potential amplitude, θ_x and θ_z are two phases depending on the boundary conditions. By assuming that the fields vanish at $x=\pm L/2$ (transverse edges of the waveguide) and $z=\pm d/2$ (film surfaces), it will be $\theta_x=\theta_z=0$.

From the MSW approximation given in Eq.(25), and by using the $\nabla\times e$ equation for the TE-mode and with the condition $\partial/\partial x=0$:

$$\begin{cases} \dfrac{\partial e_z}{\partial y} = j\omega\left(\mu h_x - jv h_y\right) \\[2mm] \dfrac{\partial e_x}{\partial z} = j\omega\left(jv h_x + \mu h_y\right) \\[2mm] \dfrac{\partial e_x}{\partial y} = -j\omega h_z \end{cases} \tag{27}$$

From the definition of the scalar field φ given in Eq.(26), the h_x and h_y components of the h field can be derived as:

$$\begin{cases} \dfrac{\partial \varphi}{\partial y} = jk_y\varphi = h_y \\[2mm] \dfrac{\partial \varphi}{\partial x} = k_x \cot\left(k_x x\right)\varphi = h_x \end{cases} \tag{28}$$

from where, after few algebraic passages:

$$\frac{\partial e_z}{\partial y} = j\omega\left(\mu + v\frac{k_y}{k_x}\tan\left(k_x x\right)\right)h_x \tag{29}$$

When the magnetic waveguide is infinite in the x-direction, k_x tends to zero, and the second term in brackets of Eq.(29) tends to $vk_y x$. In both cases, when the waveguide is finite or infinite along the direction transverse to the propagation, this term cannot be considered negligible. It means that Eq.(29) can not be furtherly simplified, and an inductor per unitary pathlength is not easily defined, unless to hypothesize a lumped element depending on the transverse direction.

In conclusion, Eq.(24) (Maxwell equations for a magnetic tensorial medium) and Eq.(27) (magnetostatic wave problem) do not allow an easy definition of the inductor per unitary length, as it is the case of a non-tensorial medium where the permeability μ is the unique information needed.

5. Conclusions

In this work, the modeling of non-linear transmission lines (NLTL) has been discussed for semiconductor and magnetic film media, to give a general description of both systems in terms of lumped elements for possible use in simulations by means of commercial software packages. The approach for obtaining the ladder network is that based on the transmission lines theory derived by the Maxwell equations, where the electric and the magnetic fields are substituted by voltages and currents, respectively, to obtain the inductance L and the capacitance C per unitary length.

Because of the constitutive L and C components of a semiconductor device, it is quite easy to define discrete components and to obtain a continuum approximation analytical treatment by means of the KdV equation for the non-linear behavior of the transmission line. Interesting properties of the semiconductor NLTL can be predicted by using different non-linear elements, like Schottky diodes or resonating tunnel diodes (RTD) which provide a non-linear response of the capacitance per unitary cell vs the applied voltage. Harmonic generation is one of the peculiar properties of such a NLTL, but pulse shaping and soliton propagation can occur by properly tailoring the elementary cell parameters and the number of the cells.

Magnetic film media, or, more generally, tensorial media for which the permeability (or the permittivity) have a matrix representation, do not have a transmission line model with the same straightforward application of the Maxwell equations to define lumped elements per unitary pathlength. Actually, also the magnetostatic problem does not allow a simple picture where L and C are easily defined. It means that a more complicated approach has to be proposed, matching it with the well known approach of the wave modulation which carries to the non-linear Schrödinger equation (NLSE). Actually, it has to be demonstrated if the NLSE and a possible ladder network approch lead to the same conclusions, and if, in the particular case of the MSW approximation, a non-linear inductor is sufficient to descibe the non-linear propagation in the tensorial medium or a non-linear capacitor has to be defined too. An alternative way of operation is to utilize experimental data to obtain the best description of the linear and non-linear behavior of the magnetic film waveguide in terms of lumped elements.

6. References

1. Remoissenet, M. (1994) *Waves Called Solitons*, Springer Verlag, Berlin.
2. Boardman, A.D., Cooper, G.S., Maradudin, A.A., Shen, T.P. (1986) Surface-polariton solitons, *Physical Review B* 34, 8273-8278.
3. Sethares, J.C., Weinberg, I.J. (1985), Circuits, Systems and Signal Processing, Special Issue on Microwave Magnetics, Boston
4. Zvezdin, A.K. and Popkov, A.F. "A contribution to non-linear spin wave theory", Sov. Phys. JETP, vol.57, p.350 (1983).
5. E. Carman, M. Case, M. Kamegawa, R. Yu, K. Giboney, M. J. W. Rodwell, "V - Band and W - Band broadband, monolithic distributed frequency multipliers", IEEE Microwave Guided Wave Lett., Vol. 2, pp. 253 - 254, June 1992.
6. E. Carman, K. Giboney, M. Case, M. Kamegawa, R. Yu, K. Abe, M. J. W. Rodwell, J. Franklin, "28 - 39 GHz Distributed harmonic generation on a soliton non linerar trasmission line", IEEE Microwave Guided Wave Lett., Vol. 1, pp. 28 - 31, Feb. 1991.
7. M. Dragoman, A. Muller, S. Iordanescu, F. Craciunoiu, S. Simion, B. Szentpali, K. Somogyi, F. Riesz, S. Varga, "Millimeter frequencies generation on a travelling MMIC Schottky diode array and applications in an automotive sensor", Proceedings of GAAS 94 Symposium, April 1994, Torino, pp 293 - 296.
8. R. Y. Yu, Y. Konishi, S. T. Allen, M. Reddy, M. J. W. Rodwell "A travelling - wave resonant tunnel diode pulse generator", IEEE Microwave Guided Wave Lett., Vol. 4, pp. 220 - 222, July 1994.
9. R. P. Smith, S. T. Allen, M. Reddy, S. C. Martin, J. Liu, R. E. Muller, M. J. W. Rodwell, "0.1 μm Schottky - collector AlAs - GaAs resonant tunnelling diodes", IEEE Trans. Electron. Device Letters, Vol. 15, n. 8, pp. 295 - 297, August 1994.

10. S. K. Diamond, E. Özbay, M. J. W. Rodwell, D. M. Bloom, Y. C. Pao, J. S. Harris, "Resonant tunneling diodes for switching applications", Appl. Phys. Lett. 54 (2) pp.153-155 (1989).

Other References on Magnetostatic Wave Devices and Non-linear Signal Processing:

Microwave Applications of Magnetic Garnets

1. Proceedings of the IEEE, Special Section on Microwave Magnetics, vol.76, No.2, pp.97-208 (1988).
2. P. Kaboš and V.S. Stalmachov: "Magnetostatic Waves and their Applications", Chapman and Hall, New York, 1994.
3. D.D. Stancil: "Theory of Magnetostatic Waves", Springer Verlag, New York, 1993.
4. J. Helszajn, YIG Resonators and Filters, John Wiley and Sons, New York (1985).
5. C. Vittoria: "Microwave properties of magnetic films", World Scientific, Singapore, 1993.
6. B. Lax and K.J. Button: "Microwave ferrites and ferrimagnetics", Mc Graw Hill, New York, 1962.

Advanced Topics (Nonlinear Effects)

1. M. Cottam: "Linear and non-linear spin waves in magnetic film and superlattices"; World Scientific, Singapore, 1994.
2. IEEE Trans. on Magnetics, Special Section on Advances in Magnetics: Nonlinear Effects in Magnetics, Vol.30, No.1, 1994
3. P.E. Wigen: "Nonlinear phenomena and chaos in magnetic materials"; World Scientific, Singapore, 1993.

NONLINEAR DYNAMICS OF OPTICAL SOLITONS

STEFAN WABNITZ

Fondazione Ugo Bordoni
Via B. Castiglione 59, 00142 Rome, Italy

1. Introduction

Optical soliton pulses or beams represent a stable balance between the broadening due to linear dispersion (or diffraction) and the self-compression (or self-focusing) induced by the intensity-dependent refractive index. Perhaps the most important property of solitons is their particle-like nature. As a consequence, the propagation of solitons in the presence of perturbations may often be analysed with the help of a simple low-dimensional approximate dynamical model.

In this work, we intend to present in a unified manner the method of analysis of optical soliton propagation in the presence of perturbations. We will first outline the derivation of the nonlinear Schrödinger equation (NLS) in a lossless optical medium. As we shall see, the same NLS equation also describes pulse propagation in periodically amplified fiber links, where a strong attenuation of soliton amplitude occurs between two lumped amplifiers. The NLS equation also describes spatial soliton formation in diffractive waveguides. As a consequence, often temporal soliton phenomena have their counterpart in the space domain.

In order to analyse the soliton dynamics in the presence of perturbations, we will not employ here the results from the inverse scattering theory. This method permits to obtain exact solutions, however since it rests on the complete integrability of the NLS equation [1], it is less general than the Lagrangian method which is adopted here. For example, the Lagrangian perturbation analysis is easily extended to the vector NLS case. The perturbation results will then be applied to specific applicative examples such as two-soliton interactions, fiber soliton lasers and soliton couplers.

R. Marcelli and S.A. Nikitov (eds.), Nonlinear Microwave Signal Processing: Towards a New Range of Devices, 101–118.
© *1996 Kluwer Academic Publishers.*

2. The Nonlinear Schrödinger Equations

In nonlinear optics, solitons represent a stable balance between nonlinearity and temporal dispersion or transverse diffraction. The two distinct cases lead to the observation of temporal or spatial solitons, respectively. In the most general case, dispersion and diffraction may act simultaneously and spatio-temporal solitons or light bullets are formed. However, in most practical situations the peak powers that lead to the compensation of diffraction and dispersion are so different that one may safely neglect either one of the two effects. In the following, we briefly sketch the derivation of the basic soliton equation or nonlinear Schrödinger equation in either dispersive or diffractive optical waveguides.

2.1. TEMPORAL SOLITONS

2.1.1. *Lossles Medium*
Let us consider the propagation of optical pulses in optical fibers. Take at first a constant polarization state and a single guided mode. The last condition permits to separate transverse $((x, y))$ and longitudinal (z) variables: the field in the fiber reads as $\mathbf{E}(x, y, z, t) = E(z, t)\mathbf{F}(x, y)$, where $\mathbf{F}$ is the transverse modal profile and

$$E(z, t) = \int \hat{E}(\omega)e^{i\{k(\omega)z - \omega t\}}d\omega =,$$

$$\left(\int \hat{E}(\omega_0 + \Delta\omega)e^{i\{[k(\omega_0 + \Delta\omega) - k_0]z - i\Delta\omega t\}}d(\Delta\omega)\right)e^{i\{k_0 z - i\omega_0 t\}} + c.c., \quad (1)$$

where k is the modal propagation constant and ω_0 is the mean frequency of the quasi-monochromatic field. One uses the following expansion for k in Maxwell equations

$$k - k_0 = k'\Delta\omega + \frac{1}{2}k''\Delta\omega^2 + \frac{\omega n_2}{cA_{eff}}|E|^2 \quad (2)$$

where n_2 is the nonlinear refractive index and A_{eff} is the transverse mode $\mathbf{F}$ cross-section. With $k - k_0 = -i\frac{\partial}{\partial z}$, $\Delta\omega = i\frac{\partial}{\partial t}$, and $\tau = (t - k'z)$, one obtains, in dimensionless form,

$$iq_Z + \frac{1}{2}q_{TT} + |q|^2 q = iR, \quad (3)$$

where $q = E\sqrt{\omega_0 n_2 z_0/(cA_{eff})}$, $Z = z/z_0$, $z_0 = -\tau_0^2/k''$, $T = \tau/\tau_0$, τ_0 is an arbitrary time unit. Here R represents the various perturbations: with $R = 0$, eq.(3) has the well-known one-soliton solution

$$q(Z,T) = \eta \text{sech}[\eta(T - \xi(Z))]e^{-i\kappa(T-\xi(Z))+i\delta(Z)},$$

$$\frac{d\xi}{dZ} = -\kappa, \tag{4}$$

$$\frac{d\delta}{dZ} = \frac{\eta^2 - \kappa^2}{2}.$$

2.1.2. Periodically Amplified Lossy Link

In a fiber link (or fiber laser), nonlinear pulse propagation is described by the perturbed NLS equation [2, 3]

$$iq_Z + \frac{1}{2}q_{TT} + |q|^2 q = -i(\Gamma - G(Z))q - iG''(Z)q_{TT} \equiv$$

$$iF(Z)q - iG''(Z)q_{TT} \tag{5}$$

where the distributed loss coefficient is $\Gamma = \gamma z_0$, and γ is the loss per unit length. G and G'' represent the periodic gain of the optical amplifiers and their bandwidth, respectively. Note that low-dispersion fibers typically yield $\Gamma \gg 1$. Whenever the amplifier distance $Z_a = z_a/z_0 \ll 1$, propagation over Z_a is linear. Whereas nonlinear and dispersive effects occur at lengths of the order of z_0. Set $q(Z,T) = a(Z)u(Z,T)$, where a(Z) represents the rapidly varying soliton amplitude and $da/dZ = F(Z)a$, or

$$\frac{da}{dZ} = -\Gamma a + [exp(\Gamma Z_a) - 1] \sum_{n=1}^{M} \delta(Z - nZ_a)a \tag{6}$$

One obtains

$$iu_Z + \frac{1}{2}u_{TT} + a(Z)^2|u|^2 u = iP[u, u^*, Z] \tag{7}$$

Here P represents small-amplitude periodic perturbations such as the excess amplifier gain and the finite gain bandwidth. Averaging eq.(2) over Z_a, with the condition that $a(Z = 0)^2 = a_0^2 = 2\Gamma Z_a/(1 - exp(-2\Gamma Z_a))$ (i.e., $< a^2(Z) > = 1$), leads to a guiding-center (or averaged) equation [3]

$$iq_Z + \frac{1}{2}q_{TT} + |q|^2 q = iR[q, q^*] + O(Z_a^2). \tag{8}$$

2.2. SPATIAL SOLITONS

The NLS equation may be also derived for the stationary propagation of light beams in thin-film planar waveguides. Let us consider here the simplest case of a transverse electric (TE) wave in a Kerr medium. Take y for the direction of the electric field and z for the propagation direction. The *scalar* electric field $\vec{E}(x, y, z)e^{i\omega t}$ satisfies the simplified Maxwell equation (or Helmoltz equation)

$$\frac{\partial^2 E}{\partial z^2} + \frac{\partial^2 E}{\partial x^2} = -n^2 \frac{\omega^2}{c^2} E \tag{9}$$

where n is the refractive index of the medium, that reads $n^2 = n_0^2 + \hat{n}_2^2 |E|^2$. We further assume to deal with light waves that are close to plane waves, and write

$$\vec{E} = \hat{y} F(x, z)e^{i(\beta k_0 z - \omega t)} + cc. \tag{10}$$

where $F(x, z)$ is the envelope of the field, βk_0 is the effective wavenumber, and $k_0 = \omega/c$. To be more specific, we consider beam propagation in the framework of the paraxial ray theory: rays perpendicular to surfaces of constant phase are nearly parallel. In other words, the field envelope is slowly varying as it propagates along the z direction ($\partial^2 F/\partial z^2 << 2i\beta k_0 \partial F/\partial z$). Whenever the diffraction term and $(\beta^2 - n^2)F$ are of the same order of the propagation term, one obtains for the envelope F

$$2i\beta k_0 \frac{\partial F}{\partial z} + \frac{\partial^2 F}{\partial x^2} - (\beta^2 - n^2)F = 0 \tag{11}$$

As well known, in linear optical waveguides the refractive index is a function of the transverse and longitudinal coordinates (x, z). In the present case, the refractive index also depends on the local light intensity: $n^2 \equiv n_0^2 + \hat{n}_2^2 |F|^2 + \delta n^2(x, z, |F|^2)$, where $n_0, \hat{n}_2$ are constants and the last term is a small perturbation. In the absence of other perturbations, the balance between diffraction and the intensity-dependent self-focusing mechanism generates a self-guided spatial mode, which is again the one-soliton solution of the NLS equation (11)

$$F(x, z) = \frac{\sqrt{2}}{n_2} 2\eta \, sech[2\eta(x - \frac{vz}{2\beta} - \bar{x})]e^{i(\frac{vx}{2} - (\frac{v^2}{4} + \beta^2 - n_0^2 - 4\eta^2)\frac{z}{2\beta})} \tag{12}$$

Here, the parameter η defines the width of the beam and is related to the power across by the relation

$$P = \int |E|^2 dx = \frac{8\eta}{\hat{n}_2^2} \tag{13}$$

The second soliton parameter v is now the x wavenumber and it also defines the angle θ that the beam makes with the z axis through the relation $\theta = tan^{-1} v/2\beta$.

2.3. VECTOR SOLITONS

The optical field in fibers and waveguides has two orthogonal polarization components. In many situations it is important to take into account the degree of freedom associated with light polarization. As we shall see, polarized solitons may be formed in nonlinear optical fibers. The envelope of the polarized field reads as $\mathbf{E}(Z,T) = q_x(Z,T)\mathbf{e_x}f_x(\mathbf{r}) + q_y(Z,T)\mathbf{e_y}F_y(\mathbf{r})$, where $\mathbf{e_{x,y}}$ are orthogonal polarization unit vectors, and $\mathbf{F}_{x,y}(\mathbf{r})$ is the transverse mode profile. With a nonlinear polarizability of the form

$$P = \chi[(1 - B)(\mathbf{E} \cdot \mathbf{E}^*)\mathbf{E} + B(\mathbf{E} \cdot \mathbf{E})\mathbf{E}^*], \qquad (14)$$

where B is the polarization anisotropy coefficient (in a silica fiber, $B - 1/3$), one obtains that the vector amplitude $\mathbf{q} = (q_x, q_y)^T$ obeys the NLS system [4]

$$i\mathbf{q}_Z + \mathbf{M}\left[i\delta\mathbf{q}_T + k\mathbf{q}\right] + \frac{1}{2}\mathbf{q}_{TT} + \mathbf{N}(\mathbf{q}) = 0, \qquad (15)$$

where δ and k represent polarization dispersion and linear birefringence. Moreover, $\mathbf{M}$ and $\mathbf{N}$ indicate the linear and the nonlinear birefringence, respectively, and read

$$\mathbf{M} = \begin{pmatrix} cos(2\psi) & sin(2\psi) \\ sin(2\psi) & -cos(2\psi) \end{pmatrix}, \qquad (16)$$

$$\mathbf{N} = \begin{pmatrix} (|q_x|^2 + (1 - B)|q_y|^2)q_x + Bq_y^2 q_x^* \\ (|q_y|^2 + (1 - B)|q_x|^2)q_y + Bq_x^2 q_y^* \end{pmatrix}, \qquad (17)$$

where $\psi = \psi(Z)$ denotes the random angle between the linear birefringence axes at any given point Z and the fixed x, y polarization axes. In order to average out the rapid variations of the linear birefringence of the fiber (the length scale for dispersive and nonlinear effects is much longer than the correlation distance of these random changes of the linear birefringence properties), one may take $\psi(Z) = \tau(\epsilon(Z))Z$, where $\epsilon(Z)$ is a slowly varying function of Z. The vector propagation equation (17) may then be rewritten in a new frame that rotates with rate τ. Whenever eq.(15) is expressed in the new basis of the linear birefringence eigenmodes, one obtains the following coupled NLS equations [5]

$$i\frac{\partial U}{\partial Z} + \frac{1}{2}\frac{\partial^2 U}{\partial T^2} + \left(\sigma(Z)|U|^2 + \rho(Z)|V|^2\right)U = 0,$$

$$\qquad (18)$$

$$i\frac{\partial V}{\partial Z} + \frac{1}{2}\frac{\partial^2 V}{\partial T^2} + \left(\sigma(Z)|V|^2 + \rho(Z)|U|^2\right)V = 0.$$

where $\sigma = (1 + 2t^2(1 - 2B) + t^4)/(1 + t^2)^2$, $\rho(Z) = (1 - B + 2t^2(1 + 3B) + (1 - B)t^4)/(1 + t^2)^2$, $t = -r + \sqrt{r^2 + 1}$, and $r = k/\tau$. We took $k > 0$, so that $1 \geq t \geq 0$. In the above equations we assumed that polarization dispersion is rapidly varying along Z so that it may be averaged out. We also dropped in eqs.(18) all the rapidly varying (i.e., oscillating over the length scale of the linear birefringence beat length $L_b = \pi/k$) nonlinear terms that are obtained from the matrix $\mathbf{N}$. As a result of the randomly rotating birefringence, the self- and cross-modulation coefficients σ and ρ are sampled from random processes of the distance Z. As a first approximation, one may then consider the average over Z of eqs.(18). In the rotating reference frame, the local birefringence may be associated to a vector on the Poincare' sphere which reads $\Omega = (2k\cos(\theta), 2k\sin(\theta), 2\tau)$, where θ is an arbitrary angle. If one assumes that $\tau = k\sin(\psi)$, where the angle ψ has an uniform distribution, one obtains the following average (over Z) values: $< \tau^2 >= k^2/2$, $< r^2 >= 2$, $< \rho/\sigma >= 1$, and $\sigma = 8/9$ (for $B = 1/3$). In conclusion, averaging the random birefringence over the propagation distance reduces eqs.(18) to the integrable Manakov equations [6], or vector NLS equation

$$i\frac{\partial U}{\partial Z} + \frac{1}{2}\frac{\partial^2 U}{\partial T^2} + \left[|U|^2 + |V|^2\right]U = iR_U \tag{19}$$

$$i\frac{\partial V}{\partial Z} + \frac{1}{2}\frac{\partial^2 V}{\partial T^2} + \left[|U|^2 + |V|^2\right]V = iR_V. \tag{20}$$

Here R_U, R_V are small perturbation terms: whenever $R_U = R_V = 0$, the one–soliton solution of equations (19-20) reads as

$$U_0(T, Z) = \eta \, \cos(\theta) \, \text{sech}\left[\eta(T - \xi)\right]e^{i(\kappa(T-\xi)+\delta_U)} \tag{21}$$

$$V_0(T, Z) = \eta \, \sin(\theta) \, \text{sech}\left[\eta(T - \xi)\right]e^{i(\kappa(T-\xi)+\delta_V)} \tag{22}$$

where $2\nu, \xi$ and μ represent the soliton amplitude, position and frequency. Whereas δ_U, δ_V are the phases of the orthogonal polarization components and θ is the polarization angle.

3. The Dynamical Structure

In order to develop a perturbation theory for the NLS equation, one may exploit its Lagrangian (or Hamiltonian) structure. In fact, the NLS equation is an infinite dimensional dynamical system: the evolution of the field $q(Z, T)$ yields an extremum of the action. This is given by the $T-$ and $Z-$integrals of the Lagrangian density $\mathcal{L}$

$$\delta \int \int \mathcal{L}(q, q^*, q_Z, q_Z^*, q_T, q_T^*, q_{TT}, \cdots) \, dTdZ = 0. \tag{23}$$

In the case of the scalar NLS equation, $\mathcal{L} \equiv \mathcal{L}_0$ and reads

$$\mathcal{L}_0(q, q^*, \cdots) = \frac{i}{2}(qq_Z^* - q_Z q^*) - \frac{1}{2}(|q|^4 - |q_T|^2) \qquad (24)$$

The variation (23) is defined as

$$\begin{aligned}
&\delta \int \int \mathcal{L}(q, q^*, \cdots)\, dT dZ \\
&\equiv \lim_{\varepsilon \to 0} \frac{1}{\varepsilon} \int \int \{\mathcal{L}(q + \varepsilon(\delta q), q^* + \varepsilon(\delta q^*), \cdots) - \mathcal{L}(q, q^*, \cdots)\}\, dT dZ
\end{aligned} \qquad , \quad (25)$$

where δq and δq^* vanish at the integration boundaries. We also define $\delta q_Z = \partial(\delta q)/\partial Z$, $\delta q_Z^* = \partial(\delta q^*)/\partial Z$, $\delta q_T = \partial(\delta q)/\partial T$, and so on. With an integration by parts, eq.(23) reduces to

$$\int \int \left[\left\{ \sum_{n=0}^{\infty} (-1)^n \frac{\partial^n}{\partial T^n} \frac{\partial \mathcal{L}}{\partial q_{nT}^*} - \frac{\partial}{\partial Z} \frac{\partial \mathcal{L}}{\partial q_Z^*} \right\} \delta q^* + \{\text{c.c.}\} \right] dT dZ = 0 \qquad (26)$$

where $q_{nT}^* = \partial^n q^*/\partial T^n$, and note $(\partial \mathcal{L}/\partial q_{nT})^* = \partial \mathcal{L}/\partial q_{nT}^*$. The variations δq and δq^* are assumed to be arbitrary and independent, and eq.(26) reduces to

$$\sum_{n=0}^{\infty} (-1)^n \frac{\partial^n}{\partial T^n} \frac{\partial \mathcal{L}}{\partial q_{nT}^*} - \frac{\partial}{\partial Z} \frac{\partial \mathcal{L}}{\partial q_Z^*} = 0 \ , \qquad (27)$$

along with its complex conjugate. We may denote eq.(27) as

$$\frac{\delta}{\delta q^*} \mathcal{L}[q, q^*] = 0 \ . \qquad (28)$$

By substituting (24) in (27), one immediately obtains the NLS equation.

4. The Lagrangian Perturbation Theory

The complete integrability of the NLS equation by means of the inverse scattering transform permits what is perhaps the most peculiar property of NLS solitons: their identity, (amplitude (or width) and velocity (or frequency)) remains constant after their mutual interaction. However, in real physical systems several perturbations to the ideal NLS equation exist. As a consequence, it is essential to analyse the stability of the NLS solitons by means of perturbation theory. We discuss here a general perturbation method which is based on the Lagrangian structure of the NLS equation. An important advantage of this method is that it may be also applied to non-integrable systems, as long as a well-defined solitary wave solution exists [7].

4.1. SCALAR NLS

By using (28) and (24), the perturbed NLS equation (8) may be rewritten as

$$i\frac{\delta \mathcal{L}_0}{\delta q^*}[q, q^*] = R[q, q^*] \equiv \varepsilon P[q, q^*] , \qquad (29)$$

The main hypothesis here is that a solution of eq.(29) may be written

$$q(T, Z) = q_0(T, Z) + \varepsilon q_1(T, Z) + O(\varepsilon^2) . \qquad (30)$$

where $|\varepsilon| \ll 1$. For a perturbed one-soliton solutions, as a leading order solution q_0 one takes the expression eq.(4). By inserting eq.(4) in the Lagrangian density (24), and integrating this density over T, a reduced finite-dimensional Lagrangian results that describes the dynamics of the soliton parameters. In fact, this reduced Lagrangian may be exploited for obtaining evolution equations for these parameters. The time averaged Lagrangian L_0 for $q = q_0$ reads

$$L_0 \equiv \int_{-\infty}^{\infty} \mathcal{L}_0[q_0, q_0^*] \, dT = 2\eta \left(\kappa \frac{d\xi}{dZ} + \frac{d\delta}{dZ} \right) - \frac{1}{3}\eta_1^3 + \eta_1 \kappa_1^2 . \qquad (31)$$

One may also compute the variations of L_0 with respect to the one-soliton parameters,

$$\frac{\delta L_0}{\delta \eta} = \frac{\partial L_0}{\partial \eta} - \frac{d}{dZ} \frac{\partial L_0}{\partial(d\eta/dZ)} = 2\left(\kappa \frac{d\xi}{dZ} + \frac{d\delta}{dZ} \right) - \eta^2 + \kappa^2 , \qquad (32)$$

$$\frac{\delta L_0}{\delta \kappa} = \frac{\partial L_0}{\partial \kappa} - \frac{d}{dZ} \frac{\partial L_0}{\partial(d\kappa/dZ)} = 2\eta \frac{d\xi}{dZ} + 2\eta\kappa , \qquad (33)$$

$$\frac{\delta L_0}{\delta \xi} = \frac{\partial L_0}{\partial \xi} - \frac{d}{dZ} \frac{\partial L_0}{\partial(d\xi/dZ)} = -2\kappa \frac{d\eta}{dZ} - 2\eta \frac{d\kappa}{dZ} , \qquad (34)$$

$$\frac{\delta L_0}{\delta \delta} = \frac{\partial L_0}{\partial \delta} - \frac{d}{dZ} \frac{\partial L_0}{\partial(d\delta/dZ)} = -2\frac{d\eta}{dZ} . \qquad (35)$$

These variations yield the Euler-Lagrange equations for the soliton parameters. Without a perturbation R, the evolution of the soliton parameters is obtained by setting the variations to zero: this yields again the one-soliton solution eq.(4). With a perturbation, the derivatives of the Lagrangian L_0 (32)-(35) may also be written in the chain-rule form, e.g.,

$$\frac{\delta L_0}{\delta \eta} = \int_{-\infty}^{\infty} \left(\frac{\delta L_0}{\delta q_0(T)} \frac{\partial q_0(T)}{\partial \eta} + \frac{\delta L_0}{\delta q_0^*(T)} \frac{\partial q_0^*(T)}{\partial \eta} \right) dT . \qquad (36)$$

Here we may express the functional derivatives of $\mathcal{L}_0[q_0, q_0^*]$ with the help of (29) and (30). In fact, by inserting (30) with $q = q_0$ into (29), one obtains (up to order ε) that

$$i\frac{\delta \mathcal{L}_0}{\delta q_0^*}[q_0, q_0^*] = \varepsilon P[q_0, q_0^*] . \tag{37}$$

By using of (37) in each of the equations similar to (36), we may determine the right-hand side of eqs.(32)-(35). This eventually yields the evolution equations for the one-soliton parameters [8]-[10]

$$\frac{d\eta}{dZ} = \int_{-\infty}^{\infty} \text{Re}\{R[q, q^*]e^{-i\varphi}\} \, \text{sech} \, \tau \, d\tau , \tag{38}$$

$$\frac{d\kappa}{dZ} = -\int_{-\infty}^{\infty} \text{Im}\{R[q, q^*]e^{-i\varphi}\} \, \text{sech} \, \tau \, \tanh \tau \, d\tau , \tag{39}$$

$$\frac{d\xi}{dZ} = -\kappa + \frac{1}{\eta^2}\int_{-\infty}^{\infty} \text{Re}\{R[q, q^*]e^{-i\varphi}\} \, \tau \, \text{sech} \, \tau \, d\tau , \tag{40}$$

$$\frac{d\delta}{dZ} = \frac{\eta^2 - \kappa^2}{2} - \kappa\frac{d\xi}{dZ} +$$
$$\frac{1}{\eta}\int_{-\infty}^{\infty} \text{Im}\{R[q, q^*]e^{-i\varphi}\} \, \text{sech} \, \tau \, (1 - \tau \tanh \tau) \, d\tau , \tag{41}$$

noindent with $\tau = \eta(T - \xi)$.

4.2. VECTOR NLS

In this section, we show that the same Lagrangian method may be applied in order to derive perturbation equations for the evolution of the perturbed vector one–soliton parameters (21-22). One first rewrites eqs.(19,20) in the form

$$\frac{\delta \mathcal{L}_0}{\delta U^*} = iR_U \quad , \quad \frac{\delta \mathcal{L}_0}{\delta V^*} = iR_V \tag{42}$$

where the Lagrangian density is

$$\mathcal{L}_0 = -\text{Im}\left[U\frac{\partial U^*}{\partial Z} + V\frac{\partial V^*}{\partial Z}\right] - \frac{1}{2}\left[|U|^2 + |V|^2\right]^2 + \frac{1}{2}\left[\left|\frac{\partial U}{\partial T}\right|^2 + \left|\frac{\partial U}{\partial T}\right|^2\right] . \tag{43}$$

The time-averaged (on the one–soliton solution $U = U_0$, $V = V_0$ (21,22)) Lagrangian reads as

$$L_0 = \int \mathcal{L}_0[U_0, U_0^*, V_0, V_0^*]dT = 2\kappa\eta\frac{d\xi}{dZ} - \frac{\eta^3}{3} + \eta\kappa^2 +$$
$$2\eta\left[\cos^2(\theta)\frac{d\delta_U}{dZ} + \sin^2(\theta)\frac{d\delta_V}{dZ}\right] \tag{44}$$

One may again evaluate as before the variations of the Lagrangian L_0 with respect to the soliton parameters. First, one may directly use eq.(44), which yields

$$\frac{\delta L_0}{\delta \kappa} = 2\eta \frac{d\xi}{dZ} + 2\eta\kappa \tag{45}$$

$$\frac{\delta L_0}{\delta \eta} = 2\kappa \frac{d\xi}{dZ} - \eta^2 + \kappa^2 + 2\left[\cos^2(\theta)\frac{d\delta_U}{dZ} + \sin^2(\theta)\frac{d\delta_V}{dZ}\right] \tag{46}$$

$$\frac{\delta L_0}{\delta \xi} = -2\left(\kappa\frac{d\eta}{dZ} + \eta\frac{d\kappa}{dZ}\right) \tag{47}$$

$$\frac{\delta L_0}{\delta \delta_U} = -2\cos^2(\theta)\frac{d\eta}{dZ} - 4\eta\sin(\theta)\cos(\theta)\frac{d\theta}{dZ} \tag{48}$$

$$\frac{\delta L_0}{\delta \delta_V} = -2\sin^2(\theta)\frac{d\eta}{dZ} - 4\eta\sin(\theta)\cos(\theta)\frac{d\theta}{dZ} \tag{49}$$

$$\frac{\delta L_0}{\delta \theta} = -4\eta\sin(\theta)\cos(\theta)\left(\frac{d\delta_U}{dZ} - \frac{d\delta_V}{dZ}\right). \tag{50}$$

Secondly, one may also write the variations of L_0 in the chain–rule form (36). Inserting in the resulting expressions both eqs.(42) and the derivatives of the one–soliton solution (21,22), yields

$$\frac{\delta L_0}{\delta \kappa} = \frac{2}{\eta}\mathrm{Re}\int\left\{\cos(\theta)R_U e^{-i\beta_U} + \sin(\theta)e^{-i\beta_V}\right\}\tau\,\mathrm{sech}(\tau)d\tau \tag{51}$$

$$\frac{\delta L_0}{\delta \eta} = \frac{2}{\eta}\mathrm{Im}\int\left\{\cos(\theta)R_U e^{-i\beta_U} + \sin(\theta)e^{-i\beta_V}\right\}\mathrm{sech}(\tau)(1 - \tau\tanh(\tau))d\tau \tag{52}$$

$$\frac{\delta L_0}{\delta \xi} = 2\eta\mathrm{Im}\int\left\{\cos(\theta)R_U e^{-i\beta_U} + \sin(\theta)e^{-i\beta_V}\right\}\tanh(\tau)\mathrm{sech}(\tau)d\tau +$$
$$-2\kappa\mathrm{Re}\int\left\{\cos(\theta)R_U e^{-i\beta_U} + \sin(\theta)e^{-i\beta_V}\right\}\mathrm{sech}(\tau)d\tau + \tag{53}$$

$$\frac{\delta L_0}{\delta \delta_U} = -2\mathrm{Re}\int\cos(\theta)R_U e^{-i\beta_U}\mathrm{sech}(\tau)d\tau \tag{54}$$

$$\frac{\delta L_0}{\delta \delta_V} = -2\mathrm{Re}\int\sin(\theta)R_V e^{-i\beta_V}\mathrm{sech}(\tau)d\tau \tag{55}$$

$$\frac{\delta L_0}{\delta \theta} = -2\mathrm{Im}\int\left\{\sin(\theta)R_U e^{-i\beta_U} - \cos(\theta)R_V e^{-i\beta_V}\right\}\mathrm{sech}(\tau)d\tau. \tag{56}$$

By comparing the right-hand sides of (45–50) and eqs.(51–56), one finally obtains the perturbation equations

$$\frac{d\eta}{dZ} = \text{Re} \int \left\{ \cos(\theta) R_U e^{-i\beta_U} + \sin(\theta) R_V e^{-i\beta_V} \right\} \text{sech}(\tau) d\tau \qquad (57)$$

$$\frac{d\kappa}{dZ} = -\text{Im} \int \left\{ \cos(\theta) R_U e^{-i\beta_U} + \sin(\theta) R_V e^{-i\beta_V} \right\} \tanh(\tau) \text{sech}(\tau) d\tau \quad (58)$$

$$\frac{d\theta}{dZ} = \frac{1}{2\eta} \text{Re} \int \left\{ \cos(\theta) R_V e^{-i\beta_V} - \sin(\theta) R_U e^{-i\beta_U} \right\} \text{sech}(\tau) d\tau \qquad (59)$$

$$\frac{d\xi}{dZ} = -\kappa + \frac{1}{\eta^2} \text{Re} \int \left\{ \cos(\theta) R_U e^{-i\beta_U} + \sin(\theta) R_V e^{-i\beta_V} \right\} \tau \text{sech}(\tau) d\tau \quad (60)$$

$$\begin{aligned}
\frac{d\delta_U}{dZ} &= \frac{\eta^2 - \kappa^2}{2} - \kappa \frac{d\xi}{dZ} + \\
&+ \frac{1}{\eta} \text{Im} \int \left\{ \cos(\theta) R_U e^{-i\beta_U} + \sin(\theta) R_V e^{-i\beta_V} \right\} \text{sech}(\tau)(1 - \tau\tanh(\tau)) d\tau + \\
&+ \frac{1}{2\eta} \frac{\sin(\theta)}{\cos(\theta)} \text{Im} \int \left\{ \sin(\theta) R_U e^{-i\beta_U} - \cos(\theta) R_V e^{-i\beta_V} \right\} \text{sech}(\tau) d\tau \qquad (61)
\end{aligned}$$

$$\begin{aligned}
\frac{d\delta_V}{dZ} &= \frac{\eta^2 - \kappa^2}{2} - \kappa \frac{d\xi}{dZ} + \\
&+ \frac{1}{\eta} \text{Im} \int \left\{ \cos(\theta) R_U e^{-i\beta_U} + \sin(\theta) R_V e^{-i\beta_V} \right\} \text{sech}(\tau)(1 - \tau\tanh(\tau)) d\tau + \\
&- \frac{1}{2\eta} \frac{\cos(\theta)}{\sin(\theta)} \text{Im} \int \left\{ \sin(\theta) R_U e^{-i\beta_U} - \cos(\theta) R_V e^{-i\beta_V} \right\} \text{sech}(\tau) d\tau \qquad (62)
\end{aligned}$$

where $\beta_U = \kappa/(\eta Z) + \delta_U$ and $\beta_V = \kappa/(\eta Z) + \delta_V$.

5. Examples of Application

In the next sections, we apply the above theory to some specific applicative examples of soliton perturbation theory to optical fiber devices such as a fiber ring soliton laser and a soliton directional coupler.

5.1. SOLITON LASER

Let us consider the generation of a stable soliton train from continuously frequency-shifted feedback and bandwidth-limited laser (see figure 1).

112

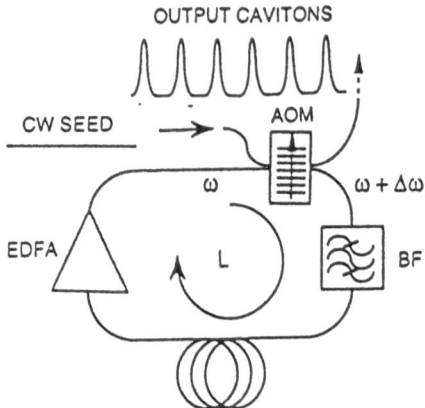

Figure 1. Schematic of laser

In this laser, the frequency shifting is introduced by an acousto-optic modulator (AOM). On the other hand, spectral broadening by self-phase modulation in the fiber leads to pulse shortening: a soliton with a fixed time width results in the presence of anomalous group-velocity dispersion and a bandpass filter [11, 12]. The principle of operation of this soliton laser is analogous to the sliding-frequency filter method of Mollenauer *et al.* [13] for the stabilization of soliton transmission systems.

The NLS equation that describes soliton propagation in this laser is derived as follows. The cw seed $E = A_0 exp\{-i\omega_0 t\}$ is injected into the ring through the zeroth-order diffraction peak of the AOM, with the amplitude transmission and reflection coefficients τ and ϱ, respectively. On each transit, the AOM shifts the circulating light by $+\Delta\omega = 2\pi f$. The ring of length $L = L_a + L_n$ is composed of a bandpass filter (BF), an erbium-doped fiber amplifier (EDFA) of length L_a, and a fiber of length L_n. Let $E_m(z,t) = A_m(z,t)exp\{-i\omega_0 t\}$ be the field after m $(m \geq 1)$ circulations through the cavity. Nectecting at first nonlinearity and dispersion, the envelope $A_m(z = 0, t)$ obeys

$$A_m(z = 0, t) = \tau A_0 + \varrho' e^{i\phi_m - i\Delta\omega(t+t_s)} A_{m-1}(z = 0, t) \qquad (63)$$

where $z = 0$ is at the AOM, t_s is an arbitrary time, $\phi_m = k_0 L + (m - 1)k_0'\Delta\omega L \equiv \phi_0 + \bar{\phi}_m$ is the linear cavity detuning, $\varrho' = \varrho\sqrt{G_0}exp\{-\alpha L_n\} > 1$, α is the fiber loss coefficient and $k_0' = \partial k_0/\partial\omega|_{(\omega=\omega_0)}$. On the other hand, the nonlinear dynamics of the field in the ring, $A_m(z,t)$, is described by the NLS equation

$$\frac{\partial A_m}{\partial z} + k_0'\frac{\partial A_m}{\partial t} + \frac{ik_0''}{2}\frac{\partial^2 A_m}{\partial t^2} = i\frac{\omega_0 n_2}{cA_{eff}}|A_m|^2 A_m. \qquad (64)$$

After rewriting eqs.(63,64) in dimensionless units, one may average the above map over a circulation through the ring, and obtain the perturbed NLS equation for the cavity field q,

$$i\frac{\partial q}{\partial Z} + \frac{1}{2}\frac{\partial^2 q}{\partial T^2} + |q|^2 q = iR \equiv$$

$$(\alpha_0(T + T_s) + i\delta)q + iV_f\frac{\partial q}{\partial T} + i\sigma\frac{\partial^2 q}{\partial T^2} - iSe^{i\Lambda Z}. \qquad (65)$$

where $\alpha_0 = \Delta\omega'/Z_l$, $\Lambda = \phi_0/Z_l$ is the linear cavity detuning, $\delta = ln(\varrho')/Z_l > 0$ is an excess gain, and S represents the injected cw beam. We also defined the filter-induced inverse group velocity shift $V_f = -2/(Bt_0Z_l)$, and $\sigma = -g''/2|k_0''| \simeq 2/(|k_0''|B^2Z_l)$, where B is the filter bandwidth in the Lorentzian approximation. Perturbation theory (38-41) leads to the equations for the soliton parameters

$$\frac{d\eta}{dZ} = 2\delta\eta - 2\sigma\eta(\eta^2/3 + \kappa^2) + S\pi sech(\beta)sin(\chi)$$

$$\frac{d\kappa}{dZ} = \alpha_0 - 4\sigma\eta^2\kappa/3 - \frac{\pi\kappa S}{\eta}sech(\beta)sin(\chi),$$

$$\frac{d\xi}{dZ} = -\kappa - V_f + \frac{\pi^2 S}{2\eta^2}sech(\beta)tgh(\beta)cos(\chi), \qquad (66)$$

$$\frac{d\theta}{dZ} = \frac{\eta^2 - \kappa^2}{2} + \xi\dot{\kappa} - \frac{\pi^2\kappa S}{2\eta^2}sech(\beta)tgh(\beta)cos(\chi).$$

where $\chi \equiv \Lambda Z + \kappa\xi - \theta - \pi/2$, and $\beta \equiv \pi\kappa/(2\eta)$. Provided that $|\alpha_0| \leq \alpha_c \simeq (2/3)^{3/2}\sigma$, $V_f \simeq 0$ and $\Lambda = 1/2$, the above equations have a stable eigensolution with $\eta = \eta^* = 1$, $\kappa = \kappa^* = 3\alpha_0/(4\sigma + 3\pi Ssin(\chi^*))$, $\xi = \xi^* = 0$ and $\chi = \chi^* = arcsin\{[2\delta - 2\sigma(k^{*2} + 1/3)]/[\pi S]\}$. This eigenmode corresponds to the stable output pulse from the sliding frequency soliton laser.

5.2. SOLITON COUPLER

The previous scalar theory may be extended to treat polarization effects in fiber lasers [14], where conservative (linear and nonlinear birefringence) and dissipative (the gain medium, filters and polarizers) perturbations balance for the stable propagation of vector solitons. As a consequence, soliton stability in fiber lasers may be analysed in terms the perturbed NLS equations (19,20) which represent the averaging of the various elements over each circulation. In a rather general form, these equations read as

$$R_U = (\alpha + \delta + i\rho)U + (\gamma + iK)V + i\Delta\frac{\partial U}{\partial T} + +\beta\frac{\partial^2 U}{\partial T^2} + i\sigma|V|^2 U,$$

$$(67)$$

$$R_V = (\alpha - \delta - i\rho)V + (\gamma + iK)U - i\Delta\frac{\partial V}{\partial T} + \beta\frac{\partial^2 V}{\partial T^2} + i\sigma|U|^2 V.$$

The meaning of the various perturbing terms in the right-hand side of eqs.(67) is the following: $\alpha > 0$ is the isotropic gain coefficient, whereas δ, γ and β represent gain dichroism and dispersion. Moreover, Δ represents polarization dispersion, ρ is the linear birefringence, K is the linear coupling, and σ the differential cross-phase modulation coefficient.

By applying the general perturbation theory results (57-62) to the above case (67) (we neglect here for simplicity the contribution of polarization dispersion, i.e., we set $\Delta = 0$), one obtains the following equations for the vector soliton parameters

$$\frac{d\eta}{dZ} = \eta\left(\alpha + \delta\cos(2\theta)\right) - 2\beta\kappa\left(\frac{\eta^2}{3} + \kappa^2\right) + 2\gamma\eta\cos(\psi)\sin(2\theta),$$

$$\frac{d\kappa}{dZ} = -\frac{4}{3}\beta\kappa\eta^2,$$

$$\frac{d\xi}{dZ} = -\kappa \tag{68}$$

$$\frac{d\theta}{dZ} = -K\sin(\psi) - \delta\sin(2\theta) + \gamma\cos(\psi)\cos(2\theta),$$

$$\frac{d\psi}{dZ} = 2\rho - 2K\cos(\psi)\mathrm{cotg}(2\theta) - \frac{2}{3}\sigma\eta^2\cos(2\theta) - 2\gamma\csc(2\theta)\sin(\psi),$$

where $\psi \equiv \delta_U - \delta_V$. The availability of a set of ordinary differential equations as eqs.(68) permits to derive the conditions for the mode locking eigenstates of the laser: these are represented by the stable fixed points of the above system (68) (i.e., were $d\eta/dZ = d\kappa/dZ = \ldots = d\psi/dZ = 0$). Clearly the limits of validity of soliton perturbation theory should be verified by comparing their predictions with the numerical solutions of the original eqs.(19,20,67)).

For example, figure (2) shows that the perturbation theory is able to reproduce (over a limited distance) the numerically observed polarization rotation or coupling of the vector solitons. Here the solid curves show the evolution of the polarization angle θ and relative phase ψ from the perturbation theory eqs.(68). Whereas the dots show the results that are obtained by computing the two angles from the direct numerical solutions (by a split-step Fourier method) of eqs.(19,20,67)). In figure (2) we took a linear coupling $K = 1/2$ and the cross phase modulation coefficient $\sigma = -1/3$. Moreover, we considered the initial values $\eta(0) = 1$, $\theta(0) = \pi/4$, and $\psi(0) = \pi/2$.

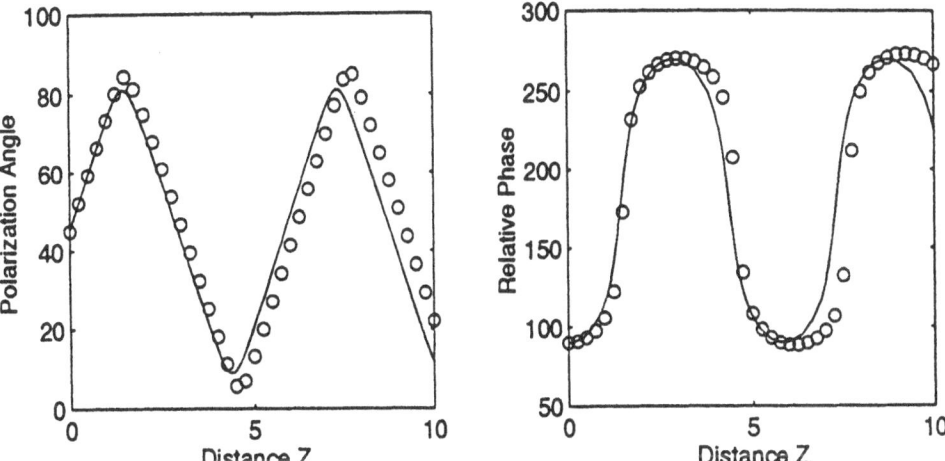

Figure 2. Soliton polarization rotation

As can be seen, in this case there is a very good qualitative agreement between perturbation theory and numerical results.

6. The Soliton Interactions

Another important application of perturbation theory is the study of the interactions between optical solitons. In particular, soliton interactions introduce an unwanted cross-talk among different bits in soliton-based transmissions. For example, figure (3) shows the collision of two in-phase solitons of the NLS equation. Therefore, it is important to discover ways to suppress or minimize the influence of these interactions: several methods have been recently proposed for the control of soliton interaction phenomena (for a review, see [15]). A simple approach to analyse the interaction of two adjacent solitons by means of perturbation theory is to consider each individual soliton as weakly perturbed by the tail of the other pulse, which is valid in the limit case of well-separated pulses. Then the dynamics of the interacting solitons reduces to the study of the nonlinear coupling between the parameters of these individual solitons. The main assumption of this quasi-particle approach, as developed by Karpman and Solov'ev [16], is that we may express the field q as

$$q(T, Z) = q_1(T, Z) + q_2(T, Z) , \tag{69}$$

where q_l are given by eq.(4) with $Z-$ dependent parameters.

The evolution equations for these parameters may be derived as follows [16]. Let us write $|q|^2 q = (|q_1|^2 q_1 + q_1^2 q_2^* + 2|q_1|^2 q_2) + (|q_2|^2 q_2 + q_2^2 q_1^* + 2|q_2|^2 q_1)$:
one may separate the first from the second term in the right hand side on

116

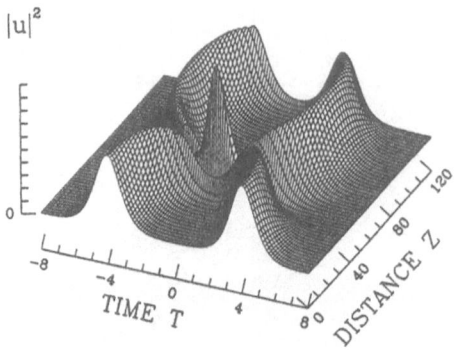

$|u|^2$

TIME T

DISTANCE Z

Figure 3. Two–soliton interaction

the basis of their degree of temporal overlapping. This permits to split the
perturbed NLS equation (8) into two equations for the evolution of $q_l(T, Z)$,

$$\frac{\partial q_l}{\partial Z} - \frac{i}{2}\frac{\partial^2 q_l}{\partial T^2} - i|q_l|^2 q_l = R[q_l, q_l^*, q_{\bar{l}}, q_{\bar{l}}^*] \equiv (q_{\bar{l}}^2 q_l^* + 2|q_l|^2 q_{\bar{l}}) , \qquad (70)$$

where $l = 1, 2$, and $\bar{l} = 3 - l$. Each of the two eqs.(70) may be analysed
by means of the general eqs.(38)-(39). Let us apply these equations to the
case of two soliton interaction as described by eq.(70). One then obtains
the following evolution equations for the variables $\eta = (\eta_1 + \eta_2)/2$, $\kappa = (\kappa_1 + \kappa_2)/2$, $p = (\eta_2 - \eta_1)/2$, $q = (\kappa_2 - \kappa_1)/2$, $\Delta = \xi_1 - \xi_2 > 0$, and
$\Psi = \theta_2 - \theta_1 - \kappa\Delta$

$$\frac{d\eta}{dZ} = 0, \frac{d\kappa}{dZ} = 0, \qquad (71)$$

$$\frac{dp}{dZ} = 4\eta^3 e^{-\eta\Delta} \sin(\Psi) \qquad (72)$$

$$\frac{dq}{dZ} = -4\eta^3 e^{-\eta\Delta} \cos(\Psi) \qquad (73)$$

$$\frac{d\Delta}{dZ} = 2q, \frac{d\Psi}{dZ} = 2p\eta. \qquad (74)$$

In the derivation of the above equations, it was assumed that $|p| \ll \eta$,
$|q| \ll 1$, $\eta\Delta \gg 1$, and $|p|\Delta \ll 1$ [16]. Eqs.(71)-(74) are exactly integrable.
In fact

$$4(q + ip)^2 - 16\eta^2 \exp(-\eta\Delta + \Psi) \equiv \Lambda^2 = \text{constant} , \qquad (75)$$

is a complex constant of motion. By using this equation, one reduces eqs.(72)-
(73) to a single equation for $Y = 2q + 2ip$,

$$\frac{dY}{dZ} = -\frac{2}{\eta}(Y^2 - \Lambda^2) \qquad (76)$$

whose explicit solution reads $Y = \Lambda \tanh(2\eta\Lambda Z + \Theta)$, where Θ is a complex integration constant. From this solution and eq.(75), one may derive explicit expressions for the soliton separation Δ. For example, in the case of equal amplitude, in-phase pulses, one obtains

$$\Delta = \Delta_0 + 2\ln\left|\cos\left(\frac{1}{2}Im\{\Lambda\}Z\right)\right|, \qquad (77)$$

which leads to the estimate for the collision distance (with $Im\{\Lambda\} = \Delta\eta$)

$$Z_c = \frac{2}{\Delta\eta}\cos^{-1}(e^{-\Delta/2}) \simeq \frac{\pi}{\Delta\eta} = \frac{\pi}{4}e^{\Delta_0/2}. \qquad (78)$$

7. Conclusions

In short summary, in this work I presented an unified approach for treating soliton propagation in optical waveguides. After deriving the scalar and the vector NLS equations that rule the propagation of temporal or spatial optical solitons, it was pointed out that the Lagrangian structure of these equations permits to derive in a relatively simple way perturbation equations for the adiabatic variation of the soliton parameters. The application of this theory to several specific examples of great interest for the applications was then described. The generality of the present approach may permit to extend the above theory and results to other physical processes such as for example in nonlinear magnetooptics.

I would like to acknowledge that the results presented here stem from my collaboration with Y. Kodama, A. B. Aceves, C. De Angelis, M. Midrio and M. Romagnoli. I also wish to thank R. Marcelli for his encouragement. This work was carried out under the agreement between the Fondazione Ugo Bordoni and the Italian Post and Telecommunication Administration.

References

1. V.E. Zakharov, and A. B. Shabat, "Exact theory of two-dimensional self-focusing and one-dimensional self-modulation of waves in nonlinear media", *Zh. Eksp. Teor. Fiz.*, Vol. **61**, 118 (1971). [*Sov. Phys. JETP*, vol. 34, 62 (1972)]
2. L. F. Mollenauer, S. G. Evangelides, and H. A. Haus, "Long-distance soliton propagation using lumped amplifiers and dispersion shifted fiber", *J. Lightwave Technol.*, Vol. **9**, 194 (1991).
3. A. Hasegawa, and Y. Kodama, "Guiding-center soliton in optical fibers", *Optics Lett.*, Vol. **15**, 1443 (1990); "Guiding-center soliton", *Phys. Rev. Lett.*, Vol. **66**, 161 (1991).
4. S. G. Evangelides, L. F. Mollenauer, J. P. Gordon, and N. S. Bergano, "Polarization multiplexing with solitons", *J. Lightwave Technol.*, Vol. **10**, 28 (1992).
5. C. R. Menyuk, "Pulse propagation in elliptically birefringent fibers", *IEEE J. Quantum Electron.*, Vol. **QE-25**, 2674 (1989).

118

6. S. V. Manakov, "On the theory of two-dimensional stationary self-focusing of electromagnetic waves", *Sov. Phys. JETP* Vol. **38**, 248 (1974) [*Zh. Eksp. Teor. Fiz.*, Vol.**65**, 505 (1973)].

7. A. Bonderson, M. Lisak, and D. Anderson, "Soliton perturbations : A variational principle for the soliton parameters", *Physica Scripta*, Vol. **20**, 479 (1979).

8. V. I. Karpman, and E. M. Maslov, "Perturbation theory for solitons", *Zh. Eksp. Teor. Fiz.*, Vol. **73**, 537 (1977) [*Sov. Phys. JETP*, Vol. **46**, 281 (1977)]

9. D. J. Kaup, and A. C. Newell, "Solitons as particles, oscillators, and in slowly changing media : a singular perturbation theory", *Proc. R. Soc. London Ser. A*, Vol. **361**, 413 (1978).

10. D. Anderson, and M. Lisak, "Bandwidth limits due to mutual pulse interaction in optical soliton communication systems" *Opt. Lett.*, Vol. **11**, 174 (1986).

11. F. Fontana, L. Bossalini, P. Franco, M. Midrio, M. Romagnoli, and S. Wabnitz, "Self-starting sliding-frequency fibre soliton laser", *Electronics Lett.*, Vol. **30**, 321 (1994).

12. Y. Kodama, M. Romagnoli, and S. Wabnitz, "Stabilization of optical solitons by an acousto-optic modulator and filter", *Electronics Lett.*, Vol. **30**, 261 (1994).

13. L. F. Mollenauer, J. P. Gordon, S. G. Evangelides, "The sliding-frequency guiding filter: an improved form of soliton jitter control", *Optics Lett.*, Vol. **17**, 1575 (1992).

14. C. De Angelis, and S. Wabnitz, "Interactions of orthogonally polarized solitons in optical fibers", *Optics Communic.*, Vol. 125, 186 (1996).

15. S. Wabnitz, Y. Kodama, and A. B. Aceves, "Control of optical soliton interactions", *Optical Fiber Technology*, Vol. 1, 187 (1995).

16. V. I. Karpman, and V. V. Solov'ev, "A perturbational approach to the two-soliton systems", *Physica D*, Vol. **3**, 487 (1981).

Chapter II
Spin Wave Instabilities

THEORY OF SPIN-WAVE INTERACTIONS IN HEISENBERG FERROMAGNETIC THIN FILMS

M.G. COTTAM and N.J. ZHU
Physics Department, University of Western Ontario,
London, Ontario N6A 3K7, Canada

1. Introduction

The spin-wave (or magnon) excitations in Heisenberg ferromagnets are not exact eigenstates of the system at nonzero temperatures below the Curie temperature T_C (see, *e.g.*, [1]). As a consequence there are interactions between the spin waves (SWs) leading to a variety of *nonlinear processes*. These include a *renormalization* (or shift) in the energy of the SWs and the occurrence of a *damping* (or reciprocal lifetime) of the SWs. It is these particular effects that we shall concentrate on in this article. Other nonlinear SW processes, involving SW instabilities, parallel pumping of SWs, auto-oscillations, transition to chaos, SW solitons, *etc.*, are reviewed. in [2] and also are discussed extensively in other parts of this book.

Mainly we shall be concerned with ferromagnets described by the Heisenberg model, *i.e.*, where the exchange terms in the Hamiltonian dominate over the long-range magnetic dipole-dipole interactions, which is typically the case for the SW wave number $q > 10^6$ cm^{-1} . However, later in this article we shall discuss the generalization of our theory to include the dipolar interactions and hence make the extension to smaller wave numbers (the dipole-exchange regime).

In infinite Heisenberg ferromagnets the theory of the interactions between volume (or bulk) SWs has received much attention (see [1] for a review) and was put on a rigorous basis by Dyson [3]. However, for a *finite* Heisenberg ferro-magnet, the presence of a surface (or surfaces) leads to a richer SW spectrum that may consist of one or more localized surface SW modes as well as the volume SW modes, which are now required to satisfy boundary conditions at any surface. Reviews of the existence conditions and dispersion relations for surface SWs *in a linear approximation* are given, for example, in [4,5]. Going beyond the linear SW approximation, the occurrence of surface and modified volume

R. Marcelli and S.A. Nikitov (eds.), Nonlinear Microwave Signal Processing: Towards a New Range of Devices, 121–138.
© *1996 Kluwer Academic Publishers.*

modes in a finite ferromagnet will give rise to more complicated schemes of SW interactions, since scattering processes may occur involving both types of modes. There have previously been a number of calculations for semi-infinite Heisenberg ferromagnets (*i.e.*, considering the effects of just one planar surface terminating the sample) at low temperatures $T \ll T_C$. In particular, results for the damping and energy renormalization, respectively, of the surface SW were obtained in some special cases by Tarasenko and Kharitonov [6] and Mazur and Mills [7]. Subsequently, more detailed results, which were applicable to both surface and volume SWs and included pinning through surface anisotropy and modified surface exchange, were derived by Kontos and Cottam [8-10]. By contrast, for ferromagnets in the long-wavelength (small q) dipole-exchange regime there have been extensive calculations of nonlinear SW properties using a continuous-medium or macroscopic type of theory (see, *e.g.*, [2] for a recent review).

In this article we present a study of SW interaction effects in Heisenberg ferromagnetic thin films, generalizing the above-mentioned previous work for the semi-infinite geometry. Compared with the semi-infinite case, two important differences for films are that there may be an additional surface SW and the volume SW spectrum is discrete (corresponding to "quantized" or "standing" volume SWs). We employ a diagrammatic perturbation method in terms of an expansion parameter 1/z (where z is the number of spins interacting with any given spin) to obtain the renormalized energy of all the SW modes of a film. This approach, which we describe later, has the advantages that the approx-imations can be introduced consistently through the expansion parameter and it is valid over a wide range of temperatures below T_C, not just $T \ll T_C$. This is important because the SW interaction effects (nonlinear effects) become more pronounced at higher temperaures. In addition we are able to obtain results for ultrathin films, where the number of atomic layers may be small, as well as the thick-film case. We employ a microscopic approach, assuming a particular lattice structure for the film, rather than a continuous-medium theory. This is important for films with relatively few atomic layers and/or for general SW wave vectors throughout the Brillouin zone.

In Section 2 we discuss the assumed geometry and model Hamiltonian for the ferromagnetic film, and we include a review of the previous work on SW inter-actions in semi-infinite media in the low-temperature limit $T \ll T_C$. Then in Section 3 the diagrammatic perturbation method, valid over a wide range of temperatures, is introduced. The Green-function results for the quantized volume and surface SWs are obtained in Section 4, and the extension to higher orders (the nonlinear SW regime) is indicated in Section 5. Explicit results for the renormalized energy and damping of all the SWs are given in Sections 6 and 7, respectively. In Section 8 we describe how the extension to include magnetic dipole-dipole interactions in the theory can be accomplished, and the overall conclusions are in Section 9. Further details of the calculations will be published elsewhere [11].

2. Theoretical Models

In this section a description is first given of the film geometry and the model Hamiltonian, including a specification scheme for the surface parameters. Then we briefly review some of the previous work on SW interactions in semi-infinite ferromagnetic media (the limit of film thickness becoming infinite) for $T << T_C$.

2.1. FILM GEOMETRY AND HAMILTONIAN

We consider a ferromagnetic film with its surfaces parallel to a family of lattice planes, as shown schematically in Figure 1. The position vector of any magnetic site may be written as $\mathbf{r} = (\rho, nc)$, with $n = 1, 2, ..., N$ for a film having N atomic

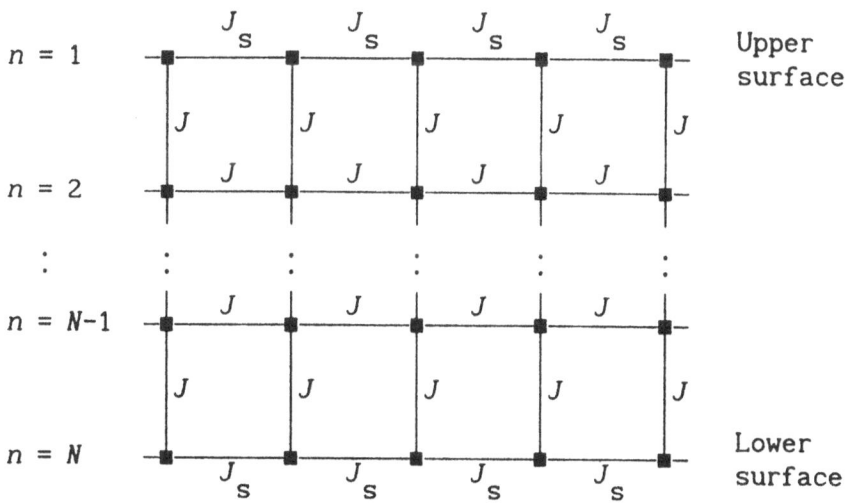

Figure 1. Schematic microscopic model for a ferromagnetic film of simple cubic structure with a pair of (001) surfaces. The film has N atomic layers (labeled by the index n) and the nearest-neighbor exchange constants J and J_S are indicated.

layers. Here ρ is a two-dimensional (2D) vector parallel to the surfaces in the xy-plane and integer n labels the layers. The interlayer spacing is denoted by c. The external magnetic field H_0 (and the static magnetization M_0) can be taken either perpend-icular or parallel to the film surfaces; here we describe the results for the perpendicular case.

The Heisenberg Hamiltonian H for the film can be expressed as

$$H = -\frac{1}{2}\sum_{r,\delta} J(r, r+\delta)S_r \cdot S_{r+\delta} - g\mu_B H_0 \sum_r S_r^z \tag{1}$$

where S_r is the spin operator at magnetic site r, and $J(r, r+\delta)$ denotes the exchange interaction between sites connected by vector δ. The spins in the surface layers have missing neighbors and, in addition, the exchange parameters may have modified values at the surfaces. Two specific models of exchange interactions have been considered, as follows:

Model 1. The exchange interactions are nonzero only between nearest-neighbor pairs of sites, having the value J_S if both the sites are in a surface layer ($n = 1$ or $n = N$) and the bulk value J otherwise. In particular, we discuss here simple cubic (sc) lattices with (001) film surfaces. Other structures and surface orient-ations that have been considered include sc (011), bcc (001), bcc (011), and fcc (001).

Model 2. The exchange interactions take their bulk values everywhere. They are equal to J_1 for nearest neighbors, J_2 for next-nearest neighbors, and zero other-wise. Results in this case have been obtained for sc (001) structures.

2.2. REVIEW OF RESULTS FOR SEMI-INFINITE FERROMAGNETS

Before describing the perturbation method and its application to finding the SW spectrum for the finite-thickness films, it is useful to mention briefly the previous results obtained for semi-infinite Heisenberg ferromagnets (the limiting case where $N \rightarrow \infty$). The references cited earlier (*i.e.*, [6-10]) all employ the Holstein-Primakoff transformation from spin operators to boson creation and annihilation operators (a^+ and a), together with an expansion treating $a^+a/2S$ as a small parameter, where S is the spin quantum number. Terms up to fourth order in the operators are then retained in the Hamiltonian, thus giving a description of the SW interactions. Because of the operator expansion and the truncation of the Hamiltonian, this approach is valid only in the low-temperature regime $T \ll T_C$.

Using the above method Tarasenko and Kharitonov [6] calculated the relaxation time for damping of the surface SW due to four-magnon scattering processes (which could involve surface or volume SW modes in any combination). Assuming Model 1 for a sc (001) structure, they concluded that the dominant contribution to the surface SW damping comes from scattering processes involving four surface waves. The energy shift of the surface SW due to the magnon-magnon interactions was studied by Mazur and Mills [7] for the case of a fcc (001) structure in Model 1 with no exchange perturbation at the surface (*i.e.*, $J_S = J$). In a low-temperature limit and when the applied field $H_0 = 0$, these authors predicted a $T^{5/2}$ correction to the $T = 0$ surface SW energy. This is the same type of temperature variation as found for the renormalization of volume SWs in infinite ferromagnets (*e.g.*, see [1-3]).

A more detailed study of SW interactions in semi-infinite Heisenberg ferromagnets was carried by Kontos and Cottam [8-10], again based on the low-temperature Holstein-Primakoff method but employing Models 1 and 2 for various different lattice structures and surface orientations. In particular, the energy shift and damping of the volume SW modes, as well as the surface SW modes, were evaluated, and the effects of pinning due to surface anisotropy were investigated, thereby generalizing [6,7]. For example, consider the case of a sc (001) structure described by Model 1 with zero surface anisotropy and $H_0 = 0$. At small wave vectors, such that $a^2Q^2 \ll 1$ where a is the lattice parameter and Q is the magnitude of the wave vector parallel to the surface, the lower edge of the volume SW region has energy $E_V(Q) \approx g\mu_B H_0 + SJa^2Q^2 + O(Q^4)$. If $J_S < J$ there is a surface SW branch with energy $E_S(Q)$ which is split off below the volume region by an amount proportional to $\sigma^2 a^4 Q^4$, where $\sigma = (1 - J_S/J)$. As a consequence the surface SW is only weakly bound to the surface in this case. On introducing the reduced variables $\tau = k_B T/SJ$ and $h = g\mu_B H_0/SJ$, the authors obtained for the energy shift $\Delta E_S(Q)$ and damping $\Gamma_S(Q)$ in leading orders:

$$\Delta E_s(Q) \approx -\left(J/32\pi^{3/2}\right)\left[\zeta(5/2)a^2Q^2\tau^{5/2} + 4\pi^{1/2}\zeta(2)(1 - 8\sigma^2)\sigma a^4Q^4\tau^2\right]$$

(2)

$$\Gamma_s(Q) \approx \left(J/32\pi S\right)\zeta(3)\sigma^2 a^4 Q^4 \tau^3$$ (3)

assuming $a^2Q^2 \ll \tau \ll 1$, where $\zeta(m)$ is a Riemann zeta function. The dominant contribution to $\Delta E_S(Q)$ in this case comes from the first term in Eq. (2) and is due to interaction with volume SWs. We note that this term has the same dependence on Q and τ as the SW energy shift in an infinite Heisenberg ferromagnet (see, *e.g.*, [1,3]) and is the analog of the result for a fcc (001) semi-infinite ferromagnet with $\sigma = 0$ found in [7]. The surface corrections show up explicitly in the second term in Eq. (2), which depends on the surface parameter σ and has a different Q and τ dependence. The requirement $a^2 Q^2 \ll \tau \ll 1$ for Eqs. (2) and (3) implies that the surface SW attenuation length ($\approx 1/\sigma a Q^2$) extends over a large number of atomic layers. Clearly, these results would become modified in the case of a thin film due to the presence of a second surface.

While it would be possible to extend the above-mentioned method, based on the Holstein-Primakoff transformation to boson operators, to apply to our models of a finite-thickness film, the results would still be applicable only in the low temperature limit of $T \ll T_C$. In order to remove this restriction and obtain results valid over a wider range of temperatures where the nonlinearities may be more important, we employ instead the diagrammatic perturbation method described in the following sections.

3. Diagrammatic Perturbation Expansion

The diagrammatic perturbation method for finite spin systems can be developed using either the Vaks, Larkin and Pikin method [12,13] or the drone-fermion method (see [14,15]). Both techniques were originally employed for infinite bulk ferromagnets and antiferromagnets, and they lead to equivalent results except that the former (henceforth referred to as VLP) method applies for general values of the spin quantum number S, whereas the drone-fermion method in its usual formulation is only for $S = \frac{1}{2}$. It is the VLP method that we shall describe in this article.

The standard application of these perturbation methods in infinite bulk media (see [12-15]) involves making a 3D Fourier transform of the Hamiltonian and other spatially-dependent variables in terms of a 3D wave vector $\mathbf{q} = (q_x, q_y, q_z)$, which characterizes the spin wave. However, in the film geometry, the diagrammatic theory must be generalized to take account of the loss of translational symmetry in the z-direction perpendicular to the film. This leads to a mixed representation in terms of a 2D in-plane wave vector $\mathbf{Q} = (q_x, q_y)$ and the layer index n (where $n = 1, 2, ..., N$).

Also, as in previous applications of the VLP and drone-fermion methods, we make a classification of the diagrams in terms of the expansion parameter $1/z$, where z is the number of spins interacting with any given spin. In lowest order $(1/z)^0$ this leads to mean-field theory, and spin fluctuation effects are obtained from the higher orders. A description of linear spin waves (non-interacting SWs) is found in order $(1/z)^1$, and various SW interaction processes occur first in orders $(1/z)^2$ and $(1/z)^3$. This type of expansion has the advantage of being valid over a wide range of temperatures below T_C, typically for T up to about $0.7\,T_C$ depending on the lattice structure and the scheme of exchange interactions.

4. Spin Waves in the Linear Approximation

To establish the diagram technique, following [12], the first step is to define the following spin-spin Green functions involving transverse and longitudinal spin components, respectively:

$$T(\mathbf{r},\mathbf{r}';\omega) = \langle\langle S_{\mathbf{r}}^+ ; S_{\mathbf{r}'}^- \rangle\rangle \tag{4}$$

$$L(\mathbf{r},\mathbf{r}';\omega) = \langle\langle (S_{\mathbf{r}}^z - \langle S_{\mathbf{r}}^z \rangle);(S_{\mathbf{r}'}^z - \langle S_{\mathbf{r}'}^z \rangle)\rangle\rangle \tag{5}$$

where $\langle\langle \,...\, \rangle\rangle$ denotes a finite-temperature Green function, defined in the standard way (see, e.g., [16]), and $\langle .. \rangle$ indicates a thermal average taken with respect to the Hamiltonian H in Eq. (1). Physically, the Green functions provide a description of the correlations between the appropriate spin components at different sites and at different times; the above quantities $T(\mathbf{r},\mathbf{r}';\omega)$ and $L(\mathbf{r},\mathbf{r}';\omega)$ denote the Green-function Fourier components at frequency ω.

The Green functions may be easily evaluated in mean-field theory $(1/z)^0$ and *then* renormalized to order $(1/z)^1$ corresponding to non-interacting SWs. Details can be found elsewhere [11], but the procedure is formally analogous to the well-known diagrammatic RPA calculation of the electron-hole propagator for a Fermi gas (see, e.g., [16]).

Specifically, the SW energies (or their corresponding frequencies) are found from the poles of the transverse Green function after using the property of translational symmetry parallel to the surfaces to make a Fourier transform with respect to an in-plane wave vector $\mathbf{Q}$. For a film with N atomic layers, this condition leads to a transcendental equation whose solutions are the N discrete branches of the linear SW spectrum in the Heisenberg ferromagnet. It is known from previous work (see, e.g., [17] for $T \ll T_C$) that the SW spectrum consists of p surface SWs and $(N-p)$ volume SWs, where p can take the values 0, 1, or 2, depending on the values of wave vector $\mathbf{Q}$, the number of layers N, and other parameters of the model.

A numerical example is given in Figure 2 for the case of $N = 5$, $J_s = 0.5\,J$ in Model 1 for a sc (001) structure, and $T \ll T_C$. The energy E vs. $\mathbf{Q}$ relationships

128

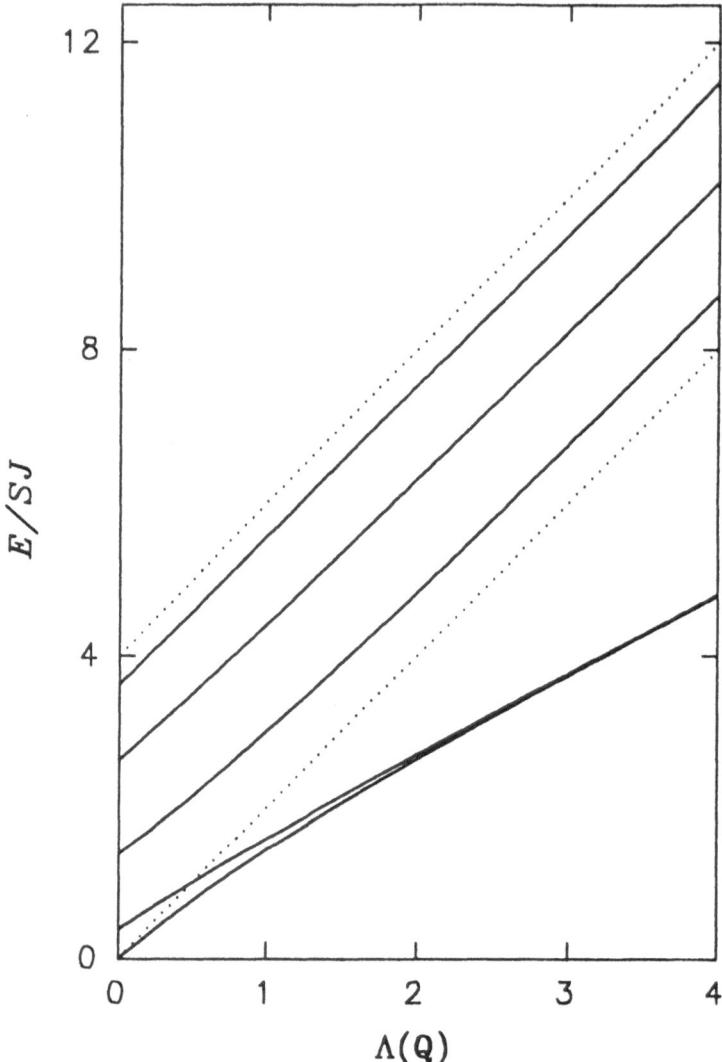

Figure 2. Dispersion relationships for the discrete SWs in a thin film in order $(1/z)^1$, *i.e.*, in the absence of SW interactions. The example is for a sc (001) film with $N = 5$, $J_S = 0.5\,J$ in the case of Model 1, and applied field $H_0 = 0$. The SW energies (solid lines) are plotted in dimensionless units E/SJ against the in-plane wave vector $\mathbf{Q}$ through the function $\Lambda(\mathbf{Q})$ in Eq. (6). The dotted lines are the upper and lower bounds of the volume SW region.

of the five SW branches are shown plotted in dimensionless form as a function of $\Lambda(\mathbf{Q})$, where

$$\Lambda(\mathbf{Q}) = 2 - \cos(q_x a) - \cos(q_y a) \tag{6}$$

The dotted lines indicate the upper and lower bounds of the volume SW region. Thus, in this example, there is just one surface SW for small wave vector (such that $\Lambda(\mathbf{Q})$ is less than about 0.5) and four SW modes within the volume region. For larger $Q \equiv |\mathbf{Q}|$ there are two surface SWs, which become almost degenerate in energy as $\Lambda(\mathbf{Q}) \to 4$ (*i.e.*, for $q_x \to \pi/a$ and $q_y \to \pi/a$ corresponding to a corner of the 2D Brillouin zone edge).

The surface modes in Figure 2 occur below the volume SW region and are often called *acoustic* surface SWs (see, *e.g.*, [4,5]). This is typically the case when the total effective exchange field on a surface spin is sufficiently less than that on a spin in the bulk (*i.e.*, $J_S < J$ in the case of Model 1, or $J_1 > 0$ and $J_2 > 0$ in the case of Model 2, assuming a sc (001) structure in each case). If the total effective exchange field at the surface is sufficiently larger than in the bulk, then *optical* surface SWs may occur with energies above the volume SW region.

We note that one of the SW branches in Figure 2 is a Goldstone mode, *i.e.*, its energy tends to zero as $\mathbf{Q} \to 0$ and $H_0 \to 0$. This is as expected on general symmetry grounds for the isotropic Heisenberg model and it applies even in the film geometry. The other SW branches have $\mathbf{Q} = 0$ gaps that depend on the exchange parameters.

5. Higher-order Perturbation Analysis

Diagrammatically we may represent the Green functions T and L in order $(1/z)^1$ by a directed solid line and a broken line, as shown in Figure 3(a) and (b) respectively. Higher-order contributions in the $1/z$ expansion for each of these Green functions can then be represented as diagrams containing these lines, together with various "*interaction vertices*". Generally, for the Heisenberg model, these interaction vertices involve $2m_1$ solid lines and m_2 broken lines, where m_1 and m_2 are zero or positive integers. This defines an m-term interaction vertex, where

$$m = (2m_1 + m_2) \geq 3 \tag{7}$$

As an example, an m-term vertex with $m = 4$ is shown in Figure 3(c). These complicated vertices occur because there is no direct analog of Wick's theorem for spin operators. The vertices are related to mean-field thermal averages of the

130

(a)

(b)

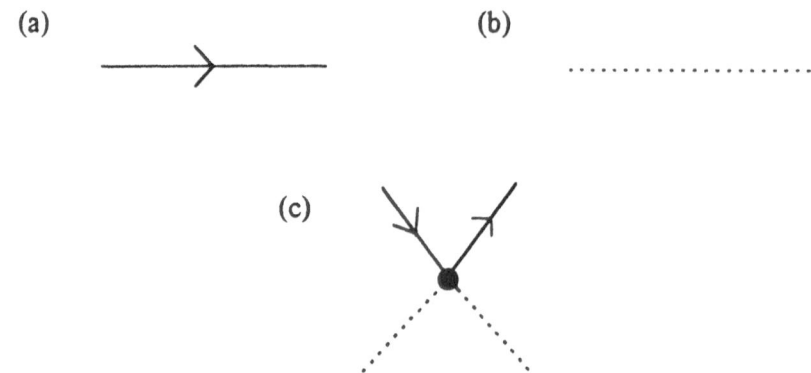

(c)

Figure 3. Form of the diagrammatic representation: (a) the transverse Green function; (b) the longitudinal Green function; (c) example of a *m*-term interaction vertex (with *m* = 4).

product of the *m* types of spin operators for the lines entering or leaving the vertex, *e.g.*, averages of the form

$$\langle S_{\mathbf{r}}^{+} S_{\mathbf{r}}^{-} S_{\mathbf{r}}^{z} S_{\mathbf{r}}^{z} \rangle$$

for the vertex in Figure 3(c). These quantities, which are nonzero only when the spin operators refer to the same site, are defined formally as in [12].

The above description is analogous to the procedure in infinite bulk magnetic systems (see, *e.g.*, [12,13]). An important difference in the present application to a finite-thickness film is that we employ the mixed layer-index – 2D-wave-vector representation mentioned in Section 3. Hence our Green function lines have layer indices n and n' at each end, where n and n' may be different due to interlayer spin-spin correlations, a single 2D wave-vector label $\mathbf{Q}$, and a single frequency label of the diagram technique. At each interaction vertex there is an overall conservation of the 2D wave vectors and frequency labels.

As discussed in Section 4, the SW excitations are associated with the transverse Green function, and thus the renormalization of the linear SW results described earlier is obtained by evaluating this Green function in higher orders of the perturbation analysis. The 1/z dependence of any diagram can be found from the simple rule that each independent momentum label in the diagram gives a 1/z factor (see, *e.g.*, [12-15]).

The appropriate set of diagrams in the next order $(1/z)^2$ consist of the one-loop diagrams shown in Figure 4. In *most* cases this order will be sufficient to obtain

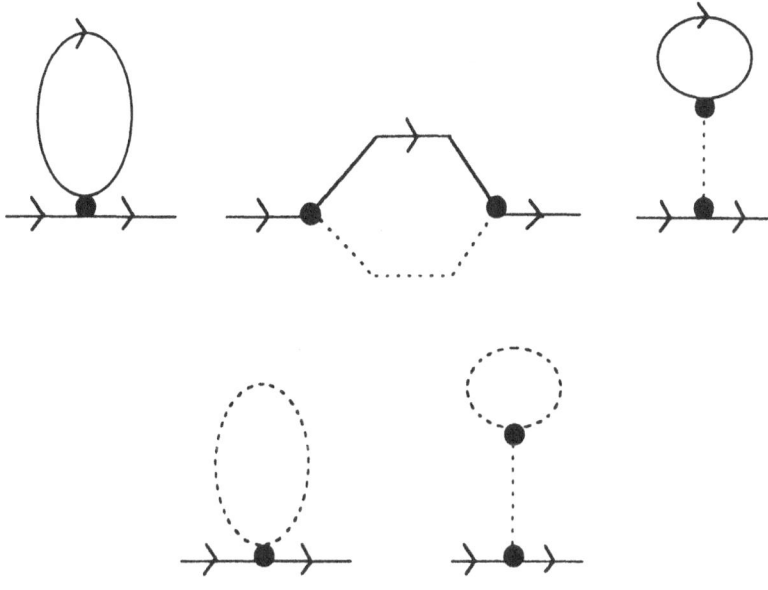

Figure 4. Diagrammatic contributions to the transverse Green function in order $(1/z)^2$: the one-loop diagrams.

the results for the SW energy renormalization and damping. However, later we shall show that for a proper desciption of the damping it is sometimes necessary to go to order $(1/z)^3$. The contributions in this order are two-loop diagrams and an example is shown in Figure 5.

Formally, the inclusion of the nonlinear SW interactions can be conveniently taken into account by evaluating proper self-energy contributions and using a Dyson-equation formalism, by extension of the treatment for infinite magnetic systems in [12-15]. The details of this application to the film geometry are given in [11]. For SW branch k (where $k = 1, 2, \ldots, N$) the excitation energy $E_k(Q)$ in the linear approximation is shifted to $E_k(Q) + \Delta E_k(Q)$ and the damping is $\Gamma_k(Q)$, where $\Delta E_k(Q)$ and $\Gamma_k(Q)$ are proportional to the real and imaginary parts, respectively, of a combination of self-energy terms.

We next discuss separately some of the results obtained for the quantities $\Delta E_k(Q)$ and $\Gamma_k(Q)$; a much fuller account, including details of the derivations and a wider range of examples, is to be found in [11].

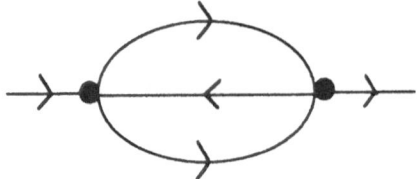

Figure 5. Example of a two-loop diagrammatic contribution to the transverse Green function in order $(1/z)^3$.

6. Results for Renormalized Energy of Spin Waves

As an example, suppose we consider the temperature region of T less than about $0.4\ T_C$. The dominant process of energy renormalization is due to four-magnon interactions, and each SW branch can interact with the thermally excited SWs corresponding to the same or different SW branch. The shift in SW energy for in-plane wave vector Q and branch k takes the form

$$\Delta E_k(Q) = \sum_{k'=1}^{N}\sum_{Q'}V_{k,k'}(Q,Q')n_B\left[E_{k'}(Q')\right] \tag{8}$$

where $V_{k,k'}(Q,Q')$ is a function of the exchange parameters, branch labels, and wave vectors; it is a generalization of the Dyson SW interaction vertex [1,3] in infinite ferromagnets. Also $n_B[E_{k'}(Q')]$ is the Bose-Einstein thermal factor for SW energy $E_{k'}(Q')$.

To illustrate the above result with a specific case for which the analytic results are relatively straightforward, we take the case of a sc (001) film with N = 3. Thus there are three discrete SW branches, and their energies in the $(1/z)^1$ non-interacting case (the linear approximation) in the region of small Q (such that $a^2Q^2 \ll 1$) are

$$E_1(Q) = D_1Q^2\ , \quad E_2(Q) = \frac{1}{2}J + D_2Q^2\ , \quad E_3(Q) = \frac{3}{2}J + D_3Q^2 \tag{9}$$

for Model 1, where the SW stiffness parameters are

$$D_1 = \frac{1}{6}(J + 2J_S)a^2 \ , \quad D_2 = \frac{1}{2}J_S a^2 \ , \quad D_3 = \frac{1}{6}(2J + J_S)a^2 \qquad (10)$$

In this case mode $E_1(Q)$ is the Goldstone mode (*i.e.*, having the property that $E_1(Q) \to 0$ as $Q \to 0$). It is an acoustic surface SW if $J_S < J$; otherwise it is a volume mode in this small Q region. The results for the energy shifts are found to be, up to order a^2Q^2,

$$\Delta E_1(Q) = -\left[\zeta(1)/6\pi\right]a^2Q^2 T \qquad (11)$$

$$\Delta E_2(Q) = \left[\zeta(2)(J_S - J)/\pi(J + 2J_S)^2\right]T^2 \\ -\left[\zeta(1)J_S/2\pi(J + 2J_S)\right]a^2Q^2 T \qquad (12)$$

$$\Delta E_3(Q) = -\left[\zeta(2)(J_S - J)/\pi(J + 2J_S)^2\right]T^2 \\ -\left[\zeta(1)(2J + J_S)/2\pi(J + 2J_S)\right]a^2Q^2 T \qquad (13)$$

Note that there is no renormalization of the gap at $Q = 0$ for SW mode 1; thus its Goldstone-mode property is correctly preserved in the perturbation analysis, as expected. For SW modes 2 and 3 there are corrections to the nonzero gaps at $Q = 0$ that are proportional to T^2 (except in the special case of $J_S = J$ when the T-dependence becomes higher order). For all three SW modes Eqs. (11)-(13) indicate that ther is a temperature renormalization of the SW stiffness parameter by terms proportional to T. The same leading dependences of $E_k(Q)$ on Q and T as in Eqs. (11)-(13) are found in the case of Model 2 with $N = 3$, but the coefficients are different.

In the case of a film with $N = 4$ there are four SW branches, one of which (mode 1) is the Goldstone mode with energy in order $(1/z)^1$ behaving as $E_1 = D_1Q^2$ in the long-wavelength region when $H_0 = 0$). We find for the energy shift due to SW interactions that ΔE_1 is proportional to Q^2T^2 for $T \ll T_C$ for both Models 1 and 2. Again this corresponds to a temperature renormalization of the SW stiffness parameter, but the proportionality to T^2 contrasts with the proportionality to T in the case of $N = 3$. It is closer to the $T^{5/2}$ dependence in an infinite ferromagnet [13].

In all the above examples the main contribution to the $E_k(Q)$ for each branch comes from the $k' = 1$ term in the summation in Eq. (8), *i.e.*, from the interaction with the zero-gap Goldstone mode. This would not necessarily be the case for larger Q and/or T, and the results for $E_k(Q)$ are then more complicated [11].

7. Results for Damping of Spin Waves

The results for the SW damping are of particular interest because there are two distinct mechanisms coming from different terms in the $1/z$ expansion. In some regions of wave vector and temperature, one of the mechanisms may turn out to be dominant, while in other cases there may be a competing effect. This behavior in the film case is broadly analogous to similar effects found in infinite ferromagnets and antiferromagnets with the $1/z$ expansion methods (see [12-15]).

Except at low temperaures where $T \ll T_C$ the dominant contribution to the SW damping $\Gamma_k(Q)$ for most Q arises in order $(1/z)^2$ from the diagrams in Figure 4. It can be interpreted physically as being due to the scattering of the SW from longitudinal spin fluctuations (in S^z). The general expression for the damping of branch k for in-plane wave vector Q takes the form

$$\Gamma_k(Q) = \sum_{k'=1}^{N} \sum_{Q'} W_{k,k'}(Q,Q') \delta \left[E_k(Q) - E_{k'}(Q') \right] \qquad (14)$$

The δ-function term describes the scattering of a SW of energy $E_k(Q)$ by another SW which has the same energy. It may be on the same or a different SW branch, and in general $Q' \neq Q$. The amplitude factor $W_{k,k'}(Q,Q')$ depends on the S^z fluctuations and typically increases sharply with increasing temperature.

As in the previous section we take the example of a sc (001) film in zero applied field $H_0 = 0$ with $N = 3$. It can be shown that the SW damping of the lowest-energy branch 1 at long wavelengths arises only due to scattering into the *same* branch ($k = k' = 1$ in Eq. (14)), implying $|Q'| = |Q|$. After using polar coordinates to sum over all directions for Q', we eventually find for the ratio of the damping to energy:

$$\frac{\Gamma_1(Q)}{E_1(Q)} = g(T)a^2Q^2 + O(a^3Q^3) \qquad (a^2Q^2 \ll 1) \qquad (15)$$

The above leading-order proportionality to Q^2 holds both for Models 1 and 2, and it is different from the case of an infinite ferromagnet [13] where the dependence is like Q^3. This difference may be attributable to the lower dimensionality in the present $N = 3$ case and to the discrete spectrum of the SWs. The factor

$g(T)$ increases monotonically with increasing T, and for temperatures less than but of order T_C it behaves as

$$g(T) \sim \left(1 - \frac{T}{T_C}\right)^{-2} \qquad \left(\text{for } T > \frac{1}{2}T_C\right) \tag{16}$$

Thus we can conclude, by analogy with the case of an infinite ferromagnet [13], that at *any* temperature below T_C the Goldstone SWs are well defined (in the sense that $\Gamma_1/E_1 \ll 1$ for sufficiently small Q).

However, in the low-temperature region $T \ll T_C$ the factor $g(T)$ is proportional to $\exp(-T_C/T)$ and is negligibly small. In this case the $(1/z)^2$ damping mechanism is "frozen out" as the fluctuations in S^z become unimportant. It is then necessary to proceed to order $(1/z)^3$ in order to obtain a proper description of the SW damping. In fact, the appropriate diagrammatic contribution corresponds to that shown in Figure 5. It describes a four-magnon scattering process, and it becomes dominant for $T \ll T_C$ and small $\mathbf{Q}$. For example, in the case of a $N = 3$ sc (001) film, we predict

$$\frac{\Gamma_1(Q)}{E_1(Q)} \propto T^2 \qquad \left(\frac{T}{T_C} \ll a^2 Q^2 \ll 1\right) \tag{17}$$

for the Goldstone mode (mode 1) when the applied field $H_0 = 0$.

8. Inclusion of Magnetic Dipole-dipole Interactions

In this section we discuss briefly how magnetic dipole-dipole interactions can be incorporated into the diagrammatic perturbation method to calculate $\Delta E_k(Q)$ and $\Gamma_k(Q)$ for the SW modes. In terms of the microscopic approach this means that an extra term H_D has to be included in the total spin Hamiltonian H, along with the Heisenberg exchange and Zeeman terms in Eq. (1), where

$$H_D = \tfrac{1}{2}g^2\mu_B^2 \sum_{\mathbf{r},\mathbf{r}'} \left\{\frac{\mathbf{S_r} \cdot \mathbf{S_{r'}}}{R^3} - \frac{3(\mathbf{S_r} \cdot \mathbf{R})(\mathbf{S_{r'}} \cdot \mathbf{R})}{R^5}\right\} \tag{18}$$

and $R = |\mathbf{R}|$ with $\mathbf{R} = (\mathbf{r} - \mathbf{r}')$. The terms with $\mathbf{r} = \mathbf{r}'$ are excluded from the summation. The effects of H_D on the SW spectrum *in the linear approximation* are well known, both for infinite ferromagnets (see, *e.g.*, [1]) and for ferromagnetic films (see, *e.g.*, [18]). The role of H_D is important for the description of SW properties at very small wave vectors, as we discussed in Section 1.

The more widely-used approach to dealing with the dipolar terms in this very small wave-vector regime has been in terms of a continuous-medium approximation where the macroscopic dipolar fields are represented in terms of Maxwell's equations, usually with retardation effects ignored, to describe the dipole-exchange and magnetostatic regions of SW behavior. The results for the SW modes in thin films in the linear approximation have been extensively reviewed elsewhere (see, *e.g.*, [4] and Chapter 2 in [2]). Recent work applying this continuous-medium approach to SW interactions (the three-wave and four-wave terms) is reviewed, for example, in Chapters 7-9 of [2]. However, these calculations can usually only be extended to higher temperatures in a phenomenological way and also the continuum approximation breaks down for ultrathin films with N small, and this is our motivation for using the microscopic approach to include H_D of Eq. (18) in the total Hamiltonian. It also allows us to examine the interesting problem of how the macroscopic and microscopic theories relate to one another in various limits. This connection is not entirely straightforward even in the linear SW approximation [19].

The extra difficulties involved in carrying out the microscopic theory for the SW energy shift and damping, with both dipolar and exchange terms included, can be summarized as follows. First, there are extra vertices that occur in the VLP diagrammatic method due to spin products like

$$S^+S^+, \quad S^-S^-, \quad S^+S^z, \quad S^-S^z,$$

etc., which are absent in the Heisenberg Hamiltonian, Eq. (1). However, these vertices are relatively easily included by generalizing the method due to Cottam [20] for infinite ferromagnets using the analogous drone-fermion diagrammatic method. Second, for a N-layer film there is an increase in the number of diagrams (by a factor of order $N/3$ compared with the exchange-only case). This is a consequence of the long-range nature of the dipolar terms which may couple spins across the entire thickness of the film. Third, there are long-range (and therefore slowly-converging) dipole-dipole sums with respect to the in-plane directions and the 2D wave vector $\mathbf{Q}$ that need to be evaluated. This can be most readily accomplished using analytic transformation methods such as in [18] or [21].

We expect to report on these calculations for the effects of the dipole-dipole interactions on $\Delta E_k(\mathbf{Q})$ and $\Gamma_k(\mathbf{Q})$ in the near future, together with applications to other nonlinear SW processes.

9. Conclusions

We have presented a perturbation expansion method, developed in terms of the small parameter $1/z$, to investigate analytically the SW interactions in thin ferromagnetic films with a finite number N of layers. The method has the advantage of being valid for a wide range of temperatures below T_C (not just the regime of $T \ll T_C$ as in some previous calculations for semi-infinite ferromagnets). Also, since a microscopic approach (using a Hamiltonian description) is employed, we are able to discuss general values of the in-plane wave vector $\mathbf{Q}$ throughout the 2D Brillouin zone. In this article the emphasis has been on the Heisenberg model where the short-range exchange coupling dominates, and we have described results for the energy shift $\Delta E_k(\mathbf{Q})$ and damping $\Gamma_k(\mathbf{Q})$ of the discrete SW branches.

However, the extension to include the longer-range magnetic dipole-dipole coupling into the formalism has been discussed, and further details of this will be reported shortly. It will supplement some of the existing theoretical work on nonlinear dipole-exchange SWs based on long-wavelength continuous-medium (or macroscopic) methods. For example, the relation between the microscopic and macroscopic approaches merits further study, particularly for ultrathin films and for a proper description of temperature dependences. Also there are interesting effects associated with including various types of surface anisotropy (pinning) and with varying the directions of the applied field H_0 and static magnetization M_0 relative to the surface planes.

A further extension to the present work would be the application of the diagrammatic perturbation formalism (either with or without the dipolar terms) to *other nonlinear processes* involving SWs in a film geometry. For example, it would be of interest to include extra terms in the Hamiltonian that describe the parallel and perpendicular components of an applied microwave pumping field, and to study within our formalism the various instabilities and SW parametric processes that are well known to arise due to nonlinearities treated within the macroscopic approaches (see, *e.g.*, Chapters 7-9 of [2] for reviews).

10. References

1. Keffer, F. (1966) Spin waves, *Handbuch der Physik* **18**, 1-273.
2. Cottam M.G. (ed.) (1994) *Linear and Nonlinear Spin Waves in Magnetic Films and Superlattices*, World Scientific, Singapore.
3. Dyson, F.J. (1956) General theory of spin-wave interactions, *Phys. Rev.* **102**, 1217-1230.

138

4. Wolfram, T., and De Wames, R.E. (1972) Surface dynamics of magnetic materials, *Progr. Surf. Sci.* **2**, 233-330.
5. Mills, D.L. (1984) Surface spin waves on magnetic crystals, in V.M. Agranovich and R. Loudon (eds.), *Surface Excitations*, North-Holland, Amsterdam, pp. 379-439.
6. Tarasenko, V.V., and Kharitonov, V.D. (1974) Spin-spin relaxation of surface waves in a Heisenberg ferromagnet, *Sov. Phys. Solid State* **16**, 1031-1032.
7. Mazur, P., and Mills, D.L. (1984) Temperature variation of surface spin-wave frequencies in the Heisenberg ferromagnet: The exchange-dominated regime. *Phys. Rev. B* **29**, 5081-5088.
8. Kontos, D., and Cottam, M.G. (1984) Effect of a surface on magnon-magnon interactions in ferromagnets, in S.W. Lovesey, U. Balucani, F. Borsa, and V. Tognetti (eds.), *Magnetic Excitations and Fluctuations*, Springer-Verlag, Berlin, pp. 153-158.
9. Kontos, D., and Cottam, M.G. (1986) Spin wave interactions in semi-infinite Heisenberg ferromagnets: I. Basic theory, *J. Phys. C* **19**, 1189-1202.
10. Kontos, D., and Cottam, M.G. (1986) Spin wave interactions in semi-infinite Heisenberg ferromagnets: II. Renormalized spin wave energy and damping, *J. Phys. C* **19**, 1203-1214.
11. Zhu, N.J., and Cottam, M.G. (1996) Spin-wave interactions in Heisenberg ferromagnetic films: I. Renormalized energy of spin waves, and II. Damping of spin waves; to be published.
12. Vaks, V.G., Larkin, A.I., and Pikin, S.A. (1968) Thermodynamics of an ideal ferromagnetic substance, *Sov. Phys. JETP* **26**, 188-199.
13. Vaks, V.G., Larkin, A.I., and Pikin, S.A. (1968) Spin waves and correlation functions in a ferromagnetic, *Sov. Phys. JETP* **26**, 647-655.
14. Spencer, H.J. (1968) Quantum field theory approach to the Heisenberg ferromagnet, *Phys. Rev.* **167**, 434-444.
15. Cottam, M.G., and Stinchcombe, R.B. (1970) Thermodynamic properties of a Heisenberg antiferromagnet, *J. Phys. C* **3**, 2283-2304.
16. Rickayzen, G. (1980) *Green's Functions and Condensed Matter*, Academic Press, London.
17. Puszkarski, H. (1972) Theoretically predictable effects and new interpretations in spin wave resonance of thin films with asymmetric boundary conditions, *Phys. Stat. Sol. (b)* **50**, 87-97.
18. Benson, H., and Mills, D.L. (1969) Spin waves in thin films; dipolar effects, *Phys. Rev.* **178**, 839-847.
19. Wolfram, T, and De Wames, R.E. (1970) Macroscopic and microscopic theories of dipole-exchange spin waves in thin films: Case of the missing surface states, *Phys. Rev. Lett.* **24**, 1489-1492.
20. Cottam, M.G. (1971) Theory of dipole-dipole interactions in ferromagnets: II. The correlation functions, *J. Phys. C* **4**, 2673-2683.
21. Tsymbal, E. (1994) Evaluation of the magnetic dipolar fields from layered systems on atomic scale, *J. Mag. Magn. Mat.* **130**, L6-L12.

KINETIC INSTABILITY AND BOSE CONDENSATION OF MAGNONS - THE SOURCES OF CONTROLLED MICROWAVE EMISSION FROM MAGNETIC CRYSTALS

G.A.MELKOV AND A.YU.TARANENKO

Faculty of Radiophysics, Taras Shevchenko Kiev University
64 Vladimirskaya Str, 252017 Kiev, Ukraine

1. Introduction

The action of a microwave magnetic field $h \cdot cos\omega_p t$ on a magnetic dielectric located in a constant magnetic field H_0 is known to lead to parametric instability of magnons (or spin waves) at some critical value $h = h_{thr}$ (see e.g. [1 to 3]). In this case wave vectors k, k' and frequencies ω_k, $\omega_{k'}$ of excited magnon pairs satisfy the condition of parametric resonance

$$\omega_k + \omega_{k'} = m\omega_p; \quad k + k' = 0, \tag{1}$$

where the number "m" determines the order of instability. Parametric magnons demonstrate complicated nonlinear behaviour as a level of excitation increases. Theoretical and experimental investigations of ferrites and antiferromagnets being carried out for about forty years revealed a diversity of nonlinear phenomena such as collective resonance of magnons, excitation of periodic and chaotic autooscillations, stepwise excitation of parametric magnons and etc.(see e.g.[4 to 9]). The theory in view of nonlinear interaction of spin waves with each other and with the pump field gives satisfactory explanation of the behaviour of a magnon system for the overthreshold region $h/h_{thr} - 1 \sim 1$[4, 5, 8, 9]. In this case a narrow packet of parametric magnons with frequencies closed to the parametric resonance (1) is considered. The population of other magnons is supposed to be the thermally equilibrium one. However the level of overheating of nonresonant magnons substantially increases with increasing the pump power. In this paper we focus our attention to the problem of excitation of magnons with frequencies far from the parametric resonance region under the pumping of strong parametric magnons.

Lavrinenko et al. predicted and detected experimentally in [10] a new phenomenon in yttrium iron garnet (YIG) at room temperature under parallel

139

R. Marcelli and S.A. Nikitov (eds.), Nonlinear Microwave Signal Processing: Towards a New Range of Devices, 139–164.
© *1996 Kluwer Academic Publishers.*

140

pumping: microwave emission from the bottom of spin-wave spectrum. This emission appeared in a threshold manner when the parallel pump microwave amplitude h exceeded a certain value h_c ($h_c/h_{thr} \sim 5 - 10$) was of a noisy nature with frequency ω_l closed to the minimal magnon frequency

$$\omega_i = gH_i; \quad H_i = H_0 - N_z M_0 + H_A . \qquad (2)$$

Here g is the gyromagnetic ratio, H_i is the static internal magnetic field, H_A is the anisotropy field, M_0 is the magnetisation, N_z is the demagnetisation factor of the sample.

The second harmonic $2\omega_l$ was also observed. Just above the threshold h_c the emitted power was $P \sim 10^{-10}$ W. The signal emitted at doubled frequency $2\omega_l$ had greater amplitude. Its power was about $P_{em}(2\omega_l) \sim 100$ mW during absorption of maximal pump power $P \sim 10$ W. The width of emission spectrum was several MHz when the threshold h_c was just reached. The emission was found to occur in a limited magnetic field region. In Figure 1 is shown the magnetic field

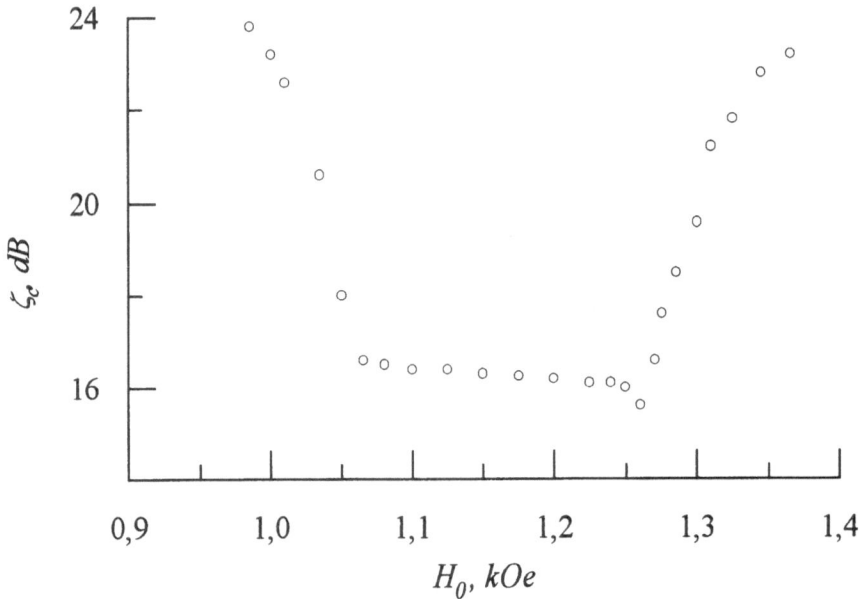

Figure 1. Magnetic field dependence of relative critical power of the onset of electromagnetic emission from the bottom of spin-wave spectrum for an yttrium iron garnet sphere 1.5 mm in diameter. The pump frequency is $\omega_p/2\pi = 9.4$ GHz, $H_0 \parallel [111]$, $T = 300$ K.

dependence for the critical pump power with respect to the threshold power of parametric resonance of magnons $\zeta_c = P_c/P_{thr} = (h_c/h_{thr})^2$. It is seen that variation in ζ_c is not significant within the limits $H_1 \leq H_0 \leq H_2$ where $H_1 = 0.9$ kOe and

$H_2 = 1.27$ kOe and outside this region increases steeply. The radiation frequency ω_l demonstrates linear dependence on static magnetic field (2) within the occurrence region. It should be also mentioned that the onset of emission always accompanied by an inflection in the dependence of nonlinear susceptibility of a ferrite on the pump power.

The appearance of noisy emission from the bottom of spin-wave spectrum means that the occupation numbers of magnons with the frequency ω_l and wave vector $k_l \sim 0$ substantially exceed their thermal level. There are known three theoretical explanation for the bottom magnon accumulation. The first is kinetic instability which was proposed and developed in [10 to 13, 5] on the base of so-called S-theory [4]. The second is based on the thermodynamic theory of parametric resonance of magnons [14 to 16] and the third concerns the theory of Bose condensation of nonequilibrium magnons [17, 18].

2. Theory of Kinetic Instability of Magnons

Strong parallel pumping can create a highly nonequilibrium distribution of spin waves in the phase space. The presence of superheated domains in the spin-wave spectrum leads to energy flows to the new, previously equilibrium, parts of the spectrum. These flows in the case of four-magnon nonlinearity are directed to the magnons with frequencies ω_1 and ω_2 and wave vectors k_1 and k_2:

$$\omega_1 + \omega_2 = \omega_3 + \omega_4; \quad k_1 + k_2 = k_3 + k_4. \tag{3}$$

Here ω_3, ω_4 and k_3, k_4 are frequencies and wave vectors of parametric magnons. In this case $\omega_3 = \omega_4 = \omega_p/2$ but ω_1 and ω_2 are far from the parametric resonance $\omega_1 < \omega_p/2$, $\omega_2 > \omega_p/2$. According to the theory of kinetic instability process (3) reduces the damping of magnons of the frequency $\omega_l \approx \omega_i$ to zero. When the parallel pump amplitude h is greater than h_c the number of secondary long-wavelength magnons (ω_l, $k_l \sim 0$) rises strongly. This gives rise to electromagnetic emission from the bottom of spin-wave spectrum.

The theory of kinetic instability has been developed by L'vov and Cherepanov in [10] for that particular case when three-magnon interaction processes are prohibited by the laws of conservation of energy and momentum. Since to the present time the concept of kinetic instability was widely used for interpretation of the majority experiments we shell consider it in detail.

The process (3) is described by the spin-wave Hamiltonian with four-magnon nonlinearity

$$\mathcal{H} = \sum_{k_1 k_2, k_3 k_4} T_{12,34} b_1^* b_2^* b_3 b_4 \Delta \left(k_1 + k_2 - k_3 - k_4 \right) \qquad (4)$$

where b_3 and b_4 are amplitudes of the pump spin waves ($b_{3,4} = A_{3,4} exp(i\omega_p t/2)$); b_1 and b_2 are amplitudes of excited spin waves ($b_{1,2} = a_{1,2} exp(i\omega_{1,2} t)$); $T_{12,34} = T$ is the interaction matrix element. The dynamic equations for the amplitude b_1 and for the envelope a_1 of the packet of excited waves can be written in form

$$\frac{db_1}{dt} + \left(\gamma_1 + i\omega_1 \right) b_1 - i \sum T b_2^* b_3 b_4 \Delta = 0, \qquad (5)$$

$$\frac{da_1}{dt} + \gamma_1 a_1 - i \sum T a_2^* A_3 A_4 \Delta = 0, \qquad (6)$$

where γ_1 is the damping of excited waves. When the pump waves are coherent, for example, in the case of uniform precession ($A_3 = A_4 = A_0$) Eq.(6) describes the second-order Suhl instability [1]. The threshold amplitude can be obtained as

$$A^2_{0\,thr} = \gamma_1/T. \qquad (7)$$

Should the pump waves are the parametric spin waves excited, for example, by parallel pumping the sum in (5) turns to zero since the phases of parametric spin waves are random ($\langle A_3 A_4 \rangle = 0$). Therefore parametric magnons can not give rise to a second parametric instability.

However scattering of parametric magnons can lead to the energy flows to some other parts of spin-wave spectrum. This can increase the number of secondary spin waves. One can write

$$\langle a_1 \rangle = 0; \langle b_1 \rangle = 0; \langle b_1 b_1^* \rangle = n_1 \neq 0. \qquad (8)$$

Thus the parametric magnons can amplify a noise in some parts of spin-wave spectrum. The magnon distribution function far from parametric resonance is described by a kinetic equation [19]:

$$\frac{d}{dt} n_1 = \frac{d}{dt} \langle b_1 b_1^* \rangle \qquad (9)$$

Substituting Eq.(5) for derivative of b_1 in the kinetic equation (9) one can obtain the kinetic equation for the numbers of secondary magnons of the frequency ω_1

$$\frac{\partial n_1}{\partial t} + 2\gamma_1 n_1 = -2\operatorname{Im}\sum T\left\langle b_1^* b_2 A_3 A_4\right\rangle \ell^{i\omega_p t}\Delta \tag{10}$$

The amplitude b_1 of secondary spin waves being acted by parametric magnons are considered to be slightly different from the thermally equilibrium one b_1^0

$$b_1 = b_1^0 + \alpha_1 \tag{11}$$

In view of Eq.(11) the kinetic equation (10) can be rewritten as

$$\frac{\partial n_1}{\partial t} + 2\gamma_1 n_1 = -2\operatorname{Im}\sum T[\left\langle \alpha_1^* b_2^{0*} A_3 A_4\right\rangle + \left\langle b_1^{0*}\alpha_2^* A_3 A_4\right\rangle]\ell^{i\omega_p t}\Delta \tag{12}$$

Since α_1 depends on A_3 and A_4 (it is easy to make certain substituting Eq.(11) in Eq.(5)) the correlators in (12) are not equal to zero and have the following form

$$\left\langle \alpha_1^* b_2^{0*} A_3 A_4\right\rangle \ell^{i\omega_p t} = 2T\frac{n_2 N_3 N_4}{\omega_p - \omega_1 - \omega_2 - i\gamma_1} \tag{13}$$

$$\left\langle b_1^0\alpha_2^* A_3 A_4\right\rangle \ell^{i\omega_p t} = 2T\frac{n_1 N_3 N_4}{\omega_p - \omega_1 - \omega_2 - i\gamma_2} \tag{14}$$

Here $n_{1,2}$ and $N_{3,4}$ - the numbers of secondary and parametric spin waves accordingly. Then one has

$$\frac{dn_1}{dt} + 2\tilde{\gamma}_1 n_1 = 0 \tag{15}$$

where $\tilde{\gamma}_1$ is the total damping constant of secondary spin waves of the frequency ω_1 in the presence of parametric spin waves

$$\tilde{\gamma}_1 = \gamma_1 - 2\pi\sum |T|^2 N_3 N_4\delta(\omega_1 + \omega_2 - \omega_p)\Delta(k_1 + k_2 - k_3 - k_4) \tag{16}$$

Turning from summation to integration in Eq.(16) one can obtain the final expression for the relaxation frequency of secondary magnons

$$\tilde{\gamma}_1 = \gamma_1 - \pi\frac{(TN)^2}{kv} \tag{17}$$

144

Here k and v are the wave vector and the group velocity of parametric magnons. From (17) it follows that if the number of parametric magnons N is greater than the critical number

$$N_c = \frac{1}{|T|}\left(\frac{\gamma_1 kv}{\pi}\right)^{1/2}$$ (18)

at which $\tilde{\gamma}_1 = 0$ then the amplitude of secondary spin waves of the frequency ω_l will grow exponentially. The magnons with the smallest intrinsic damping γ_l will be excited first. According to [10] $\omega_l \approx \omega_i$ and $k_l \sim 10^3$ cm^{-1}. The secondary magnons (ω_2, k_2) carries away the excess energy and momentum relaxing rapidly. The distribution function for excited magnons beyond the kinetic

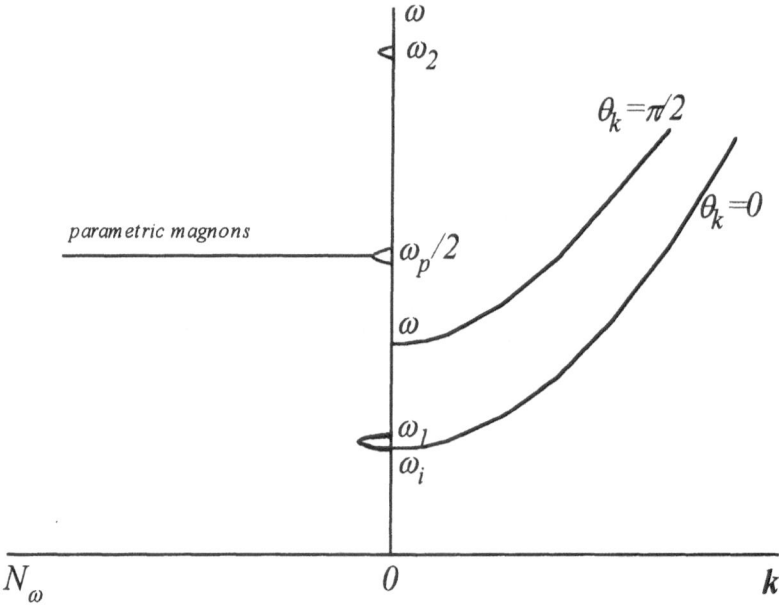

Figure 2. Spectrum and distribution function of excited spin waves beyond the kinetic instability; $\omega_\perp$ and ω_i are upper and lover boundaries of spin-wave spectrum correspondingly. instability is shown in Figure 2 [20].

It is obvious that secondary spin waves with small wave vectors can be transformed into microwave emission from the sample. The power $P_{em}(\omega_l)$ of the emission from the bottom of spin wave spectrum is weak as far as it is governed by magnetodipole emission of magnons with wave vector $k_l \sim 10^3$ cm^{-1} considerably exceeded the wave vector of electromagnetic waves of the frequency $k_0 = \omega_r/c \sim 1$ cm^{-1}. The emission at doubled frequency arises as a result of confluence of two secondary magnons with opposite wave vectors into a photon.

The threshold of kinetic instability can be found from Eq.(18) by determining the number of parametric magnons N from S-theory[4];

$$\zeta_c = \frac{|S|^2 \gamma_l k \mathrm{v}}{\pi \gamma_p^2 |T|^2} \tag{19}$$

Here S is the interaction matrix element for parametric magnon pairs, γ_p is the relaxation frequency of parametric magnons. The numerical estimate made using (19) for YIG under parallel pumping gives $\zeta_c \sim 15$ dB ($h_c/h_{thr} \sim 5.5$) that shows good agreement between theory and experiment (see Figure 1). Existence of upper boundary H_2 or the minimal wave vector of parametric magnons finds natural explanation in the framework of the above theory and it is just due to energy and momentum conservation (3). Increase in ζ_c in the field region lower than H_1 is a consequence of the fact that three-magnon scattering processes are allowed for short parametric spin waves when $H \leq H_1$ [21]. As a result nonlinear attenuation of parametric magnons arises. Therefore greater pump power is required to attain the threshold level N_c.

The inflection in the pump power dependence of the nonlinear susceptibility demonstrates a significant influence of secondary spin waves on parametric waves. A theory of interaction of secondary and parametric magnons beyond the kinetic instability threshold has been developed in [11,13].

3. Experimental Observation of Secondary Magnons in Ferromagnetic Materials

3.1 KINETIC INSTABILITY OF MAGNONS AT ARBITRARY PUMPING

As it was noted kinetic instability was observed in [10] under parallel pumping, i.e., in the presence of parametrically excited spin waves with polar angle $\theta_k = \pi/2$ (see Figure 2). An excitation of secondary spin waves by parametric waves with θ_k in the wide interval $\pi/4 \leq \theta_k \leq \pi/2$ on the resonance surface $\omega_k = \omega_p/2$ was reported in [22]. Parametric spin waves with different polar angles θ_k and correspondingly wave vectors k were excited as the pump angle θ_h was varied from parallel pumping ($H_0 \parallel h$, $\theta_h = 0$) to perpendicular one ($H_0 \parallel h$, $\theta_h = \pi/2$) [23].

It was established in [22] that parametric magnons irrespective to their polar angles always excited the secondary magnons of the frequency closed to the lower limit of spin-wave spectrum. The secondary magnons were detected in the same manner as in [10]. Figure 3 shows the kinetic instability threshold curve for the pump angles $\theta_h = 0$, $\theta_h = \pi/4$, $\theta_h = \pi/2$. The threshold $P_c \sim h_c^2$ is seen to demonstrate weak dependence on the pump angle θ_h while the region where kinetic instability occurs is shifted to the high-field side as the pump angle increases. Such a behaviour is due to broadening the parametric magnon packet with increasing the pump angle θ_k [22, 23] and conservation laws as well.

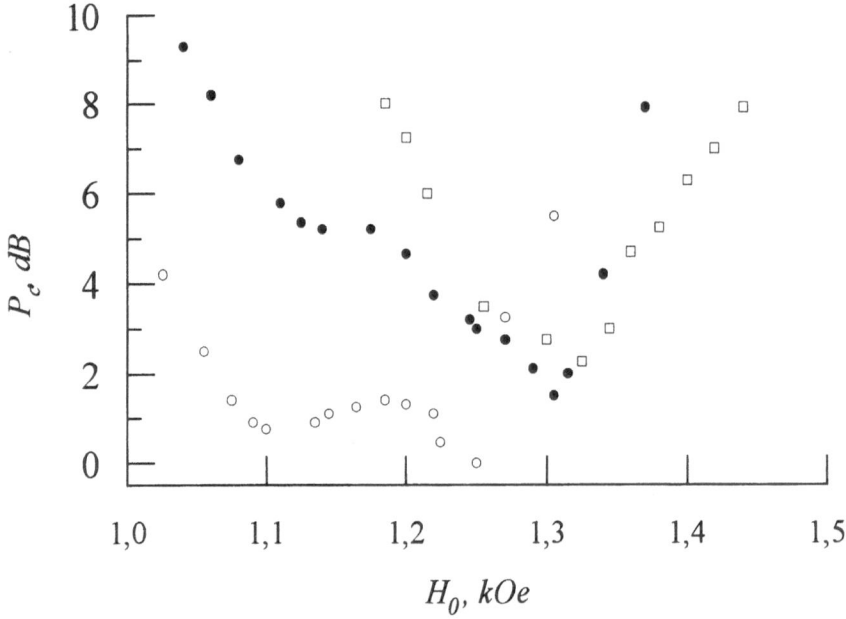

Figure 3. Kinetic instability threshold as a function of static magnetic field at various pump angles θ_h between H_0 and h. $\theta_h = 0$ (whitecircle), $\theta_h = \pi/4$ (blackcircle), $\theta_h = \pi/2$ (whitebox). The origin for the decibel scale on the ordinate axis corresponds to the minimal kinetic instability threshold under parallel pumping ($\theta_h = 0$). The sample is a single crystal sphere of yttrium iron garnet 1.5 mm in diameter. The pump frequency is $\omega_p/2\pi = 9.34$ GHz, $H_0 \parallel [111]$, $T = 300$K.

3.2. NONLINEAR EMISSION FROM A FERRITE IN FERROMAGNETIC RESONANCE

Electromagnetic emission was also detected from single crystal YIG spheres in ferromagnetic resonance (FMR) at the frequency in the X-band range under second-order Suhl instability [24]. The emission was noisy with frequencies concentrated in two regions below the pump frequency ω_p. It arises when the threshold ζ_c is reached in the reduced pump power $\zeta = P/P_{thr}$. The value of ζ_c varies over the range $\zeta_c = 6 - 8$ dB for various samples. The emitted power depends in a resonant fashion on the static magnetic field, the maximum of the emission is usually shifted a few oersteds above ferromagnetic resonance field. Immediately after the threshold ζ_c is exceeded the maximal emitted power P_{em}

amounts to few tens of microwatts and reaches values on the order of 1 mW at a pump power $P \sim 10^3 W$.

Figure 4 shows the spectrum of the emission from the ferrite. It is seen to consist of two components with frequencies ω_{r1} and ω_{r2} satisfying $\omega_{r1} < \omega_{r2} < \omega_p = 2\pi \cdot 9.37$ GHz. The frequencies ω_{r1} and ω_{r2} depend on the diameter of the sphere, the static magnetic field H_0 and the reduced pump power ζ. There is no harmonic relationship of any sort of these frequencies with each other and with the pump frequency. For the case shown in Figure 4 we have $\omega_{r1}/2\pi \approx 8.8$ GHz,

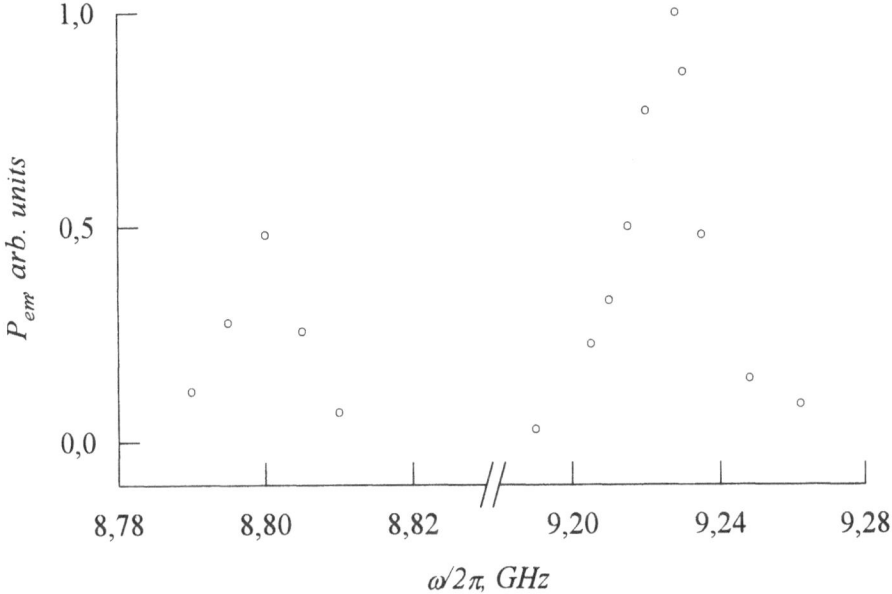

Figure 4. The spectral density of the non-linear emission versus the frequency for a single crystal sphere of yttrium iron garnet 2.89 mm in diameter (H_{FMR} = 3380 Oe, H_0 = 3383 Oe, ζ_c = 12 dB, H_0 [111], T = 300K).

$\omega_{r2}/2\pi \approx 9.22$ GHz, and $\omega_p/2\pi \approx 9.37$ GHz. Figure 5 shows the positions of the emission frequencies with respect to the characteristic frequencies of the ferromagnetic sample. The two spectral components of the emission at ω_{r1} and ω_{r2} appear simultaneously in the case $\zeta \geq \zeta_c$ but when the pump power is just above the threshold the spectral density of the low-frequency component (ω_{r1}) is significantly lower than that of the high-frequency component. With a further increase in the pump power, however, the two spectral densities become equal. The width of each component Δf_{r1} and Δf_{r2} is 10 - 20 MHz when $\zeta = \zeta_c$. The width of high-frequency peak Δf_{r1} is greater than Δf_{r2} in all cases. At the highest values of the reduced pump power the width Δf_{r2} reaches 100 MHz while Δf_{r1} is less than 30 MHz.

148

It was suggested in [24] that the reason for this phenomenon is a kinetic instability of secondary spin waves driven by the primary waves parametrically excited under second-order Suhl instability. The second-order parametric instability gives rise to spin waves with a frequency equal to the pump power frequency ($m = 2$ in (1)) and with polar angles $\theta_k = 0$ and π [1] (see Figure 5)

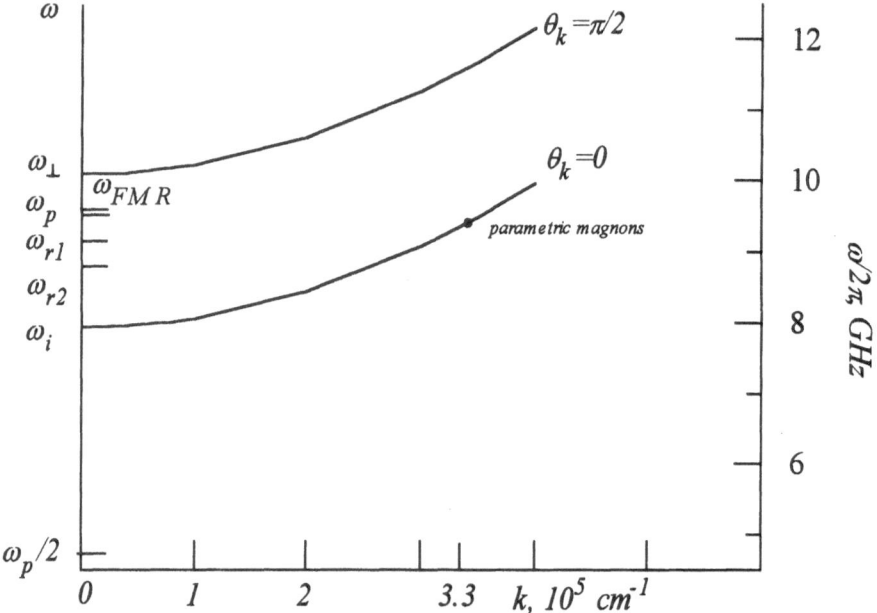

Figure 5. Position of the emission frequencies ω_{r1} and ω_{r2} in the spectrum of spin wave oscillations of the ferrite for the condition of the experiment whose result are shown in Figure 4. $\omega_\perp$ - upper boundary of spin-wave spectrum in the limit $k \rightarrow 0$; ω_p -frequency of the pumping and of the parametric magnons; ω_{FMR} - FMR frequency; ω_i - minimal spin wave frequency.

In view of such a distribution of spin waves the only allowed process which is capable of driving a kinetic instability of spin waves in the present experimental situation is a process in which we have

$$|k| = |-k'|; \quad |k_1| \approx |k_2| \approx 0; \quad \omega_1 + \omega_2 = \omega_k + \omega_{k'}, \tag{20}$$

where k, k' and ω_k, $\omega_{k'}$ are the wave vectors and frequencies of parametric magnons and k_1, k_2 and ω_1, ω_2 are the wave vectors and frequencies of secondary magnons. For energy conservation the frequencies ω_1 and ω_2 ($\omega_2 > \omega_1$) must lie within the spin-wave spectrum in the limit $k \rightarrow 0$:

$$\omega_2 \leq g[H_i(H_i + 4\pi M_0]^{1/2} = \omega_\perp; \quad \omega_1 \geq \omega_i; \quad \omega_1 + \omega_2 = 2\omega_p. \tag{21}$$

Under the condition of present experiment process (3) is forbidden by energy and momentum conservation. The decay of parametric magnon into magnon and phonon is allowed by conservation laws but because of the weak magneto-elastic interaction in YIG this decay has significant probability at room temperature only at $\zeta > 80$ dB [25] - well above the values reached experimentally.

The secondary spin waves being formed by the process (20) are distributed over a broad frequency interval below the pump frequency. As a result of two-magnon scattering process the secondary spin waves work through the magnetostatic precession modes which are degenerate with them to excite electromagnetic wave corresponding to the eigenfrequencies of these precessions ω_ν. The frequencies of the (210) and (300) magnetostatic precessions modes are founded to agree within 10 MHz with the frequencies ω_{r1} and ω_{r2}.. The width of the emission spectrum should be determined by the width of the line of the magnetostatic modes. The behaviour of the emission frequency and spectral width as a function of the magnetic field and pump power is determined by the same dependencies of the frequency and width of the magnetostatic precession modes.

The excitation of secondary spin waves leads to a decrease in the susceptibility of the ferrite as well. This is revealed experimentally as an appearance of a dip in the resonant curve[24, 26].

3.3 KINETIC INSTABILITY OF MAGNONS IN THIN FERRITE FILMS

Kinetic instabilities in magnetic materials represent a fairly wide class of phenomena which occur at high excitation rate. Such an instability can be excited in ferrite films which differ from solid samples in the structure of the spin-wave spectrum in the range of low values of the wave vector. Observation of a kinetic instability in thin ferrite films under parallel pumping was reported in [27]. It was found that by virtue of the characteristics of the spin-wave spectrum of the films there are at least two different regions in which a kinetic instability exists, not one, as in solid samples.

Normally and tangentially magnetised YIG films 16 - 30 μm thick grown on a gallium-gadolinium garnet substrate of (111) orientation were studied under parallel pumping of the frequency $\omega_p/2\pi = 9.5$ GHz. Kinetic instability was detected by recording the electromagnetic emission from the film. The emitted power varied from $P_{em} \sim 10^{-11}$ W to $P_{em} \sim 10^{-14}$ W depending on the experimental condition.

Figures 6 and 7 show typical results of the study of kinetic instability in an YIG film 20 μm thick; Figure 6 corresponds to normally magnetised film, and Figure 7 to tangentially magnetised film. These Figures depict the reduced threshold ζ_c at which kinetic instability is formed and the frequencies of the electromagnetic radiation from the film at the threshold ζ_c as a function of the

150

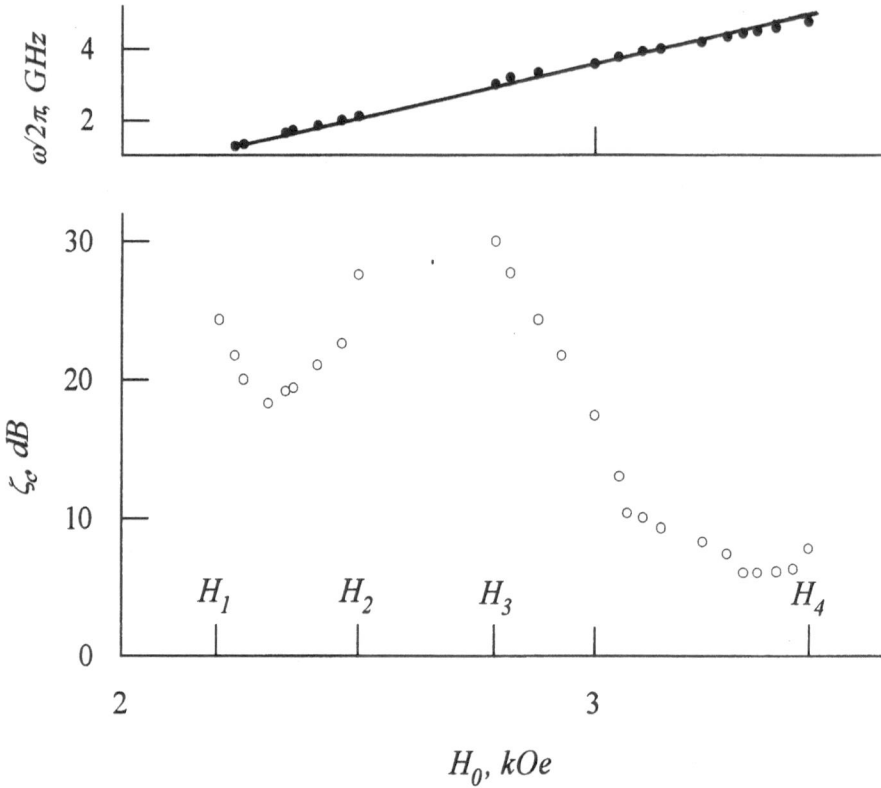

Figure 6. Reduced threshold of kinetic instability (whitecircle) and frequency of the emission (blackcircle) arising at the threshold as a function of constant magnetic field for perpendicularly magnetised yttrium-iron garnet film 20 μm thick. Straight line - frequency of the bottom of spin-wave spectrum; H_1 and H_2 , H_3 and H_4 are boundaries of experimental regions within the confines of which kinetic instability is observed.

magnetic field H_0. The kinetic instability exists in two regions of the magnetic field bounded by fields H_1, H_2 and H_3, H_4. in regard to the characteristic fields H_c and H_i corresponding to the coincidence of the frequency of parametric spin waves $\omega_p/2$ with $\omega_\perp$ (21) and ω_i (2) they are arranged as follows:

for normally magnetised film H_1, $H_2 < H_c - H_1 \approx 0.6$ kOe, $H_c - H_3 \approx 0.3$ kOe, $H_3 \approx H_c = 2.8$ kOe, $H_4 \approx H_i = 2.45$ kOe;

for tangentially magnetised film $H_c - H_1 \approx 0.57$ kOe, ($H_c = 1.03$ kOe), $H_4 \approx H_i$ ($H_i = 1.66$ kOe), $H_c - H_2 \approx 0.24$ kOe, $H_3 \approx H_c + (200 \div 300)$ Oe.

In the first region $H_1 < H_0 < H_2$ for both films the sources of secondary spin waves as in solid samples were short parametrically excited plane spin waves whose path length $l_\perp$ along the normal to the surface of the film was small

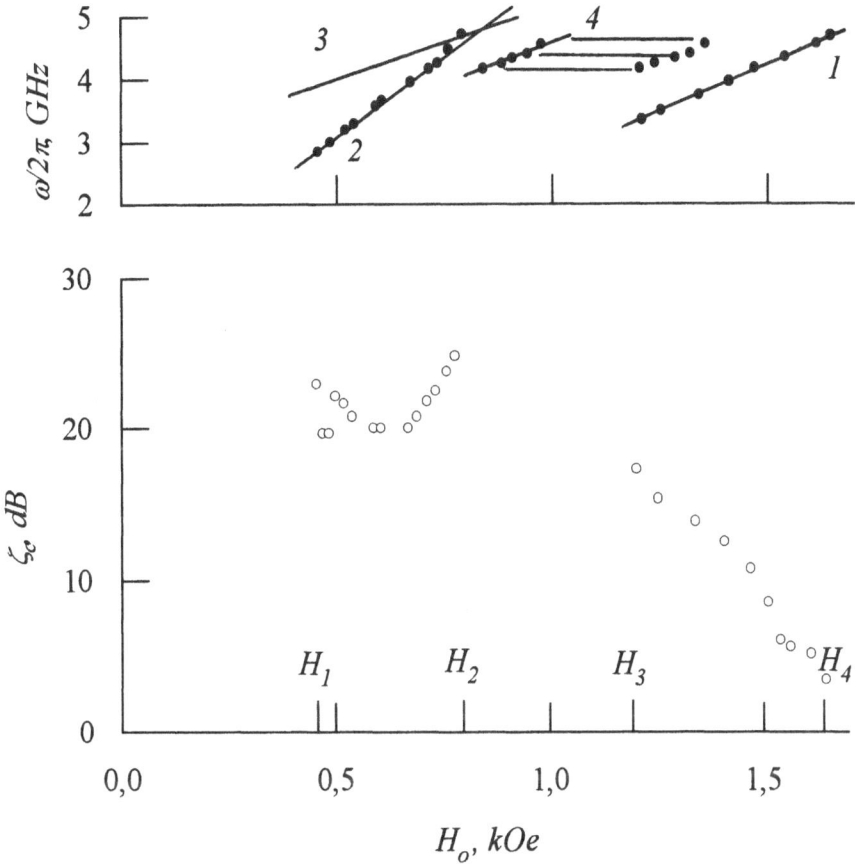

Figure 7. Reduced threshold of kinetic instability (whitecircle) and frequency of the emission (blackcircle) arising at the threshold as a function of constant magnetic field for tangentially magnetised yttrium-iron garnet film 20 μm thick. Continuous lines correspond to; $1 - f = gH_i /2\pi$, $2 - f = 2gH_i/2\pi$, $3 - f = (gH_i + g2p_{M0})/2\pi$, $4 - f = g[H_i(H_i + 4\pi M_0)]/2\pi$. H_1 and H_2, H_3 and H_4 are boundaries of the regions where kinetic instability occurs; region shaded at the top - nonthreshold emission.

compared to its thickness d. Consequently the boundaries of the first region do not differ appreciably in the magnitude of the static internal magnetic field from the case of kinetic instability in solid samples. The boundaries H_1 and H_2 are determined by the lows of conservation energy and momentum when a kinetic instability occurs in an unbounded medium.

In the second region $H_3 < H_0 < H_4$ spin waves with $l_\perp > d$ are excited, i.e., waves travelling in the plane of the film and standing along its thickness. This means that the low (3) should now be obeyed only for projections of the wave vectors on the film plane

$$k_1 + k_2 = k_3 + k_4; \quad \omega_1 + \omega_2 = \omega_3 + \omega_4 \tag{22}$$

As a result of (22) kinetic instability in films can also be observed at fields greater than H_2. In contrast to solid samples the inequality of the transverse wave-numbers before and after interaction now affects only the magnitude of the threshold ζ_c but does not prohibit the occurrence of the process.

As a result of kinetic instability in both regions of magnetic fields secondary spin waves are excited which lie near the bottom of spin-wave spectrum and whose frequencies determined by the magnitude of the internal magnetic field H_i (2). By virtue of the characteristics of the spectral structure for a normally magnetised film the wave number of secondary spin waves satisfies $k \to 0$ while for a tangentially magnetised we have $k = 10^4$ - 10^5 cm^{-1}, the lower value being reached in the range of the high values of the magnetic field H_0, i.e., in the second region $H_3 < H_0 < H_4$. Therefore for a normally magnetised film excitation of secondary waves gives rise to electromagnetic emission at frequency ω_i in both regions of kinetic instability. For a tangentially magnetised film this emission is observed only in the second existence region of kinetic instability. However , for such magnetisation emission also take place in the first existence region of kinetic instability at doubled frequency $2\omega_i$ as a result of confluence of two secondary waves travelling in opposite directions into a long-wave ($k \to 0$) magnetostatic oscillation and this oscillation reradiates electromagnetic energy.

Very far above the threshold ($\zeta_c \sim 25$ - 30 dB) in tangentially magnetised film in the intermediate region of magnetic field between first and second existence regions of kinetic instability (see Figure 7) additional emission was produced in a nonthreshold manner whose power was appreciably lower than in the two threshold regions. This emission corresponded to excitation of secondary waves lying near the upper boundary of the spectrum of exchange-free backward internal magnetostatic waves where their wave vector $k \to 0$.

It should be mentioned that a kinetic instability can be excited on excitation of a spin wave by an antenna in a ferrite film [28, 29]. Here the source of parametric waves and hence also the source of secondary spin waves consists of magnetostatic waves propagating through the film, i.e. a primary volume, backward volume and surface waves. This method however because of condition of excitation of a particular magneto-static wave can not produce packets of spin waves of arbitrary magnitude k.

3.4. FIRST-ORDER KINETIC INSTABILITY OF MAGNONS

Kinetic instability discussed so far is due to the second-order processes (3), (20) and (22) and described by the Hamiltonian (4). Kinetic instability of first order is also possible when a decay of parametric magnons is allowed by energy and momentum conservation. In this case one parametric magnon of frequency $\omega_p/2$

and wave vector k can form two secondary magnons of the frequencies ω_1, ω_2 and wave vectors k_1, k_2

$$k = k_1 + k_2; \qquad \omega_k = \omega_1 + \omega_2 \qquad (23)$$

First-order kinetic instability can be excited in two regions of spin-wave spectrum (see Figure 8). In the first region (see figure 8a) parametric magnons are of exchange nature ($k \geq 10^5$ cm^{-1}), in the second region (see Figure 8b) magneto-dipole magnons with wave vector $k \sim 10^4$ cm^{-1} are excited [30, 31].

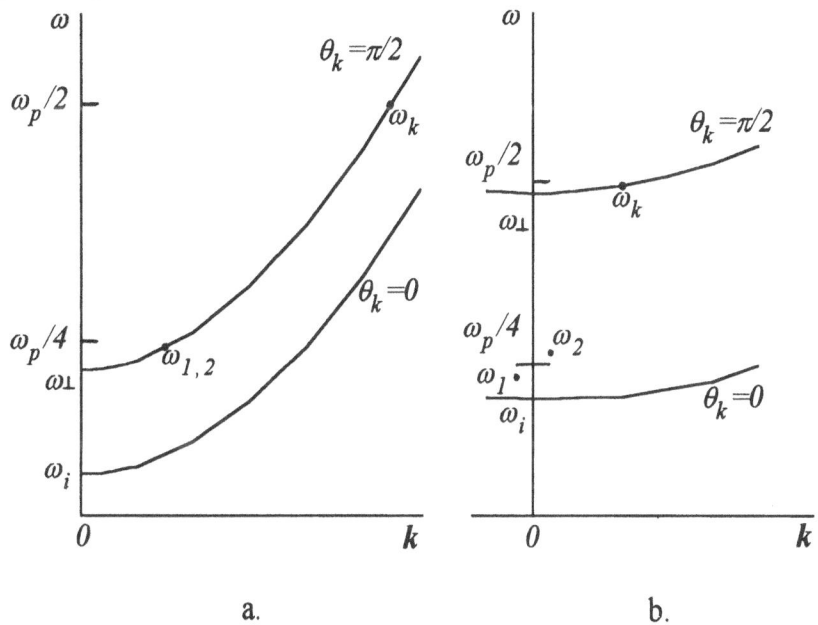

Figure 8.   Position of parametric magnon frequency ω_k and secondary magnon frequencies ω_1 and ω_2 in the short-wavelength (a) and long-wavelength (b) parts of spin-wave spectrum; ω_p is the pump frequency.

Theory of first-order kinetic instability under exchange parametric magnons was developed in [31]. As a result the formula for the threshold number of parametric magnons was obtained:

$$n_p^c = \frac{\gamma_1 S}{\omega_M} \left(\frac{2gDk^2}{\omega_M} \right)^{1/2} \left(\frac{4g\Delta H}{2\omega_M + gDk^2} \right)^{1/2} = \frac{\gamma_p S}{\omega_M} \left(\frac{2gDk^2}{\omega_M} \right)^{1/2} \frac{1}{\chi_m} \qquad (24)$$

Here k and γ_p are the wave number and equilibrium damping of parametric magnons of frequency $\omega_k = \omega_p/2$ (see Figure 8a), D is the exchange constant, S is

154

the spin of magnetic ion, $\omega_M = 4\pi g M_0$, $\Delta H = H_{3s} - H_0$, where H_{3s} is the static magnetic field at which three-magnon splitting processes start. There is nontrivial moment which is the fact that the amplitude of the process (23) turns to zero for the parametric magnons with polar angles $\theta_k = \pi/2$. So, the threshold of first-order kinetic instability is found to be sensitive to the angle structure of the parametric spin wave packet. The Eq.(24) is correct unless χ_m gets lesser than angle width of parametric magnons $\Delta\theta_k$ which is $\Delta\theta_k \sim 0.1 - 0.15$ under experimental condition [23].

Direct experimental observation of first-order kinetic instability fails because, first, the process (23) is only allowed in the part of phase space where nonlinear processeses are attenuated by strong damping and, second, short wavelength of secondary spin waves prevents their electromagnetic emission from the sample. Therefore two independent pump method was offered in [31] to detect first-order kinetic instability. The first pumping of the frequency ω_p excited parametric magnons with the frequency $\omega_p/2$ which produced secondary magnons with frequencies closed to $\omega_p/4$ (see Figure 8a). The second pumping of the frequency $\omega_p' \approx \omega_p/2$ was probing and served to record the deviation of magnon density N from the thermodynamically equilibrium one N_c at the frequency $\omega_p/4$ under the first pumping. The onset of a kinetic instability was registered experimentally at that moment when the time to the trailing edge of the second pump pulse [2] decreases abruptly. Figure 9 shows the dependence of the

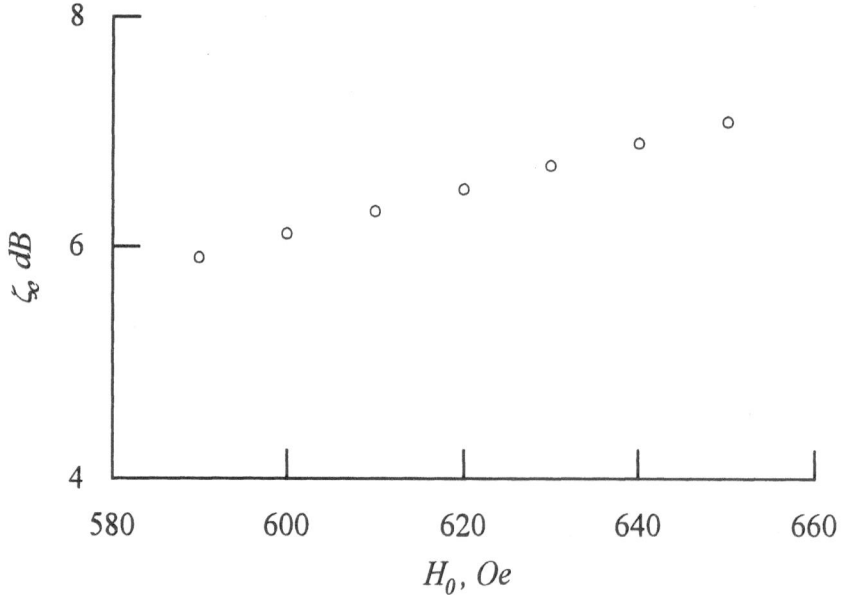

Figure 9.   Dependence of reduced threshold of first-order kinetic instability on the static magnetic field in the exchange part of spin-wave spectrum for a single crystal yttrium iron garnet sphere 2 mm in diameter; $\omega_p/2\pi = 9.4$ GHz , $H_0 \parallel [111]$, $T = 300$K.

reduced threshold ζ_c of the first-order kinetic instability on the magnetic field H_0. Kinetic instability is seen to occur in the region $H_S < H_0 < H_{3S}$ where H_S is the saturation magnetic field. The values of n_p^c obtained from the theory and experiment are of the same order and show the similar behaviour as magnetic field H_0 varies. The frequency interval where secondary magnons are excited was detected by changing the frequency of the probe pumping. It was established that magnons of the frequency closed to $\omega_p/4$ have a minimal threshold.

The first-order kinetic instability which results from the action of magneto-dipole parametric magnons was detected in [32] by using the above experimental method. It should be noted that the necessity to have the frequency $\omega_p/4$ in the spin-wave spectrum (see Figure 8b) puts the limitation on both the pump frequency ω_p and magnetisation M_0. According to the conservation laws (23) kinetic instability can be driven by parametric magnons with $k \to 0$ if $\omega_p < 4\omega_M/3$. Therefore the experiment [32] was carried out with YIG samples in the X-band range at liquid helium temperature $T = 4.2K$ when the magnetisation of YIG single crystal reaches 200 Gs instead 140 Gs at room temperature. The magnetic field dependence of the reduced threshold ζ_c of kinetic instability in the long-wavelength part of spin-wave spectrum is shown in Figure 10. For the pump frequency $\omega_p/2\pi = 8535$ GHz kinetic instability is observed within the limits 1334.6 Oe $\leq H_0 \leq 1370$ Oe. According to (23) such an instability should exist

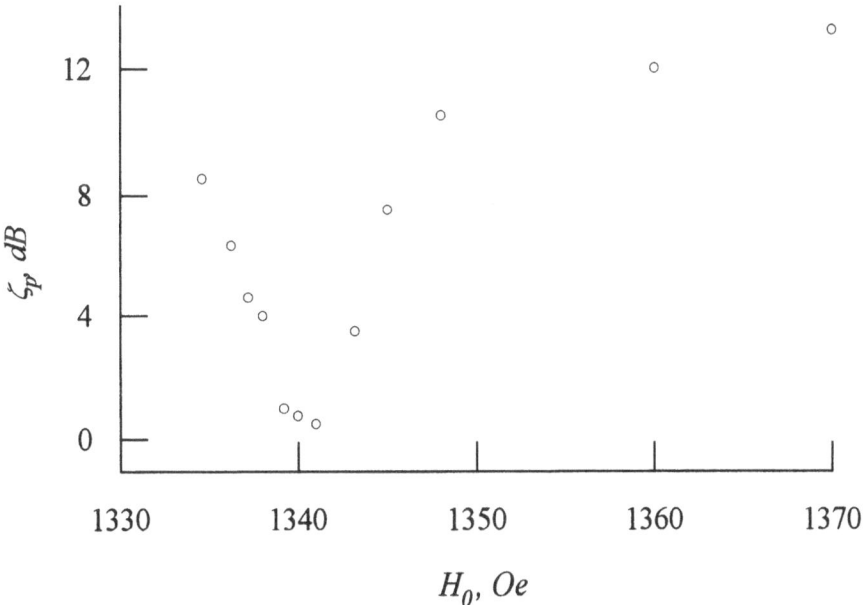

Figure 10. Dependence of the reduced threshold of first-order kinetic instability on the static magnetic field in the magneto-dipole part of spin-wave spectrum for single crystal yttrium iron garnet sphere 2 mm in diameter; $\omega_p/2\pi = 8535$ MHz, $H_0 \parallel [111]$, $T = 4.2K$

within the region 1339 Oe $\leq H_0 \leq$ 1376 Oe that agrees well with the experimental results. Secondary magnons with frequencies lower than $\omega_p/4$ were found to be excited first. This fact as well as the threshold dependence which is understood quantitatively only can be explained by developing the corresponding theory.

4. Thermodynamic Description of Strongly Driven Magnon System

4.1 THERMODYNAMIC THEORY OF PARAMETRIC EXCITATION OF MAGNONS

The thermodynamic approach in a rotating coordinate frame has been developed in the theory of parametric resonance of magnons by Kalafati and Safonov [14 to 16] in order to describe strong parametric pumping of spin-wave system. Models describing energetic flows in this case are impracticable ones due to a vary complex structure of kinetic equations. Much more simple to analyse a state of saturation. According to the theory [14 to 16] all excited magnetic vibrations can be divided into two parts. The first is the coherent with the pump field magnons. The second is the noisy deviations being described by an equilibrium quasiparticle gas of inductons [16] with an effective temperature T_{eff}. Accumulation of low-frequency magnons in this model arises due to specificity of occupation numbers N_ω

$$N_\omega = \frac{1}{exp\left[\hbar(\omega - \omega_i)/k_B T_{eff}\right] - 1} \tag{25}$$

Note that in this case there is no necessity at all for direct realisation of Eq.(3). Gpovorkov and Tulin In [33] studied parametric pumping of so-called nuclear magnons in antiferromagnet $CsMnF_3$ in which the Eq.(3) are not valid. They detected an emission from the bottom of nuclear spin-wave spectrum at $h/h_{thr} \approx$ 10. So this fact testifies in favour of applicability of thermodynamic theory of parametric resonance of magnons.

4.2. NON-EQUILIBRIUM BOSE CONDENSATION

Thermodynamic reasons are also used in an idea of Bose condensation of non-equilibrium magnons. It is well known that the phenomenon of Bose-Einstein condensation can not take place in a thermodynamically equilibrium magnon system as far as the magnon density decreases with the decreasing of temperature. However sufficient density of magnons and their quasiequilibrium condition can be supported by an external noise microwave pumping with the help of spin-wave nonlinearity. This idea has been proposed and developed by Kalafati and Safonov in [17, 18].

Parametrically excited magnons being concentrated in a narrow frequency range can play for other points of spectrum a role of incoherent pumping by means of four-magnon scattering processes. As a result a quasi-equilibrium distribution function of the form

$$N_\omega = \frac{1}{exp\left[(\hbar\omega - \mu)/k_B T_0\right] - 1} \tag{26}$$

is settled. Effective chemical potential increases from $\mu = 0$ in absence of the external pumping to $\mu = \hbar\,\omega_i$ at some critical pump power. Then a coherent magnon state appears on the bottom of spin-wave spectrum and a coherent emission should arise.

The critical magnon density is defined by following equation

$$n_c = \int \left\{ exp\left[\hbar(\omega_k - \omega_i)/k_B T_0\right] - 1 \right\}^{-1} dk / (2\pi)^3. \tag{27}$$

In order to estimate Eq.(27) one can consider the spin-wave spectrum in the simplest form $\omega_k = \omega_i + \omega_{ex}(ak)^2$ where ω_{ex} is the exchange frequency and a is the cell size. Then one can obtain

$$n_c \approx 0.06 \cdot a^{-3} (k_B T_0 / \hbar\,\omega_{ex})^{3/2} \tag{28}$$

or for YIG

$$n_c[\text{cm}^{-1}] = 1.7 \cdot 10^{17}\,(T_0[\text{K}])^{3/2}.$$

So Bose condensation of magnons can be excited at low (helium) temperatures as far as values $n \sim 10^{18}$ cm^{-1} are typical for strongly pumped spin waves.

5. Experimental Attempts to Choose an Adequate Model of the Origin of Bottom Frequency Emission

5.1. EMISSION OF THE FREQUENCY HIGHER THAN THE PARAMETRIC RESONANCE FREQUENCY FROM A FERRITE FILM

Experimental investigation was carried out in [34] in order to separate the validity regions of strong nonequilibrium and quasiequilibrium theoretical models considered above. An attempt was made to find out an accumulation of magnons with the frequency $\omega_2 > \omega_p/2$ (see Eq.(3)). The substantial growth of the magnon population in this region may be treated in favour of the theory of kinetic instability. The above magnon accumulation could be detected by the corresponding microwave emission. However a great value of wave vector k_2 prevents such a detection. In order to avoid this prevention the experiment has been carried out with tangentially magnetised YIG film in which the frequency $\omega_p/2$ is less than the upper spin-wave frequency $\omega_\perp$ (see Figure 11). The emission from (ω_2, k_2)-waves can be detected as a result of their two-magnon scattering on imperfections of the sample. This leads to an excitation of new waves with the lesser value of wave vector (in Figure 11 these are the waves of the frequency ω_2 with the indexes n_1, n_2, n_3 at $k \to 0$). Such a scattering has already considered in [35] for studying parametric magnons with $k \sim 10^5$ cm^{-1}.

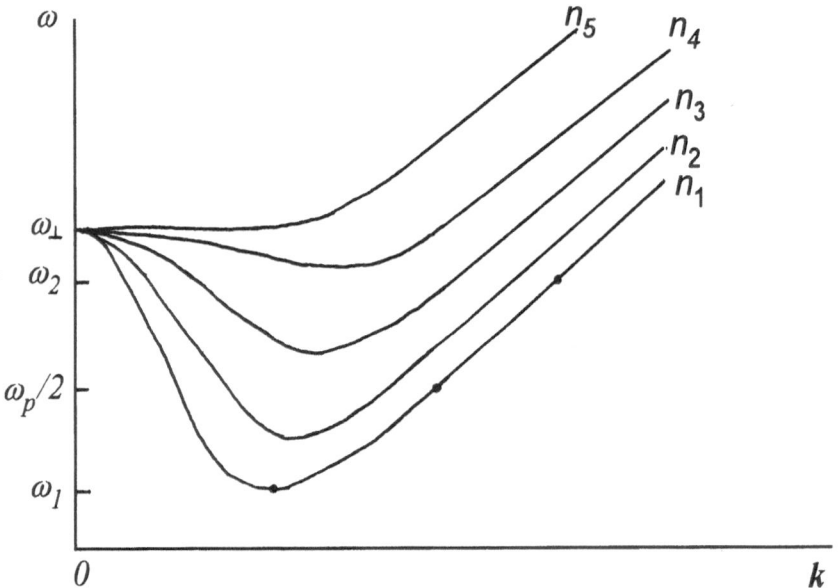

Figure 11. Spectrum of backward volume spin waves (qualitatively) for the tangentially magnetised ferrite film. Solid lines correspond to branches with different distribution of magnetisation in transverse plane. Points indicate waves excited beyond the kinetic instability.

Film 29.9 μm thickness under parallel microwave power of the frequency $\omega_p/2\pi = 9.52$ GHz was studied at room temperature. As a result the weak emission of ω_2 has been detected from the film. Its power was $P_{em}(\omega_2) \sim 10^{-14}$W. The results of the experiment are given in Figure 12. At every field H_o there are two emitted frequencies, one with ω_1 near the bottom of spin-wave spectrum

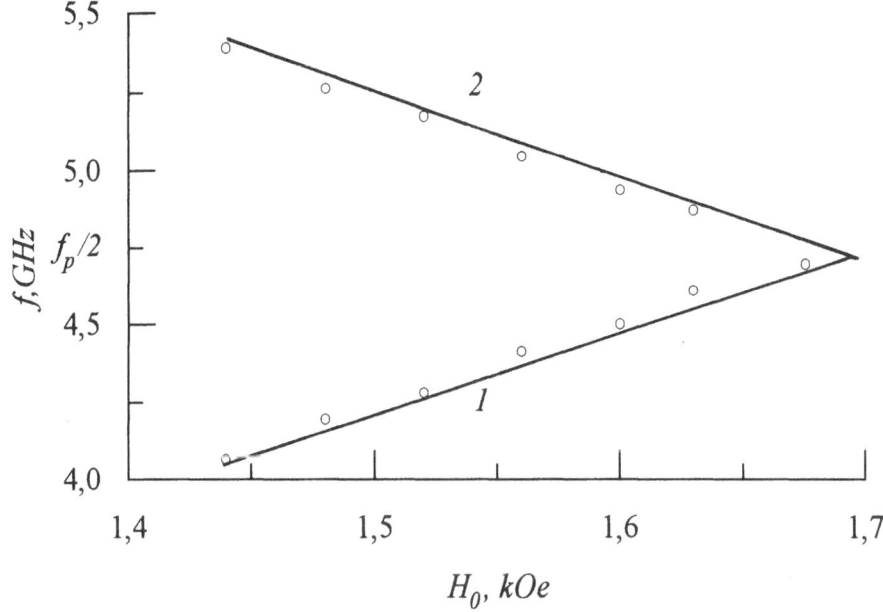

Figure 12. The magnetic field dependence for frequencies of electromagnetic emission $f = \omega/2\pi$, (1) ω_i, (2) $\omega = \omega_p - \omega_i$. The circles are experimental data.

and other with ω_2 near the frequency $\omega = \omega_p - \omega_i$ as it follows from the theory of kinetic instability. According to this theory the intensity of emission of ω_2 must be 2 - 3 orders weaker than that of ω_1. So, it has been observed when the pump power was 2 - 3 dB greater than the kinetic instability threshold ζ_c.

Thus the results of the experiment can be explained in the framework of the theory of kinetic instability. However one can propose a quite different and natural interpretation of the origin of emission with the frequency ω_2. It means that two uniform oscillation with frequencies ω_p and ω_l being mixed on nonlinearity give a harmonic $\omega_2 = \omega_p - \omega_l$. Such a process can take place at any mechanism of accumulation of magnons with $\omega_l \approx \omega_i$.

5.2. ELECTROMAGNETIC EMISSION FROM A SINGLE CRYSTAL AT LOW TEMPERATURES.

As it follows from the theory of nonequilibrium Bose condensation of magnons one can expect an appearance of a coherent magnon state on the bottom of spin wave spectrum and consequently a coherent emission at liquid helium temperature. It should be noted that low-temperature investigation of second-order kinetic instability [36] revealed the emission at doubled bottom frequency

$2\omega_l$ the onset of which was consider to be the kinetic instability threshold. Therefore the conversion of the noisy emission of the frequency $\omega_l = \omega_i$ into the coherent one at low temperatures could be treated in favour of the thermodynamic theory.

The microwave emission from the bottom of spin wave spectrum under parallel pumping of the frequency $\omega_p/2\pi = 9.6$ GHz has been detected in [37] at temperatures $T = 77$K, 4.2K and 1.4K. The sample under the study was single crystal yttrium iron garnet sphere 1 mm in diameter . The second and the third harmonics were also observed .The emission was found to appear in the same manner as in experiments at room temperature. Just above the critical pump power the emitted power was $P_{em}(\omega_l) \sim 2 \cdot 10^{-16}$ W. In Figure 13 is shown the magnetic field dependence for the reduced power ζ_c for two temperatures. Both experimental curves demonstrate a similar character in a restricted region of magnetic fields but ζ_c is greater for the liquid helium temperature than that for the case of liquid nitrogen temperature. Note that the values of ζ_c at $T = 77$K are comparable with those obtained at room temperature [10] and results for $T = 4.2$K agree with the experimental results [36].

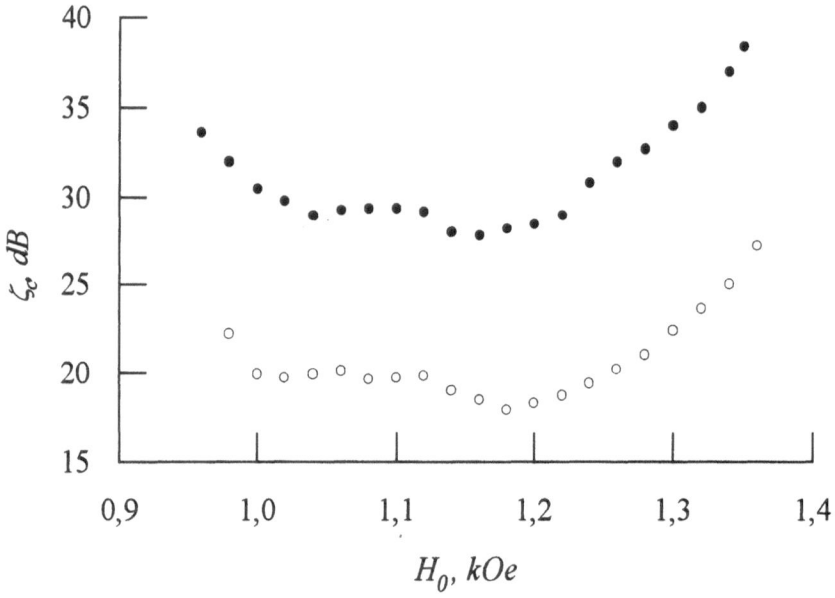

Figure 13.   Magnetic field dependence of relative critical power ζ_c of the onset of electromagnetic emission form the bottom of spin-wave spectrum for yttrium iron garnet sphere 1 mm in diameter; $T = 77$K (whitecircle), $T = 4.2$K (blackcircle), $H_0 \parallel$ [111]

The spectrum of electromagnetic emission at various levels of pump power is shown in Figure 14. One can see a significant broadening of the lineshape

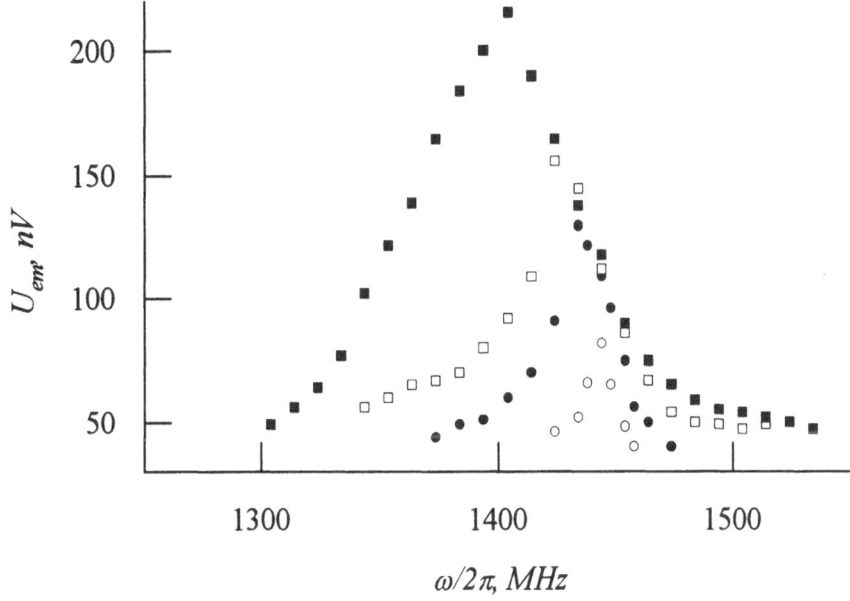

Figure 14. Spectrum of electromagnetic emission from the spin-wave bottom (ω_l). H_0 = 1170 Oe, T = 4.2K; P/P_c = 1 dB (whitecircle), 5 dB (blackcircle), 10 dB (whitebox), 15 dB (blackbox).

and the negative frequency shift with increasing the P/P_c. The second and the third harmonics represent proportional changes of the lineshapes. The analogous pump power dependencies for the emitted spectra were observed at all temperatures of experiment [37]. It should be also mentioned that the linewidth of the emitted spectrum decreases with decreasing the temperature.

From the above experimental results is not possible yet to choose an adequate model of an origin of emission from the bottom of spin-wave spectrum. The fact that the linewidth of the emission decreases on decrease in the temperature could be treat in favour of the theory of nonequilibrium Bose condensation of magnons however there is no tendency to narrowing the emission spectrum with increasing the pump power. So, experimental and theoretical investigations in this field are supposed to be continued.

Microwave emission from the ferrite reported in this paper is very attractive from both fundamental standpoint and from applied standpoint. In fact this effect makes it possible to create controlled noise emission generators. Then as it follows from the thermodynamic theory in case of Bose condensation of magnons it could be possible to obtain a kind of transformer from noise emission into coherent emission. Anyway magnon excitation well above the parametric instability threshold that is an additional channel of energy dissipation in a ferrite should be taken into account when constructing ferrite devices working at high-

162

level microwave power in the wide range of frequencies, magnetic fields and temperatures.

6. References

1. Suhl, H. (1957) The theory of ferromagnetic resonance at high signal power, *J. Phys. Chem. Solids* 1, 209-227.
2. Morgenthaler, F.R. (1960) Survey or ferromagnetic resonance in small ferromagnetic ellipsoids, *J. Appl. Phys.* **31**, 95S-97S.
3. Shlomann, E., Green, J.J., and Milano, U. (1960) Recent developments in ferromagnetic resonance at high power levels, *J. Apple. Phys.* **31**, 386S-395S.
4. Zakharov, V.E., L'vov, V.S., and Starobinets, S.S. (1974) Turbulence of spin waves beyond the threshold of their parametric excitation, *Usp. Fiz. Nauk* **114**, 609-654.
5. L'vov, V.S. (1987) *Non-linear Spin Waves [in Russian]*, Nauka, Moscow; (1992) *Wave Turbulence Under Parametric Excitation*, Springer Verlag, Berlin.
6. Gibson, G. and Jeffries, C. (1984) Observation of period doubling and chaos in spin-wave instabilities in yttrium iron garnet , *Phys. Rev. A* **29**, 811-818.
7. Yamazaki, H. and Mino, M. (1989) Chaos in spin-wave instabilities, *Progr. Theor. Phys. Suppl.* **98**, 400-419.
8. Cherepanov, V.B. and Slavin, A.N. (1993) Collective spin wave oscillations in finite size ferromagnetic samples, *Phys.Rev. B* **47**, 5847-5852.
9. Safonov, V.L. and Yamazaki, H. (in press) Radiation damping of parametrically excited spin waves , *J. Magn. Magn. Mater.*
10. Lavrinenko, A.V., L'vov, V.S., Melkov, G.A., and Cherepanov, V.B. (1981) "Kinetic" instability of a strongly nonequilibrium spin wave system and the tuneable radiation from a ferrite, *Zh..Eksp. Teor. Fiz.* **81**, 1022-1036 [(1982) *Sov. Phys. JETP* **54**, 542-554].
11. L'vov, V.S. and Cherepanov, V.B (1981) Non-linear theory of "kinetic" excitation of waves, *Zh. Eksp. Teor. Phys.* **81**, 1406-1422 [(1982) *Sov. Phys. JETP* **54**, 764-778].
12. Taranenko, A. Yu. and Cherepanov, V.B. (1989) Energy absorption in a ferrite beyond the of kinetic instability threshold, *Zh. Eksp. Teor. Fiz.* **95**, 1810-1819 [(1989) *Sov. Phys. JETP* **68**, 1046-1051].
13. Lavrinenko, A.V., Melkov, G.A., and Fal'kovich, G.E. (1984) Mutual influence of the kinetic and parametric methods of exciting spin waves, *Zh. Eksp. Teor. Fiz.* **87**, 205-211 [(1984) *Sov. Phys. JETP* **60**, 118-121].

14. Kalafati, Yu.D. and Safonov, V.L. (1989) Thermodynamic theory of saturation of the impulse magnon excitation, *J. Phys. France* **50**, 1157-1161.

15. Kalafati, Yu.D. and Safonov, V.L. (1989) Thermodynamic approach in the theory of parametric resonance of magnons, *Zh. Eksp. Teor. Fiz.* **95**, 2009-2020 [(1989) *Sov. Phys. JETP* **68**, 1162-1168].

16. Safonov, V.L. (1992) Thermodynamic description of strongly excited Bose systems, *Physica A* **188**, 675-686.

17. Kalafati, Yu.D. and Safonov, V.L. (1991) Theory of quasiequilibrium effects in magnon system excited by incoherent pumping, *Zh. Eksp. Teor. Fiz.* **99**, 1511-1521 [(1991) *Sov. Phys. JETP* **73**, 836-844].

18. Kalafati, Yu.D. and Safonov, V.L. (1993) Theory of Bose condensation of magnons excited by noise, *J Magn. Magn. Mater.* **123**, 184-186.

19. Akhiezer, A.I., Bar'yakhtar, V.G., and Peletminskii, S.V. (1968) *Spin Waves*, North-Holland, Amsterdam; Willey, New York.

20. Cherepanov, VB. (1986) Nonequilibrium magnons beyond the second-order kinetic instability in a ferrite, *Preprint No.318, Institute of Automation and Electrometry, Siberian Branch of the Academy of Sciences of the USSR, Novosibirsk.*

21. Gottlib, P. and Suhl, H. (1962) Saturation of ferromagnetic resonance with parallel pumping, *J. Appl. Phys.* **33**, 1508-1514.

22. Lavrinenko, A.V., Melkov, G.A., and Taranenko, A.Yu. (1984) Kinetic instability of spin waves under arbitrary pumping, *Fiz. Tverd. Tela.* **26**, 1499-1500 [(1984) *Sov.Phys. Solid State* **26**, 910-911].

23. Bakai, A.S., Krutsenko, I.V., Melkov, G.A., and Sergeeva, G.G. (1979) Study of the distribution of parametrically excited spin waves under parallel pumping of spin-wave instability in a ferrite, *Zh. Eksp. Teor. Fiz.* **91**, 231-237 [(1979) *Sov. Phys. JETP* **49**, 118-123].

24. Melkov, G.A. and Taranenko, A.Yu. (1986) Nonlinear emission from a ferrite in ferromagnetic resonance, *Zh. Eksp. Teor. Fiz.* **91**, 1007-1015 [(1986) *Sov. Phys. JETP* **64**, 592-597].

25. Fal'kovich, G.E. (1983) On kinetic excitation of sound by spin waves, *Preprint No.172, Institute of Automation and Electrometry, Siberian Branch of the Academy of Sciences of the USSR, Novosibirsk.*

26. Starobinets, S.S. and Gurevich,A.G. (1968) Nonlinear effects above the threshold of spin-wave instability, *J. Apple. PhDs.* **39**, 1075-1077.

27. Melkov, G.A. and Sholom, S.V. (1991) Kinetic instability of spin waves in thin ferrite films, *Hz. Esker. Tear. Fizz.* **99**, 610-618 [(1991) *So. PhDs. JET* **72**, 341-346].

28. Zil'berman, P.E., Nikitov, S.A., and Temiryazev, A.G. (1985) Four-magnon decays of magnetostatic wave in yttrium iron garnet films, *Pis'ma Zh. Eksp. Teor. Fiz.* **42**, 92-94 [(1985) *JETP Lett.* **42**, 110-112].

164

29. Dudko, G.M., Kazakov, G.T., Kozhevnikov, A.V., and Filimonov, Yu.A. (1987) Kinetic instability and pulse propagation under three-magnon decays of surface magnetostatic waves SMW in YIG films, *Authors Abstract of Reports of the Regional Conference, Spin-Wave Phenomena in Microwave Electronics, Krasnodar.*

30. Lemeire, B., Le Gall, H., and Dormann, J.C. (1967) Splitting of parametric magnons in ferromagnetic crystals, *Solid State Comm.* **5**, 499-502.

31. Lutovinov, V.S., Melkov, G.A., Taranenko, A.Yu., and Cherepanov, V.B. (1989) First-order kinetic instability of spin waves in a ferrite, *Zh. Eksp. Teor. Fiz.* **95**, 760-768 [(1989) *Sov. Phys. JETP* **68**, 432-437].

32. Zablotskii, I.L., Melkov, G.A., and Taranenko, A.Yu. (1990) Kinetic instability of spin waves at low temperatures, *Ukr. Fiz. Zh.* **35**, 238-243.

33. Govorkov, S.A. and Tulin, V.A. (1989) Heating and energy diffusion in the electronic-nuclear magnetic system of antiferromagnet $CsMnF_3$ at powerful parametric excitation, *Zh. Eksp. Teor. Fiz* **95**, 1398-1403 [(1989) *Sov. Phys. JETP* **68**, 807-810].

34. Melkov, G.A., Safonov, V.L., Taranenko A.Yu., and Sholom, S.V. (1994) Kinetic instability and Bose condensation of nonequilibrium magnons, *J. Magn. Magn. Mater.* **132**, 180-184.

35. Melkov, G.A. and Sholom, S.V. (1989) Parametric excitation of spin waves by the surface magnetostatic wave, *Zh. Eksp. Teor. Fiz* **96**, 712-719 [(1989) *Sov. Phys. JETP* **69**, 672-678].

36. Krutsenko, I.V., Lavrinenko, A.V., and Melkov, G.A. (1983) Kinetic instability of spin waves at 4.2K, *Fiz. Nizk. Temp.* **9**, 1289-1292 [(1983) *Sov. Phys. Low Temp.* **68**, 664-666].

37. Taranenko, A.Yu., Mino, M., Yamazaki, H., and Safonov, V.L. (in press) Electromagnetic emission by a system of nonequilibrium magnons in a ferromagnet, *Phys. Rev. B.*

EXCHANGE SPIN WAVES IN NONUNIFORM FILMS OF YTTRIUM IRON GARNET

A. G. TEMIRYAZEV, M. P. TIKHOMIROVA. and
P. E. ZILBERMAN
Institute of Radioengineering & Electronics,
Russian Academy of Sciences
Vvedenskii sq. 1, Fryazino, Moscow region, 141120, Russia.

Excitation and propagation of short-wavelength exchange-dominated spin waves in garnet films with nonuniformity of saturation magnetisation or anisotropy field along the thickness of the film is considered. The paper includes experiments on propagation of pulses of spin waves through the film thickness, conversion of spin waves into acoustic waves, and three-magnon interaction of high-amplitude exchange spin waves.

1. Introduction

The aim of this paper is to show that YIG films having smooth variations of magnetic parameters through the film thickness are of interest because of their potential as a basis for a new type of microwave signal processing technology using very short-wavelength exchange-dominated spin waves (ESW). Along with elastic and magnetostatic waves, exchange spin waves are "slow" waves, i.e., their phase and group velocities are small compared to the velocity of an electromagnetic wave. Since ESW have small wave length and slow speed, it is possible to get a group delay of tens or hundreds of nanoseconds in a distance of a few micrometers. Thus, the desired delay can be achieved under ESW propagation across the film thickness. On the one hand, that is why exchange spin waves are promising candidates for use in making small-size microwave engineering elements. On the other hand, the wavelength of exchange spin waves is comparable to that of acoustic waves; therefore, these spin waves could be important instruments for excitation of acoustic waves at microwave frequencies. There are two main difficulties encountered in reaching the ESW application. First, we need to obtain an

R. Marcelli and S.A. Nikitov (eds.), Nonlinear Microwave Signal Processing: Towards a New Range of Devices, 165–212.
© *1996 Kluwer Academic Publishers.*

effective conversion from electromagnetic waves to the short-wavelength spin waves. Second, it is important to find the ways and means of controlling delay-versus-frequency characteristics. It will be shown that both of the problems can be solved using the nonuniformity of the film.

The experimental study of exchange spin waves, which started more than thirty years ago, has essentially involved investigations of standing ESW modes and parametrically excited spin waves. Three basic experimental methods have been used: study of spin-wave resonance spectra in thin ferromagnetic films [1], measurement of frequency and field dependencies of the threshold for parametric excitation of spin waves [2], and

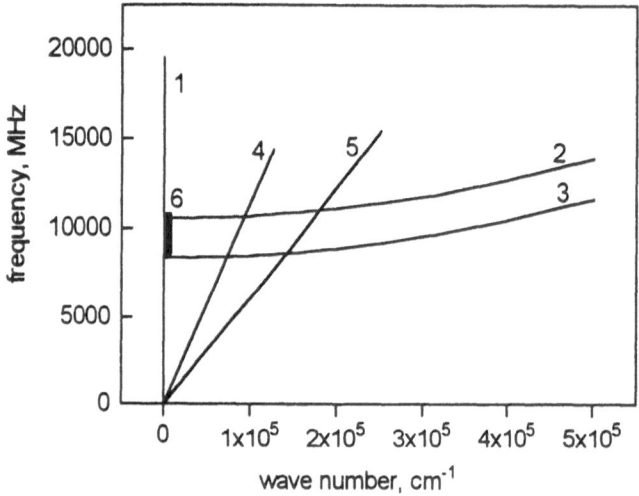

Figure 1. Dispersion diagram for the waves propagating in a ferromagnetic medium. 1 - electromagnetic wave; 2 - spin wave propagating perpendicular to the direction of the bias magnetic field (H_{int} = 3000 Oe was used for the calculation); 3 - spin wave propagating along the magnetic field; 4 and 5 - longitudinal and transversal elastic waves; painted area 6 shows the region where magnetostatic waves exist.

investigation of the scattering of light by thermal or parametrically excited spin waves [3]. In this paper we would like to make a change in emphases - from standing modes to running waves and from parametrical excitation to linear excitation of coherent spin waves. That is why considerable attention will be given to the study of ESW pulse propagation. The organisation of this paper is as follows: Section 2 provides an overview of the main methods for spin wave excitation. Section 3 describes a theoretical model that will be used for experimental data handling. Section 4 describes experiments on spin wave resonance. The aim of these experiments is to select films suitable for the excitation and propagation of running ESWs. The experimental results on the propagation of ESW pulses are presented and discussed in Section 5. Section 6 shows that ESWs can be used for the excitation of acoustic waves at microwave frequencies. Some nonlinear properties of ESW are considered in Section 7.

It is intended that the present paper should give the reader the basic understanding of the feasibility of ESWs. We do not pretend to give the complete set of the literature. The extensive literature may be found in cited reviews and monographs [1-8]. Some of the topics discussed in this paper were treated by the present authors in previous publications [9-15].

2. Methods for Exciting Spin Waves

Fig. 1 shows a dispersion diagram for the basic types of waves that can propagate in ferro- or ferrimagnetic materials at microwave frequencies, namely: electromagnetic waves, longitudinal and transversal elastic waves, and spin waves. The energy transfer by spin waves arises from two types of interaction between spins: magnetodipole and exchange. The magnetodipole interaction, as a rule, plays a fundamental role in the propagation of relatively long-wavelength spin waves with wavelength $\lambda \sim 100$ μm, where λ can be comparable to the characteristic size of the ferromagnetic sample. Such waves are customarily referred to as magnetostatic spin waves (MSW) [5]. For shorter-wavelength spin waves (with $\lambda < 1$ μm) it is the exchange

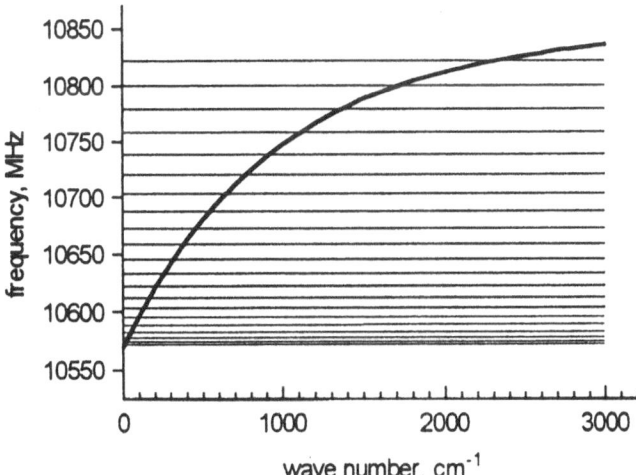

Figure 2. Bold line shows the dispersion curve for the surface magnetostatic wave in 5 μm-thick YIG film at $H = 3000$ Oe. Thin lines show the frequencies of spin-wave resonance.

interaction that plays the fundamental role. In order to emphasise this distinction, we will refer to such waves as exchange spin waves. Note that both MSWs and ESWs are slow waves and their characteristics may be obtained by use of Maxwell's equations in the magnetostatic limit, i.e., $\nabla \times \mathbf{H} = 0$, $\nabla \cdot \mathbf{B} = 0$. Nevertheless, it is common to use term "magnetostatic waves" for dipole-dominated waves only. This kind of spin waves has received the most study. An intensive investigation of MSWs for microwave signal-processing applications has started since high-quality, narrow-linewidth single-crystal thin films of yttrium iron garnet (YIG) became available. Such films with film thickness L between 0.3 and 150 μm are grown epitaxially on gadolinium gallium garnet (GGG) substrates with in-plane dimensions up to 3 inches. MSWs propagate along the plane of the film in a distance of a few millimetres or centimetres. The properties of MSWs are usually defined in terms of dispersion curve of ω vs. k, where ω is the frequency, and k is the in-plane component of the wave vector. Fig. 2 shows the calculated dispersion curve for so-called surface magnetostatic wave which propagates perpendicularly to the bias magnetic field $\mathbf{H}$ in the in-plane magnetised YIG film [16]. MSWs are easy to excite by a microwave current in a microstrip line placed near a ferrite film (see Fig. 3). The microstrip line creates a spatial nonuniformity in the rf magnetic field with a characteristic size of the same order as the width of the microstrip line w. This gives strong coupling between an electromagnetic wave and MSWs and allows us to excite magnetostatic spin waves with values of the wave number $k < \pi/w$. For $w > 10$ μm it means that $k < 3 \cdot 10^3$ cm^{-1}. It is not possible to use a microstrip antenna to excite exchange spin waves having $k \sim 10^5$ cm^{-1} because if w is less than a micrometer, it is difficult to provide good matching between an antenna and feeding microwave circuits. Thus we cannot excite the ESW propagating along the plane of the film and we shall consider the possibilities to generate waves propagating through the film thickness. The wave lengths of such waves should be much smaller than L, this is true for ESWs.

Let us consider the main properties of exchange spin waves. For the simplest case of uniform saturated isotropic ferromagnetic media the relation between frequency ω and wave number q of the ESW has a form:

$$\left(\frac{\omega}{\gamma} \right)^2 = \left[H_{int} + Dq^2 \right]\left[H_{int} + Dq^2 + 4\pi M_{sat} \cdot \sin^2 \alpha \right], \qquad (1)$$

where H_{int} is the internal magnetic field, M_{sat} is the saturation magnetisation, α is the angle between the wave vector and the magnetisation direction, γ is the gyromagnetic ratio ($\gamma = 2\pi \cdot 2.8$ MHz/Oe), D is the exchange constant ($D = 4.6 \cdot 10^{-9}$ Oe·cm^2 in YIG).

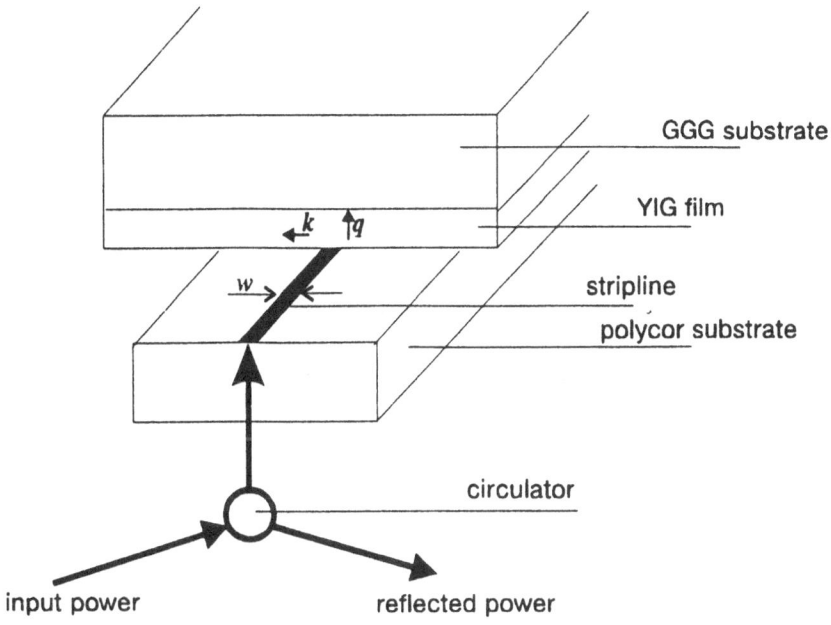

Figure 3. Schematic diagram of experimental setup.

A dispersion law of the ESW propagating along the film thickness in the external magnetic field H can be found from (1) by substituting $H_{int} = H - 4\pi M_{sat}$, $\alpha = 0$, in perpendicularly magnetised films, and $H_{int} = H$, $\alpha = \pi/2$ for in-plane magnetised films. For $\alpha = 0$, it is easy to estimate the group velocity of ESW $V_{ESW} = \partial\omega/\partial q = 2\gamma Dq$. For $q = 1 \cdot 10^5$ cm^{-1}, one obtains $V_{ESW} = 1.6 \cdot 10^4$ cm/s. Such value of group velocity is less by two orders of magnitude than one for magnetostatic waves ($V_{MSW} = 10^6 - 10^7$ cm/s). Thus, the propagation of ESW across the film thickness in a distance of, say, 50 μm gives the same group delay as the propagation of MSW in a distance of 0.5 - 1 cm. Note that only large-q-value ESWs can be considered as running waves. One can calculate that if $q < 1.3 \cdot 10^4$ cm^{-1}, the group velocity is too slow for the wave to propagate, since the delay time in a distance of a wave length exceeds the relaxation time $T_r = 227$ ns. This value of T_r corresponds to an equivalent linewidth of about 0.5 Oe.

The ESWs manifest themselves mainly in experiments on spin-wave resonance (SWR). Resonance excitation of exchange spin waves is possible if the spins at the surface of the film are pinned, that is, dynamic magnetisation falls to zero at the surface [17]. The resonant wave numbers are defined by $q = \pi n/L$, where n is the mode number. The resonant frequencies (shown in Fig. 2) can be calculated from the dispersion relation (1). Note that the thickness of the magnetically uniform film must be sufficiently small for the SWR to be observed. It follows from two features of SWR in uniform films. First, it is only the first few spin-wave resonance modes that are excited, since the amplitude of the resonances falls off as n^2 as the resonant mode number n increases. Second, the resonant fields depend on n as n^2. Therefore, in thick uniform films ($L > 10$ μm), the distance between resonances exceeds the linewidth only for high-number resonances, however such resonances are not excited.

Resonant excitation of exchange spin waves has also been observed in certain experiments involving the propagation of magnetostatic waves in thin films of YIG. MSWs and ESWs are coupled at frequencies of SWR when the surface spins are pinned [18-20]. It leads to resonant increasing the transmission loss of MSW at SWR frequencies.

The above-mentioned methods for exciting ESWs are resonant in nature, they do not allow running waves to be excited. At the same time that are running waves that should be excited in order to realise the ESW application.

The authors of Ref. 21 pioneered the effective linear excitation of travelling exchange spin waves in YIG film. They observed the ESW pulse propagation across the film thickness. An ion-implanted YIG film was used. The implantation was carried out to create a thin nonuniform layer at the surface of the film, which acts as the spin wave antenna. The effective conversion ($\sim$ 10-30%) of electromagnetic wave into spin waves has been achieved inside a field band of $\sim$ 30 Oe [22]. Note that no excitation of ESW pulses was observed in non-implanted films, even though the films demonstrated SWR spectra. The theoretical analysis conducted in [23] has shown that very poor ($\sim$ 0.03%) conversion into travelling spin waves takes place when the surface spins are pinned and dynamic magnetisation varies between zero and in-volume value in a distance of the order of exchange length $l_{exch} \sim 10^{-6}$ cm. At the same time, spin pinning makes it possible to observe SWR, because the accumulation of energy in resonator provides the conditions for the spectrum detection even under weak coupling.

As indicated above, the problem of generation of running ESWs can be solved to some extent with the use of ion-implanted YIG films. Nevertheless, such way gives rather narrow-band excitation. Another approach to obtaining short-wavelength spin waves is one in which rather long waves are excited at the first stage. In the next stage, these are converted into short waves as they propagate through a spatially nonuniform waveguiding structure.

The idea of using nonuniform magnetic media to excite short-wavelength exchange spin waves was proposed even in the 1960's, and was rather thoroughly discussed in the papers of Schloemann [24, 25]. In essence, this method makes use of the fact that if the parameters of the medium vary slowly in space, the wavelength of the spin wave will depend on the coordinates. Then it is possible to create conditions under which the wave number of the spin wave will be small in a certain region of space, which ensures effective coupling with an electromagnetic wave. At the same time, as the wave propagates its wavelength decreases, which leads to the appearance of short-wavelength spin waves. In practice, this method was first implemented in experiments involving excitation of spin wave pulses that propagated along a rod of YIG placed in an external magnetic field (see, e.g., Refs. 26 and 27). The nonuniformity of the medium arises from the demagnetising field near the end of the rod. Spin waves were excited in this experiment with $q \sim 10^4$ cm^{-1}. This excitation mechanism manifests itself in experiments on spin wave resonance in ferrite films as well when the film under study is nonuniform along the thickness. It has been shown that effective excitation of exchange spin waves with high mode numbers is possible in such films. The authors of Refs. 28-33 investigated spin-wave resonance theoretically and experimentally in nonuniform magnetic films, setting as their goal the resolution of two fundamental questions: (1) how does nonuniformity affect the resonance spectrum, and (2) is it possible to recover the nonuniformity profile from the spectrum. A possibility of using inhomogeneities to control the dispersion of magnetostatic waves was investigated in [34, 35]. In that case the theory was developed without considering exchange interaction.

We will look at nonuniform films from a different point of view - as media in which it is possible to excite and propagate travelling exchange spin waves. As we have already noted, travelling exchange spin waves show promise for fabricating new devices. In what follows, we will discuss YIG films, since the propagation loss for spin waves in this material is minimal. The films should be sufficiently nonuniform to ensure effective wavelength conversion. We will start from theoretical analysis of spin waves in nonuniform films. Because the main part of this paper is of experimental nature, only a short summary of relevant theoretical results will be given.

3. Theoretical Analysis

Let us consider a magnetic film of thickness L placed in an external magnetic field **H**, so that **H** is directed along the x-axis perpendicular to the plane of the film. We will assume that the parameters that change significantly over the film thickness are the saturation magnetisation $M_{sat}(x)$ and the uniaxial anisotropy field $H_a(x)$. We will assume that the remaining film parameters, i.e., γ and D are constant, and we will ignore the cubic anisotropy. Let us introduce a certain effective parameter - the effective magnetisation $M(x)$ - which we define by the relation $4\pi M(x) = 4\pi M_{sat}(x) - H_a(x)$. Let us consider uniform in-plane precession of rf magnetisation $m(x,t) = m(x) \cdot exp(i\omega t)$. Then we find from the Landau-Lifshitz equation that $m(x)$ should satisfy the following equation [32]:

$$D\frac{\partial^2 m}{\partial x^2} + \left(\frac{\omega}{\gamma} - H + 4\pi M - \frac{D}{M_{sat}}\frac{\partial^2 M_{sat}}{\partial x^2}\right)m = 0 .$$ (2)

The effect of the last term in the brackets can be shown to be small. Neglecting this term, we rewrite (2) in the form

$$\frac{\partial^2 m(x)}{\partial x^2} + q^2(x) \cdot m(x) = 0 ,$$ (3)

where

$$q^2(x) = \frac{\omega/\gamma - H + 4\pi M(x)}{D} .$$ (4)

Recall that in a homogeneous film the general solution to Eq. (2) has the form $m(x) = a \cdot exp(-iqx) + b \cdot exp(+iqx)$, where a and b are constants and q is the wave number of a spin wave propagating along the normal to the film.

Exact solutions to Eq. (2) for nonuniform films were found in Ref. 28 and 32 for two special cases of the profile $M(x)$ - parabolic and linear. However, we will investigate approximate solutions for an arbitrary monotone profile $M(x)$. We shall follow the WKB method. Then $q(x)$, determined by the dispersion relation (4), depends on the local magnetisation and has the physical meaning of a local wave number. We will assume that $M(x)$ varies monotonically through the thickness of the film, reaching its maximum value $M = M_0$ at $x = 0$, and its minimum value $M = M_L$ at $x = L$. One can see from Eq. (4) that if the frequency ω lies in the range

$$\omega_0 < \omega < \omega_L ,$$ (5)

where $\omega_L = \gamma(H - 4\pi M_L)$ and $\omega_0 = \gamma(H - 4\pi M_0)$, then there is a point within the film where the local wave number q turns to zero. We denote by d the coordinate x of that point. The eigenmode is a sinusoidal wave (with varying wave number) within the layer $0 < x < d$, where $q^2 > 0$, and changes to an exponentially decaying excitation in the region $d < x < L$, where $q^2 < 0$ (Fig. 4). By analogy with the problems of particle motion in a potential well, we will refer to the point $x = d$ as a turning point. Note that just one turning point can exist if $M(x)$ varies monotonically. The coordinates of the turning point depend both on frequency and on the field magnitude, which is illustrated by Fig. 5b, where we show the form of the computed values of

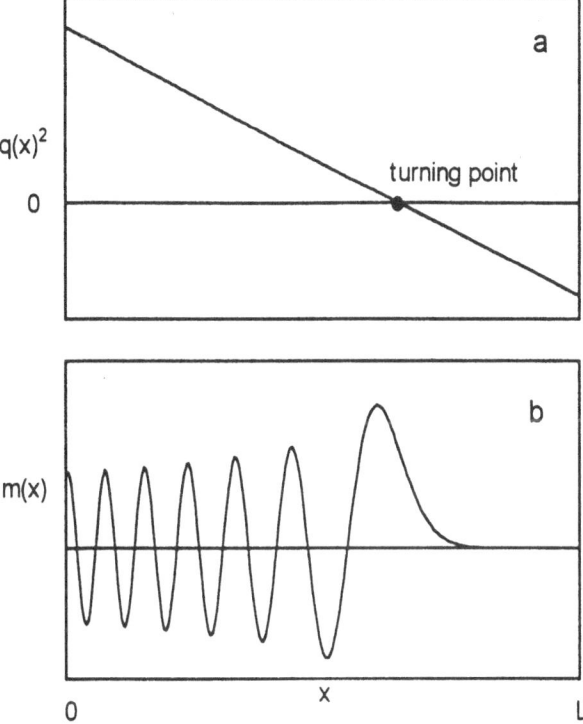

Figure 4. Variation of the wave number squared (a) and variation of rf magnetisation (b) over the film thickness.

$q(x)$ for the case of a parabolic profile $M(x)$ shown in Fig. 5a. When $\omega \cong \omega_0$, the turning point appears near the boundary with large magnetisation $x = 0$; as the frequency increases, it shifts into the bulk of the film. When $\omega \cong \omega_L$, the value of d reaches the second boundary of the film (curve 3 in Fig. 5b).

The existence of a turning point within the film causes the spin-wave resonance modes to be localised in the layer $0 < x < d$, in which $q^2 > 0$. As was shown in Ref. 32, these "local" modes are excited very efficiently. This latter circumstance is connected with the fact that the rf magnetisation varies most smoothly with x near the turning point, giving strong coupling to the external electromagnetic field [36]. Therefore, the layer where $q \sim 0$ acts as an intrinsic spin-wave antenna.

An effective excitation of normal modes can be expected only at frequencies satisfying the condition (5). We apply the WKB method to obtain the approximate solutions of Eq. (3). Then the condition for the resonance frequencies ω_n inside the above mentioned frequency band (5) has the form

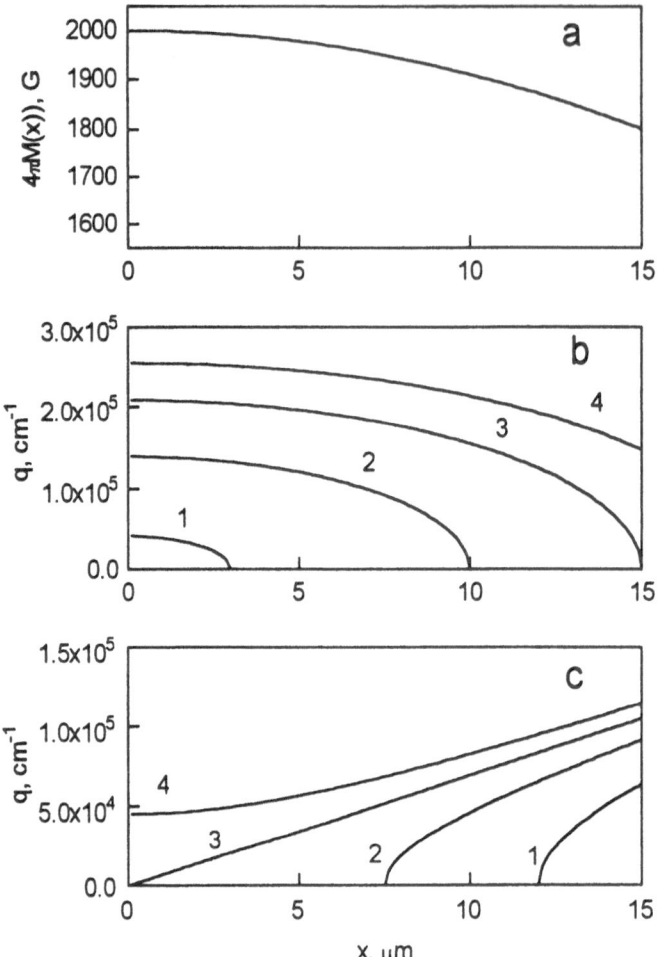

Figure 5. Profile of magnetisation variation (a) and the corresponding computed variations of the wave number across the thickness of the film for the case of normal (b) and tangential (c) magnetic fields. As an example, we show in (a) a parabolic variation in magnetisation. Curves 1 - 4 in Figs. (b) and (c) reflect the character of the change in wave number for successive increases in frequency.

$$\varphi(\omega_n, H) + \varphi_0 = 2\pi n, \quad n = 1,2,3\ldots, \tag{6}$$

where

$$\varphi(\omega, H) = 2 \int_0^{d(\omega,H)} |q(x, \omega, H)| dx . \tag{7}$$

The physical meaning of the quantity $\varphi(\omega,H)$ is quite evident - it is the phase shift accumulated by the exchange spin wave as it propagates from the turning point to the boundary of the film and back. The quantity φ_0 can be treated as a phase jump that the wave acquires as it is reflected from the film boundary and from the turning point. The value of φ_0 depends on the spin pinning at the film surface $x = 0$. For films with free spins we have $\varphi \cong 3\pi/2$ [32]. Relation (6) applies to whatever the boundary condition at $x = L$, since it is assumed that the waves have decayed before reaching that surface. By comparing the frequencies ω_n defined by Eq. (6) with the exact solutions of Eq. (2), reported in Ref. 32 for a linear variation of H_a, we can conclude that condition (5) gives us rather precise values of the resonance frequencies for all ω_n within the band determined by condition (5). The error is appreciable only for the last resonance ($\omega_n \cong \omega_L$), where the turning point is located near the film surface.

The theoretical description of spin-wave resonance in a tangential field is somewhat more complicated, since in place of Eq. (2) we have a differential equation of fourth order with variable coefficients. Analysis [37] shows that under conditions where the WKB approximation is applicable, this equation can be transformed into a system of weakly-coupled differential equations of type (3), in which the quantities $q(x)$ coincide with the two different roots of the equation

$$\left(\frac{\omega}{\gamma}\right)^2 = \left[H + Dq(x)^2\right]\left[H + Dq(x)^2 + 4\pi M(x)\right], \qquad (8)$$

which is in essence the "local" dispersion relation for exchange spin waves propagating perpendicularly to the magnetisation. In the spectral range of interest to us, these roots differ significantly in their character and magnitude. Only one of them, specifically

$$q(x)^2 = \frac{1}{D}\left\{\sqrt{\left(\frac{\omega}{\gamma}\right)^2 + \left[2\pi M(x)\right]^2} - H - 2\pi M(x)\right\}, \qquad (9)$$

can have a turning point and a segment on which $q^2 > 0$. It is this root that is of primary interest to us. The second root describes the magnetisation oscillation in time that decays rapidly in space ($q^2 < 0$, $|q| \sim 10^6$ cm^{-1}).

From Eq. (9) it is clear that the turning point, and consequently the spin-wave resonance spectrum, exists in the frequency range

$$\omega_{TL} < \omega < \omega_{T0} , \qquad (10)$$

where $\omega_{TL} = \gamma\sqrt{H(H + 4\pi M_L)}$, $\omega_{T0} = \gamma\sqrt{H(H + 4\pi M_0)}$.

In contrast to the case of a normally magnetised film, as the frequency increases the turning point shifts from the boundary with the smaller effective magnetisation ($x = L$) to that with the larger magnetisation ($x = 0$). The wave number increases with increasing x; see Fig. 5c. The phase shift is determined by

$$\varphi(\omega, H) = 2 \int\limits_L^{d(\omega, H)} |q(x, \omega, H)| dx \quad . \tag{11}$$

The resonant frequencies for a given profile $M(x)$ can be found from condition (6) provided that φ_0 is defined by spin pinning at surface $x = L$.

4. Experimental Study of Spin-Wave Resonance Spectra of Nonuniform YIG Films

This Section is concerned with experiments aimed to select the film suitable for the excitation and propagation of running ESW.

Our experimental setup (Fig. 3) consisted of a segment of 50 Ω asymmetric stripline of length $\sim$ 1 cm, fabricated on a polycor substrate of thickness 0.5 mm. The width of the stripline was 0.5 mm. The film and electrode were pressed together with a gasket between them that provided a gap of $\sim$ 0.5 mm and decreased the coupling between the antenna and the sample. The external magnetic field **H** was directed either along the normal to the film plane (a "normal" field) or within the film plane (a "tangential" field).

A microwave signal with frequency $\omega/2\pi$ was applied in the setup through a circulator. The power P of the reflected microwave signal was measured as a function either of frequency (at fixed magnetic field) or of the magnitude of the external magnetic field (at fixed frequency). The samples under investigation were YIG films grown by liquid-phase epitaxy on substrates of gadolinium-gallium garnet with orientations (111) or (100). The samples were grown for us by A. V. Maryakhin and A. S. Khe. The film thickness was normally 10-30 μm, with dimensions in the plane of order several centimetres. For uniform films with these dimensions, the width of the absorption spectrum did not exceed 15 - 30 MHz, or 5 - 10 Oe, since high-number spin-wave resonance modes are not excited in uniform films with this thickness; a broad spectrum of magnetostatic modes is not excited due to the large dimensions in the plane. Under these conditions, the width of the absorption spectrum for uniform films is determined by the region of existence (see Fig. 2) of magnetostatic waves with $k < \pi/W \sim 60$ cm^{-1}, i.e., those waves that can be excited by the wide electrode of the stripline we are

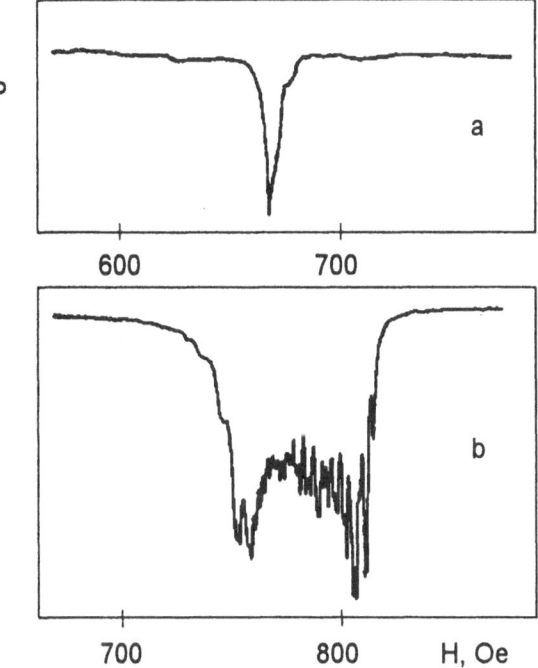

Figure 6. Dependencies of the reflected microwave power on the magnitude of the external magnetic field for two films demonstrating narrow-band (a) and wide-band (b) spectra.

using. However, among the samples we investigated we observed some films in which the absorption spectrum was much wider. Fig. 6 shows the absorption spectra for two films. While the first film demonstrates rather narrow absorption band, the second one has a wide spectrum within which one can observe a number of resonance peaks.

Among the films with the wide spectra we have selected the films whose spectra have a regular structure of the resonance peaks. A typical spectrum of this kind is shown in Fig. 7 for a tangential field. Spectra in a normal field were broader than in a tangential field, and could include up to 120 resonances. Enumerating the peaks in their order of increasing frequency, we measured the dependence of the resonance frequencies ω_n on their mode number n. The results of these measurements are shown in Fig. 8 for two samples and two orientations of the external field. In the text that follows, and in the captions for the figures, we will refer to these films as samples "1" and "2".

Using these films, we carried out some experiments and evaluations to show that the films have nonuniformity of magnetic parameters that can be described in terms of the model discussed in the preceding section. We determined the effect of the planar dimensions of the sample on the

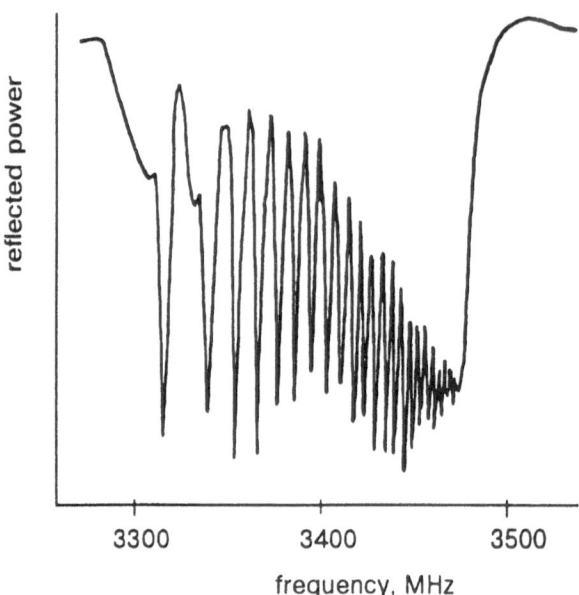

Figure 7. Frequency dependence of the microwave power reflected from the sample, which was a tangentially magnetised YIG film with orientation (111) and thickness $L = 12.5$ μm in external magnetic field $H = 563$ Oe.

absorption spectrum. Our studies showed that the form of the functions ω_n does not change as the in-plane dimensions were varied from 1 mm to several centimetres. This allows us to believe that the absorption peaks are associated with thickness resonances, and not resonances over the sample width, i.e., we are observing spin-wave resonances.

In uniform films, the distance between adjacent spin-wave resonance frequencies increases with the mode number as n^2. This clearly does not correspond to the dependencies shown in Fig. 8. Consequently, we can assume that the films are nonuniform, i.e., their parameters change along the thickness. Then we can expect that the effective excitation of high-number ESW modes is associated with the existence of the turning point. If this is the case, the absorption spectra should occupy the frequency band defined by (5) for a normal field and by (10) for a tangential field. We will introduce the effective magnetisation drop $4\pi\Delta M = 4\pi M_0 - 4\pi M_L$. From the definition of the frequencies ω_L and ω_0 [see (5)] it is clear that $4\pi\Delta M = (\omega_L - \omega_0)/\gamma$, i.e., the quantity $4\pi\Delta M$ is directly determined by the width of the spin-wave resonance spectrum in a normal field. On the other hand, the value of $4\pi\Delta M$ can be found independently from the frequencies that bound the spectrum in a tangential field, if we make use of the definitions of ω_{TL} and ω_{T0} in condition (10). When we make the corresponding estimates for

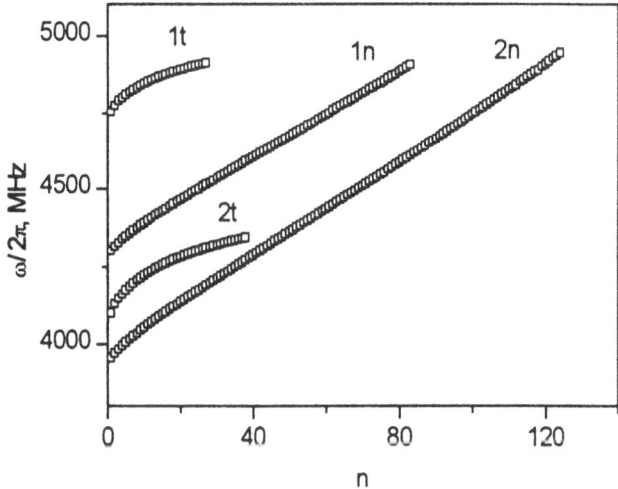

Figure 8. Dependence of the resonance frequency on mode number. (1t) - sample 1, tangential field $H = 895$ Oe; (1n) - sample 1, normal field $H = 3588$ Oe; (2t) - sample 2, tangential field $H = 808$ Oe; (2n) - sample 2, normal field $H = 3505$ Oe [the YIG films had (111) orientation, $L = 15$ μm (sample 1) and $L = 17.2$ μm (sample 2)].

the values of $4\pi\Delta M$ (starting from the spectra shown in Fig. 8), we observe that the values of the drop obtained from the tangential and normal spectra agree to good accuracy. This agreement is confirmation of the correctness of our original assumption that the observed spin-wave resonance spectra are caused by nonuniformity of the effective magnetisation over the film thickness. The value of $4\pi\Delta M$ was ~ 200 G for sample 1 and ~ 330 G for sample 2.

The next step is to examine whether the variation in effective magnetisation is monotone. Suppose such is the case. Then setting $q = 0$ in (4), we can use the experimentally obtained frequencies ω_n to compute a series of values M_n corresponding to values of the effective magnetisation at the turning points for frequencies ω_n. Now, in order to compute the profile $M(x)$, it is sufficient to find the coordinates of the turning points d_n for those values of the frequency. This can be done by assuming that the effective magnetisation varies linearly on the segments $d_n < x < d_{n+1}$ (i.e., between turning points) and also on the segment $0 < x < d_2$ adjacent to the film boundary. Then by solving Eqs. (4), (6), and (7) together, we determine from the frequency difference $(\omega_2 - \omega_1)$ the slope of the function $M(x)$ on the segment $0 < x < d_2$, and consequently the coordinates d_1 and d_2. We then substitute the value ω_3 into Eqs. (4), (6), and (7) to find d_3. Using ω_4 we then find d_4 and so on. The results of this processing of the spectrum of a normally magnetised film are shown in Fig. 9a.

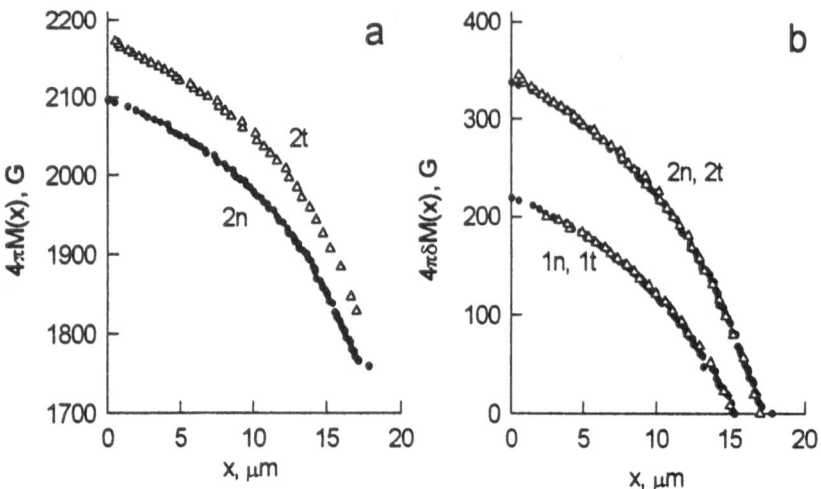

Figure 9. (a) Profile of effective magnetisation variation $4\pi M(x)$ for sample 2; (b) Profile of relative change of the effective magnetisation $4\pi\delta M(x)$ for samples 1 and 2. These data were obtained by processing the spectra shown in Fig. 8. The points correspond to data calculated from spectra of normally magnetised films, the triangles are from spectra of tangentially magnetised films.

By analogy with the case of a normally magnetised film, we can construct a profile for the variation of the effective magnetisation based on values of the resonance frequencies in a tangential field. The primary difference in procedure from the previous case is the fact that in place of (4) and (7) we make use of Eqs. (9) and (11), and the process of recovering the profile goes in the opposite direction - from the boundary $x = L$ (with the smaller magnetisation) to $x = 0$. The results of this calculation are shown in Fig. 9a. From this figure it is clear that the profiles recovered from the tangential and normal spectra are very similar. The primary difference is a disagreement in the absolute values of $4\pi M$, which are shifted by an amount $\sim$ 70 G. This shift may be explained by the effect of the cubic anisotropy field, which we have not included in the calculation. In Fig. 9b we show the relative variation of the effective magnetisation $4\pi\delta M(x) = 4\pi M(x) - 4\pi M(L)$ for samples 1 and 2. The good agreement of the profiles obtained from these spectra for various orientations of the magnetising field shows that for these two films the assumption that the variation of $M(x)$ is monotone is correct.

We note that some of our samples had extremely nonmonotone variation of $M(x)$. When their spectra were processed in accordance with the methods described above, significantly different results were obtained from the normal and tangential spectra. Apparently the profiles of these films cannot be recovered by the method of nondestructive testing. In order to solve this problem, one can use the method described in [31], i.e., analysis of spin-wave spectra measured during layer-by-layer etching.

Note that in assuming that the maximum of $4\pi M(x)$ lies at the boundary $x = 0$, we did not specify whether the point $x = 0$ is a film-air boundary or a film-substrate boundary. In order to clarify this question, it is necessary to investigate further. Let the point $x = 0$ lie at the film-substrate boundary. Then etching the film will remove a portion of the sample (near $x = L$) with low effective magnetisation. This implies that the spin-wave resonance spectrum will become narrower and the number of resonances will decrease. In a normal field, this narrowing of the spectrum occurs because the last resonances disappear. However the frequencies of the remaining resonances should remain unchanged. Our previous analysis shows why this happens. The frequency of the nth resonance is determined by the profile of variation of the magnetisation on the segment from $x = 0$ to $x = d_n$.. Therefore, as long as the thickness of the remaining film exceeds d_n, the frequency ω_n does not change. Then by etching the film we can attach the measured profiles to the proper sample boundary. These experiments showed that for our two films, whose profiles are shown in Fig. 8b, the maximum magnetisation is located at the film-air boundary i.e., the real situation is exactly the opposite what we have considered - successive etching shows that it is the spectrum in a tangential field that becomes narrower without changing the resonance frequencies of the remaining peaks.

In processing the experimental data, we made a number of assumptions. Let us discuss the question of how correct these assumption are.

First of all, we assumed that the gyromagnetic ratio γ does not change over the film thickness. From condition (5) it is clear that if the value of γ is not constant, the width of the spectrum of a normally magnetised film depends on the value of the external magnetic field. We did not observe any such dependence in our experiments.

Secondly, we assumed that the exchange constant was indeed a constant. We claim that even if D varies over the film thickness, it cannot be the sole and basic cause of the nonuniformity of the properties. The fact is that variation of D by itself does not lead to the appearance of a turning point, nor to the existence of the latter within a rather broad spectral interval. Consequently, variation of D can not explain the presence of the extended spin-wave spectrum.

Thirdly, we assumed that spins at the film surface are free, and used a value of φ_0 equal to $3\pi/2$. For completely pinned spins, the value of φ_0 would be $\cong \pi/2$ [32]. Note that, in any case, the value of φ_0 lies within the interval from 0 to 2π, and the error in determining the resonance frequencies can not exceed the distance between adjacent resonances. The experimental spectra numbered from 30 to 120 resonances; therefore, the form of the profile of *M(x)* does not change essentially, regardless of the value of φ_0 used.

Thus, we conclude that there exist YIG films that are significantly nonuniform in their magnetic parameters. The main features of spin-wave resonance spectra in such films can be explained in terms of variation of effective magnetisation through the film thickness. We will not separate the

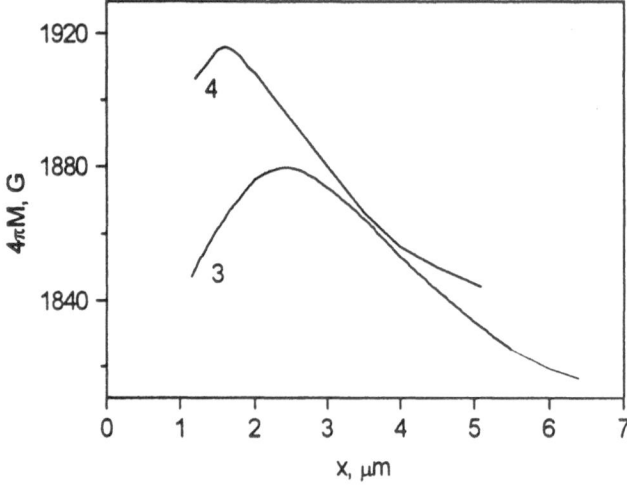

Figure 10. Profiles of effective magnetisation variation $4\pi M(x)$ for samples 3 and 4. The profiles were measured by L.V. Lutzev and Yu.M. Yakovlev with step-by-step etching the films.

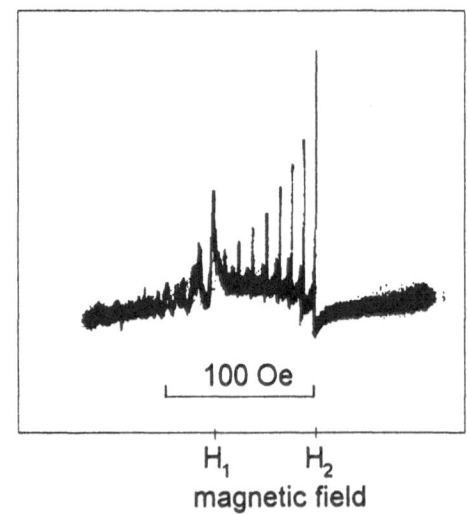

Figure 11. Absorption spectrum of 1-mm resonator made from sample 3.

contributions from the anisotropy field and from the saturation magnetisation to the change of effective magnetisation, but, it is worth noting that, as it was shown in [33], the anisotropy field in particular can be expected to vary. The reason for such variation is the change in the amount of Pb incorporated into the YIG film during the growth process.

The next step is to show that nonuniformity of magnetic parameters does not cause the dissipation to increase. Two films with nonmonotone variation of $M(x)$ (Fig. 10) were used. The films were placed at our disposal by Yu. M. Yakovlev and L. V. Lutzev who also measured the profiles of $4\pi M(x)$. We studied absorption spectra of small resonators made from these films. The films were cut to be 1-mm or 5-mm squares and were placed into the X-band cavity.

Fig. 11 shows the absorption spectrum for one of the films (sample 3). There are two regions in the spectrum. At $H < H_1$ the form of the spectrum is determined by the lateral dimensions of the sample. An increase of dimensions makes this part of the spectrum die out. In the field band $H_1 < H < H_2$ one can see a number of narrow resonances. These resonances are associated with SWR since the peak positions do not depend on the lateral dimensions. The resonances are spaced equidistantly. The dependency of the resonant frequency on mode number (Fig. 12) agrees nicely with the linear dependency predicted by Portis's theory [28] for the parabolic variation in the anisotropy field, when the vertex of parabola corresponds to the middle of the film. Note that in theory the even modes with antisymmetrical distribution of the rf magnetisation should not be excited. The deviation of

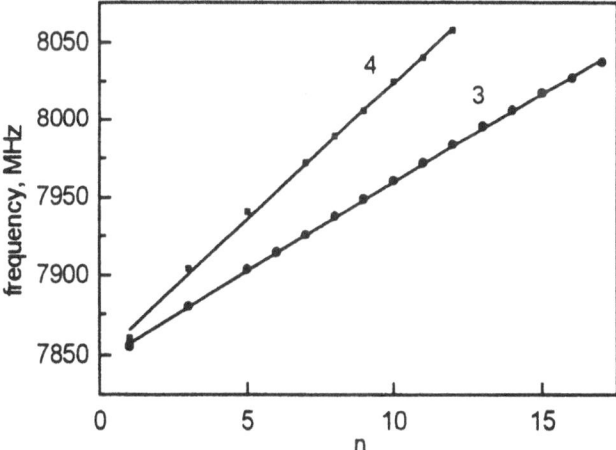

Figure 12. Dependence of the resonance frequency upon mode number. Points - experimental data. Lines - dependencies calculated in accordance with Portis's model for the parabolic approximations of the profiles of $4\pi M(x)$ shown in Fig. 10.

the real profile from the parabolic form causes the antisymmetrical modes to be slightly excited. Nevertheless, the theory by Portis adequately depicts the actual film structure. Hence, the first (and the most intensive) resonance corresponds to the mode localised in the vicinity of the maximum of $M(x)$.

An important point is that the linewidth of this resonance is found to be around 0.2 Oe. Thus, nonuniform YIG films can have low loss and it is possible to expect that exchange spin waves can propagate in such media.

5. Excitation and Propagation of Exchange Spin Wave Pulses

In the preceding section, we showed that there exist YIG films whose effective magnetisation varies smoothly with thickness over a range of several hundred gauss. As is clear from Fig. 5, spin waves propagating along the thickness of these films can have wave numbers of order 10^5 cm^{-1}, i.e., they are exchange spin waves. Experiments on spin-wave resonance show that *resonant* excitation of such waves is possible. The existence of a turning point within the film allows us to assume that the efficiency of excitation of spin-wave will remain quite high for travelling spin waves as well. In order to verify this, we carried out a number of experiments in the pulsed regime.

Our measurements setup was analogous to that described earlier; the main difference was the fact that the gap between the film and the stripline antenna did not exceed 100 μm. This increased the coupling between the film and the antenna. We used a microwave switch in order to fed microwave pulses with duration 10 -100 ns to this antenna. The pulses reflected from the film were detected, amplified, and displayed on an oscilloscope.

For $H = 0$, a microwave pulse fed to the setup was totally reflected. Within a certain range of magnetic fields (determined by (5) and (10) for normal and tangential fields, respectively) the amplitude of the reflected signal dropped sharply, and a second pulse appeared that was delayed by a time τ. For a tangential field, the value of τ depended strongly on H, as demonstrated by traces (a)-(e) in Fig. 13.

For a normal field, delayed pulses were observed over a considerably broader range of magnetic fields. The delay time for the sample under study, which was independent on the value of H within a field interval of width ~ 120 Oe, was ~ 140 ns. Traces (f) and (g) of Fig. 13, which correspond to this case, show that the shape of the delayed pulse suffers little if any dispersive distortion.

Note that the photographs shown in Fig. 13 were all taken with the same amplification coefficient over the receiver path, but with different amplitudes fed to the sample. The values of K given in the figure caption show the relative change in the power of the signal fed to the sample. Because the amplitude of the detected pulses is proportional to the microwave signal power, we can estimate the power ratios of the original and delayed signals by comparing the amplitudes of the pulses shown on the photographs.

The observation of a delayed pulse can be interpreted as evidence for excitation of exchange spin waves that then propagate along the film thickness. The excitation takes place near the turning point. The delay time is due to the time for propagation of the exchange spin waves from the turning point to the boundary of the film and back.

Let us first discuss the dependence of the time delay τ on the field H at a fixed frequency ω. It is clear from traces (b) to (e) in Fig. 13, that the delay time increases as H decreases. This result directly contradicts the data obtained in Ref. 21, where delayed exchange spin wave pulses were observed in films that were uniform (except for a near-surface layer). Furthermore,

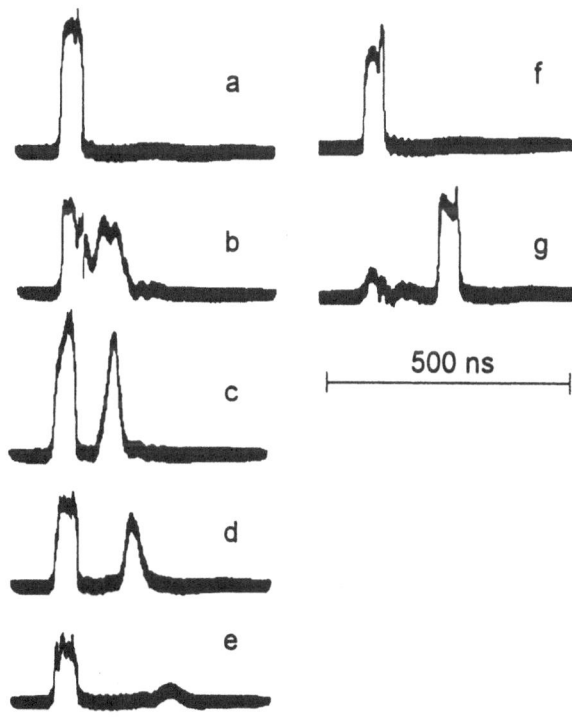

Figure 13. Time dependencies of the envelope of the reflected signal for sample 1. Traces (a)-(e) are for a tangential field with $\omega/2\pi = 5124$ MHz; (a) - $H = 0$, $K = 0$; (b) - $H = 1084$ Oe, $K = 10$ dB; (c) - $H = 1079$ Oe, $K = 10$ dB; (d) - $H = 1071$ Oe, $K = 10$ dB; (e) - $H = 1058$ Oe, $K = 10$ dB. Traces (f) and (g) are for a normal field with $\omega/2\pi = 5322$ MHz; (f) - $H = 0$, $K = 0$; (g) - $H = 3790$ Oe, $K = 7.6$ dB.

theoretical considerations based on use of the exchange spin wave dispersion law for a uniform medium imply that the exchange spin wave group velocity V_{ESW} increases as H decreases, and consequently the delay time τ must decrease with H. The increase in τ with H, observed in our experiments, is a direct consequence of the nonuniformity of the film properties. The fact is that in a nonuniform film the value of τ depends not only on V_{ESW} but also on the path length traversed by the exchange spin wave pulse within the film. On this path the condition $q^2(x) > 0$ is satisfied. Because the coordinate of the turning point changes as the frequency and magnetic field vary, the increase in group velocity can be compensated by an increase in path length, so that $\tau(H)$ will increase. The rate of displacement of the turning point as H or ω vary is determined by the profile $4\pi M(x)$. For a given profile, the time delay can be calculated from the relation

$$\tau(\omega) = \frac{\partial \varphi}{\partial \omega}, \tag{12}$$

where φ is the phase shift of the wave determined from (7) and (4) for a normal field or from (11) and (9) for the case of tangential field. On the

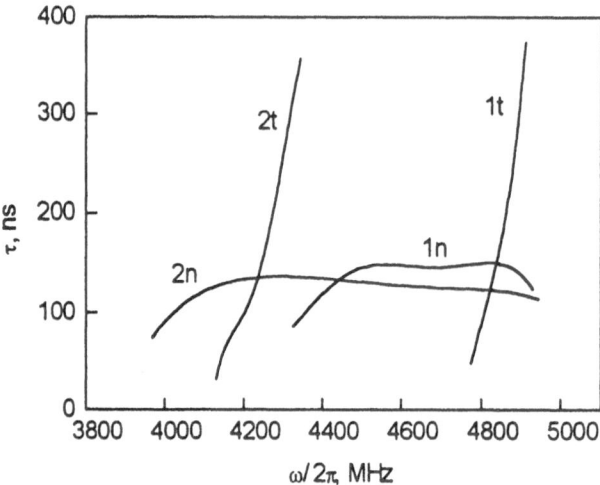

Figure 14. Dependence of time delay on frequency for samples 1 and 2, computed from the corresponding spectra shown in Fig. 8.

other hand, without recovering the profile we can estimate the delay time directly from the spin-wave resonance spectrum. From (12) it follows that

$$\tau \cong \frac{\varphi_{n+1} - \varphi_n}{\omega_{n+1} - \omega_n} = \frac{2\pi}{\omega_{n+1} - \omega_n} = \frac{1}{\Delta f}, \tag{13}$$

where Δf is the frequency difference between adjacent spin-wave resonance peaks. The functions $\tau(\omega)$ constructed in this way are shown in Fig. 14. Experiments in which the pulses were observed confirm these estimates. The only exceptions are certain data obtained in a tangential field. Thus, in Fig. 14 the segments of the curves for $\tau > 250$ ns are associated with a sharp drop in the amplitude of the delayed pulse (trace (e) of Fig. 13) and eventually its disappearance. This is probably associated with strong dispersive pulse-spreading. In normally magnetised films, the delay time depends weakly on frequency (see Fig. 14); consequently no significant distortion of the pulse shape is observed, as trace (g) in Fig. 13 shows.

Thus, the frequency dependencies of the time delay that we obtain can be explained using the model discussed above. This confirms our assertion that the delay is due to propagation of exchange spin wave pulses. Let us discuss the function $\tau(\omega)$ in more detail for various profiles of the change in effective magnetisation.

Let us choose $M(x)$ in the form of a power-law function: $4\pi M(x) = 4\pi M_0 - 4\pi \Delta M \cdot (x/L)^r$. Then for the case of a normally magnetised film, we obtain from (7) and (4):

$$\varphi(\omega) = C(\delta\omega)^{(r+2)/2r}, \quad \tau(\omega) = C\frac{r+2}{2r}(\delta\omega)^{(2-r)/2r}, \quad (14)$$

where $\delta\omega = (\omega - \omega_0)$,

$$C = \frac{2L}{\sqrt[r]{\gamma \cdot 4\pi\Delta M} \cdot \sqrt{\gamma \cdot D}} \cdot \int_0^1 \sqrt{1 - y^r}\, dy \;. \quad (15)$$

It is clear that within the frequency band (5) we can obtain frequency dependencies of various types. For example, when $r = 2/3$ the delay τ increases linearly with $\delta\omega$, when $r = 2$ the delay time is independent of frequency, and for $r > 2$ the delay decreases with increasing frequency. Our experiments confirm these conclusions. Figure 15a shows the profile $4\pi M(x)$ for three samples, and also the measured functions $\tau(\omega)$. From the figure it is clear that all three basic types of characteristics for $\tau(\omega)$ - increase, decrease, and constancy of the delay - can be realised in practice using films with various profiles. Note also that the band of frequencies in which the delayed pulse is observed can exceed 1 GHz. This is due to the large value of the drop $4\pi\Delta M$, which in one film reached 500 G. From (4) we estimate that the wave numbers of the exchange spin waves in such a sample can be around $\sim 3 \cdot 10^5$ cm^{-1}.

The function $\tau(\omega)$ can be analysed analogously for various profiles $4\pi M(x)$ when the film is placed in a tangential magnetic field. In this case, the analytic expressions for $\tau(\omega)$ obtained are quite cumbersome; however, the calculations are easily performed numerically for any specific form of

$M(x)$, by starting from Eqs. (9), (11) and (12). The results of these calculations are in good agreement with the experimental data. We will not pause to discuss them, but note only that, as follows from Fig. 14, the behaviour of $\tau(\omega)$ is a strong function of what sort of field (normal or tangential) the film is placed in. This shows that it is possible in principle to control the frequency characteristics of the time delay by choosing the direction of the bias field. Meanwhile, from the point of view of practical use of exchange spin waves, it would also be interesting to study the possibility of controlling the absolute value of τ for a nondispersive delay line (i.e., for which $\tau(\omega) = $ const).

As it was seen from Fig. 15, when one of the films we investigated was placed in a normal field, it exhibited a roughly constant value of the time delay over a frequency band with a width of order 1 GHz. Fig. 16a shows that the absolute value of τ can be varied within certain limits by applying an external magnetic field **H** at a rather small angle θ to the film normal. As we changed θ we adjusted the magnitude of the magnetic field, which was necessary in order to ensure a roughly constant value of the frequency at which the pulse delays are observed. As is clear from Fig. 16a, the increase in time delay is accompanied by a significant narrowing of the frequency band in which the delayed pulses exist. In order to explain these

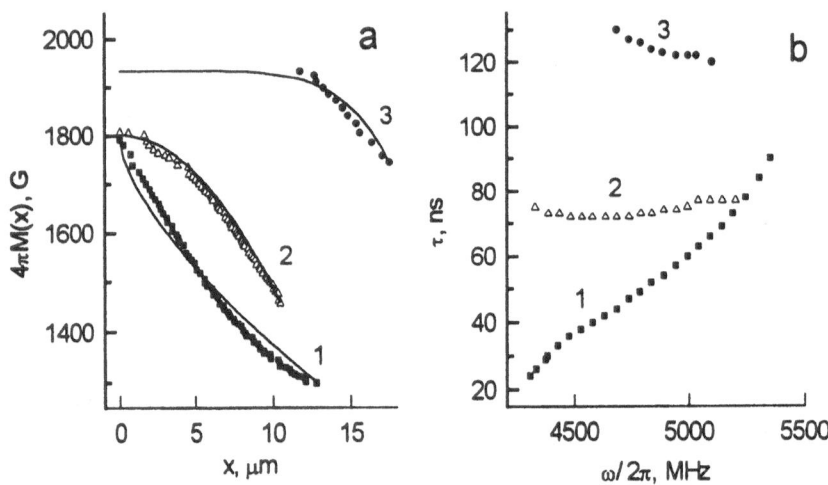

Figure 15. (a) Profiles of effective magnetisation variation for three films with orientation (100). (b) Frequency dependencies of the time delay measured on the same films. The solid curves 1-3 in Fig.(a) are attempts to approximate the experimental values using power-law functions with exponents $r = 2/3$ (1), $r = 2$ (2), and $r = 6$ (3).

experimental data qualitatively, let us discuss a simple model example.

Let an isotropic magnetic film with saturation magnetisation M_{sat} be placed in an external magnetic field **H** directed at an angle θ to the normal. Then the equilibrium position of the magnetisation makes an angle β with the normal, determined by the relation [38]

$$H \sin(\beta - \theta) = 2\pi M_{sat} \sin 2\beta \ . \tag{16}$$

The frequency for uniform resonance ω is found from

$$\left(\frac{\omega}{\gamma}\right)^2 = H_{int}\left[H_{int} + 4\pi M_{sat}(x)\sin^2 \beta\right] \ , \tag{17}$$

where H_{int} is the internal magnetic field, and $H_{int} = H \cdot \sin(\theta)/\sin(\beta)$. As a model of nonuniform film, let us consider a structure containing three layers with saturation magnetisations M_{s1}, M_{s2}, M_{s3}, such that $M_{s1} = M_{s2} - \Delta M/2$, $M_{s3} = M_{s2} + \Delta M/2$. Let us use Eq. (16) to find that dependence of external field on θ for which the frequency of uniform resonance (17) in the second layer is constant and equal, let us say, to $\omega_2/2\pi = 5$ GHz for all values of the angle θ. We then construct (Fig. 16b) the angular dependence of the frequency of uniform resonance in layers 1 and 3, using the quantity $H(\theta)$

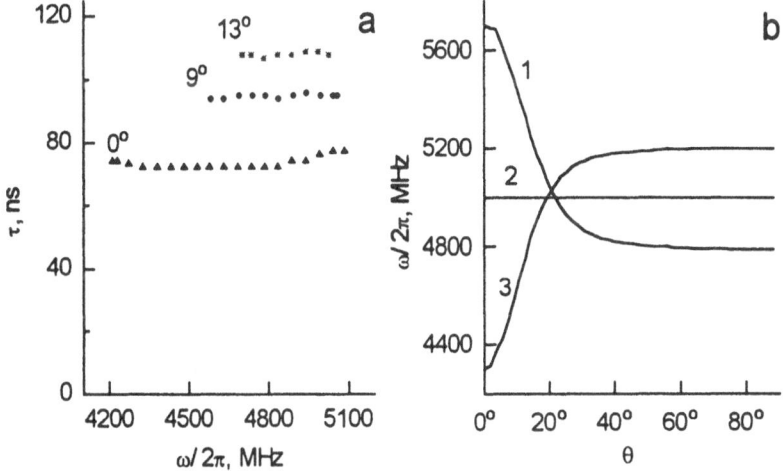

Figure 16. (a) Measured frequency dependencies of the time delay as the direction of the external magnetic field is varied; $H = 3355$ Oe for $\theta = 0^0$, $H = 3140$ Oe for $\theta = 9^0$, $H = 2995$ Oe for $\theta = 13^0$. The sample was a YIG film with orientation (100) and the magnetisation profile shown in Fig. 15a, curve 2.
(b) Computed values of the ferromagnetic resonance frequency of a three-layer structure as the angle θ varies. The numbers on the curves correspond to the layer labels.

we have found as a parameter. We will see that for small angles θ the condition $\omega_1 > \omega_2 > \omega_3$ is fulfilled, while for large θ the frequency relation is reversed: $\omega_1 < \omega_2 < \omega_3$. . There exist a range of angles for which all three frequencies are close; furthermore, in this range the resonance frequency of the second layer is the smallest: $\omega_2 < \omega_{1,3}$. This example allows us to understand qualitatively why the band of frequencies where delayed ESW pulses are observed narrows. Excitation of exchange spin waves is possible only under conditions where a turning point exists. Near this point, the condition $q = 0$ is fulfilled, hence the excitation frequency of the wave can be found by using the equation for uniform resonance in a layer with a magnetisation equal to that at the turning point. Thus, Fig. 16b shows that in oblique fields the region of excitation of exchange spin waves turns out to be small even for large effective magnetisation drops. Furthermore, it is interesting to note that the low-frequency boundary of the exchange spin wave spectrum corresponds to localisation of the turning point in the region with large magnetisation for a normal field, and one with small magnetisation for the tangential field, and in the region with intermediate magnetisation for an oblique field.

From the standpoint of ESW application it would be useful to study the possibility to shift in frequency the delayed ESW pulse. One way of getting the frequency shift is to vary the magnitude of the external magnetic field in a time while the ESW propagates. Similar effects were observed previously in experiments with spin waves in YIG rods [39-42] and MSWs propagated in YIG films [43, 44]. The reasons for the frequency shift can be explained qualitatively if we consider spin-wave pulse propagation in uniform magnetic medium. Then, each spectral component ω_l corresponds to its own specific wave number q_l . When an external magnetic field H changes to $H+\delta H$, the value of the wave number remains constant since the medium is still spatially uniform; the frequency changes so that the new frequency corresponds to the value q_l at $H+\delta H$. If the wave propagates along the direction of magnetisation, the change in frequency will be equal to $\gamma \cdot \delta H$. The frequency shift for spatially nonuniform medium is attributable to the same reasons since the variation in magnetic field does not cause the profile of nonuniformity to vary. It is worth noting that the change of magnetic field must be rather quick, since the time of field variation should be less than the time of the wave propagation.

We have conducted some experiments aimed to demonstrate the possibility to shift the frequency of the ESW pulse delayed. The film was placed in a tangential magnetic field. An additional wide stripline mounted near the film surface allowed us to obtain a quick variation in the bias field when a current pulse was fed to that stripline. The current pulse with rather long duration was adjusted so that the field variation took place between two

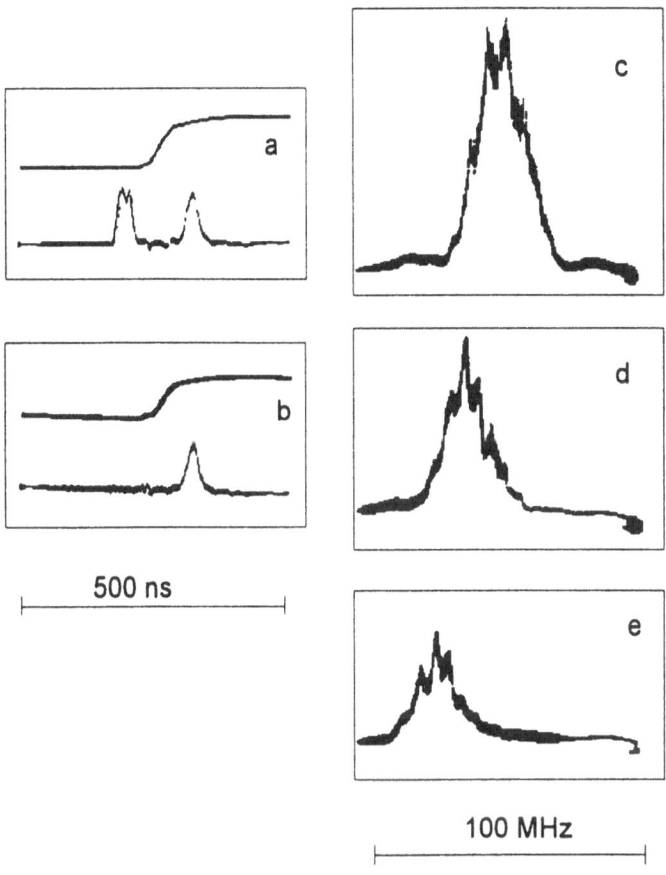

Figure 17. Upper traces in (a) and (b) show variation of magnetic field. Lower traces in (a) and (b) are envelopes of the reflected microwave signal. (c)-(e) - spectra of delayed ESW pulses: (c)- in the absence of magnetic field pulse; (d) and (e) - with consecutive increasing magnitude of the magnetic field pulse.

reflected microwave pulses (that is, during the ESW propagation) - Fig. 17a. The first reflected microwave pulse, which has no time delay, was suppressed (Fig. 17b) with the use of an additional microwave switch. The delayed pulse was fed to a spectrum analyser. Fig. 17c-e shows that the spectrum of this pulse moves with increasing the magnitude of the field pulse. Frequency translations up to 30 MHz have been achieved. Note that the main reason for the decreasing the amplitude of the shifted pulse appears to be the spatial nonuniformity of the pulse magnetic field. The value of 30 MHz is not the upper limit of the frequency shift, it can be increased by improving the magnetic system.

To summarise this section, we conclude that by using nonuniform YIG films we can bring about wide-band excitation of travelling exchange spin wave pulses. As is clear from the ratio of pulse amplitudes in Fig. 13, this excitation is quite efficient - the overall losses for excitation, propagation, and reception of the exchange spin waves are quite small, about 10 dB, for delay times of 100 to 140 ns. Note that the configuration of the rf field created by the microstrip antenna near the turning point can have a considerable influence on the excitation efficiency. A theory that would allow us to optimise the construction of the transducer does not exist at this time, and this question requires additional investigation. In this paper we note only that our experiments show that the efficiency of excitation of exchange spin waves depends both on the distance between the film and the transducer and on the antenna dimensions.

In the next section, we will discuss the main features of the interaction of spin waves with acoustic waves.

6. Excitation of Acoustic Waves by Exchange Spin Waves

It is well known that the magnetoelastic interaction can be used for the generation of acoustic waves in solids at microwave frequencies. The interaction is the most strongly pronounced when phase velocities of spin waves and acoustic waves are coincident. At the point of synchronism (see Fig. 1), where the phase velocities and hence the wave lengths of ESWs and AWs are equal, an effective transformation of magnetic oscillations energy into the energy of elastic waves can occur. Since both the spin waves and the acoustic waves in the yttrium iron garnet have rather small losses, this material is very attractive for realisation of the mentioned way of AW generation.

Note that coupling between AWs and spin waves in YIG films have been observed in previous experiments with magnetostatic waves [45-52], although there was no complete synchronism in that case. The forward and reverse conversion of spin and acoustic waves at the point of synchronism was observed in experiments with YIG rods (see, e.g., Ref. 53). Nonuniformity, whose origin was the demagnetisation field that exists at the end of a magnetised YIG rod, led to a gradual increase in the wave number of the spin waves to a value $q \sim 10^4$ cm^{-1}, which is sufficient to excite sound at frequencies of $\sim$ 500 MHz. As we noted in the previous section, in our experiments on excitation of exchange spin waves in nonuniform films we achieved values of q that are an order of magnitude higher than this. Consequently, in our case synchronism between spin and acoustic waves is possible at frequencies of the order of 10 GHz.

Let us first consider some data that suggest that the magnetoelastic interaction manifests itself in experiments on SWR. In Fig. 18 we show two spin-wave resonance spectra for the same YIG film placed in a normal field **H**. The spectra, which were taken while scanning the magnetic field, differ in their excitation frequencies ($\omega/2\pi$ = 7854 MHz for the first spectrum, and $\omega/2\pi$ = 10300 MHz for the second). For purposes of convenient comparison of these two spectra, we show the relative variation of H on this figure; the absolute values of the field H_0 corresponding to the right hand boundary of the spectrum are given in the figure caption. It is clear that the spectra are very similar, and have roughly the same width. However, within each spectrum we can identify two distinct regions. The boundary between the regions is shown by an arrow. To the right of the arrow, the dependence of the absorbed power P on magnetic field has the form of regular oscillations, while on the left the oscillation patterns are distorted. Let us denote the distance between the arrow and the right-hand boundary of the spectrum (i.e., the difference between the fields H_0 and the field corresponding to the

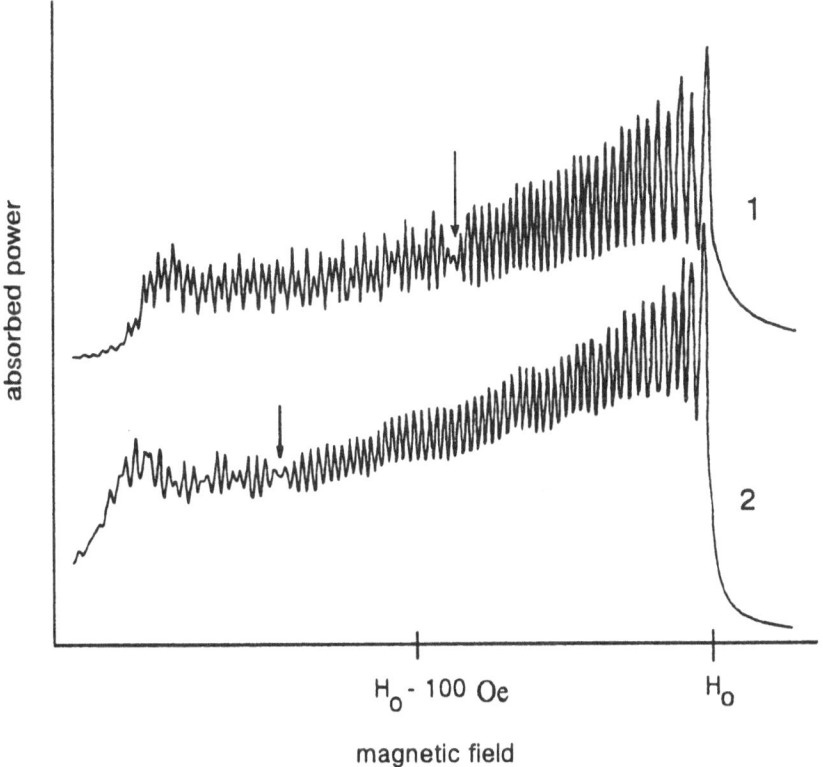

Figure 18. Dependence of the power absorbed by a YIG film (sample 1) on the magnitude of the external magnetic field for (1) $\omega/2\pi$ = 7854 MHz, H_0 = 4814 Oe; (2) $\omega/2\pi$ = 10300 MHz, H_0 = 5587 Oe.

beginning of the distortion) by ΔH. Then the basic difference between curves 1 and 2 is a change in the value of ΔH.

Recall that the spectra differ by their excitation frequency. Note also that the analysis given in Section 3 shows that for a normally magnetised film an increase in this frequency should lead only to a parallel shift of the spectrum toward higher fields. A change in the shape of the function $P(H)$ for the model under discussion is not predicted. Hence, we can assume that the distortion of the spectrum is caused by the magnetoelastic interaction, which we have not taken into account in our model. The magnetoelastic interaction should most strongly manifests itself when the spin and acoustic waves are synchronous, i.e., when their wave numbers are equal. Let us analyse the conditions under which it is possible to achieve this synchronism, and compare the results obtained with the experimental data.

The film we used in these experiments (sample 1) was the one with the effective magnetisation profile shown in Fig. 9b. Accordingly, our discussion will be based on the model described in Section 3, i.e., we will assume that $M(x)$ varies monotonically with thickness. We will include the fact that the maximum effective magnetisation is located at the film-air boundary, and assume that the elastic parameters of the film do not change with thickness. Then the wave number of the acoustic waves is coordinate-independent and determined by the wave frequency: $q_a = \omega/V_a$, where V_a is the acoustic wave velocity. The exchange spin wave spectrum, as clear from (4), lies within the range of fields

$$H_L < H < H_0 , \tag{18}$$

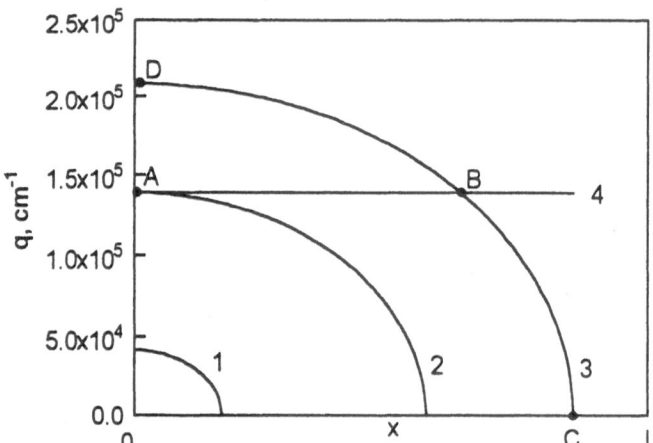

Figure 19. Variation of the wave number for exchange spin waves along the film thickness. 1 - $\delta H < \delta H^*$; 2 - $\delta H = \delta H^*$; 3 - $\delta H > \delta H^*$. Line 4 - wave number for acoustic waves.

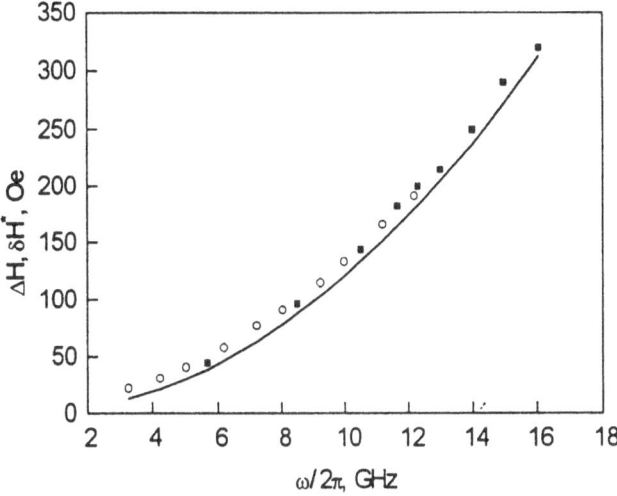

Figure 20. Computed curve $\delta H^{\bullet}(\omega)$ and experimental values of $\Delta H(\omega)$ for sample 1 (○) and sample 2 (■).

where $H_L = \omega/\gamma + 4\pi M_L$, $H_0 = \omega/\gamma + 4\pi M_0$. As the magnetic field decreases, the turning point shifts toward the boundary $x = L$, while Figure 19 shows that the magnitude of the local wave number for exchange spin waves $q(x)$ increases in the opposite direction. For a fixed magnetic field H, the maximum of $q(x)$ is reached at the boundary $x = 0$, and corresponds to

$$q(0) = \sqrt{\delta H / D} \quad , \tag{19}$$

where δH is the detuning in field from the right-hand boundary of the spectrum: $\delta H = H_0 - H$. When $q(0) = q_a$, a point of synchronism for spin and acoustic waves appears in the film (point A in Fig. 19). This will occur when $\delta H = \delta H^*$, where

$$\delta H^* = D\left(\frac{\omega}{V_a}\right)^2 \quad . \tag{20}$$

As the detuning δH increases, the point of synchronism shifts into the bulk of the film (point B).

Recall that the width of the exchange spin wave spectrum equals the drop in effective magnetisation $4\pi\Delta M$. This determines the maximum frequency ω_{max} for which we can simultaneously satisfy the conditions for the existence of a turning point and a point of synchronism in the film for a given magnetisation drop:

$$\delta H^* < 4\pi\Delta M, \quad \omega_{max} = V_a\sqrt{\frac{4\pi\Delta M}{D}} \quad . \tag{21}$$

The experimental value of ΔH corresponds to the width of that portion of exchange spin wave spectrum that is free from distortion, and .the computed value of δH^* is the width of that part of the spectrum where there is no point of synchronism between exchange spin waves and acoustic waves. Let us compare the frequency dependencies of these quantities. From Fig. 20 it is clear that they are in good agreement when $V_a = 3.85 \cdot 10^5$ cm/s, which corresponds to the velocity of transverse acoustic waves in YIG. Consequently, the distortion of the spectrum is due to the interaction of exchange spin waves with transverse acoustic waves. Note ·also that, as it is seen from Fig. 20, in a film with $4\pi\Delta M \cong 200$ G, the irregular portion of the spectrum is observed up to frequencies ~ 12 GHz, while in a film with $4\pi\Delta M \cong 340$ G it is observed up to 16 GHz. This agrees with (21).

Especially noteworthy is the fact that all the experimental values in Fig. 20 are shifted relative to the computed values in the direction of large ΔH. This shows that distortion (decreased amplitude) of the oscillations of $P(H)$ arises when the point of synchronism is located not at the boundary of the film but at a certain distance from its surface. Let us discuss this question in more detail by examining the structure of the spectral distortion. In Fig. 21,

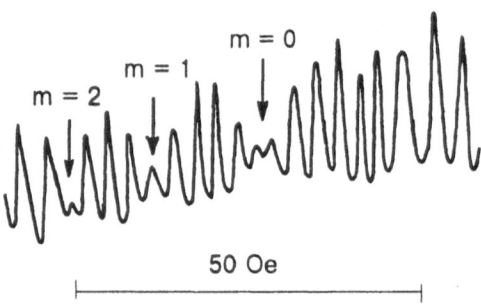

Figure 21. Fragment of absorption spectrum. Sample 2, $\omega/2\pi = 8521$ MHz.

a fragment of the absorption spectrum is shown on an expanded scale. It is clear that the distortion of the spectrum is itself regular in character. The regions where the oscillations are attenuated (shown by arrows) are spaced regularly and separated by bands of intense oscillations of $P(H)$. This behaviour of the spectrum can be explained as follows. At the point of synchronism (point B in Fig. 19) partial conversion of the exchange spin wave into an acoustic wave takes place, and also the reverse conversion of the acoustic wave into an exchange spin wave. We obtain information about processes that occur within the film from the amplitude of the signal at the turning point (point C). This amplitude is determined by the interference of two signal. One is a spin wave which propagates from the turning point to the surface (point D) and back. The second wave is a spin wave over the segment C-B and an acoustic wave over the segment B-A. At point A the acoustic wave is reflected, since $x = 0$ is the film-air boundary. At point B, the reflected wave is once again converted back into an exchange spin wave. If we omit the phase jump upon reflection, we can describe the phase shift of the first signal as $\varphi_1 = \int |q(x)dx| = 2S_{CD}$, where S_{CD} is the area under the curve CD. For the second signal $\varphi_2 = 2S_{ABC}$, where $2S_{ABC}$, is the area under the curve ABC. We may assume that the total signal amplitude decreases when the signals are opposite in phase, i.e.,

$$\Delta\varphi = \varphi_1 - \varphi_2 = \pi + 2\pi m, \quad m = 0,1,2... \quad (22)$$

When $\delta H = \delta H^*$, the phase difference equals zero, since the points A and B coincide. As δH increases, the area ABD increases, i.e., $\Delta\varphi$ increases as well. We can calculate the function $\Delta\varphi(\delta H)$ only by knowing the profile of the magnetisation change. We performed this calculation for samples 1 and 2, whose profile of saturation magnetisation was shown in Fig. 9b. In these films, we measured the magnetic fields corresponding to the portions of the signal curves where the signal amplitudes decreased. We have labelled these zones (as shown in Fig. 21) starting with the label number $m = 0$. Then, using Eq. (22), we plotted the experimental values of $\Delta\varphi$ for various

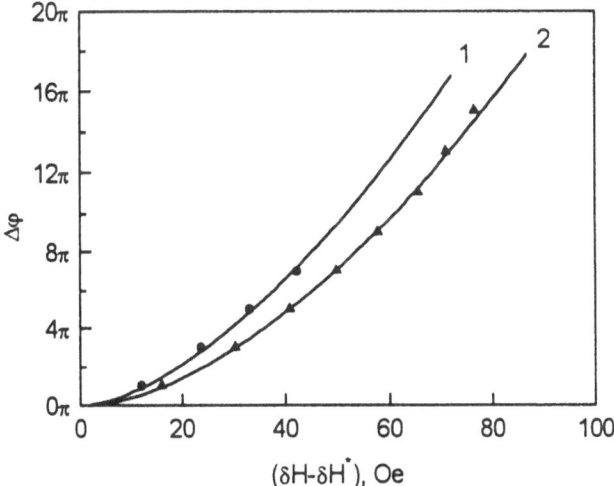

Figure 22. Dependencies of the phase difference on magnetic field. The solid curves are computed values of $\Delta\varphi(\delta H)$; the numbers on the curves correspond to sample number. Experimental values for sample 1 (•) and sample 2 (▲) are shown for $\Delta\varphi$ determined using Eq. (22).

magnetic fields and compared them with the computed function $\Delta\varphi(\delta H)$, using the first experimental point as one end of the calculated curve for $\Delta\varphi$

$= \pi$, and then plotting the relative change in measured values of the magnitude of the magnetic field (Fig. 22). For both films there is good agreement between the experimental and computed values at $\Delta\varphi = 3\pi$, $5\pi,\ldots$, i.e., when the waves are in anti-phase, which validates the correctness of our assertion regarding the interference nature of the spectrum distortions. We emphasise that in this discussion we have assumed that local conversion of exchange spin waves into acoustic waves takes place, that is, the waves interact only in a very narrow layer located near the point of synchronism. In the rest of the film volume, the exchange spin waves and acoustic waves propagate independently of one another. The good agreement between experiment and computed data shows that the use of this model is correct in the films we used, which have large gradients of $M(x)$. Thus, in these films there is no broad region where hybridised magnetoelastic waves exist. At first sight, it seem unlikely that it is possible to have efficient wave transformation in this situation. Nevertheless, the experiments show that the conversion efficiency of exchange spin waves to acoustic waves is quite high. This is revealed with particular clarity in the

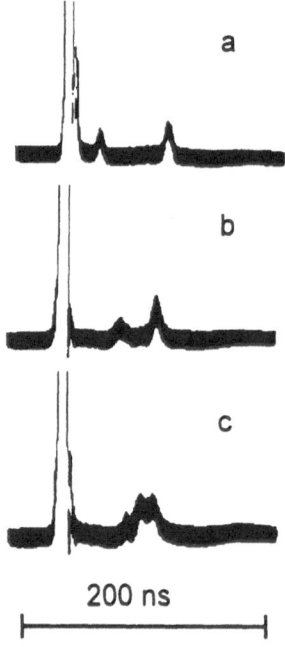

Figure 23. Time dependence of the envelope of reflected signal for sample 1, $\omega/2\pi$ = 3695 MHz, normal field; (a) - H = 2933 Oe; (b) - H = 3018 Oe; (c) - H = 3031 Oe.

pulsed regime.

Our procedure here was analogous to what we described in Section 5: we fed a microwave pulse into a normally magnetised YIG film. In Fig. 23 we show the envelope of the pulses reflected from the sample. Two delayed pulses are clearly seen, separated by a time that depends on the value of the external magnetic field. For a certain value of the field, the pulses merge; see Fig. 23c. We may assume that the presence of the additional pulse is connected with excitation of an acoustic wave. Recall that the group velocity of ESW is less than one of acoustic wave. Then the signal with the large time delay is naturally associated with the exchange spin wave propagation. The pulse with the small time delay is associated with a wave that propagates as an exchange spin wave on the segment between the turning point and the point of synchronism. and as an acoustic wave in the layer between the film surface and the point of synchronism. Knowing the effective magnetisation profile of this film (Fig. 9b), we can compute the field dependence of the phase shift of these waves $\varphi_1(H)$ and $\varphi_2(H)$, and use Eq. (12) to find the dependence of the time delay on magnetic field. Comparing the experimental data with the computed data (Fig. 24), we confirm this interpretation of the experiment. However, it is clear from Fig. 23 that the amplitudes of the delayed pulses are comparable. Consequently, a significant fraction of the energy of the exchange spin wave pulses is converted into

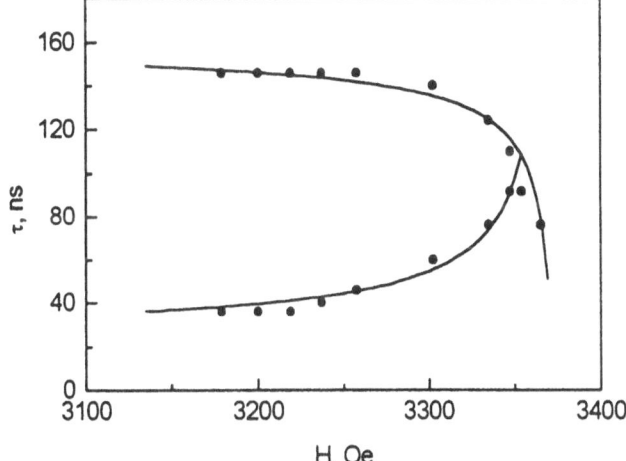

Figure 24. Dependence of the time delay on magnetic field for sample 1, $\omega/2\pi=$ 3698 MHz, normal field. The solid curves are computed values of τ, the dots are experimental values.

acoustic waves.

It should be noted that the effective reflection of acoustic waves takes place only at the film-air boundary. The surface $x = 0$ which corresponds to

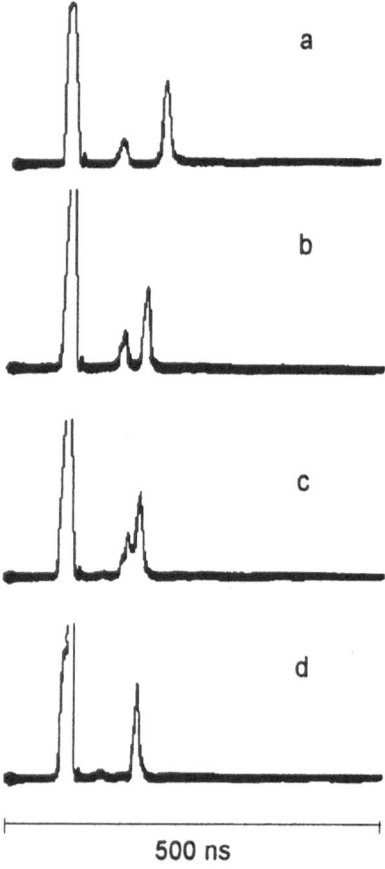

Figure 25. Time dependence of the envelope of reflected signal in a tangential field; $\omega/2\pi$ = 4996 MHz; (a) - H = 1121 Oe; (b) - H = 1171 Oe; (c)- H = 1181 Oe; (d) - H = 1190 Oe.

the maximum effective magnetisation, could also have been located at the film-substrate boundary. (This situation is realised for films with the magnetisation profiles shown in Fig. 15a). Then the acoustic wave radiates into the bulk of the substrate, and we do not observe an additional delayed pulse having time delay less than one for ESW pulse. However, such pulse appears (Fig. 25) when these films are placed in a tangential magnetic field, since the exchange spin waves and acoustic waves propagate toward the boundary with the smaller value of $4\pi M(x)$ in this case [see Fig. 5c], that is, toward the film-air boundary.

If AW penetrates to the interior of substrate, it reflects from the second boundary of the substrate and comes back to the YIG film. A typical value of the substrate thickness is 0.5-0.7 mm. A velocity of transverse sound wave in GGG equals to $3.57 \cdot 10^5$ cm/s. Thus, the acoustic wave comes back in

0.3-0.5 µs. The inverse conversion of sound into spin wave may cause the delayed pulse to appear, if the surfaces of the substrate are parallel. The time delay of such pulse should exceeds the delay of initial ESW pulse. Fig. 26 shows the reflected signal for a 30 µm thick YIG film on 700 µm thick GGG substrate. We used a heterodyne amplifier in order to observe the pulses of small amplitude. In Figures 26a-c one can see a number of delayed pulses due to several passes of the AW through the film-substrate structure. The amplitudes of the pulses may change nonmonotonically. The reasons for this is most likely to be nonperfect parallelism of the surfaces but this question invites for further investigations. Note that when we damaged the surface of the substrate to make it rough, just one delayed ESW pulse was observed (Fig. 26d).

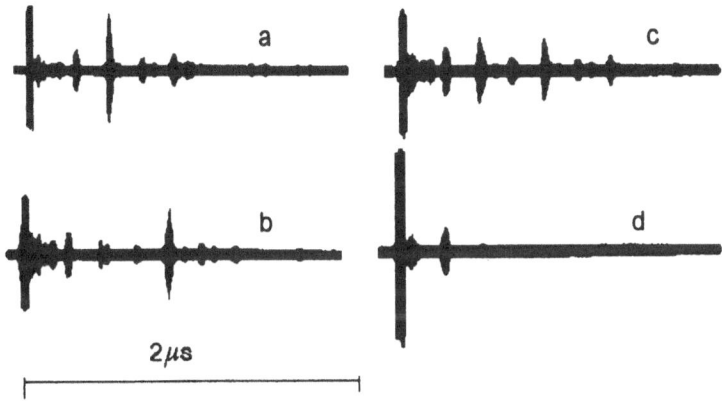

Figure 26. Time dependence of the reflected signal for a 30 µm thick YIG film on 700 µm thick GGG substrate with orientation (100) in a normal field; (a) - $H = 2960$ Oe, $\omega/2\pi = 3257$ MHz; (b) - $H = 3025$ Oe, $\omega/2\pi = 3435$ MHz; (c) - $H = 3000$ Oe, $\omega/2\pi = 3383$ MHz; (d) - $H = 2900$ Oe, $\omega/2\pi = 3227$ Mhz.

We estimated the insertion loss for the delayed pulses. It was found that the insertion loss was around 15 dB when the AW pulse reflected from the film-air surface and it increased up to 23-25 dB when the elastic wave travelled through the substrate and reflected from its boundary. Since the insertion loss includes the loss for two conversions, as well as transmission loss and reflection loss, we conclude that the loss for AW excitation is less than 10 dB. Thus, nonuniform ferrite films can be used to efficiently excite very high-frequency acoustic waves.

7. Three-Magnon Decay of Exchange Spin Wave

Up to this point we have considered linear properties of exchange spin waves. This section is concerned with high-power ESWs. We will examine decay of ESW into pairs of spin waves of half the initial frequency.

It worth noting that effects which arise because of an instability of spin oscillations in magnetic materials have attracted research interest for several reasons. One is that several processes, which are typical for various nonlinear media, can be realised by comparatively simple experimental methods. Examples of these processes are parametric wave excitation, self-modulation and formation of envelope solitons, and routes to chaos. Using of nonuniform films for the ESW excitation provides the new means of experimental investigation of nonlinear phenomena. Thus the study of ESWs may have a basic interest. We will consider three-magnon interactions which lead to a first-order instability and satisfy the conservation laws

$$\omega_1 = \omega_2 + \omega_3,$$
$$\mathbf{q}_1 = \mathbf{q}_2 + \mathbf{q}_3,$$

$$(23)$$

where ω_1 and $\mathbf{q}_1$ are the frequency and wave vector of the original wave (the pump wave), while waves 2 and 3 are parametrically excited waves. The values of $q_{2,3}$ are typically quite high, on the order of 10^4 - 10^5 cm^{-1}. Previous experiments on the first-order parametric instability have used as the pump either forced oscillations of the magnetisation or various types of magnetostatic waves and oscillations with $q_1 \leq 10^3$ cm^{-1} (including a uniform mode, $q_1 \cong 0$). There has also been a study of the parametric excitation of spin waves under parallel pumping. In all these cases, the experiments have been restricted to processes in which the condition $q_1 << q_{2,3}$ holds. Consequently, such wave properties of the pump as its wavelength and the direction of the wave vector have been unimportant. The study we are reporting here demonstrates that these characteristics turn out to be exceedingly important for the decay of an exchange spin wave.

In general, conditions (23) allow a parametric excitation of waves which may differ in both frequency and propagation direction. If the wave vectors $\mathbf{q}_1$, $\mathbf{q}_2$, and $\mathbf{q}_3$ are parallel, we say that the process is a "collinear decay". The frequencies ω_2 and ω_3 must satisfy the relations $\omega_2 = \omega_1/2 + \Delta\omega$ and $\omega_3 = \omega_1/2 - \Delta\omega$, where $\Delta\omega$ is the detuning from the half-frequency. The process with $\Delta\omega = 0$ is a "frequency-degenerate decay". Let us analyse the conditions which would be necessary for experimental realisation of the decay of an exchange spin wave. Analysing dispersion relation (1), together with conditions (23), one can show that if the initial wave is directed along the magnetisation ($\alpha = 0$) a decay is possible at frequencies [7] $\omega_1 \geq 3\gamma H_{int}$.

The limiting frequency $\omega_1 = 3\gamma H_{int}$ corresponds to the case of degenerate collinear decay, with $\omega_2 = \omega_3 = \omega_1/2$, and $q_2 = q_3 = q_1/2$. We can thus draw two conclusions. First, it can be seen from (1) that only sufficiently short waves, with $q_1^2 \geq 2\omega_1/3\gamma D$, can decay. Even at a comparatively low frequency, $\omega_1/2\pi = 1000$ MHz, the wave number q_1 must exceed $2.2 \cdot 10^5$ cm^{-1}. Second, the parametrically excited waves must also have short wavelengths.

It follows that an experimental study would require the capability to excite and receive exchange spin waves with $q \sim 10^5$ cm^{-1}, in order to (first) create the pump wave and (second) detect the decay products. This problem can be solved by working with nonuniform YIG films. To explain the

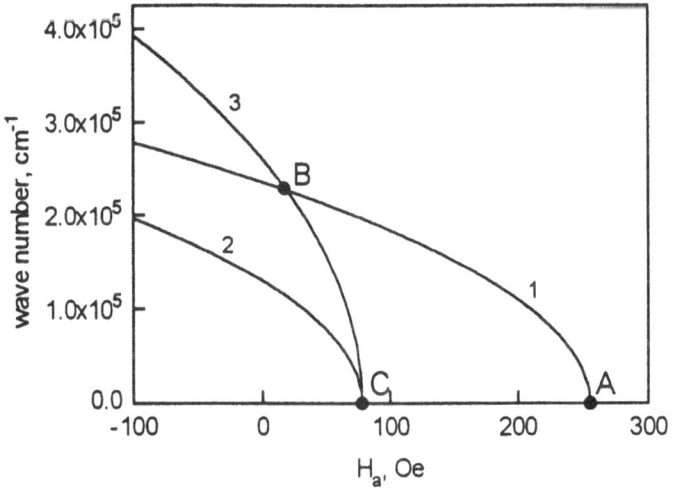

Figure 27. Wave number versus the anisotropy field for a film with $4\pi M_{sat} = 1750$ G in a normal magnetic field of $H = 1850$ Oe. (1) - $q(H_a)$ at $\omega/2\pi = 1000$ MHz; (2) - $q(H_a)$ at $\omega/2\pi = 500$ MHz; (3) - $2q(H_a)$ at $\omega/2\pi = 500$ MHz.

experiment, we assume a YIG film in an external magnetic field **H** which is directed along the normal to the surface. For the sake of simplicity assume that anisotropy field H_a depends on x, while the other parameters of the film are constants. We ignore the effect of a cubic anisotropy. The profile of the wave number along the coordinate x or the dependence of this wave number on the anisotropy field H_a can then be found from (1) substituting $\alpha = 0$ and $H_{int}(x) = H - 4\pi M_{sat} + H_a(x)$. Figure 27 shows plots of $q_1(H_a)$ at $\omega_1/2\pi = 1000$ MHz (curve 1), of $q_2(H_a)$ at $\omega_2 = \omega_1/2$ (curve 2), and $2q_2(H_a)$ (curve 3). At point A, where $q_1 \approx 0$, the pump wave is excited. The conditions for degenerate collinear decay are satisfied at point B. When the products of this decay reach point C, they have a q_2 on the order of zero, so we can expect radiation of electromagnetic waves with a frequency ω_2 from the sample. To

204

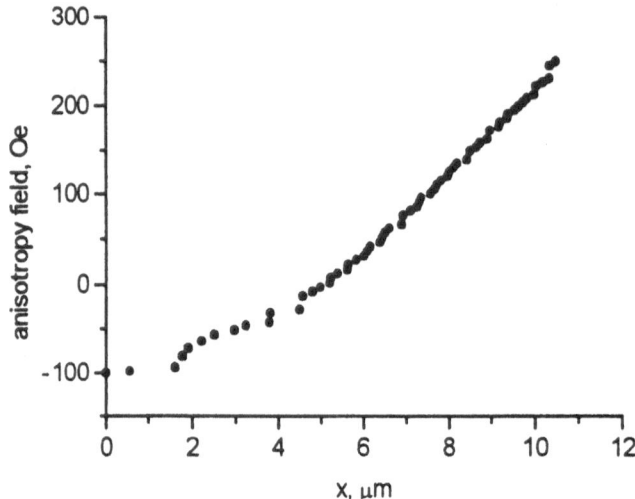

Figure 28. Profile of variation in the anisotropy field across the thickness of the film.

the left of point B, where we have $2q_2 > q_1$, there can be a parametric excitation of frequency-nondegenerate waves. The occurrence of this process can be inferred from the occurrence of emission at frequencies differing from $\omega_1/2$ by an amount $\pm \Delta\omega$.

For the experiments we used a (100) YIG film with a thickness of 11 μm. The drop of the anisotropy field over the thickness of the film was about 350 Oe. The field $H_a(x)$ varied monotonically from 250 Oe at one surface to -100 Oe in a uniform layer 2 μm thick adjacent to the second surface - Fig. 28. The spectrum of the spin-wave resonance of this test sample consists of several tens of intense absorption lines. It occupies a frequency band about 1 GHz wide. The film is pressed against a stripline 50 - 500 μm wide, to which a microwave signal with a frequency $\omega_p/2\pi = 1000$ MHz is applied. The external field is adjusted to strength such that the frequencies ω_p and $\omega_p/2$ lie within the spectrum of the spin-wave resonance. Emission from the sample is detected by the same stripline and then sent through a directional coupler to a spectrum analyser. The spectrum of this emission was studied for various strengths and directions of the external magnetic field.

It was found that, as the power applied to the sample is raised to ~ 1 mW, signals with frequencies close to $\omega_p/2$ appear in the spectrum (Fig. 29). This emission is presumably due to a parametric excitation of spin waves. Let us look at some of the experimental results.

1. Figure 29, a and b, demonstrates the possibility of a parametric excitation of spin waves with a frequency equal to precisely half the pump frequency. The emission spectrum can be extremely narrow (Fig. 29a). In

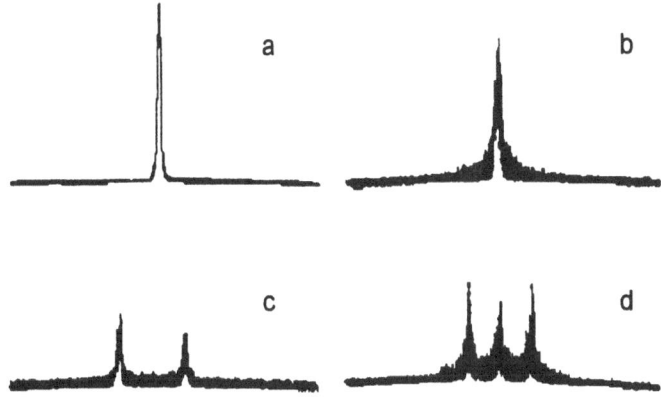

Figure 29. Typical emission spectra. The power applied to the sample was 80 mW. The various spectra correspond to various strengths and directions of the external field. For all spectra, the horizontal sweep is 100 kHz, and the central frequency is 500 MHz. The gain of the receiver for spectrum (a) is 20 dB lower than that for spectra (b)-(d).

this case the observed linewidth is determined by the resolution (1 kHz) of the spectrum analyser, and there are no significant traces of noise in the emission. The emission intensity is high; it can reach 100 μW at an input power of 80 mW.

2. The spectra in Fig. 29, c and d, show that the decay can be frequency-nondegenerate (Fig. 29c), and that it can also occur simultaneously in several channels (Fig. 29d). In this case the emission intensity is much lower than in the case of the degenerate decay. These spectra have some noise. The maximum value of the frequency detuning $\Delta\omega/2\pi$ was no more than 2-3 MHz.

3. Emission was observed only when the external magnetic field was oriented at some angle θ from the normal to the film. The interval of values $\theta_{min} \leq \theta \leq \theta_{max}$ in which the emission was detected was extremely narrow, about 1^0. Figure 30 shows values of θ_{min} and θ_{max} for various values of H. By varying θ or H slightly, one can cause a pronounced restructuring of the emission spectrum, to the point that the spectrum disappears completely. Accordingly, emission was not actually observed at every single point between θ_{min} and θ_{max} in Fig. 30.

Three-magnon processes were analysed theoretically in Ref. 7 for a spatially homogeneous medium. It is difficult to directly compare that theory with the results of the present experiments with an inhomogeneous film. Still, some of the conclusions found in [7] provide a qualitative explanation of why the emission is observed in only a narrow interval of angles θ.

First, as it was shown in Section 4, the exchange spin waves excited and received by stripline propagate along the normal to the surface of the film, since the maximum possible projection of **q** onto the plane of the film is determined by the width of the strip. This maximum possible value is 10^2 - 10^3 cm^{-1}, much smaller than the wave number of the exchange spin wave ($\sim 10^5$ cm^{-1}). It can thus be concluded that only the products of a collinear decay can be detected. Specifically, if the angle between $\mathbf{q_2}$ and $\mathbf{q_1}$ exceeds 0.01 rad, then the component of $\mathbf{q_2}$ in the plane of the film is greater than 10^3 cm^{-1}.

A deviation of **H** from the normal to the surface causes the magnetisation to rotate in a nonuniform way over thickness; i.e., the angle β, between the magnetisation and the normal, becomes a function of the coordinate. In this case the angle α, between the wave vector and magnetisation, also depends on x: $\alpha = \beta(x)$. To analyse this behaviour we first need to find $\beta(x)$. We simplify the problem, assuming that the film consists of n homogeneous layers which differ in the value of the uniaxial-anisotropy field. We ignore the exchange interaction between this layers. We can then find the angle β_i, in layer i, by numerically solving the static problem of finding the equilibrium magnetisation direction [8]. Replacing (1) by the dispersion relation for exchange spin waves in an anisotropic medium [6], we can then construct the dependence $\theta_i(H)$ for which conservation laws (23) hold in the given layer for the degenerate collinear decay of a spin wave which is propagating at an angle β_i with respect to the magnetisation. Figure 30 shows that these conditions hold over a broad

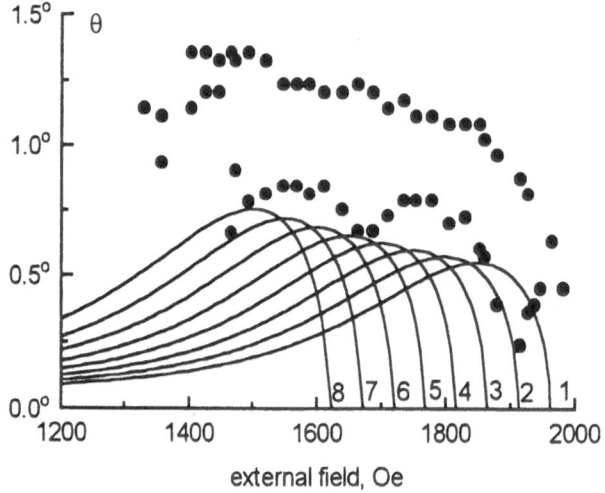

Figure 30. Points - experimental values of θ_{max} and θ_{min}; curves - theoretical values of $\theta_i(H)$ for layers with $H_a = -100 + 50(i-1)$ Oe. The curve label is the value of i.

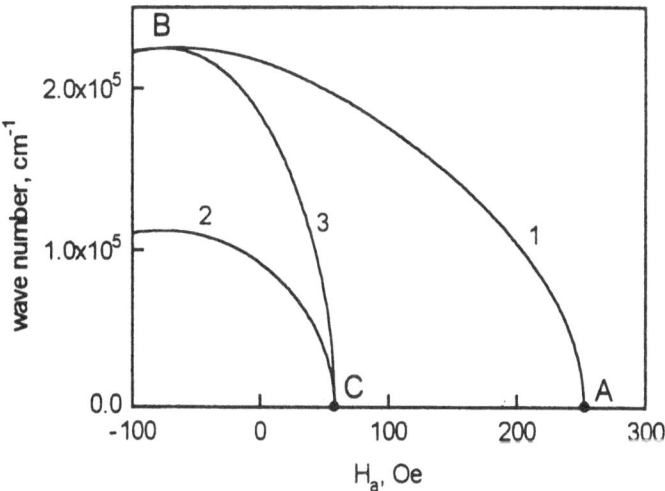

Figure 31. The same as in Fig. 27 but in an oblique magnetic field; $\theta = 0.56^0$.

range of the magnetic field H, but the maximum values of θ_i do not exceed 0.8^0. Consequently, the absence of emission at large values of θ can be explained on the basis that conservation laws (23) are violated under these conditions. We do not find a quantitative agreement here: the experimental values of θ_{max} reach 1.3^0. Nevertheless, the agreement between theory and experimental can be judged completely satisfactory, since we ignored the cubic anisotropy and the exchange interaction in solving the static problem.

It can be seen from Fig. 27 that conditions (23) also hold in the case $\theta_i = 0$. Nevertheless, an emission is not observed in this case. The apparent reason for this result is that, as follows from Ref. 7, the threshold for the collinear decay goes off to infinity if all three wave vectors are oriented parallel to the magnetisation. In a normally magnetised film, the noncollinear decay has a lower threshold, but experimentally we can not detect this process.

It can also be seen from Fig. 30 that the maximum value of the field H at which emission does arise agrees well with the calculation. At lower values of H at which the external field is much smaller than the saturation field, a theoretical analysis runs into difficulties because of the possible formation of domains.

Note that, if $\theta \neq 0$, the curves $q(H_a)$ may be nonmonotone. Then, as it is seen from Fig. 31, one can take such values of θ and H that the lines $q_1(H_a)$ and $2q_2(H_a)$ may be tangent to one another. Hence, the conditions of the frequency-degenerate collinear decay are fulfilled not just at a point, as it was when $\theta = 0$ (see Fig. 27), but in a rather wide range of H_a. Thus, such decay can take place in a spatially wide layer. At the same time, other channels of decay are forbidden. It is reasonable that the observed strong radiation at frequency $\omega = \omega_p/2$ results from this fact.

In summary, it can be concluded that the basic experimental results can be explained in the model discussed here. It follows that the emission which is observed is indeed due to a parametric excitation of waves by a short spin wave. We wish to stress that the emission spectra have several features which distinguish them from the parametric excitation of spin waves in YIG films and spheres which have been studied previously [54-58]. On the one hand, the spectrum reveals no emission at frequencies significantly different from $\omega_p/2$. On the other hand, the emission may be unusually stable and narrowband even when the pump level is substantially above the threshold. The apparent reason for the latter result is a spatial localisation of the process in the inhomogeneous medium. The propagation of the decay products out of the instability zone and the subsequent emission of these products can stabilise the amplitude of the parametrically excited wave.

Acknowledgements

We gratefully acknowledge financial support by the International Science Foundation (ISF Grant No. MSZ000 and Grant No. MSZ300), and Russian Foundation for Basic Research (Grant No. 94-02-04928-a). We are thankful to A. V. Maryakhin and A. S. Khe for providing us with the YIG films.

References

1. Wigen, P.E. (1984) Microwave properties of magnetic garnet thin films, *Thin Solid Films* **114**, 135-186.
2. Patton, C.E. (1984) Magnetic excitation in solids, *Physics Reports* **103**(5), 251-315.
3. Kottam, M.G. and Lockwood, D.J. (1986) *Light Scattering in Magnetic Solids*, John Wiley & Sons, New York.
4. Gurevich, A.G. and Melkov, G.A. (1994) *Magnetic Oscillations and Waves*, Nauka, Moscow.
5. Adam, J.D., Daniel, M.R., Emtage, P.R., and Talisa, S.H. (1991) *Magnetostatic Waves*, Academic Press.
6. Akhiezer, A.I., Baryakhtar, V.G., and Peletminskii, S.V.(1967) *Spin Waves*, Nauka, Moscow.
7. L'vov, V.S., (1987) *Nonlinear Spin Waves*, Nauka, Moscow.

8. Salanskii, N.M. and Erukhimov, M.Sh. (1975) *Physical States and Applications of Magnetic Films*, Nauka, Novosibirsk.

9. Zilberman, P.E., Temiryazev, A.G., and Tikhomirova, M.P. (1992) Dependence of the phase on the frequency of exchange spin waves in yttrium iron garnet films with magnetic properties distributed nonuniformly in depth, *Pis'ma Zh. Tekh. Fiz.* **18**(14), 79-83. [*Sov. Tekh. Phys. Lett.* **18**(7), 467-469].

10. Gulyaev, Yu.V., Zilberman, P.E., Temiryazev, A.G., and Tikhomirova, M.P. (1993) Generation of hypersound by an yttrium-iron garnet film of nonuniform thickness, *Pis'ma Zh. Tekh. Fiz.* **19**(2), 33-37. [*Sov. Tekh. Phys. Lett.* **19**(1), 50-].

11. Zilberman, P.E., Temiryazev, A.G., and Tikhomirova, M.P. (1993) Propagation of pulses of exchange spin waves in yttrium iron garnet films with magnetic properties distributed nonuniformly in thickness, *Pis'ma Zh. Tekh. Fiz.* **19**(11), 15. [*Sov. Tekh. Phys. Lett.* **19**(6), 330-332].

12. Gulyaev, Yu.V., Temiryazev, A.G., Tikhomirova, M.P., and Zilberman, P.E. (1994) Magnetoelastic interaction in yttrium iron garnet films with magnetic inhomogeneities through the film thickness, *J. Appl. Phys.* **75**(10), 5619-5621.

13. Temiryazev, A.G., Tikhomirova, M.P., and Zilberman, P.E. (1994) "Exchange" spin wave in nonuniform YIG films, *J. Appl. Phys.* **76**(9), 5586-5588.

14. Zilberman, P.E., Temiryazev, A.G., and Tikhomirova, M.P. (1995) Excitation and propagation of exchange spin waves in films of yttrium iron garnet, *Zh. Eksp. Teor. Fiz.* **108**(1), 281-302. [*Sov. Phys. JEPT* **81**(1), 151-162].

15. Temiryazev, A.G. and Tikhomirova, M.P. (1995) Three-magnon decay of exchange spin wave, *Pis'ma Zh. Eksp. Teor. Fiz.* **61**(11), 910-915. [*Sov. Phys. JEPT Lett.* **61**(11), 930-935].

16. Damon, R.W. and Eshbach, J.R. (1961) Magnetostatic modes of a ferromagnetic slab, *J. Phys. Chem. Sol.* **19**(3/4), 308-320.

17. Kittel, C. (1958) Excitation of spin waves in a ferromagnet by a uniform of field, *Phys. Rev.* **110**, 1295-1297.

18. Schilz, W. (1973) Spin-wave propagation in epitaxial YIG films, *Philips Res. Reports* **28**(1/2), 50-65.

19. Adam, J.D., O'Keeffe, T.W., and Patterson, R.W. (1979) Magnetostatic wave to exchange resonance coupling, *J. Appl. Phys.* **50**(3), 2446-2448.

20. Gulyaev, Yu.V., Bugaev, A.S., Zilberman, P.E. et al.(1979) Giant oscillations in the transmission of quasi-surface spin waves through a thin yttrium-iron garnet (YIG), *Pis'ma Zh. Eksp. Teor. Fiz.* **30**(9), 600-603. [*Sov. Phys. JEPT Lett.* **30**(9), 565-568].

210

21. Gulyaev, Yu.V., Zilberman, P.E., Sannikov, E.S., Tikhonov, V.V., and Tolkachev, A.V. (1988) Linear excitation of pulses of exchange spin waves in the films of yttrium iron garnet, *Pis'ma Zh. Tekh. Fiz.* **14**(10), 884-888. [*Sov. Tekh. Phys. Lett.* **14**, 391].

22. Tikhonov, V.V. and Tolkachev, A.V. (1994) Linear excitation of exchange spin waves in implanted YIG films, *Fiz. Tverd. Tela*, **36**(1), 185-193. [*Sov. Phys. Solid State*].

23. Zilberman, P.E. and Shishkin, V.G. (1990) Efficiency of excitation of running exchange spin waves in ferrite films with the use of microwave current, *Radiotekhnika i electronika* **35**(1), 204-206. [*Sov. Phys. Radioengineering and Electronics*].

24. Schlömann, E. (1964) Generation of spin-waves in nonuniform magnetic fields. I. Conversion of electromagnetic power into spin-wave power and vice versa, *J. Appl. Phys.* **35**, 159.

25. Schlömann, E. and Joseph, R.I. (1964) Generation of Spin Waves in Nonuniform Magnetic Fields. III Magnetoelastic Interaction, *J. Appl. Phys.* **35**, 2382; Erratum: Schlömann, E. and Joseph, R.I. (1965), *J. Appl. Phys.* **36**(3), 875-876.

26. Eshbach, J.R. (1963) Spin-wave propagation and the Magnetoelastic Interaction in Yttrium Iron Garnet, *J. Appl. Phys.* **34**(4), Part 2, 1298-1304.

27. Strauss, W. (1964) Magnetoelastic waves in yttrium iron garnet, *J. Appl. Phys.* **35**, 1022.

28. Portis, A.M. (1963) Low-lying spin wave modes in ferromagnetic films, *Appl. Phys. Lett.* **2**, 69-71.

29. Schlömann, E. (1965) Theory of spin-wave resonance in thin films, *J. Appl. Phys.* **36**, 1193-1194.

30. Wigen, P.E., Kooi, C.F., and Shanabarger, M.R. (1964) Evidence of unpinned surface spins from parallel spin-wave resonances in permalloy films, *J. Appl. Phys.* **35**, 3302-3311.

31. Wilts, G.H. and Prasad, S. (1981) Determination of magnetic profile in implanted garnets using ferromagnetic resonance, *IEEE Trans. Magn.* **MAG-17**, 2405.

32. Hoekstra, B., Van Stapele, R.P., and Robertson, J.M. (1977) Spin-wave resonance spectra of inhomogeneous bubble garnet films, *J. Appl. Phys.* **48**, 382-395.

33. Lutsev, L.V., Shcherbakova, V.O., and Fedorova, G.Ya. (1993) Magnetostatic waves, spin-wave resonance, and mechanism of formation of inhomogeneity of magnetic parameters in epitaxial garnet films with composition variation along the thickness, *Fiz. Tverd. Tela* **35**, 2208-2224. [*Sov. Phys. Solid State* **35**, 1098].

34. Buris, N.E. and Stansil, D.D. (1985) Magnetostatic surface- wave propagation in ferrite thin films with arbitrary variations of ·the magnetization through the film thickness, *IEEE Trans. on MTT.* **33**(6), 484-491.

35. Buris, N.E. and Stansil, D.D. (1985) Magnetostatic volume modes of ferrite thin films with magnetization inhomogeneities through the film thickness, *IEEE Trans. on MTT.* **33**(10), 1089-1096.

36. Gulyaev, Yu.V., Zilberman, P.E., and Temiryazev, A.G., (1995) Mechanism of effective electromagnetic excitation of exchange spin waves in nonuniform ferrite films, *Pis'ma Zh. Tekh. Fiz.* **21**(19), 27-31. [*Sov. Tekh. Phys. Lett.*].

37. Zilberman, P.E., Lugovskoi, A.V., and Sharafatdinov, A.A. (1995) Spin-wave resonance and propagation of exchange spin waves in ferrite thin films with inhomogeneity through the film thickness, *Fiz. Tverd. Tela* **37**(7), 2010-2020. [*Sov. Phys. Solid State*].

38. Mikhailovskaya, L.V. and Khlobopros, R.G. (1969) Magnetostatic spectrum of ferromagnetic film, *Fiz. Tverd. Tela* **11**(10), 2854-2857. [*Sov. Phys. Solid State* **11**, 2135]..

39. Webb, D.C. and Moore, R.A. (1966) YIG magnetostatic mode serrodyne, *Proc. of IEEE.* **54**(4), 685-686.

40. Auld, B.A., Collins, J.H., and Zapp, H.R. (1967) Spin wave-frequency conversion by adiabatic field pulsing, *Electr. Lett.* **3**(1), 35-36.

41. Resende, S.M. and Mongenthaler, F.R. (1967) Frequency conversion of spin waves in pulsed magnetic fields, *Appl. Phys. Lett.* **10**(6), 184-186.

42. Resende, S.M. and Mongenthaler, F.R. (1969) Magnetoelastic waves in time-varying magnetic fields, *J. Appl. Phys.* **10**, 524-545.

43. Preobrazhenskii, V.L. and Fetisov, Yu.K. (1988) Magnetostatic waves in nonstationary medium, *Izv. Vysch. Uchebn. Zaved. Fiz.* **31**(11), 54-66.

44. Ogrin, Yu.F. and Tarasenko, V.V. (1984) Magnetostatic converter of radio signal spectrum, *Pis'ma Zh. Tekh. Fiz.* **10**(15), 940-945. [*Sov. Tekh. Phys. Lett.*].

45. Dötsch, H., Röschmann, P., and Schilz, W. (1978) Ferrimagnetic resonance spectra of magnetic bubble films at low microwave frequencies, *Appl. Phys.* **15**, 167-173.

46. Strotmann, M., Dötsch, H., Sure, S., Löhrmann, B., and Staas, O. (1994) Coupled magnetoelastic resonances in garnet crystals studied by light scattering, *Phys. Stat. Sol. (b)* **182**, 461-469.

47. Gulyaev, Yu.V., Zilberman, P.E., Kazakov, G.T., Sysoev V.G., Tikhonov, V.V., Filimonov, Yu.A., Nam, B.P., and Khe, A.S. (1981) Observation of fast magnetoelastic waves in thin slabs and epitaxial films of yttrium iron garnet, *Pis'ma Zh. Eksp. Teor. Fiz.* **39**(9), 500-504.

48. Kazakov, G.T., Tikhonov, V.V., and Zilberman, P.E. (1983) Resonant interaction of magnetodipole and elastic waves in slabs and films of yttrium iron garnet, *Fiz. Tverd. Tela* **25**(2), 2307-2312.

49. Zilberman, P.E. (1989) Magnetoelastic waves in ferromagnetic plates and films, in M. Borissov, L. Spassov, Z. Georgiev, and I. Avramov (eds.), *II International symposium on surface waves in solids and layered structures and IV International scientific technical conference acoustoelectronics '89*, World Scientific, Singapore, pp. 74-88.

50. Ishak, W.S. (1983) Magnetostatic surface wave devices for UHF and L band applications, *IEEE Trans. on Magn.* **19**(5), 1880-1882.

51. Andreev, A.S., Gulyaev, Yu.V., Zilberman, P.E., Kravchenko, V.B., Ogrin, Yu.F., Temiryazev, A.G., and Filimonova, L.M. (1985) Magnetoelastic effects in tangentially magnetized yttrium iron garnet films., *Radiotekhnika i electronika* **30**(10), 1992-1998. [*Sov. Phys. Radioengineering and Electronics*].

52. Gorskii, V.B. and Pomyalov, A.V. (1989) Excitation of hypersound waves by magnetostatic oscillations in the structure: YIG film on GGG substrate, *Pis'ma Zh. Tekh. Fiz.* **15**(7), 61-64. [*Sov. Tekh. Phys. Lett.*].

53. LeCroy, R. and Comstock, R. (1965) Magnetoelastic interaction in ferromagnetic dielectric, in E. Mason (ed.), *Physical Acoustics. Lattice Dynamics*, Academic Press, New York and London.

54. Melkov, G.A. and Sholom, S.V. (1989) Parametric excitation of spin waves by surface magnetostatic wave, *Zh. Eksp. Teor. Fiz.* **96**(2), 712-719. [*Sov. Phys. JEPT* **69**, 403].

55. Mednikov, A.M. (1981) Nonlinear phenomena under propagation of surface magnetostatic waves in YIG film, *Fiz. Tverd. Tela* **23**(1), 242-245. [*Sov. Phys. Solid State* **23**, 136].

56. Temiryazev, A.G. (1987) Conversion of the frequency of a surface magnetostatic wave under three-magnon decay conditions *Fiz. Tverd. Tela* **29**(2), 313-319. [*Sov. Phys. Solid State* **29**(2), 179-182].

57. Zilberman, P.E., Nikitov, S.A., and Temiryazev, A.G. Four-magnon decay and the kinetic instability of a magnetostatic travelling wave in yttrium garnet ferrite films. *Pis'ma Zh. Eksp. Teor. Fiz.* **42**(3), 92-94. [*Sov. Phys. JETP Lett.* **42**(3), 110-113].

58. Krutsenko, I.V., L'vov, V.S., and Melkov, G.A. (1978) Spectral density of parametrically excited waves, *Zh. Eksp. Teor. Fiz.* **75**, 1114-1131. [*Sov. Phys. JEPT* **48**, 561].

PARAMETRIC INSTABILITY OF SPIN WAVES IN FERROMAGNETS UNDER A SPATIALLY LOCALIZED LONGITUDINAL MAGNETIC PUMP FIELD

YU.V. GULYAEV, A.V. LUGOVSKOI and P.E. ZILBERMAN.

Institute of Radioengineering & Electronics,

Russian Academy of Sciences,

Vvedensky sq. 1, Fryazino, Moscow region,

141120, Russia.

Abstract

The equation of motion for slowly varying amplitudes of spin waves parametrically excited by means of a spatially localized longitudinal magnetic pump field was derived. The initial points of the derivation were the precession equation for the magnetization vector with the relaxation term in the form of Landau and Lifshitz and the magnetostatic equations. A solution of the derived equations satisfying certain given initial and boundary conditions was obtained. Starting from some fixed level of the pump, called below as a "regeneration" threshold, the dissipative loss of the waves decreases with the pump amplitude increasing. The loss becomes equal to zero and changes its sign at some other fixed value of the pump amplitude - at the instability threshold. It was found that the two mentioned thresholds depend strongly on the pump action zone linear dimension L. Namely, both the thresholds increase, if L decreases. Because of the presence of parametric regeneration the attenuation length for excited waves increases infinitely near the instability threshold. Therefore the influence of the boundaries manifests itself in any inner point of the zone, whatever large L may be. As the result all the picture of the parametric instability depends significantly on the existence of the zone boundaries: 1) the wave amplitude spatial distributions become nonuniform everywhere inside the zone, 2) the rate of the wave amplitude growth in

R. Marcelli and S.A. Nikitov (eds.), Nonlinear Microwave Signal Processing: Towards a New Range of Devices, 213–251.
© 1996 *Kluwer Academic Publishers.*

any point of the zone becomes slower after wave distortions from the boundaries have reached this point, 3) the instability threshold becomes independent on the value of deviation of the frequency of excited waves from the parametric resonance frequency. Form of the 'butterfly" curve is found to be strongly dependent on L , this result being in agreement with experimental data available.

1. Introduction

The phenomenon of spin wave parametric instability, discovered and initially investigated in ferromagnets in 50th and 60th [1-8], is also the object of investigations in our days. As it was noted in the comparatively recent review article [9], the two stages in the development of these investigations may be outlined. At the first stage the boundaries of a specimen or a pump action zone were not taken into account in the theoretical treatment of the phenomenon [4,8]. This assumption was based on the fact, probably seemed to be obvious, that in inner points of a sufficiently large and uniform in properties system the amplitudes of parametrically excited waves should not depend on spatial coordinates. We shall refer to such approach as to the 'uniform model". The theoretical results obtained in the frames of the uniform model, were in good agreement with then experiments.

The second stage of the investigations, according to [9], was stimulated by the technological developments and by the appearance of high quality spherical yttrium-iron garnet (YIG) specimens. Fine structure of subsidiary absorption spectra had been observed for the first time in the paper [10]. This structure appeared due to the influence of the ferrite sphere boundaries - reflections of parametrically excited spin waves from the boundaries and a resonant standing wave series formation in the specimen. The influence of boundaries on parametric absorption spectra was also observed in experiments with antiferromagnets [11]. Analogous effect in ferrite films was investigated experimentally and theoretically in a series of works [12-14].

Having taken into account recent works we can, following the paper [9], continue the classification of parametric instability investigations and point out the beginning of a new (the third) stage. This stage is stimulated by a progress in the field of monocrystalline YIG films growing by means of a liquid phase epitaxy method. Due to this progress, starting from 70th, it became possible to investigate experimentally travelling spin waves in YIG films. These waves may be excited artificially by means of adjacent localized transducers, for example, by conductors with microwave current or dielectric resonators. Typical experimental situation is shown on the fig.1. Specific peculiarities of the situation are the following: 1) the exciting microwave field is localized near the transducer in a spatial region with a characteristic linear dimension of the order of L and 2) the excited spin waves propagate and may leave this region of localization. If the amplitude of the exciting magnetic field is large enough, the parametric instability may occur inside the region of localization. So we face here a new situation, where a dimension of a specimen along some direction may be unlimitedly large, but a dimension of a pump action zone along the same direction is limited.[1]

Parametric instability in such a new situation was investigated experimentally in the papers [19,20]. The instability threshold at different values of L was for the first time investigated in details in the paper [19]. Open rectangular dielectric resonators were chosen as transducers. The resonators were pasted with the film in the center of

---

[1] The following comments should be done here. A conductor with microwave current, adjacent to a film surface, may, in principle, radiate sufficiently long dipole or dipole-exchange spin waves in a linear regime at the current frequency ω_p. These waves propagate in the film along its surface and may go far from the transducer and then may play the role of a pump themself, initiating the parametric instability at the frequencies in the vicinity of the value $\omega_p/2$. Such processes have been investigated both experimentally [15] and theoretically [16-18]. But speaking here about the parametric instability in a localized pump field, we shall mean the instability in the zone under or near the transducer only. In many cases only such an instability is possible. For example, that is the case for a magnetic pump field h_p parallel to a static bias field H_0 (longitudinal pump) or for values of ω_p significantly higher than the limiting frequencies of the film dipole spectrum.

its surface. The dimensions of the resonators and modes excited may be changed to obtain various localization length L values and various magnetic pump field vector $\mathbf{h}_p$ directions. An external magnetic bias field $\mathbf{H}_0$ was applied normally to the film or in

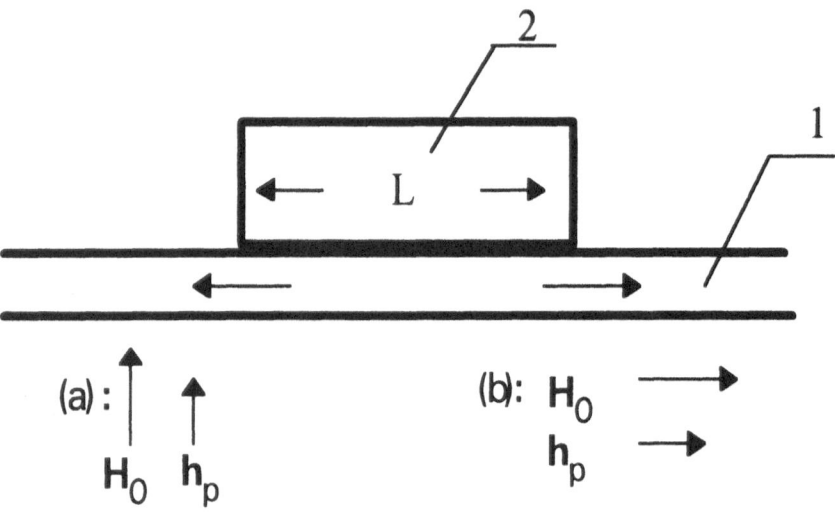

Figure 1. Typical experimental situation on travelling spin waves excitation by a localized pump: 1 - YIG film, 2 - a localized spin wave transducer; L - an approximate dimension of a pump localization zone. The arrows inside the film 1 show directions of excited spin wave propagation. At the bottom of the figure the two types of orientations are shown for a bias magnetic field $\mathbf{H}_0$ and a longitudinal pump magnetic field $\mathbf{h}_p$ relative to the pump localization zone:
(a) - transversally localized pump,
(b) - longitudinally localized pump.

the plane of the film. In all cases two vectors $\mathbf{h}_p$ and $\mathbf{H}_0$ were provided to be parallel. At the normal orientation of the vectors we have the near transversal type of the pump localization, as it is shown in the fig.1(a). The instability threshold was determined from the specific deformation of the microwave pulse reflected from the resonator. The dependencies of the threshold on $\mathbf{H}_0$ were measured. It was shown, that the dependencies are significantly different for different values of L. Typically, the threshold increases if L decreases. The following explanation of the fact was suggested. Parametrically excited spin wave propagates inside the specimen and therefore may carry its energy away from the pump action zone. Such a process manifests itself as an effective loss energy mechanism additional to the ordinary dissipation processes and results, obviously, in increasing of the instability threshold.

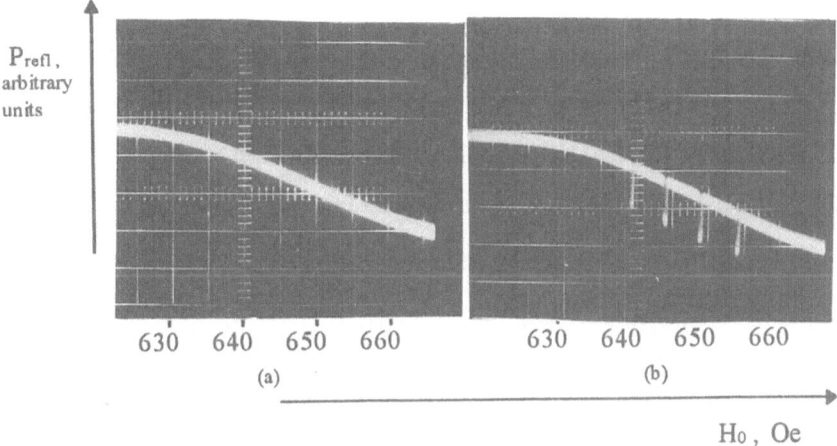

Figure 2. Experimentally obtained in [20] dependencies of a reflected from the transducer power P_{refl} on magnetic bias field H_0 under various levels of an incident power P_{inc}:
(a) - $P_{inc} = 69\mu W$,
(b) - $P_{inc} = 79\mu W$.
YIG film [111] with thickness $30\mu m$; pump frequency $\omega_p = 2\pi \cdot 5090 \cdot 10^6 \, s^{-1}$.

In the second experimental work [20] the parametrical excitation of spin waves was realized in accordance with the fig.1(b) configuration. Consequently we have here an example of a near longitudinal type of pump localization. Metallic stripes with microwave current were used as transducers. Width of the stripes was of the order of 100 μm. Dependency of the microwave power reflected from the stripes, P_{refl}, on a bias field H_0 value was investigated in the regime of a continuous signal at a fixed frequency ω_p and various levels of the incident microwave power, P_{inc} . Typical experimental dependencies are shown at the fig.2(a,b). The smooth curve at the fig.2(a) transforms to the curve with drops at the fig.2(b) if the power P_{inc} increases . Appearance of the drops was considered in [20] as an evidence that the parametric instability threshold is achieved. It is significant, that initially the threshold was achieved for some discrete values of the field H_0 only, say for $H_0 = H_{01}, H_{02}, H_{03}, \ldots$. It was shown, that for a fixed wave frequency, equal to $\omega_p /2$, these fields correspond exactly to those points at dispersion curves, where components of group velocities in plane are equal to zero. It means, that initially only those waves may be parametrically excited, which can not propagate along the surface and therefore can not carry the energy away from the pump action zone. As it was shown in the paper [20], the width of the drops increases with power P_{inc} increasing. So we may conclude, that other waves became unstable too, but the threshold for these waves became higher.

Theoretical paper [21] had been published before the performing the experiments [19,20]. In this paper the threshold of the parametric instability in a localized pump field was calculated. The authors saw their main task in the derivation of master equations for wave amplitudes describing spectra of a weak turbulence in nonuniform media. The parametric instability threshold calculation was made in the paper for the illustration of the equations usage. A stationary state solution of the equations was obtained and the threshold was found from the condition that this solution grows to infinity. An explicit expression for the threshold was obtained in limiting cases only - a strong and weak localization. It was shown, that the threshold increases if the localization degree increases (that is L decreases). This result was explained for the first time in [21], as a consequence of carrying the wave energy away from the pump action zone.

Nevertheless, many principal questions of the theory have remained undiscussed up to now, for example : 1) how amplitudes of parametrically excited spin waves depend on spatial coordinates and time, 2) how these amplitudes behave in the limit $L \rightarrow \infty$, 3) how an instability threshold depends on the value of deviation of a frequency of spin waves from the parametric resonance frequency $\omega_p /2$ and 4) how the dependency of minimum instability threshold on bias field H_0 ('butterfly" curve) looks like at finite values of L. The task of this work is to answer the indicated questions..

Moreover, there exists a very general problem we want to discuss here. This problem had been discussed for the first time in the theory of plasma wave parametric instability (see, for example [22]). However , a principal clarification of the problem, as it seems, have not given up to now. As it may be seen from the theory [22], the threshold not simply depends on L, but also in some cases does not go to its limiting value for uniform model if L grows to infinity.

What is the physical reason for such a behaviour? We try to show in this paper, that this behaviour is the direct consequence of the parametric instability. In an unstable system any perturbations , inserted by the boundaries, cannot be localized near these boundaries, but effectively penetrate to all the inner points of the system. This idea may explain the fact that the boundaries influence all the picture of parametric instability (not only the threshold value!) whatever large the length L may be.

2. System Investigated, Initial Equations and Assumptions

A typical experimental situation (see fig.1) is somewhat idealized below. We consider a bulk infinite ferromagnet instead of an infinite ferromagnetic film. It should be expected that such a model may reasonably correspond to the experimental situation only if the boundary conditions at the film surface provide either complete absorption or diffusive scattering for spin waves.

Let us suppose, beside of this, that our ferromagnet is in an uniform saturating bias field H_0 and in a pump field $h_p(r, t)$ depending on time t and radius-vector r. Let the pump be localized in a region of a layer with boundaries which are perpendicular to some vector L ("localization vector" with $|L|$ =L).

A mutual orientation of the vectors H_0 , h_p and L has a significant meaning. The following two orientations were realized in the experiments [19,20]: 1) $H_0||h_p \perp L$ and 2) $H_0||h_p,||L$. Therefore the pump was always longitudinal, but the localization might be either transversal or longitudinal. For the sake of definiteness we consider in details below the case of the transversal localization only. The main peculiarities of the longitudinal localization will be discussed in the end of the paper.

Going now to the equations of motion, let us note, that, in essence, we deal here with the description of spin wave turbulence in an nonuniform medium. A general derivation of corresponding equations of motion, starting from the system hamiltonian, was performed in the articles [23,24]. According to these articles, dynamic equations for statistical average wave amplitudes are valid if the nonuniformity is smooth enough and wavelengths for excited waves are short enough[2] . The equations for average amplitudes also may be derived from the standard Landau and Lifshitz equations. Therefore these latter equations may be and will be taken as the starting point for the further calculations.

For simplicity, let us neglect the influence of ferromagnets anisotropy and represent vectors of the total magnetic field $\mathbf{H}(\mathbf{r}, t)$ and the magnetization $\mathbf{M}(\mathbf{r}, t)$ as the sums

$$\mathbf{H}(\mathbf{r},t) = \mathbf{H}_0 + \mathbf{h}_p(\mathbf{r},t) + \delta\mathbf{H}(\mathbf{r},t) \quad \text{and} \quad \mathbf{M}(\mathbf{r},t) = \mathbf{M}_0 + \delta\mathbf{M}(\mathbf{r},t) , \qquad (1)$$

where we have the static parts $\mathbf{H}_0 = -H_0 \cdot \mathbf{e}_z$ and $\mathbf{M} = -M_0 \cdot \mathbf{e}_z$ with $H_0 > 0$, $M_0 > 0$, - unit vector for z-axis and alternating parts $\delta\mathbf{H}(\mathbf{r}, t)$, $\delta\mathbf{M}(\mathbf{r}, t)$ produced by a pump action. Let us suppose, that deviations from the static (equilibrium) quantities are sufficiently small and the pump is small also, so that the following conditions are valid

$$|\delta\mathbf{H}|, |\delta\mathbf{M}| \ll |\mathbf{h}_p| \ll M_0 . \qquad (2)$$

These conditions allow us to write a linearized system of the equations of motion (precession and magnetostatics) in the form

$$[\nabla, \delta\mathbf{H}] = 0, \quad \nabla(\delta\mathbf{H} + 4\pi\delta\mathbf{M}) = 0,$$

$$\frac{1}{\gamma}\frac{\partial\delta\mathbf{M}}{\partial t} + [\mathbf{M}_0, \delta\mathbf{H}_{\text{eff}}] + [\delta\mathbf{M}, (\mathbf{H}_0 + \mathbf{h}_p)] = 0,$$

---

[2] The formal condition of validity coincides with (7).

$$\delta H_{eff} = \delta H + \alpha \Delta \delta M - \frac{\omega_d}{(\gamma M_0)^2} \frac{\partial \delta M}{\partial t}, \qquad (3)$$

where $\gamma > 0$ - gyromagnetic ratio, α - nonuniform exchange constant and ω_d - relaxation frequency [25]. The latter quantity is considered here to be small enough,

$$\varepsilon_d \equiv \frac{\omega_d}{\gamma M_0} \ll 1. \qquad (4)$$

The pump field, varying in time with the frequency ω_d, is written in the form

$$h_p(r,t) = \frac{1}{2} \left\{ h(r) e^{-i\omega_p t} + h^*(r) e^{i\omega_p t} \right\} e_z. \qquad (5)$$

Solution of the system (3) is found in the form of Bloch functions

$$\delta H(r,t) = e^{i(qr-\omega t)} \sum_{n=0}^{\infty} \delta H_n(r,t) e^{-in\omega_p t},$$

$$\delta M(r,t) = e^{i(qr-\omega t)} \sum_{n=0}^{\infty} \delta M_n(r,t) e^{-in\omega_p t}, \qquad (6)$$

where the frequency ω and the wave vector q are connected with each other by the dispersion equation for spin waves. Let us suppose that the pump amplitude $h(r)$ may vary in space sufficiently smoothly, namely

$$\varepsilon_L \equiv \frac{|\nabla h|}{q|h|} \ll 1. \qquad (7)$$

The interaction parameter is also small because of (2), that is

$$\varepsilon_h \equiv \frac{|h|}{M_0} \ll 1. \tag{8}$$

In that case the complex harmonic amplitudes δM $(\mathbf{r}, t)$ and δH $(\mathbf{r}, t)$ may be sufficiently slow varying functions of $\mathbf{r}$ and t. Without loss of generality we may suppose, that the complex pump amplitude $h(y)$ depends only on coordinate y. According to our definition of the vector $\mathbf{L}$, it means that $\mathbf{L} \parallel \mathbf{e}_y$. Vector $\mathbf{q}$ lays in the plane (y, z), that is $\mathbf{q} = q_y \mathbf{e}_y + q_z \mathbf{e}_z$, and $q_y / q_z = \mathrm{tg}\theta$. Then, if we substitute the expressions (5) and (6) into (3) and put coefficients at the exponents equal to zero, we obtain the following system of equations for δM and δH:

$$a_n \delta M_{nx} + b \delta M_{ny} - \delta H_{ny} = \frac{1}{\gamma M_0} \frac{\partial \delta M_{nx}}{\partial t} + 2i\alpha q_y \frac{\partial \delta M_{ny}}{\partial y} + i\varepsilon_n \delta M_{ny} +$$

$$\frac{h(y)}{2M_0} \delta M_{n-1,y} + \frac{h*(y)}{2M_0} \delta M_{n+1,y} \, ,$$

$$b \delta M_{nx} - a_n \delta M_{ny} = \frac{1}{\gamma M_0} \frac{\partial \delta M_{ny}}{\partial t} + 2i\alpha q_y \frac{\partial \delta M_{nx}}{\partial y} + i\varepsilon_n \delta M_{nx} +$$

$$+ \frac{h(y)}{2M_0} \delta M_{n-1,x} + \frac{h*(y)}{2M_0} \delta M_{n+1,x} \, , \tag{9}$$

$$iq_y \delta H_{nz} - iq_z \delta H_{ny} = \frac{\partial \delta H_{nz}}{\partial y} \, , \delta M_{nz} = 0 \, , \delta H_{nx} = 0 \, ,$$

$$iq_y \left(\delta H_{ny} + 4\pi \delta M_{ny} \right) + iq_z \delta H_{nz} = -\frac{\partial \left(\delta H_{ny} + 4\pi \delta M_{ny} \right)}{\partial y} \, ,$$

where the following notations are introduced

$$a_n = \frac{i\left(\omega + n\omega_p \right)}{\gamma M_0}, \quad b = \frac{H_0}{M_0} + \alpha q^2, \quad \varepsilon_{dn} = \frac{\omega_d \left(\omega + n\omega_d \right)}{\left(\gamma M_0 \right)^2} \ll 1. \tag{10}$$

As it may be seen, harmonic amplitudes having indexes (n+1) and (n-1) enter the system of equations (9), obtained for a harmonic with index n. It shows, that, in principle, all the harmonics are dependent on each other. But the dependence is weak due to the condition (8).

The terms of the first order with respect to small parameters ε_h (8), ε_L (7) and ε_d (4) are collected in the right hand side of the equations (9). Neglecting this terms in zero order with respect to our small parameters, we obtain a series of noninteracting harmonics. Let us choose the parameter q at a given ω from the condition that the determinant of equations (9) should be equal to zero at number n=0. It means that this harmonic will be considered from the very beginning as the eigenfunction, quantities q and ω being connected through the dispersion relation for this wave

$$\omega^2 = \left(\omega_H + \tilde{\alpha}q^2 \cdot \omega_m\right)\left(\omega_H + \tilde{\alpha}q^2 \cdot \omega_m + \omega_m \sin^2\theta\right) \quad , \tag{11}$$

where $\omega_H = \gamma H_0$, $\omega_M = 4\pi\gamma M_0$ and $\tilde{\alpha} = \alpha/4\pi$. Let us now choose the frequency ω from the condition that another determinant for a harmonic with a number $n \neq 0$ should be equal to zero. It means that this harmonic becomes an eigenfunction too. This choice may be done at the parametric resonance frequency $\omega = -n\omega_p/2$. Only two harmonics among all the harmonics appearing in (6) may become eigenfunctions simultaneously.

Coupling between the two eigenharmonics having indexes 0 and n arises due to the pump action and realizes through , generally speaking, a number of excited forced harmonics with intermediate indexes from ± 1 to $\pm(n -1)$. The strength of the coupling appears to be of the order of the parameters $\varepsilon_h^{|n|}$, $\varepsilon_L^{|n|}$ and $\varepsilon_d^{|n|}$. The most strong coupling occurs for $n = \pm 1$. We restrict ourselves further to only this case of the most strong coupling between the eigenharmonics having indexes 0 and (-1) at the frequencies close to $\omega_p/2$.

In the first order with respect to parameters ε_h, ε_L and ε_d we may neglect: 1) terms with (+1)-th harmonic amplitudes, δM_{1x} and δM_{1y}, in the equations for 0-th harmonic and 2) terms with (-2)-th harmonic amplitudes, δM_{-2x} and δM_{-2y}, in the equations for (-1)-th harmonic. We obtain then a closed system containing 8 equations in partial derivatives for determination of 8 unknown functions.

3. Derivation of Equations for Amplitudes of Parametrically Coupled Spin Waves

The equations (9) become too complicated even after the indicated simplifications are made. That is not surprising because the equations contain nearly all the information about the structure of spin waves. Meanwhile in zero approximation with respect to parameters ε_h, ε_L and ε_d the structure of the waves may be considered as well known from the literature [26]. The only things remained to be done are, the first , how amplitudes of the spin waves vary in space and time under the interaction through the pump and, the second, what the small corrections induced by pump to the structure of the waves are. To answer these questions it is enough to obtain (in the spirit of the "secondary quantization" method) equations for the wave amplitudes only.

A standard way to obtain these equations was based on the introduction of the amplitudes by means of Holstein-Primakoff transform [23,24]. It is possible , in principle, another more elementary way - the direct transformation of equations (9). We shall use just this second way.

Let us write the equations for amplitudes of 0-th and (-1)-th harmonics , which follow from the system (9), in the form

$$\hat{\Omega}\frac{\partial \vec{e}}{\partial t} + \hat{Q}\frac{\partial \vec{e}}{\partial y} + \hat{R}\vec{e} + \hat{E}\vec{e} = \hat{P}\vec{f} \quad ,$$

$$\hat{\Omega}\frac{\partial \vec{f}}{\partial t} + \hat{Q}\frac{\partial \vec{f}}{\partial y} - \hat{R}\vec{f} + \hat{F}\vec{f} = \hat{P}^{*}\vec{e} \quad ,$$

(12)

where unknown functions $\vec{e}$ and $\vec{f}$ are four component vectors-columns

$$\vec{e} = \begin{Vmatrix} \delta M_{0x} \\ \delta M_{0y} \\ \delta H_{0y} \\ \delta H_{0z} \end{Vmatrix} \qquad \text{and} \qquad \vec{f} = \begin{Vmatrix} \delta M_{-1x} \\ \delta M_{-1y} \\ \delta H_{-1y} \\ \delta H_{-1z} \end{Vmatrix} \qquad (13)$$

and the coefficients in the equations (12) are the matrixes 4 x 4

$$\hat{E} = \begin{Vmatrix} a_0 & b & -1 & 0 \\ b & a_0 & 0 & 0 \\ 0 & 0 & -iq_z & iq_y \\ 0 & 4\pi iq_y & iq_y & iq_z \end{Vmatrix}, \quad \hat{F} = \begin{Vmatrix} a_{-1} & b & -1 & 0 \\ b & -a_{-1} & 0 & 0 \\ 0 & 0 & -iq_z & iq_y \\ 0 & 4\pi iq_y & iq_y & iq_z \end{Vmatrix},$$

$$\hat{\Omega} = \frac{1}{\gamma M_0}\hat{A}, \quad \hat{Q} = -2i\alpha q_y \hat{B}, \quad \hat{R} = -i\varepsilon_{d0}\hat{B}, \quad \hat{P} = \frac{h}{2M_0}\hat{B}, \qquad (14)$$

$$\hat{A} = \begin{Vmatrix} -1 & 0 & 0 & 0 \\ 0 & 1 & 0 & 0 \\ 0 & 0 & 0 & 0 \\ 0 & 0 & 0 & 0 \end{Vmatrix}, \qquad \hat{B} = \begin{Vmatrix} 0 & 1 & 0 & 0 \\ 1 & 0 & 0 & 0 \\ 0 & 0 & 0 & 0 \\ 0 & 0 & 0 & 0 \end{Vmatrix}.$$

The frequency ω may deviate slightly from the parametric resonance frequency, that is $\omega = [(\omega_p/2) + \Delta\omega]$, where $\Delta\omega/\omega_p \sim \varepsilon_h$, ε_L, ε_d. In such a case the matrix $\hat{F} = (\hat{F}_0 + \delta\hat{F})$, where

$$\hat{F}_0 = \hat{F}(\omega)\Big|_{\omega = \omega_p/2} \qquad \text{and} \qquad \delta\hat{F} = -\frac{i\Delta\omega}{\gamma M_0}\hat{A}. \qquad (15)$$

Let us find solutions of (12) in the form

$$\vec{e} = C\vec{e}_0 + \delta\vec{e} \quad \text{and} \quad \vec{f} = D\vec{f}_0 + \delta\vec{f}, \qquad (16)$$

where $\vec{e}_0$ and $\vec{f}_0$ are solutions for the unperturbed problem

$$\hat{E}\vec{e}_0 = 0 \quad \text{and} \quad \hat{F}_0\vec{f}_0 = 0, \qquad (17)$$

C and D are the spin wave amplitudes, slowly varying with **r** and t. These amplitudes, in general, are not small. Terms, $\delta\vec{e}$ and $\delta\vec{f}$ are small and slowly varying. These terms describe corrections to the wave "structure". Determinants of the matrixes $\hat{E}$ and $\hat{F}_0$ are equal to zero. Therefore to obtain a single solution of equations (16) we may choose one component for each of the vectors $\vec{e}_0$ and $\vec{f}_0$ to be equal to 1. Moreover, to conserve the total number of the unknown functions we may put equal to zero one component for each of the vectors $\delta\vec{e}$ and $\delta\vec{f}$. After this we may substitute expression (16) into system (12) and see that the zero order terms vanish because of the equations (17). If we confine ourselves by the 1-st order with respect to our small parameters ε_h, ε_L, and ε_d, we should neglect all the derivatives of the weak and slowly varying functions $\delta\vec{e}$ and $\delta\vec{f}$. Then, after excluding these functions from the system (12), we obtain finally two equations for amplitudes C and D

$$
\begin{aligned}
\frac{\partial C}{\partial t} + V_1 \frac{\partial C}{\partial y} + \gamma_1 C &= g_1(y)D + \gamma_1 C_0 \quad, \\
\frac{\partial D}{\partial t} + V_2 \frac{\partial D}{\partial y} + \gamma_2 D &= g_2(y)C + \gamma_2 D_0 \quad,
\end{aligned}
\tag{18}
$$

where the parameters $V_{1,2}$, $\gamma_{1,2}$ and $g_{1,2}$ have the following meaning respectively: the projections of group velocities on y-axis, the dissipation coefficients and the parametric coupling coefficients. For these parameters we obtain the following formulae

$$V_1 = -V_2 \equiv V, \quad V = 2k \frac{\omega_m \sin\theta}{q}\left[\tilde{\alpha}q^2 + \frac{\cos^2\theta}{2k\left(k + \frac{\omega_m}{\omega_p}\sin^2\theta\right)}\right],$$

$$\gamma_1 = \gamma\Delta Hk, \quad \gamma_2 = \gamma\Delta Hk - 2i\Delta\omega, \quad g_1(y) = -i\gamma h(y)\frac{\sin^2\theta}{2}\frac{\omega_m}{\omega_p}, \tag{19}$$

$$g_2(y) = g_1^*(y), \quad k = \sqrt{1 + (\frac{\omega_m}{\omega_p}\sin^2\theta)^2}, \quad \Delta H = \omega_d\omega_p\big/2\gamma^2 M_0.$$

Subsidiary terms $\gamma_1 C_0$ and $\gamma_2 D_0$ ("noise sources") are added to the right hand sides of equations (18). These terms manifest the fact, that in the absence of the pump some initial ("noise") amplitudes exist: $C=C_0$ and $D=D_0$. We require that the following initial conditions were fulfilled at the pump switching on at the moment t=0

$$C(y, t = 0) = C_0, \quad D(y, t = 0) = D_0. \tag{20}$$

The pump h(y) is considered to be localized within the interval 0<y<L - fig.3. At the ends of this interval the pump varies on the distance of the order of $\Delta L \ll L$, satisfying the conditions

$$\Delta L\left|\frac{\partial C}{\partial y}\right| \ll |C|, \quad \Delta L\left|\frac{\partial D}{\partial y}\right| \ll |D|. \tag{21}$$

The pump is constant inside this interval and equal to h_0. Note that the previously imposed condition (7) is equivalent to $q\Delta L \gg 1$. Therefore the values of ΔL are restricted from above and from below in our calculation. Let us take into account now that the waves enter the pump action zone, propagating in opposite directions, since $V_1 = -V_2$ (19). Therefore the amplitude of the wave entering the zone must be of the "noise" type at the input. That is why we impose here the following boundary conditions (see also [21]):

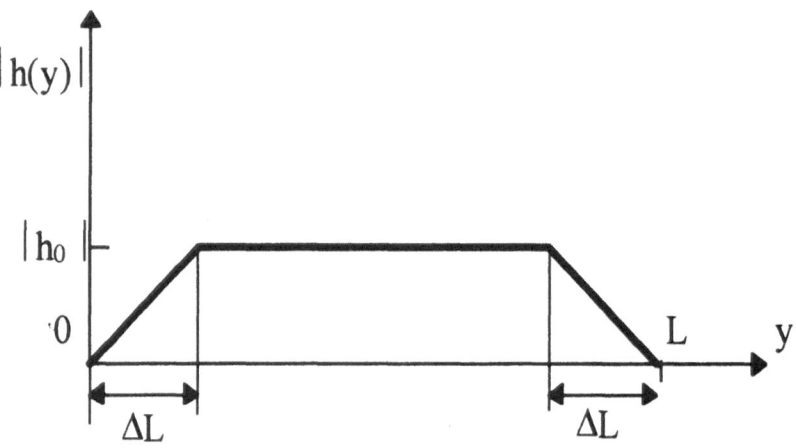

Figure 3. Spatial profile of the pump amplitude.

$$C(y = 0, t) = C_0, \quad D(y = L, t) = D_0. \tag{22}$$

If we restrict ourselves to the frames of the uniform model, we should take the derivatives $\partial C/\partial y \equiv \partial D/\partial y \equiv 0$. Then the equations (18) have the solution in the form of the linear combinations of the functions $\exp(p_1 t)$ and $\exp(p_2 t)$, where

$$p_{1,2} = -\frac{(\gamma_1 + \gamma_2)}{2} \pm \sqrt{(\frac{\gamma_1 - \gamma_2}{2})^2 + g_1 g_2} \tag{23}$$

and the coupling parameters g_1 and g_2 are those inside the pump action zone. Instability condition for this solution, $\mathrm{Re}\, p_1 > 0$, may be written in the form

$$\sqrt{g_1 g_2} > \sqrt{(\mathrm{Re}\, \gamma_1)^2 + (\Delta\omega)^2}, \tag{24}$$

where we have used the expressions (19). Let us suppose now, that $\mathrm{Re}\gamma_1 \neq \mathrm{Re}\gamma_2$. According to (19) such an inequality cannot be fulfilled. But it is fulfilled, however, in a bit more complicated situations. For example, it is fulfilled at parametric wave generation in plasma or in ferromagnets, if waves generated are different in nature (say, spin and acoustic waves). If we formally suppose that the indicated inequality is fulfilled, we obtain immediately at $\Delta\omega = 0$ the following instability condition

$$\sqrt{g_1 g_2} > \sqrt{\mathrm{Re}\gamma_1 \, \mathrm{Re}\gamma_2}. \tag{25}$$

It will be useful to compare the conditions (24) and (25) with the instability conditions which will be obtained further in the paper outside the frames of the uniform model, that is for the solution of equations (18), satisfying boundary conditions (22).

4. Solution of Equations for Wave Amplitudes, Instability Conditions

Let us write a Laplace transform in time t for equations (18) and take into account the conditions (20). Then, we obtain a system of ordinary differential equations for images $C(y,p)$ and $D(y,p)$. The solution of these equations, satisfying conditions (22), is

$$C(y,p) = \frac{U(y,p)}{p(p - p_1)(p - p_2)\Delta(p)}, \tag{26}$$

where

$$\Delta(p) = \mathrm{ch}\frac{\beta L}{V} + \frac{\alpha}{\beta}\mathrm{sh}\frac{\beta L}{V}, \tag{27}$$

$$\alpha = p + \frac{\gamma_1 + \gamma_2}{2}, \quad \beta = \sqrt{\alpha^2 - g_1 g_2} \quad \text{and}$$

$$U(y,p) = \left[g_1 D_0 + (p + \gamma_1) C_0 \right] \cdot$$

$$\left\{ (p + \gamma_2) \Delta(p) - \frac{g_1 g_2}{\beta} \sh \frac{\beta y}{V} \exp\left[\frac{(\gamma_1 - \gamma_2)}{2} \frac{(L - y)}{V} \right] \right\} -$$
(28)

$$g_1 \left[g_2 C_0 + (p + \gamma_2) D_0 \right] \left[\ch \frac{\beta(L - y)}{V} + \frac{\alpha}{\beta} \sh \frac{\beta(L - y)}{V} \right] \exp\left[-\frac{(\gamma_1 - \gamma_2)}{2} \frac{y}{L} \right].$$

The expression for $D(y,p)$ may be obtained from (26) - (28) by means of the following replacements $\gamma_1 \rightarrow \gamma_2$, $C_0 \rightarrow D_0$, $g_1 \rightarrow g_2$ and $y \rightarrow L - y$.

To calculate the originals $C(y,t)$ and $D(y,t)$ we should clarify the analytical properties of the images $C(y,p)$ and $D(y,p)$ in the complex plane of Laplace variable p. This question, in principle, have been discussed already in the literature (see [22,27]). But some details, however, require further consideration. First of all, note that $C(y,p)$ and $D(y,p)$ have no poles in the points $p = p_{1,2}$ at any, even large, length L. To make it obvious we may carry out direct but somewhat cumbersome calculation, starting from the expression (28), and show, that the function $U(y,p)$ is regular in p-plane and in the vicinity of $p = p_{1,2}$ it may be represented as $U(y,p) = (p - p_{1,2}) U'(y,p)$, where $|U'(y,p)| < \infty$ at any L. Consequently, the images $C(y,p)$ and $D(y,p)$ may have poles only at the point $p = 0$ and at the points p, which are the roots of the equation

$$\Delta(p) = 0.$$
(29)

To find all the roots of (29), we introduce, following the paper [22], a new complex variable φ instead of p, the introduction being made by

$$\ch\varphi = \frac{\alpha}{\sqrt{g_1 g_2}}, \quad \sh\varphi = \frac{\beta}{\sqrt{g_1 g_2}},$$
(30)

where $\varphi = u + i(v + \pi v)$, u and v are real quantities, v is an integer. Now to find u and v we obtain from (30) the following equations

$$u + \lambda(-1)^v \mathrm{shu} \cdot \cos v = 0,$$
$$v + \lambda(-1)^v \mathrm{chu} \cdot \sin v = 0,$$

(31)

where parameter $\lambda = \sqrt{g_1 g_2} \cdot L/V$. We need to find only the solutions (u,v), having u ≥ 0 and v ≥ 0. The other solutions will be:(-u,v), (u,-v) and (-u, -v). The nonnegative solutions of (31) were found numerically. These solutions are shown at fig.4 as functions of λ. Different branches of the solutions are convenient to numerate by means of a number v: $u_v(\lambda)$ and $v_v(\lambda)$. Then the roots found, $p = p_v(\lambda)$, are

$$p_v(\lambda) = -\frac{(\gamma_1 + \gamma_2)}{2} + (-1)^v \sqrt{g_1 g_2} \mathrm{ch}\big(u_v(\lambda) + i v_v(\lambda)\big), \quad v = 3,4,5....$$

(32)

Let us now go on the analysis of the roots (32). Let the quantity u≠0. Then the equation (32) with the aid of (31) may be rewritten in the form

$$p_v(\lambda) = -\frac{(\gamma_1 + \gamma_2)}{2} - \frac{1}{\lambda} \sqrt{g_1 g_2} \left(\frac{u}{\mathrm{thu}} + i v \cdot \mathrm{thu} \right),$$

(32')

from which it is clear, that $\mathrm{Re} p_v(\lambda) < 0$ always. It means, that all the roots with u≠0 describe stable, that is attenuating in time, contributions to the amplitudes C(y,t) and D(y,t). This conclusion coincides with the one obtained previously in the paper [22].

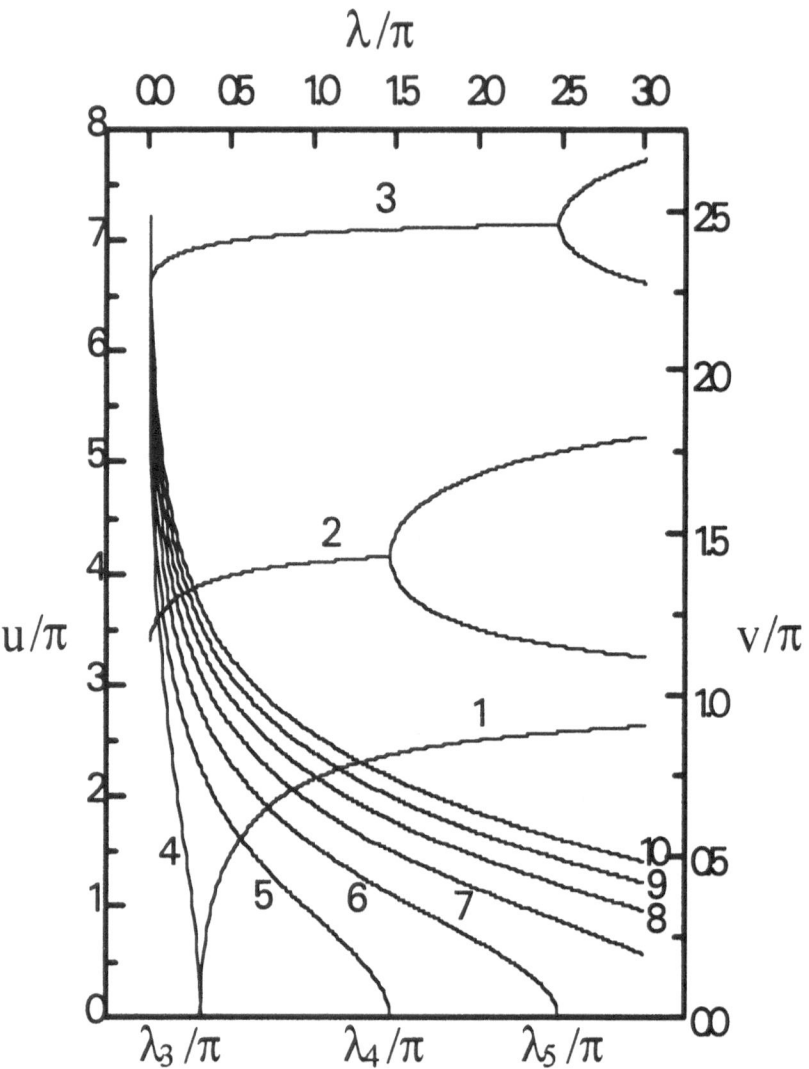

Figure 4. Results of equation (31) numerical solution.

Curves 1,2,3.....represent the functions $v_\nu(\lambda)$ for the numbers $\nu=3,4,5.....$ respectively with $v_3(\lambda)\equiv0$ at $\lambda<\lambda_3$ and the bifurcation takes place for each function $v_\nu(\lambda)$, having $\nu\geq4$, at the point $\lambda=\lambda_\nu$.

Curves 4,5,.....10 represent the functions $u_\nu(\lambda)$ for the numbers $\nu=3,4,5.....$respectively, with $u_\nu(\lambda)\equiv0$ at $\lambda>\lambda_\nu$.

Let us consider now the roots with u=0. The first equation (31) is satisfied identically. Excluding v from the second equation (31) and from (32), we obtain finally

$$\text{Im} p_v(\lambda) = -\frac{1}{2} \cdot (\text{Im} \gamma_1 + \text{Im} \gamma_2),$$

$$\text{Re} p_v(\lambda) = \frac{1}{2} \cdot (\text{Re} \gamma_1 + \text{Re} \gamma_2) \cdot \Gamma_v(\lambda),$$

(33)

the dimensionless increment being

$$\Gamma_v(\lambda) = -1 + \frac{\zeta_v(\lambda)}{|\zeta_v(\lambda)|} \cdot \frac{\eta}{\sqrt{1 + \zeta_v^2(\lambda)}},$$

(34)

where $\zeta_v(\lambda)$ - real roots of the equation

$$\lambda = \frac{\sqrt{1 + \zeta_v(\lambda)^2}}{|\zeta_v(\lambda)|} \cdot [\pi(v-3) - \text{arctg}\zeta_v(\lambda)],$$

(35)

v=3,4,5 ..., |arctg $\zeta_v(\lambda)| \le \pi/2$ and the dimensionless pump $\eta = 2\sqrt{g_1 g_2} / (\text{Re}\gamma_1 + \text{Re}\gamma_2)$ is introduced. It is convenient to introduce the dimensionless length of the pump action zone, $\xi \equiv L(\text{Re}\gamma_1 + \text{Re}\gamma_2)/2V$, so that parameter $\lambda = \xi\eta$.

It is seen from (34), that the instability may take place only if the roots with $\zeta_\gamma(\lambda) > 0$ exist. A family of curves in the plane (ξ, η) may be obtained for such roots which are, in essence, the lines of demarcation between the regions of stability and instability. If we take $\Gamma_v(\lambda)=0$ in (34), that should be valid at the frontiers of stability, we have $\zeta_v(\lambda) = \sqrt{\eta^2 - 1} > 0$. Substituting it into (35), we obtain the following equations for the curves of the mentioned above family

$$\xi = \frac{\left[\pi(v-3) - \mathrm{arctg}\sqrt{\eta^2 - 1}\right]}{\sqrt{\eta^2 - 1}}, \qquad v = 4,5,....$$ (36)

The shape of the curves (36) is shown at fig.5. The curve having $v=4$ play the specific role, because it lies lower than the other curves and therefore it represents the instability threshold. In essence, the expression (36) at $v=4$ gives us a quantitative connection between the threshold pump value and the degree of its localization.

According to (36) and the fig.5, the threshold value of the pump increases as the localization increases (or parameter ξ decreases). This conclusion is in agreement with the previous papers [19-21]. It should be stressed here that the threshold pump value $\eta \rightarrow 1$ at $\xi \rightarrow \infty$. It means that instability condition takes the form

$$\sqrt{g_1 g_2} > \frac{1}{2} \cdot (\mathrm{Re}\,\gamma_1 + \mathrm{Re}\,\gamma_2)$$ (37)

at any permissible $\Delta\omega$ and $\mathrm{Re}\gamma_{1,2}$. This condition differs from the conditions (24) and (25) in two respects: 1) the threshold does not depend on $\Delta\omega$, which is in contrast to (24) and 2) at $\mathrm{Re}\gamma_1 \neq \mathrm{Re}\gamma_2$ and $\Delta\omega=0$ the threshold coincides with the dissipation coefficients arithmetical mean value rather than with the geometrical one, as it follows from (25). At the fig.5 the threshold (37) is presented by the line $\eta=1$, while the threshold (25) - by the line $\eta =\eta_0 \leq 1$. We see, that these lines do not coincide. It means that the true threshold may differ from the value given by the uniform model, even if $L \rightarrow \infty$. The reason of such an inequality were not clarified previously. We shall try to clarify them in the next chapter.

In conclusion of this chapter let us examine in more detail the dependency of the increment $\Gamma_4(\eta)$ on the pump η, that follows from (34) and (35). The family of such dependencies for various values of localization lengths are presented at the fig.6. It may be seen that a parametric regeneration of the system, that is a partial

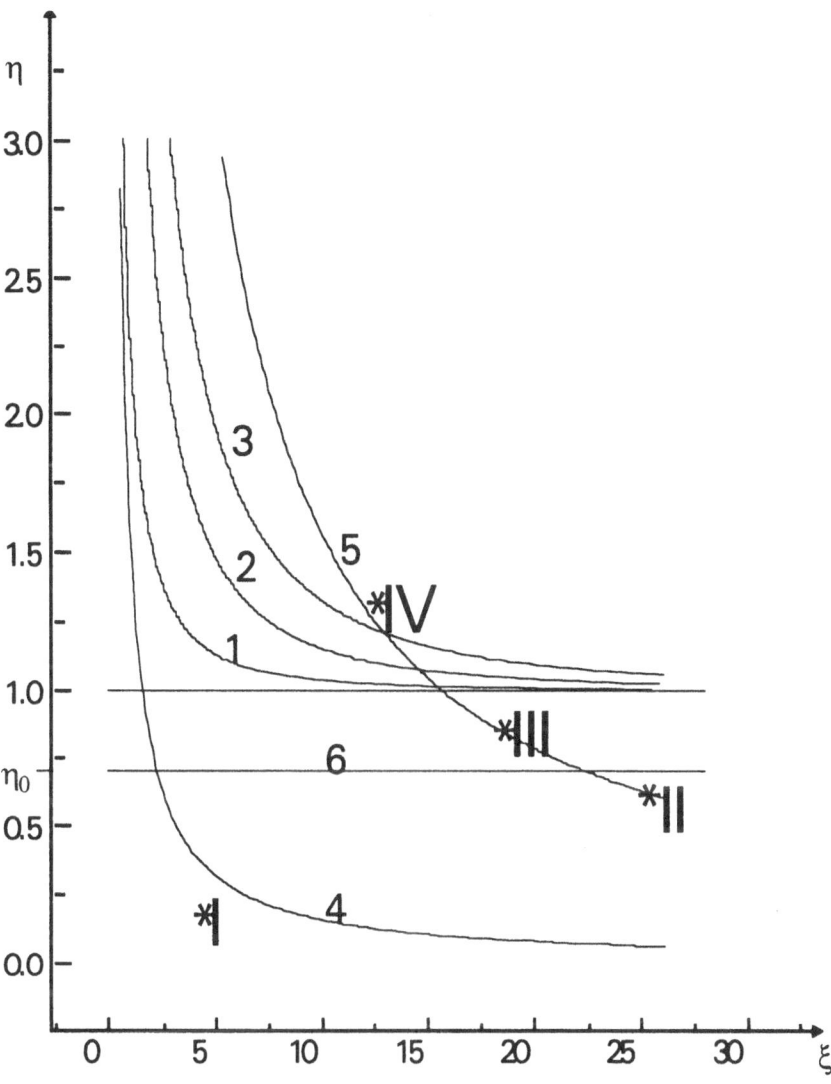

Figure 5. Plane (η,ξ) - "pump - localization length":
1,2,3 - the curves family (36) with the numbers $\nu=4,5,6$ respectively;
4 - hyperbola $\xi\cdot\eta=\pi/2$; 5 - hyperbola $\xi\cdot\eta=\lambda$ at $\lambda = 5\pi$;
6 - line $\eta=\eta_0\equiv 2\cdot\sqrt{\text{Re}\gamma_1\cdot\text{Re}\gamma_2}\big/(\text{Re}\gamma_1 + \text{Re}\gamma_2)$ for an arbitrary $\text{Re}\gamma_1/\text{Re}\gamma_2$.

compensation of a dissipative loss due to the pump has a threshold for finite values of ξ. At the regeneration threshold we may see, that the increment $\Gamma_4=-1$ and this quantity grows with η. According to (34), the regeneration threshold is achieved at $\lambda=\pi/2$, since at this value the positive root $\zeta_4=\infty$ appears. Let us deal with the curve

$\xi\eta=\pi/2$, shown at the fig.5. This curve, in essence, represents the dependence of the regeneration threshold on the localization length. Note that at the regeneration threshold the increment derived above grows with the pump faster than in the uniform model. Really, the line 1 at the fig.6 has the derivative $d\Gamma/dy=1$ at $\Gamma=-1$, while all the lines 2,3,4 and others have the same derivatives $d\Gamma_4/dy=\lambda=\pi/2>1$ at $\Gamma_4=-1$. Consequently, although the initial slopes $\Gamma_4(\eta)$ do not depend on the degree of the pump localization, they do not coincide with the slope, calculated in the frames of the uniform model. Only as the pump increases the dependence $\Gamma_4(\eta)$ tends asymptotically to the one following from the uniform model, $\Gamma(\eta)$. Therewith, the smaller is the localization length, the larger should be the pump η, providing the dependencies $\Gamma_4(\eta)$ and $\Gamma(\eta)$ to be close to each other.

5. Dependence of Wave Amplitudes on Spatial Coordinates and Time, Discussion

The images $C(y,p)$ and $D(y,p)$ (see (26)-(28)) have no other singular points in p-plane, except the poles in the point $p=0$ and in the points $p=p_v(\lambda)$ (32). Therefore, we may calculate the reverse Laplace transform integrals for the amplitudes $C(y,t)$ and $D(y,t)$ using the residue theorem. Therewith we should take into account, that the second order poles occurs in the two cases only: 1) in the point $p=0$, if $\Delta\omega=0$ and one of the roots lies at a frontier of the stability region and 2) at $p=p_v(\lambda)$, if parameter λ corresponds to the point of the bifurcation at the fig.4. In all the other cases the poles are of the first order. At small times t one should take into account too many poles. In this case the most convenient way is the direct numerical calculation of the integrals for amplitudes.

The results of such calculations of amplitudes are presented at the fig.7. The calculation were fulfilled for four characteristic points of the type I,II,III and IV, shown at the fig.5. At the first stage of the calculation it was supposed, for simplicity, that $Re\gamma_1=Re\gamma_2$ and there is no frequency deviation, that is $\Delta\omega=0$.

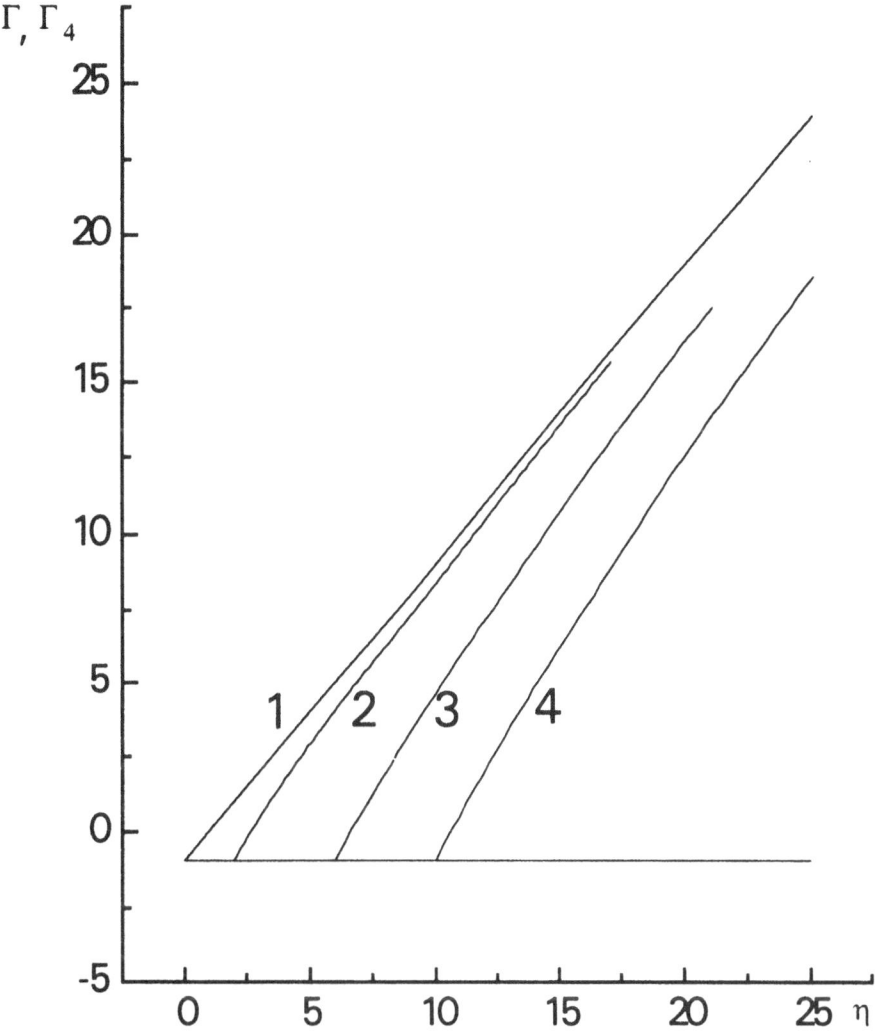

Figure 6. Dependencies of increments on the pump:
1 - the increment, calculated in the uniform model according to the formula (23) at $Re\gamma_1 = Re\gamma_2$ and $\Delta\omega = 0$, namely, $\Gamma(\eta) \equiv Rep_1 / Re\gamma_1 = (-1 + \eta)$;
2,3,4 - the increments $\Gamma_4(\eta)$, calculated for the system with boundaries according to the formulas (34) and (35) at $\xi = \pi/4$, $\pi/12$, $\pi/20$ respectively.

The point I lies below both the threshold of instability and the threshold of regeneration, that is below the curves 1 and 4 at the fig.5. It is seen from the curves at the fig.7, plotted for the point I (curves at the fig.7(I)), that after certain time $t > L/V$

have passed a stationary distribution in space for amplitudes is achieved. This distribution coincides with the one, obtained in the frames of the uniform model almost through the whole length of pump action zone except only a boundary regions having dimensions of the order of a "dissipation length", $L_d=2V/(Re\gamma_1+Re\gamma_2)$. In the point I, as one may see, the uniform model describes the situation fairly well, if $L\gg L_d$.

The point II lies above the regeneration threshold (the curve 4 at the fig.5), but below the instability threshold for a realistic situation (system with boundaries) as well as for the uniform model (the curve 6 at the fig.5). It is seen from the fig.7(II), that the influence of the boundaries penetrates with time inside the excitation zone much more farther than at the fig.7(I). It is, naturally, explained by a partial compensation of the loss and corresponding increasing in an effective dissipation length. It is interesting to stress here, that the stationary distribution is significantly nonuniform along the whole excitation zone and do not coincide in any point with the distribution, obtained in the frames of the uniform model. The point III differs from the point II only by the fact that it lies above the curve 6 at the fig.5. Respectively, at this point a stationary distribution is not achieved in the uniform model at all, but it is achieved in the system with boundaries (see curve (4)). Finally, the point IV lies above the instability threshold. According to the fig. 7(IV), the amplitude distribution is not uniform and stationary distribution is not achieved in this case. Let us point out, that for all the curves maximum $|C/C_0|$ is shifted to the right from the centre of the excitation zone. For quantity $|D/D_0|$ the maximum is shifted to the left.

Let us discuss now a dependency of the wave amplitude on time in a fixed spatial point. It is convenient to take the centre of the excitation zone, that is the point $y=L/2$. The results of the calculation are shown at the fig.8 for three points in the plane of the fig.5, namely, for points II, III and IV. We see, that there exists some time interval

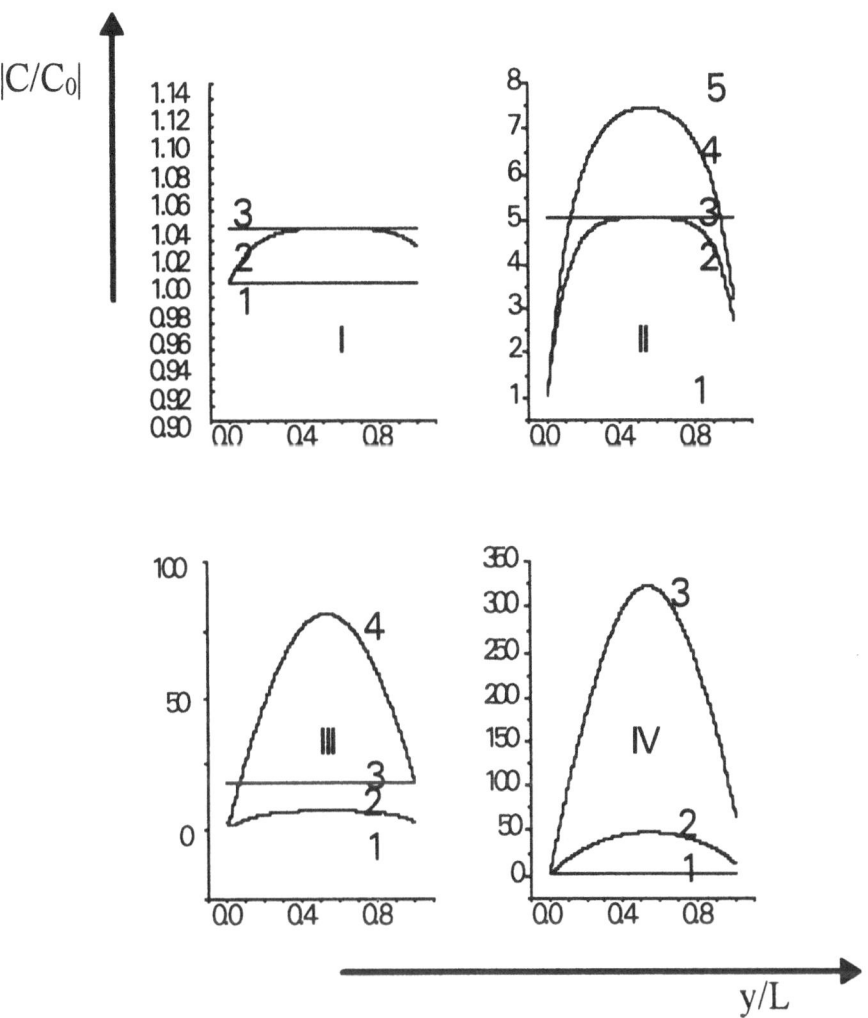

Figure 7. Relative wave amplitude module $|C/C_0|$ versus the dimentionless coordinate (y/L) at the various moments in time for characteristic points of the type I, II, III, and IV at the fig.5. We have taken the following parameters for a point of the type:

I - $\xi=3\pi$, $\eta=0.16$, Vt/L=0 (curve 1) and Vt/L=∞ (curves 2 and 3);

II - $\xi=5.5\pi$, $\eta=0.9091$, Vt/L=0 (curve 1), Vt/L=0.6 (curves 2 and 3) and Vt/L=∞ (curves 4 and 5);

III - $\xi=4.95\pi$, $\eta=1.0101$, Vt/L=0 (curve 1), Vt/L=1.4 (curves 2 and 3) and Vt/L=∞ (curve 4);

IV - $\xi=4.75\pi$, $\eta=1.0526$, Vt/L=0 (curve 1), Vt/L=2 (curve 2) and Vt/L=5 (curve 5).

For all the figures we have taken $\Delta\omega=0$, $\gamma_1=\gamma_2$, $g_{1,2}=g^*_{1,2}$ and $C_0=D_0$. The horizontal lines show for comparison the relative amplitude modules, calculated in the frames of the uniform model.

$0 \le t \le L/2V$, for which the dependency does not differ from the one obtained in the uniform model - dotted and solid lines merge at fig.8. It occurs due to the finiteness of the perturbations velocity V. As a result, a narrowing with time region, where the influence of the boundaries does not manifest itself, appears around the central point y=L/2.. However, the boundaries influence the whole excitation zone, if $t > L/2V$ and the degree of regeneration is sufficiently large, the latter condition providing absence of attenuation of the waves when they propagate along the zone. The situation is illustrated as the splitting of dotted and solid lines at the fig.8. The both lines tends with time to certain (different) stationary levels in the point II. A stationary level is achieved only due to the influence of boundaries in the point III. A stationary level is not achieved in the point IV at all, since the parametric instability appears. So, we may conclude, that the existence of the pump action zone boundaries influence significantly on the whole picture of parametric instability at any value of L, even at $L\to\infty$. But the value of the instability threshold itself for spin waves coincides with the one obtained in the uniform model at $L\to\infty$, if $\Delta\omega=0$. Really, according to (19), the equality $\mathrm{Re}\gamma_1$ $=\mathrm{Re}\gamma_2$ is valid for the spin waves and that is why lines $\eta=1$ and $\eta=\eta_0$ at the fig.5 coincide. Since the instability threshold is just the experimentally measured value, it is clear, that many important peculiarities of the real picture are masked.

It is interesting to understand, why the two thresholds mentioned above may coincide in spite of very different pictures for the instability development in space and time. Fig.9 helps us to understand the reason for such a coincidence.

If the pump localization length L is not too large, as it is shown at the fig.9(a,c), the distributions of modules and phases for the derived amplitudes C and D turned out to be different but symmetrical with respect to the centre of the zone y=L/2. Under this conditions, at least one of the two possible inequalities, $\partial C/\partial y\neq0$ and $\partial D/\partial y\neq0$, is fulfilled at any point inside the pump action zone. As the length L increases, the deformation of amplitude distributions occurs in the directions pointed out at the fig.9(a,c) by arrows. As a result, at sufficiently large L the distributions of the amplitudes C and D, shown at the fig.9(b,d), merge and flat section arises in the central part of the zone. Inside this section the equalities $\partial C/\partial y=0$ and $\partial D/\partial y=0$ may be fulfilled simultaneously. The formation of such a flat section is a consequence of the wave parameter symmetry and the fact, that the wave amplitude distributions cannot depend on the direction of propagation at L→∞. Inside this section we may neglect the

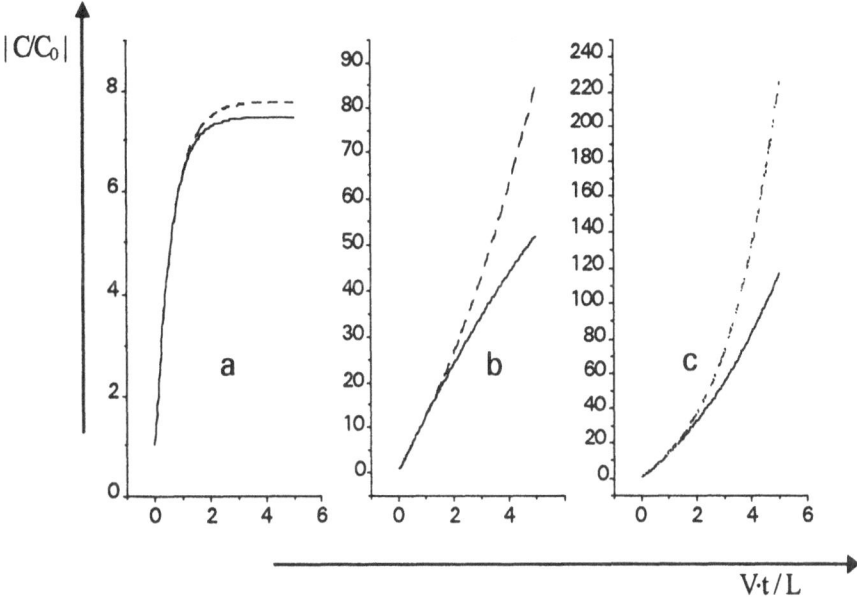

Figure 8. Relative wave amplitude module |C/C₀| versus the dimensionless time (Vt/L) in the centre of the excitation zone y=L/2:
(a) - for a point of the type II, $\xi=5.5\pi$, $\eta=0.9091$;
(b) - for a point of the type III, $\xi=4.95\pi$, $\eta=1.0101$;
(c) - for a point of the type IV, $\xi=4.75\pi$, $\eta=1.0526$.
It is supposed here that $\Delta\omega=0$, $\gamma_1=\gamma_2$, $g_{1,2}=g^{\cdot}_{1,2}$ and $C_0=D_0$.
Solid lines were calculated for the system with boundaries, dotted lines - for the uniform model.

coordinate derivatives in the equations (18). These equations lead then immediately to the instability threshold, coinciding with the threshold in the uniform model.

It should be stressed here once more, that the previous derivation is valid for a symmetrical situation, when $\text{Re}\gamma_1=\text{Re}\gamma_2$ and $\Delta\omega=0$. If $\Delta\omega\neq0$, the symmetry of the parameters is disturbed, since $\gamma_1\neq\gamma_2$ (19). The calculation shows then, that the wave amplitude distribution becomes asymmetrical, even if L→∞, and at any point inside the zone we cannot neglect the derivatives $\partial C/\partial y$ and $\partial D/\partial y$. The instability conditions

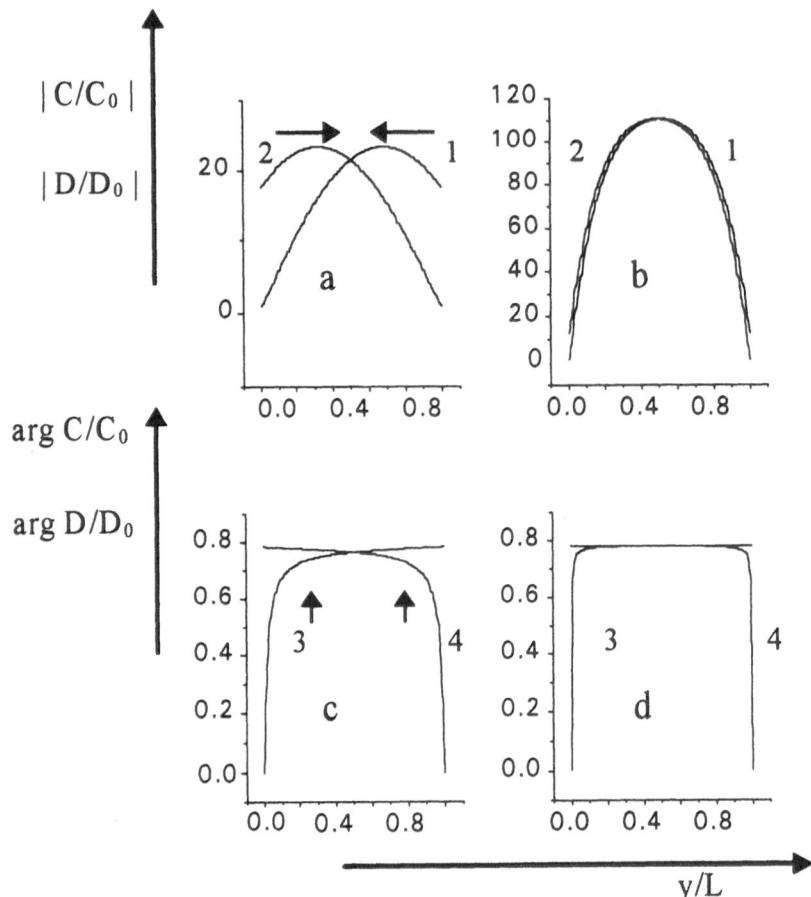

Figure.9. Stationary dependencies of complex wave amplitudes on spatial coordinate at $\Delta\omega=0$, $\gamma_1=\gamma_2$, $g_{1,2}=g^*_{1,2}$ and $C_0=D_0$.
Lines 1 and 2 represent $|C/C_0|$ and $|D/D_0|$ respectively at:
(a) - $\xi=0.7\pi$, $\eta=1.43$ and
(b) - $\xi=14.98\pi$, $\eta=1.001$.
Lines 3 and 4 represent $|\arg C/C_0|$ and $|\arg D/D_0|$ respectively at:
(c) - $\xi=0.7\pi$, $\eta=1.43$ and
(d) - $\xi=14.98\pi$, $\eta=1.001$.

takes then the form (37), where $\mathrm{Re}\gamma_1=\mathrm{Re}\gamma_2$. Therefore, the threshold does not depend on $\Delta\omega$ in the system with boundaries contrary to the uniform model. It is clear, of course, that $\Delta\omega$ may vary in a permissible limits only (remember, that $\Delta\omega/\omega\sim\varepsilon_h$, ε_L, $\varepsilon_d\ll1$).

6. Peculiarities of Longitudinal Localization

As it was pointed out in the chapters 1 and 2 the two types of localization were investigated in the experiments with a localized pump in the papers [19,20]: transversal localization, at which $L \perp H_0$, and longitudinal localization, at which $L \parallel H_0$. Up to now we have confined ourselves to the consideration of the transversal localization only. Let us consider briefly now peculiarities of the longitudinal localization.

The way the theory have been developed, outlined in the previous chapters, is not in essence changed. Following this way, we come to the equations for the wave amplitudes, having the form (18), in which, however, the following replacements should be made

$$V_1 \cdot \frac{\partial C}{\partial y} \rightarrow V_1' \cdot \frac{\partial C}{\partial z} \qquad \text{and} \qquad V_2 \cdot \frac{\partial D}{\partial y} \rightarrow V_2' \cdot \frac{\partial D}{\partial z},$$

where V_1' and V_2' are the projections of the group velocities on z - axis and, as before, the bias field H_0 is considered to be directed along z - axis. Formulae (19) for coefficients remain valid, but for newly appeared velocities we have $V_1' = -V_2' \equiv V'$, where

$$V' = 2k \frac{\omega_m \cos\theta}{q} \left(\tilde{\alpha} q^2 - \frac{\sin^2\theta}{2k\left(k + \frac{\omega_m \sin^2\theta}{\omega_p}\right)} \right). \qquad (38)$$

As it may be seen from (38), at $\theta=\pi/2$ the velocity $V'=0$. According to the dispersion relation (11), waves with $\theta=\pi/2$, travelling perpendicular to a field H_0, exist really at a sufficiently small H_0,

$$H_0 < H_{0c} = \frac{\omega_p}{2\gamma}\left(k - \sqrt{k^2 - 1}\right). \tag{39}$$

Wave numbers q for waves under consideration at such fields H_0 are not equal to zero ($q \neq 0$) and they may be large enough.[3] Therefore to estimate the instability threshold for the waves we may apply the relationships, derived above in this paper. The dimensionless localization parameter ξ, introduced in chapter 4, in this situation is equal to infinity, that is $\xi \equiv L\gamma_1/2V'=\infty$, because of $V_1'=0$. According to (36) and to the fig.5, the instability threshold reaches its minimum value at $\xi=\infty$ and coincides with the value resulting from the uniform model ($\eta=\eta_0=1$).

This conclusion is based on the expression for the velocity (38) and specific only for the longitudinal localization. Really, the velocity $V=0$ only at $\theta=0$ for the transversal localization (see (19)). The waves with $\theta=0$ travel along the field H_0 and therefore, as it is well known [8], cannot interact with the longitudinal pump. Formally it also may be seen from our formulae (19) - at $\theta=0$ the parametric coupling coefficient $g_1=0$. Thus, at the strictly transversal localization a longitudinal pump cannot excite waves, which do not carry the energy out of the pump action zone. On the strictly longitudinal localization such waves may be excited and that is why just these waves determine the minimum instability threshold.

The explanation of the minimum threshold outlined above is in a good agreement with the experimental results obtained in the paper [20]. In this experimental work a nearly longitudinal pump localization was organized. The threshold oscillations versus

---

[3] Recall that the present theory , according to (7) and to condition $L \gg \Delta L$, is valid at $qL \gg q\Delta L \gg 1$.

H_0 was observed in [20], the minima of the oscillations corresponding to the excitation of the waves with a zero projection of the group velocity on the film plane. These waves do not leave the pump action zone and therefore do not carry the energy out.

7. "Butterfly" Curve

In this chapter we shall discuss a dependency of the parametric instability minimum threshold on a bias field H_0 (so called, "butterfly" curve [28]). It is interesting to clarify, how the boundaries of the pump action zone influence this dependency. As it was mentioned already in chapter 6, this influence can be observed in the whole field H_0 interval for the case of the transversal localization. We consider further only this type of localization.

Our calculation is based on the equation for the instability threshold (36) under the index $v=4$. In this equation the parameters ξ and η may be written in the form

$$\xi = \frac{\gamma \Delta H}{\omega_p} \cdot \frac{qL}{\sqrt{k^2-1} \cdot \left\{ 2\tilde{\alpha}q^2 + \frac{\cos^2\theta}{k\left(k+\sqrt{k^2-1}\right)} \right\}},$$

$$\eta = \frac{|h_0|}{2\Delta H} \cdot \frac{\sqrt{k^2-1}}{k}, \tag{40}$$

where we took into account (19).

Thus, the mentioned parameters are written as functions of q and θ. The wave number q itself may be expressed as a function of the angle θ by means of the dispersion relation (11), if we put in this equation $\omega=\omega_p/2$. By doing so, we obtain, that our equation (36) determines the module of the threshold field $|h_0|$, as a function of θ, at fixed values of the following quantities: M_0, H_0, L, ω_p and ΔH. We take the values of the parameters corresponding to the experimental situation in the paper [19]:

$4\pi M_0 = 1750$ G, $\omega_p = 2\pi \cdot 9{,}37 \cdot 10^9 \cdot s^{-1}$. Starting from (36), we determine numerically those values of the angle θ, which correspond to the minimum threshold value of $|h_0|$ at a fixed H_0. If H_{oc} is determined by the formula (39), there exist only one such an optimal value of the angle at $H_0 < H_{oc}$, namely $\theta_{opt} = \pi/2$. The angle θ_{opt} decreases to zero as H_0 rises at $H_0 > H_{oc}$. We substitute into (36) the value of the angle $\theta = \theta_{opt}$ and calculate the minimum threshold field, min $|h_0|$, versus H_0. The family of the dependencies obtained for various lengths L is shown at the fig.10. It is seen, that as L decreases the threshold increases at any H_0. However, the absolute variations of the threshold turn out to be somewhat lower at $H_0 < H_{oc}$, than at $H_0 > H_{oc}$.

It is interesting to compare the curves at the fig.10 with the curves obtained experimentally in the paper [19] for a normally magnetized YIG film. The experiment geometry in [19] corresponds to that shown schematically at the fig.1, the orientation of the fields being similar to the case (a). Therefore the nearly transversal localization was realized. It may be seen, that at L=(0,25-0,87) mm theoretical and experimental curves for $H_0 > H_{oc}$ are correlated with each other. For $H_0 < H_{oc}$ the dependence of a minimum threshold on L was not observed experimentally at all. At the same time such a dependence is present at the fig.10.

We propose, that this discrepancy may be caused partly by the fact, that the calculated model does not completely correspond to the experiments in the paper [19], because the surfaces of the film were rather polished, than rough and diffusively scattering for spin waves. Another cause of the discrepancy may be connected with the fact that no dependency of the line width $2\Delta H$ on q was taken into account in the theory. The necessity of taking into account such a dependency was indicated in the literature [8,25]. The threshold is weakly dependent on ΔH at the most small L, which are of the order or less than the dissipative length. At the small L the threshold is determined only by carrying out the wave energy from the pump action zone. Therefore as ΔH rises with q its value does not affect significantly the "butterfly" curve at $H_0 < H_{oc}$. The influence is enhanced at not so small L and as a result the "butterfly" curve moves up

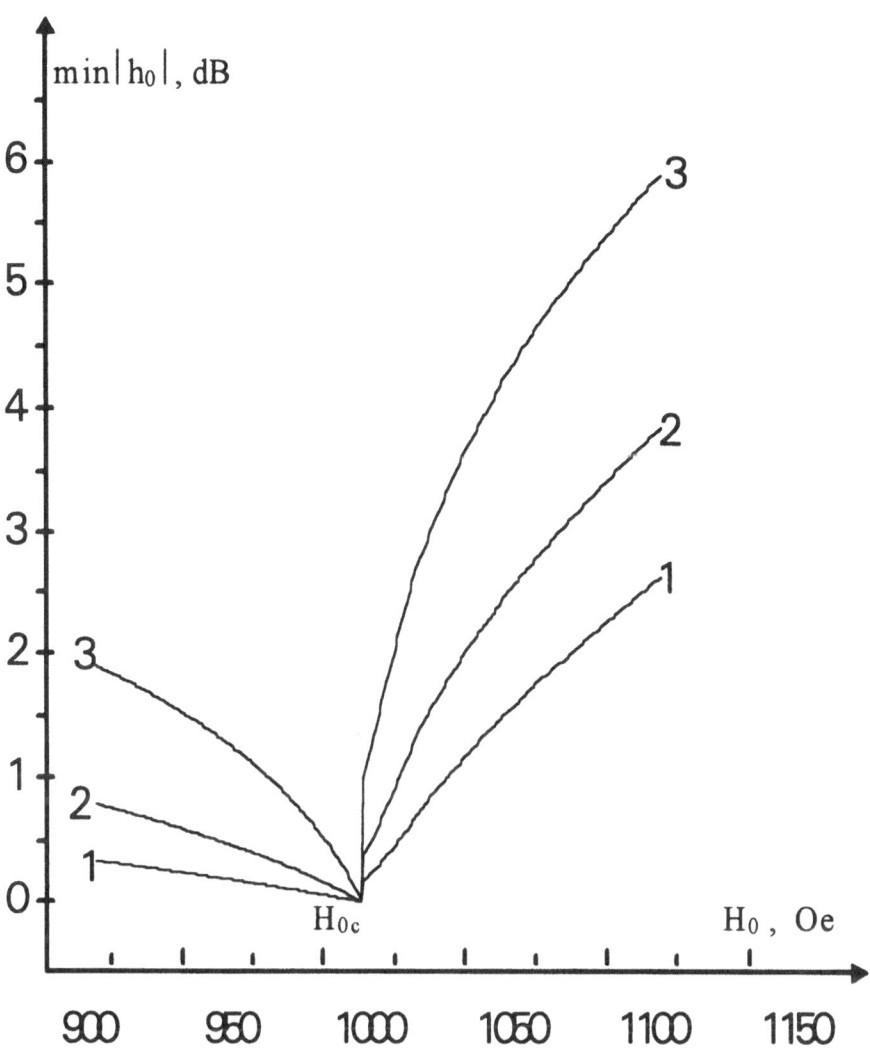

Figure 10. Calculated minimum instability threshold, $\min|h_0|$, versus H_0, expressed in dB, with respect to the threshold at a bias field $H_0=H_{0c}$. The pump $\mathbf{h_p}$ is longitudinal ($\mathbf{h_p}||H_0$) and transversally localized ($L{\perp}H_0$). The curves are obtained at the various pump localization lengths L:
1 - 3,5mm, 2 - 2mm, 3 - 1mm.
The system parameters are: $4\pi M_0=1750$ Oe, $\omega_p=2\pi{\cdot}9.37{\cdot}10^9$ s^{-1}, $2\Delta H=0.4$ Oe.

due to increasing of ΔH with q. Therefore, the lower curves at fig.10 approach from below to the upper curve at $H_0 < H_{oc}$ and the agreement with the experiment becomes better.

8. Conclusion

The parametric instability of spin waves under a localized pump field is of interest both from fundamental and applied points of view. From the fundamental position it is interesting to follow how boundaries influence the processes in an unstable system. That is exactly what we try to do in this paper, but in the frames of linearized problem only, neglecting the interaction between parametrically excited spin waves. Clarification of the role of boundaries at the nonlinear stage of the instability - that is a very attractive direction for further development of the theory.

From the position of possible applications the instability under a localized pump is of interest because it determines often the highest available power in linear devices. Therefore it is desirable in future investigations to take into account a number of factors, in particular: 1) the influence of ferrite film surfaces, which are often rather reflecting than absorbing or diffusively scattering for spin waves, 2) the possibility of parametric generation of not only purely exchange but also dipole spin waves, if the frequency $\omega_p / 2$ lies inside the dipole spectrum of the film, 3) the real structure of the localized pump field.

Acknowledgment

This work was partly supported by *International Science Foundation* (grant MSZ000) and by *Russian Foundation for Basic Researches* (grant 94-02-04928a). We are greatly indebted to Prof. S.A. Nikitov, Dr. A.G. Temiryazev and Dr M.P. Tikhomirova for helpful discussions of the work.

References

1. Damon, R.W. (1953) Relaxation Effects in the Ferromagnetic Resonance, *Reviews of Modern Physics* **25**, 239-245.

2. Bloembergen, N. and Wang, S. (1954) Relaxation Effects in Paro- and Ferromagnetic Resonance, *Physical Review* **93**, 72-83.

3. Suhl, H.J. (1957) The Theory of Ferromagnetic Resonance at High Signal Powers, *J. Physics and Chemistry of Solids* **1**, 209-227.

4. Suhl, H.J. (1958) Origin and Use of Instabilities in Ferromagnetic Resonance, *J. Applied Physics* **29**, 416-421.

5. Morgenthaler, E.R. (1960) Survey of Ferromagnetic Resonance in Small Ferrimagnetic Ellipsoids, *J. Applied Physics* **31**, 95S-97S.

6. Schloemann, E., Green, J.J. and Milano, U. (1960) Recent Developments in Ferromagnetic Resonance at High Power Levels, *J. Applied Physics* **31**, 386-395.

7. Schloemann, E. (1960) Generation of Phonons in High-Power Ferromagnetic Resonance Experiments, *J. Applied Physics* **31**, 1647-1656.

8. Schloemann, E. and Joseph, R.J. (1961) Instability of Spin Waves and Magnetostatic Modes in a Microwave Magnetic Field Applied Parallel to the dc Field, *J. Applied Physics* **32**, 1006-1014.

9. Patton, Carl E. (1984) Magnetic Excitation in Solids, *Physics Reports* **103**, 251-315.

10. Jantz, W. and Schneider, J. (1971) High Resolution Spin Wave Spectroscopy in YIG, *Solid State Communication* **9**, 69-71.

11. Kotyuzhansky, B.Ya. and Prozorova, L.A. (1981) Parametric excitation of spin waves in the $FeBO_3$ antiferromagnetic, *Zh.Eksperimental'noi i Teoreticheskoi Fiziki* **81**, 1913-1924. [Sov. Phys. JETP]

12. Berezin, I.L., Kalinikos, B.A., Kovshikov, N.G., Orobinski, S.P. and Chartorizhski, D.N. (1978) Determination of exchange interaction parameter by

spin wave parametric resonance spectrum in thin ferromagnetic films, *Fizika Tverdogo Tela* **20**, 2101-2103. [Sov. Phys. Solid State]

13. Wiese, G. (1993) Theory for the First-Order Spin-Wave Instability Threshold in Ferromagnetic Insulating Thin Films, *Z.Phisik* **B91**, 57-64.

14. Wiese, G., Kabos, P. and Patton C.E. (1995) Subsidiary-Absorption Spin-Wave-Instability Processes in Yttrium Iron Garnet Thin Films: Coupled Lateral Standing Modes, Critical Modes and the Kink Effect, *Physical Review* **B51**, 15085-15102.

15. Chivileva, O.A., Gurevich, A.G., Anisimov, A.N.,Gusev B.N., Vugal'ter G.A. and Sher,E.S. (1987) Threshold fields and magnetizations in parametric excitation of spin waves by a surface magnetostatic wave, *Fizika Tverdogo Tela* **29**, 1774-1782. [Sov. Phys. Solid state]

16. Kalinikos, B.A. (1983) Decay instability threshold for spin waves in ferromagnetic films under a localized excitation, *Pis'ma v Zh. Tekhnicheskoi Fiziki* **9**, 811-814. [Sov. Phys. Techn. Phys. Lett.]

17. Kalinikos, B.A. and Slavin, A.N. (1991) First Order Parametric Instability of Quasi-Surface Spin Waves in Ferromagnetic Films, *IEEE Transactions on Magnetics* **27**, 5444-5446.

18. Melkov, G.A. (1988) Parametric excitation of spin waves by a surface magnetostatic wave, *Fizika Tverdogo Tela* **30**, 2533-2534. [Sov. Phys. Solid State]

19. Melkov, G.A. and Sholom, S.V. (1987) Parametric excitation of spin waves by localized pump, *Fizika Tverdogo Tela* **29**, 3257-3261. [Sov. Phys. Solid State]

20. Zilberman, P.E., Golubev, N.S. and Temiryazev, A.G. (1990) Parametric excitation of spin waves by spatially localized pumping in tangentially magnetized iron-yttrium garnet films, *Zh. Eksperimental'noi i Teoreticheskoi Fiziki* **97**, 634-643. [Sov. Phys. JETP]

21. L'vov, V.S. and Rubenchik, A.M. (1976) Quasihydrodynamic description of spatially nonuniform singular spectra of weak turbulence, Report N1, Institute of

Automatics and Electrometrics, Academy of Sciences of USSR (Siberian Branch).

22. Gorbunov, L.M. (1974) Development of parametric instability in a restricted region of space, *Zh. Eksperimental'noi i Teoreticheskoi Fiziki* **67**, 1386-1400. [Sov. Phys. JETP]

23. Zakharov, V.E. and L'vov, V.S. (1972) Parametric excitation of spin waves in ferromagnets with magnetic inhomogenities, *Fizika Tverdogo Tela* **14**, 2913-2922. [Sov. Phys. Solid State]

24. L'vov, V.S. and Rubenchik, A.M. (1977) Spatially inhomogeneous singular spectra of weak turbulence, *Zh. Eksperimental'noi i Teoreticheskoi Fiziki* **72**,127-139. [Sov. Phys. JETP]

25. Gurevich,A.G. (1973) *Magnetic Resonance in Ferrites and Ferromagnets*, Nauka, Moscow.

26. Akhiezer, A.I., Bar'yakhtar, V.G. and Peletminski, S.V. (1967) *Spin Waves*, Nauka, Moscow.

27. Bobroff, D.L. and Haus, H.A. (1967) Impulse Response of Active Coupled Wave Systems, *J.Applied Physics* **38**, 390-403.

28. Monosov, Ya.A. (1971) *Nonlinear Ferromagnetic Resonance*, Nauka,Moscow.

SPATIAL NONUNIFORMITY AND SPIN WAVE TURBULENCE IN ANTIFERROMAGNET

A.I. SMIRNOV
P.L.Kapitza Institute for Physical Problems
Russian Academy of Sciences
Kosygin str.2 117334 Moscow, Russia

Abstract.
The experimental investigation of the nonuniform distribution of the energy of the excited spin system in the antiferromagnet at the parametric excitation of spin waves is described. The condensation of excited magnons near the central part of the specimen is ascribed to the nonlinear frequency shift of the magnons spectrum.

The observed condensation of magnons is unstable in time and realizes in the periodic or chaotic manner depending on magnetic field, microwave power and temperature.

The routes to chaotic regimes of magnons distribution are characteristic for the deterministic chaos. These are the period doubling cascades and intermittency. On the diagram in the axes microwave power - magnetic field, the islands of motion with the multiplied period are imbedded in the area of chaotic regime. The dimension of the reproductive phase space for the chaotic regimes was found to be not higher then 5. The embedding dimension equal to 3 is characteristic for a family of turbulent regimes. For these family of regimes the hybrid types of strange attractors, constructed from the elements of Rössler- and Lorenz- type attractors were found in experiments.

1. Introduction

Spin waves or magnons are the elementary excitations of magnetically ordered crystals, i.e to a first approximation magnons are noninteracting quasiparticles. The consideration of the higher order terms in the spin-

R. Marcelli and S.A. Nikitov (eds.), Nonlinear Microwave Signal Processing: Towards a New Range of Devices, 253–273.

wave Hamiltonian leads one to the idea that a magnetically ordered crystal is a nonlinear wave media. The interaction of magnons of different types results not only in the decay of magnons but also in the variation of the magnons eigenfrequency. The dispersion low depends on the amplitude of the spin wave or on the number of excited magnons. One of the very interesting points here is the nondissipative interaction between magnons, in most cases it is analogous to the attraction between magnons. As shown by Dyson in 1956 [1] this interaction gives rise to some corrections to the Bloch's law concerning magnetization dependence on temperature.

We looked for another consequence from this interaction, namely for the effect of the condensation of magnons.

To overview the physical concepts of the described phenomena we consider the classical Hamiltonian following [2]

$$
\begin{aligned}
\mathcal{H} = \; & \sum_{\mathbf{k}} \hbar \omega_{\mathbf{k}} b_{\mathbf{k}} b_{\mathbf{k}}^* \\
& + \sum_{\mathbf{k}_1 \mathbf{k}_2 \mathbf{k}_3 \mathbf{k}_4} (\hbar W_{\mathbf{k}_1 \mathbf{k}_2 \mathbf{k}_3 \mathbf{k}_4} b_{\mathbf{k}_1}^* b_{\mathbf{k}_2}^* b_{\mathbf{k}_3} b_{\mathbf{k}_4} \Delta(\mathbf{k}_1 + \mathbf{k}_2 - \mathbf{k}_3 - \mathbf{k}_4) + c.c) \\
& + \frac{\hbar}{2} \sum_{\mathbf{k}} (h e^{-i\omega_p t} V_{\mathbf{k}} b_{\mathbf{k}}^* b_{-\mathbf{k}} + c.c.) + \mathcal{H}\prime_{int}
\end{aligned}
\tag{1}
$$

Here $b_{\mathbf{k}}$ is the amplitude of the normal mode of spin oscillations with the wavevector $\mathbf{k}$ corresponding to the magnon branch of the excitation spectrum of the crystal. $\omega_{\mathbf{k}}$ is the eigenfrequncy of magnons in the unexcited crystal. For the antiferromagnet of the easy plane type anisotropy (like CsMnF$_3$ described in this paper) $\omega_{\mathbf{k}}$ for magnons of the lower branch of the spin wave spectrum is the following function of the in-plane magnetic field $\mathbf{H}$ and temperature T:

$\omega_{\mathbf{k}} = g(H^2 + \frac{H_\Delta^2}{T} + \alpha k^2)^{\frac{1}{2}}$,

g=2.8 GHz/kOe is magnetomechanic ratio, H_Δ^2=6.3 kOe^{2}K is hyperfine constant, α=9.5·10−6 kOecm is the exchange constant.

$W_{\mathbf{k}_1 \mathbf{k}_2 \mathbf{k}_3 \mathbf{k}_4}$ is the entire probability of the four magnon process with the convertion of "two into two", $V_{\mathbf{k}}$ is the coupling coefficient of the microwave magnetic field $h e^{-i\omega_p t}$ and the pairs of magnons with the wavevectors $\mathbf{k}$ and $-\mathbf{k}$. The magnitude of $V_{\mathbf{k}}$ depends on the nonlinear coupling of two magnons with a photon. $\mathcal{H}\prime_{int}$ is the Hamiltonian involving three magnons interactions, four magnons interactions of the type "one to three", magnon-phonon interactions. c.c. is the complex conjugated term. The population numbers of magnons are $n_{\mathbf{k}} = b_{\mathbf{k}}^* b_{\mathbf{k}}$.

The first sum corresponds to noninteracting magnons with the eigenvalues of energy $\hbar \omega_{\mathbf{k}}$. The second sum resulted from the four magnons interaction gives a contribution to the decay of spin wave with the defined wavevector, but also to the renormalization of spin wave spectrum and partially to the term analogous to the additional pumping [2]. The renormalized magnon eigenfrequency in the presence of the other magnons in the sample is given by the expression:

$$\tilde{\omega}_{\mathbf{k}} = \omega_{\mathbf{k}} + \sum_{\mathbf{k}'} T_{\mathbf{k},\mathbf{k}'} n_{\mathbf{k}'} \tag{2}$$

Here $T_{\mathbf{k},\mathbf{k}'} = W_{\mathbf{k}\mathbf{k},\mathbf{k}'\mathbf{k}'}$

The resulting pumping, consists from the action of microwave magnetic field of the frequency ω_p and from the self-action of the pairs of magnons with the opposite wavevectors. Taking into account that for the parametrically excited waves $b_{-\mathbf{k}} = b_{\mathbf{k}}$ the expression obtained for the resulting pumping $P_{\mathbf{k}}$ is [2]:

$$P_{\mathbf{k}} = hV_{\mathbf{k}} + \sum_{\mathbf{k}'} S_{\mathbf{k}\mathbf{k}'} n_{\mathbf{k}'} e^{i\psi_{\mathbf{k}}} \tag{3}$$

Here $S_{\mathbf{k}\mathbf{k}'} = W_{\mathbf{k}-\mathbf{k},\mathbf{k}'-\mathbf{k}'}$, $\psi_{\mathbf{k}}$ is the sum of phases of the spin waves constituting the pair with respect to the phase of the pumping field $he^{-i\omega_p t}$. This phase $\psi_{\mathbf{k}}$ is strictly coupled with the uniform oscillation of the magnetization with the pumping frequency (the magnons excited parametrically are oscillations of the magnetic moment with the half pumping frequency).

The part of the Hamiltonian $\mathcal{H}_{int}$ corresponds to the different types of magnons interaction resulting in the relaxation rate of spin wave amplitude γ. γ is the inverse characteristic time of the free decay of the amplitude of the spin wave. For the described experiments this time is about $2\mu s$.

Many magnetic crystals, both ferro- and antiferromagnets show the possibility to excite spin waves parametrically. To perform the parametric excitation the microwave magnetic field of the pumping is to be larger then the threshold value $h_c = \frac{\gamma}{V_{\mathbf{k}}}$. In the case of our experiment the characteristic value of h_c is 0.1 Oe, it is variable depending on magnetic field, temperature or pumping frequency.

Under the action of the microwave magnetic field of about $10h_c$ the magnons in the antiferromagnet may be excited parametrically to the population numbers $n_{\mathbf{k}} = 10^6$ [3]. This value is six orders of magnitude larger then the thermal equilibrium population numbers for the magnons of microwave frequencies at liquid helium temperatures and corresponds to the magnons density of about $10^{18} cm^{-3}$.

In the case of antiferromagnets the coupling coefficient $V_{\mathbf{k}}$ does not depend essentially on the direction of the wavevector with respect to the magnetic field or crystal axes, therefore the spin-wave ensemble with the isotropic distribution of wavevectors is excited. The absolute value of $\mathbf{k}$ is fixed by the parametric resonance condition

$$\tilde{\omega}_{\mathbf{k}} = \frac{\omega_p}{2} \qquad (4)$$

For many magnetic crystals the part of the pumping arising from the self-action of spinwaves according to eq (3) leads to the restriction of the amplitude of parametrically excited spin waves before the nonlinear damping restricts it [2]. The vector sum of the terms from the equation (3) on the complex plane with the account on the angle $\psi_{\mathbf{k}}$ is shorter then the microwave pumping vector $hV_{\mathbf{k}}$. For the stationary state based on this restriction mechanism the absolute value of $P_{\mathbf{k}}$ equals the threshold value γ. Therefore only the spinwaves for which the parametric resonance condition (4) is valid are to be excited.

This fact was proved in experiments [4, 5] both for ferro and antiferromagnets, the width of the frequency spectrum of the excited magnons was found much smaller then the relaxation rate. The experiments on the stationary state of the parametric excitation proved the validity of the "phase mechanism" of restriction in antiferromagnet [6].

It was shown experimentally [3] that the nonlinear frequency shift

$$\Delta = \tilde{\omega}_{\mathbf{k}} - \omega_{\mathbf{k}} \qquad (5)$$

is of the order of 10^{-4} of the eigenfrequency value at the density of magnons described above. This nonlinear frequency shift was found to be negative. This fact indicates the attractive character of the magnons interaction: the crystal energy is lower, when the magnons are collected in some restricted area in comparison with the uniform distribution case.

The negative nonlinear shift of the frequency of the wave with the defined wavevector is known to result in selffocusing. It occurs due to the difference in phase velocities of the different part of the wavefront with different amplitudes of the wave. This difference causes the outer parts of the beam which have the lower amplitude to move faster then the inner part, and the focusing may take place if the divergency of the beam is small enough. But what can we expect in the case when the excited magnons propagate in all possible directions and the divergency is therefore large? The condensation of a group of magnons would possible if the dispersion part of the frequency, $\omega_{\mathbf{k}} - \omega_0$, would be equal to the absolute value of the nonlinear shift Δ. The number of excited magnons which is necessary to satisfy this condition is far beyond the possible level of excitation. Nevertheless, in the experiments with the action of pumping, the magnons intersecting the boundary between the strongly excited and weakly excited parts of the sample, will enlarge the component of the wavevector which is directed inside the area with the larger density of magnons. By the other words, it is the refraction of spin waves at the boundary. If we assume the originating area wit the higher density of magnons, the effect of refraction at the boundary of this area will result in some additional energy flow into this region. If this flow will be comparable with the flow of the energy into lattice, the instability may develop. To test this scenario, the following numerical model is constructed.

The enlarging of the group velocity for the magnon with the direct incidence on the boundary is

$$u_\perp = \frac{\Delta}{k}(1 - \frac{\alpha^2 k^2}{g^2 \omega_{\mathbf{k}}^2})$$

For the tangential incident this gain in the group velocity is

$$u_\| = \sqrt{\frac{2\Delta}{\omega_{\mathbf{k}}}}$$

For all the incident angles not close to 90° the increase in velocity is proportional to Δ, i.e to the density of excited magnons N. For the typical values of $k = 10^5 \text{cm}^{-1}$, $\Delta = 2\pi 2.5 \cdot 10^6 \text{s}^{-1}$ these values are:
$u_\perp = 1.5 \cdot 10^2 \text{cm/s}$, $u_\| = 3.8 \cdot 10^3 \text{cm/s}$

The average value $\bar{u}$ for all incident angles is taken to be $3 \cdot 10^2 \text{cm/s}$. Supposing the originating area of the large density to be a sphere of the radius a, the additional flow of magnons inside the sphere will be described by the expression

$$\Phi = 4\pi a^2 \mathcal{N} \bar{u}$$

Here $\mathcal{N}$ is the density of the excited magnons. This flow will exceed the dissipation for the originating sphere of the radius of 0.05 mm, i.e 20 times smaller than the typical sample size.

The equations for the evolution of the number N and phase of the parametrically excited magnons, derived from the Hamiltonian (1) with some simplifying conditions may be taken from [2]. Adding to the N - term the additional flow Φ we obtain approximately for the total number N of magnons within the originating area:

$$\begin{aligned}
\tfrac{1}{2}\dot{N} &= N h V_{\mathbf{k}} \sin \psi_{\mathbf{k}} - \gamma N + \frac{C}{2} N^2 \\
\tfrac{1}{2}\dot{\psi}_{\mathbf{k}} &= h V_{\mathbf{k}} \cos \psi_{\mathbf{k}} + S_{\mathbf{kk}} N
\end{aligned} \qquad (6)$$

The numerical experiment with this equations showed, that N grows in an unlimited manner if $\frac{C}{2} > \frac{0.8\gamma}{N_0}$, here N_0 is the stationary value of N for $C=0$. So the spatial type instability is probable due to the "pumping-out" of the magnons from the low excitation areas to the high excited regions

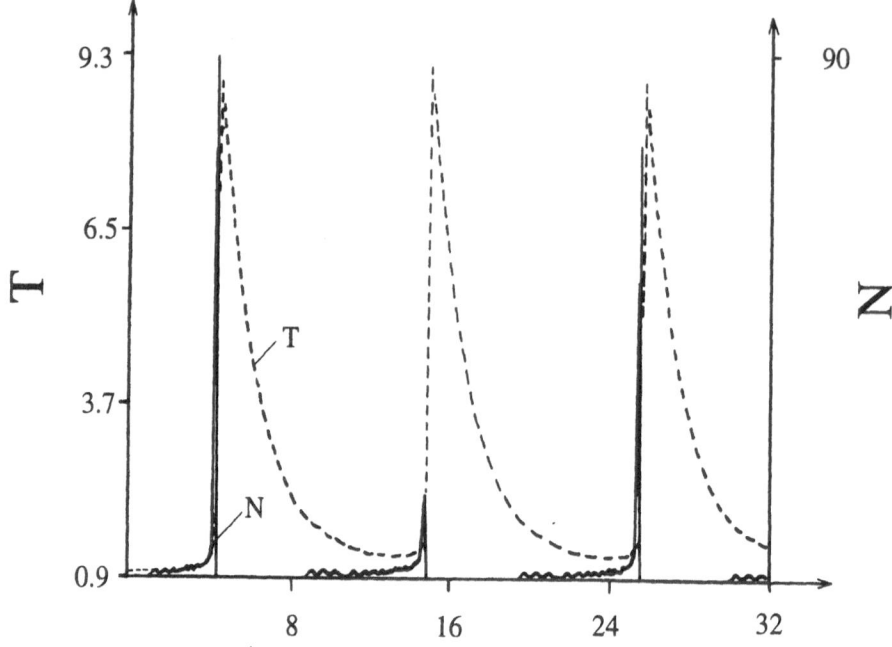

Figure 1. The evolution of the temperature T (dashed line) and of the total number of magnons N (solid line) within the originating area of the high density of magnons.

of the sample. Furthermore, the overheating of the condensation area is proposed as a mechanism for the restriction of this secondary instability at the final level of N. The relaxation rate of magnons in antiferromagnetic is known to be depending on temperature as the seventh power of T [7].

Adding to (6) the equation describing the temperature dependence of the relaxation rate and the equation for temperature evolution with the heating term due to the decay of magnons and cooling term due to the natural thermal relaxation of the sample to the helium bath temperature $T_0 = 1.4K$, and keeping in mind that at the absence of pumping the number of magnons equals the temperature equilibrium number N_{T0} we derive:

$$\begin{aligned}
\tfrac{1}{2}\dot{N} &= NhV_{\mathbf{k}}\sin\psi_{\mathbf{k}} - \gamma(N - N_{T0}) + \frac{C}{2}N^2 \\
\tfrac{1}{2}\dot{\psi}_{\mathbf{k}} &= hV_{\mathbf{k}}\cos\psi_{\mathbf{k}} + S_{\mathbf{kk}}N \\
\gamma &= \gamma_0(1 + DT^7) \\
\dot{T} &= 2A\gamma(N - N_{T0}) - 2B\gamma(T - T_0)
\end{aligned} \qquad (7)$$

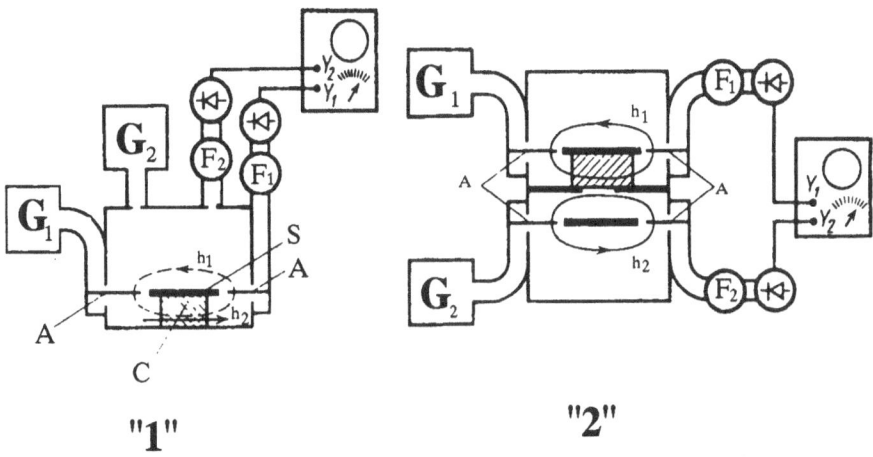

Figure 2. Microwave units for two pumpings experiments. G_1, G_2 - microwave generators of the powerful and testing pumpings, h_1, h_2 - microwave magnetic fields of these pumpings, F_1, F_2- microwave band filters, C-crystal of CsMnF$_3$, A-antennas for the strip resonators excitation, S- strips of the strip-type resonators, Y_1, Y_2-inputs of the oscilloscope.

For numerical solving of this equation we took the parameters: $S_{kk}/\gamma_0 = 10$ to have N normalized to unit at $\frac{h}{h_c} = 10$; $D = 0.01$ from the results of [7]; $A = 0.1$, $B = 0.5$ are taken from the measurements of the overheating of the sample at fixed power and from the known value of the heat capacity. C/γ_0 is taken equal to 1, it proposes the dimension of the considered originating area as large as estimated above.

The result of the numerical experiment with (8) is shown on fig 1. It is to see here that the instability is restricted due to the heating and relaxation rate growth. It is followed by the terminating of the process of parametric excitation because of the increase of the threshold and by the decay of magnons. Thereafter the cooling of the lattice proceeds till the moment when the parametric excitation again can occur, and so on - the situation becomes unstationary and quasi periodic. The regime of autooscillations with the large pauses between the spikes of the absorption was observed in CsMnF$_3$ for the parametric excitation of spin waves at large pumping power [8]. Therefore we were going to look for the change of the distribution of magnons along the sample during this short intervals of large absorption power.

2. Detection of the nonuniform distribution of magnons

We made attempts to measure the density of parametrically excited magnons near the middle of the sample surface and the density averaged over the whole sample. To measure the density of magnons we used the phenomenon of the nonlinear frequency shift described above. The frequency shift Δ is proportional to the entire density of excited magnons. The measurement is made by means of two microwave pumpings following [3]. The microwave field of the first pumping with the frequency ω_{p1} is used for the excitation of the first group of magnons denoted as PM1. The second pumping with the frequency ω_{p2} has the much smaller power just above the threshold of the parametric instability of the magnons PM2. This testing pumping is used to measure the nonlinear frequency shift produced by the group of magnons PM1.

In the process of the measurements both pumping excite magnons. The parametric excitation is detected when the appropriate diminishing of the microwave signal transmitting through the resonator is observed. This diminishing corresponds to the absorption of microwave power by the parametrically excited magnons.

The testing pumping is working in a continuous way. The high power pumping is switched on and after some level of excitation is reached, this pumping is switched off. The magnons from the group PM1 start to decay from this moment. Due to the decay of the magnons PM1 the shift of the eigenfrequency of PM2, provided by PM1, diminishes. Hence the condition of the parametric resonance (4) for the testing pumping becomes unvalid and the transition process occurs. During this transition process the magnons arisen out of resonance decay and new magnons are being excited. In the case when Δ is large in comparison with γ the new magnons are excited from the thermal level and there is a long period of silence, when the power absorbed by the new magnons is below the sensitivity of the apparatus. The magnetic moment associated with the magnons PM2 oscillates with the frequency $2\bar{\omega}_{k2}$. Due to the change of this frequency the beats between the oscillations of magnetic moment and the pumping field may be observed during the life time of PM2, as it was shown in [3, 9]. The oscillogram of such beats is shown below in fig.4a. The beats frequency is to be approximately equal to 2Δ and therefore is to be proportional to the density of magnons PM1.

Fig2. shows the experimental set-up. Two modifications "1" and "2" of the microwave cell were used. The microwave unit "1" provides two microwave magnetic fields h_1 and h_2 acting along the whole sample which is 1mm thick and 1.5 mm in diameter. The two-mode microwave resonator is used for this purpose. The mode used as the powerful source for the

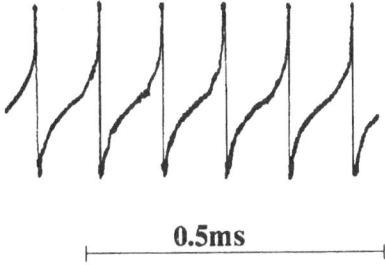

Figure 3. The oscillogram of the transmitted microwave power for the regime of the periodic spiking of the absorption, $\frac{h}{h_c}=12, H=2.0\text{kOe}, T=1.4\text{K}$

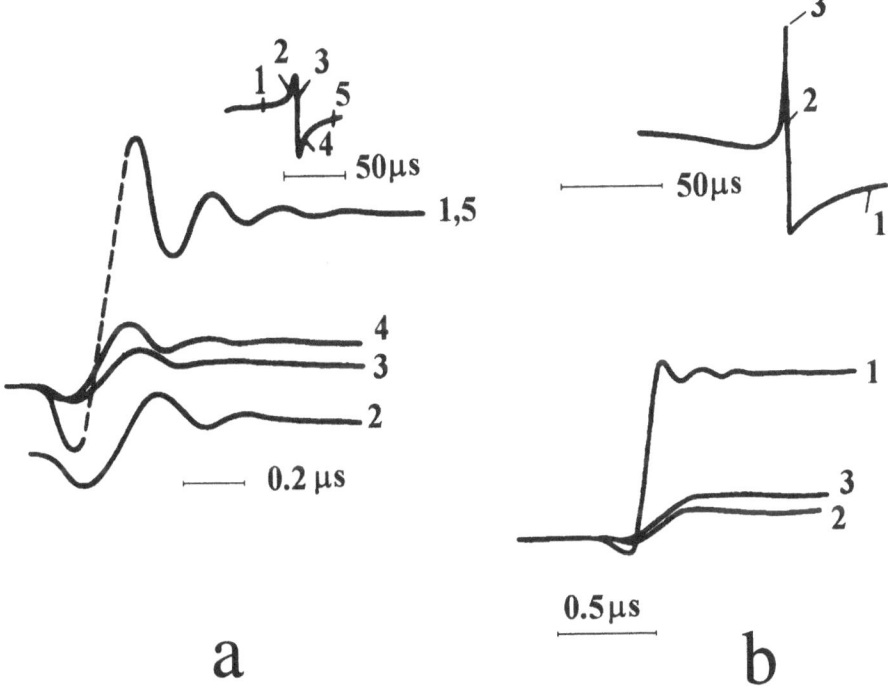

a   b

Figure 4. The oscillograms of the transmitted microwave signal for the powerful pumping in the vicinity of the interrupting moment (upper trace) and for the testing pumping(lower trace). The numbers indicate the moments of interruption. a)- microwave unit "1" with the microwave fields acting over the whole sample. b)- microwave unit "2"with the testing pumping acting near the surface of the sample, high power pumping field covers the whole sample. $\frac{h}{h_c}=12$, H=2.0 kOe,T=1.4K.

magnons PM1 is the strip-type mode of the strip placed within the rectangular cavity. The frequency of this mode is $\omega_{p1}=18$ GHz. The mode TE_{014} of the rectangular cavity with dimensions 7.2x3.4x20 mm serves as the source of testing pumping of the frequency $\omega_{p2}=36$ GHz.

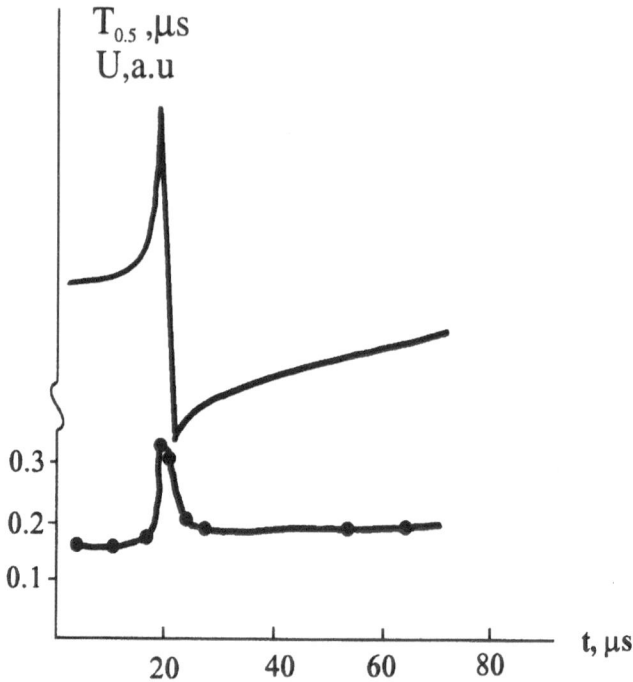

Figure 5. The dependence of the half of the beats period on the oscillogram of the testing pumping during the transition process on the moment of the powerful pumping interrupting (lower curve). Upper curve - the oscillogram of the transmitted signal of the powerful pumping.

Microwave unit "2" shown on fig.2 is constructed to measure the density of magnons in the restricted region of the sample near the center of the bottom end of the cylindrical specimen. We used here the same strip type resonator as in unit "1" and the additional strip resonator, acting on the sample in the restricted area near the center of the bottom of the sample via a small hole in the wall. This last resonator is the source of the microwave field of the testing pumping. For this microwave cell ω_{p1}=18 GHz, ω_{p2}=24GHz. As it was already mentioned, at the high power of parametric excitation, the nonstationary regime with the periodical spiking of absorption is characteristic for $CsMnF_3$ at low temperatures. The oscillogram of the microwave power of the powerful pumping transmitted through the resonator demonstrating this periodic spiking is shown on fig3.

We performed the interrupting of the first pumping in the vicinity of spiking moments, as well as between spikings. The oscillograms of the transition processes, occurred in the channels of testing pumping are presented in fig 4. The upper traces are the oscillograms of the powerful pumping power transmitted through the resonator, with the indicating of moments of the interrupting of the powerful pumping for the corresponding transition processes oscillograms, given in the lower traces. Note the sufficient difference in the horizontal scale for upper and lower traces. For the case when the both pumpings are acting over the whole volume of the sample (fig.4a), the beats have approximately the same period for all moments of the powerful pumping interrupting. The presence of higher frequency oscillations of small amplitude is also observable in this oscillogram.

There is a difference in experiments with the second microwave unit, where the testing pumping has the area of action only near the center of the upper surface of the sample. We see, that the period of beats (fig. 4b) becomes more then two times larger when the interrupting is performed during the peak of absorption. The dependence of the half of the period of the beats on the moment of interruption is shown in fig.5. These results show that the density of magnons averaged over the sample does not differ sufficiently during the spikes, but at the same time there is the area near the sample surface where the density is dropped for more than two times. It means that the redistribution of parametrically excited magnons takes place when the absorption is spiking nonstationary. The total absorbed power increases during the spikes, it proves the increase of the total number of magnons. Together with the diminishing of the magnons density near the surface it means, that the condensation of magnons takes place.

3. Routes to turbulence.

The periodic spiking becomes chaotic when pumping power or the magnetic field is changed. We observed a rich variety of different chaotic and periodic regimes. Periodic regimes arise at enlarging the pumping power from the stationary absorption. With the further enlarging of power the period of spiking changes slowly, but at some point the period of spiking becomes doubled by originating of spiking of two different amplitudes. Periods multiplied not only by 2, but also multiplied by 3,4 and 7 were observed. The diagram of the stationary, periodic and chaotic states in the coordinates microwave power - magnetic field is shown in fig.6. The oscillograms of the transmitted signal power, corresponding to some of regimes, shown in the diagram, are shown in fig 7.

The period doubling cascade is well known as a way to turbulence for the deterministic chaos. Many types of nonlinear dynamical systems of finite and infinite dimension were found which showed the period doubling as the way to chaotic behavior [10]. More fine characteristics of the deterministic chaos were found in our experiments with spin wave turbulence in antiferromagnet. These subtle chaotic phenomena are not visible at first sight, but are observable when one is looking for it in accordance with

264

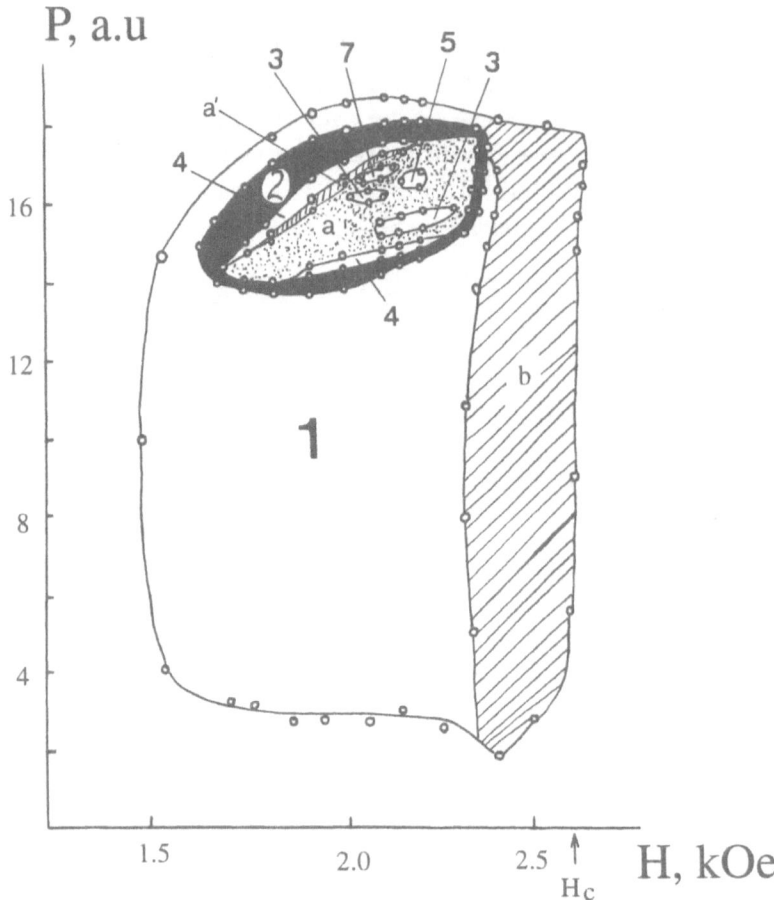

Figure 6. Diagram of the periodic and chaotic regimes. Numbers are indicating the multiplied periods, chaos 'a' is developed after period-doubling bifurcations, chaos 'b' - via intermittency

the predictions of the general theory. The interesting question is about the transition to chaos from the periodic regime with the period multiplied by 3. The general consideration of the chaotic or periodic behavior of the dissipative system moving in the 3-dimensional phase space deals with the Poincaré section. This section is constructed from the points of the intersection of the phase trajectory with a plane. Since the phase portrait of the dissipative system is to be of the zero volume, the intersection points are to lie on the line or on the set of lines. The position of the n-th intersection point may be given as a distance x_n from some reference point on this line. Because the motion is suggested to be deterministic, the position of every successive intersection point is to be the function of the preceding point:

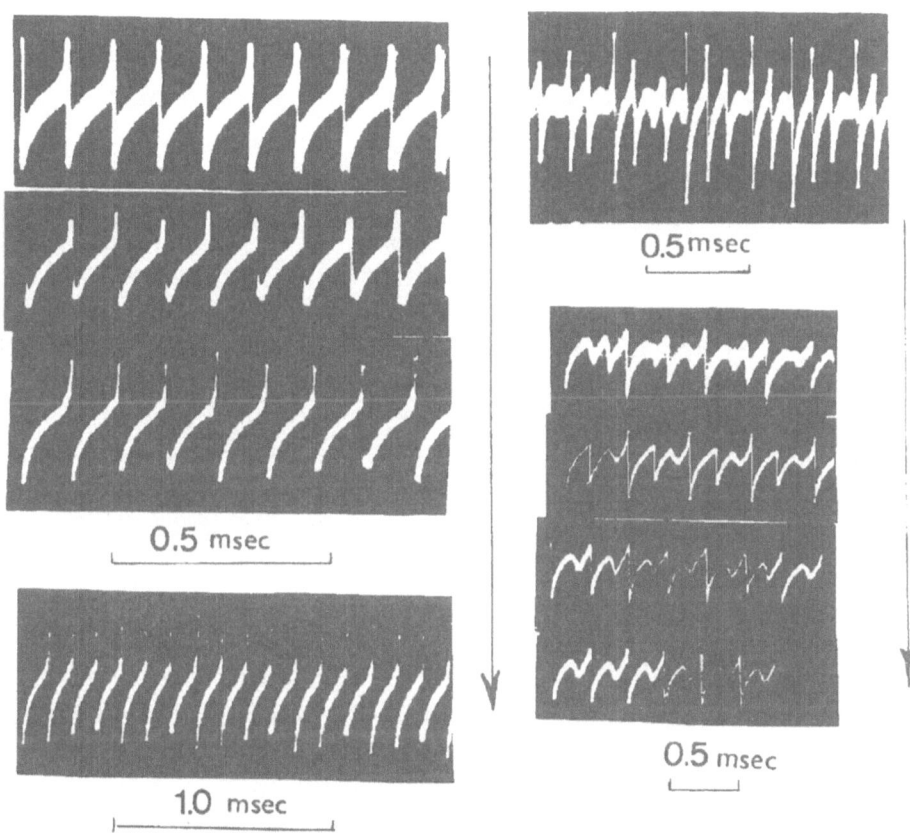

Figure 7. Oscillograms of the transmitted signal for the different regimes: in the left part, up to down: main period cycle, doubled period cycle, period multiplied by four, chaotic regime. In the right part: chaos, period multiplied by 6, period multiplied by 7, period multiplied by 4, period multiplied by 2. The arrows show the direction of the enlarging of the pumping power

$x_{n+1} = f(x_n)$. The chaotization of the motion (which is associated with the positions of points x_n in some interval, but not in the fixed number of points) arises when the function f is nonmonotonic, i.e. has an extremum. The investigation is performed for the extremum of the quadratic type, when the function f may be expressed as:

$$x_{n+1} = 2C x_n + x_n^2 \tag{8}$$

The elegant numerical and analytical investigations are made of the motion, controlled by the mapping of the preceding point on the succes-

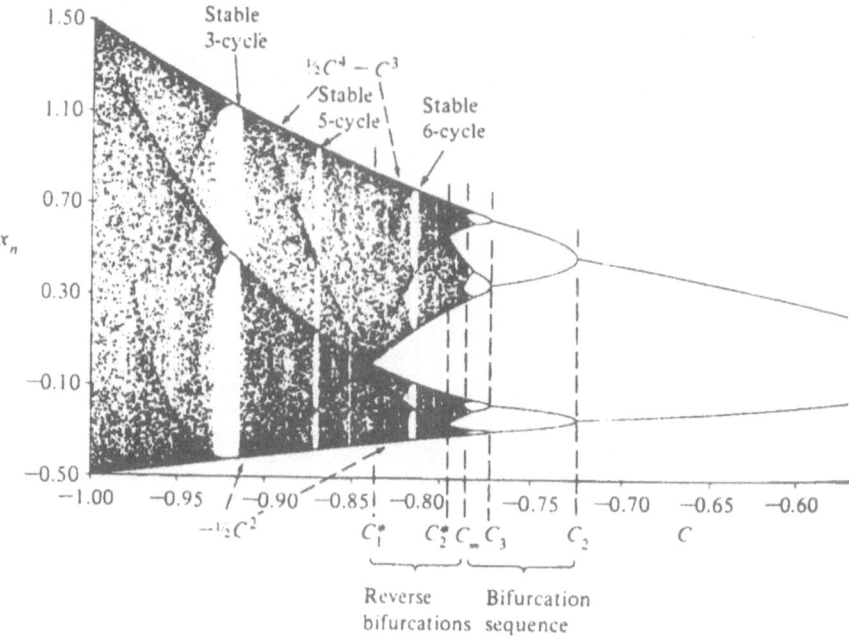

Figure 8. Results of the numerous performance of the mapping (8) at the different values of C (the figure is given after[10])

Figure 9. Phase portraits on the plane transmitted power- reflected power for the cycle 3 regime and for the cycle 6, developed from cycle3 via period doubling

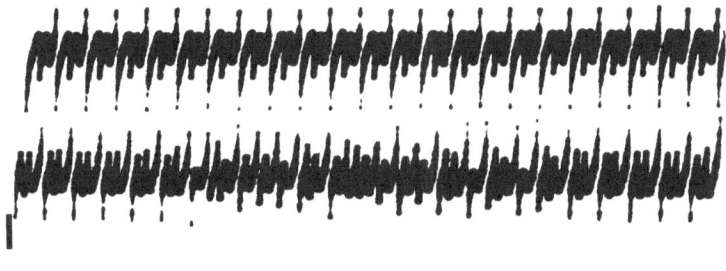

Figure 10. Oscillograms of the transmitted microwave power, illustrating the intermittency of cycle 3

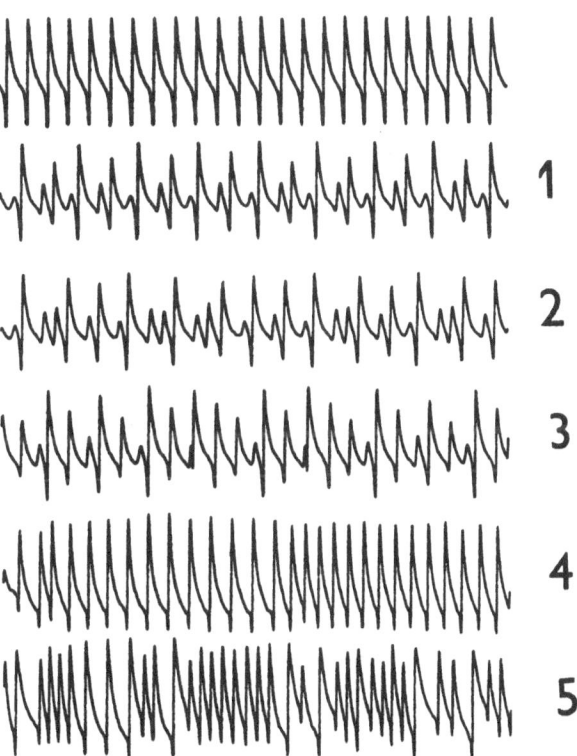

Figure 11. The oscillograms of the transmitted microwave power for the periodic regime (unnumbered), chaotic regimes from chaos "a" area in the fig.6 (curves 1,2,3, the total time base is 2 ms) and from chaos "b"(curves 4,5, the total time base 5 ms) H=2.0 kOe, T=1.4 K

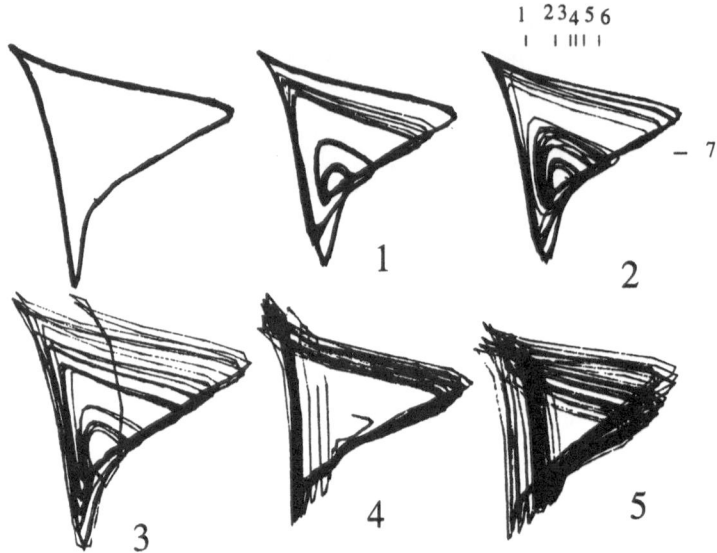

Figure 12. Phase portraits on the plain X_1, X_2 for the regimes 1-5 presented in fig.11. The numbered straight lines show the directions of sections displayed in fig.14

sive position by this and other functions (for the reviews see, for example [10, 11]). The conclusion of this consideration may be illustrated on the fig.8, taken from [10]. Here the values of x_n for large n numbers, obtained by the repeatedly numerical realization of the transformation (8) are plotted along the vertical axes for different values of the parameter C, which is plotted along the horizontal axis. The periodic regime with the main period is presented by only one possible asymptotic value of x_n (is beyond the right edge of the fig.), the period multiplied by 2 - by two different values of x_n, which are being obtained in turn, and so on. The period doubling is performed during the diminishing of C. Beyond the limit point of this period doubling procedure the chaotic regime takes place. The chaotic regime is interrupted by the intervals of the periodic motion with the periods, multiplied by 3,5,6. The exit from the period-3 regime to chaos is a special question. As it can be seen from fig.8, the diminishing of C leads to the period doubling, i.e to the period multiplied by 6 motion. The enlarging results in chaos via intermittency (Pomeau-Manneville) scenario [11]. Intermittency is the motion with the long periods of the coherent motion interrupted by the intervals of irregular oscillations.

Examining the evolution of the regime near the cycle-3 island on the diagram fig.6, we found both these two regimes. The diminishing in microwave power results in the cycle 6, which is illustrated by the phase portrait on the plane "transmitted signal - reflected signal", fig.9. The cycle 3 corresponds here to the 3-loop portrait, the cycle 6 is demonstrated by the splitted middle loop. The enlarging of the microwave power results in the intermittency of cycle 3, which is illustrated on the oscillogram fig.10.

4. Diagnostic of developed chaos

Many turbulent systems show the turbulent, or chaotic behavior with only low number of degrees of freedom which are involved into motion. These physical systems may be infinitedimensional, as the spin-wave system under the action of parametrically exciting pumping - the state of the sample is determined by the value of density of magnons in every point of the sample. The main question of the diagnostic of turbulence is whether the chaos considered is of deterministic nature. The deterministic chaos in the distributed physical system may be characterized by changing in time of the final number of sufficient variables. The number of this variables is called the embedding dimension of the phase space. We shall follow the procedure of constructing the reproductive phase space which is proposed in [12]. The arbitrary number of phase coordinate may be obtained from only one measured value $X(t)$ by calculating the so-called delayed coordinates:
$X_1(t) = X(t)$, $X_2(t) = X(t - \tau)$ ... , $X_n(t) = X(t - (n - 1)\tau)$ ...
τ must have the reasonable value not to short to provide the difference between coordinates and not too large to remain the correlation in the motion of the deterministic system at short intervals of time.

To determine the maximum value of the degrees of freedom, involved into motion, one has to determine, how many of these delayed coordinate is enough to determine all other coordinates. The simple geometric criterion may be used for this purpose [13]. The variable
$a_{n+1}(t) = |X_{n+1}(t) - X_{n+1}(0)|$,
module of the difference of the $n + 1$-th coordinates of some running point and of a fixed point of the phase portrait, is plotted vs the distance $d_n(t)$ between these points in the space of n first coordinates. If $a_{n+1}(t)$ is close to zero when $d_n(t)$ is close to zero, then X_{n+1} is the function of $X_1, X_2, ...X_n$, and n coordinates determine the motion. The coordinates with the greater numbers would be surplus coordinates.

We apply the described method for the determination of the embedding dimension of the strange attractors of the spin wave turbulence. The transmitted microwave power is taken as the basic value $X(t)$ for the phase space construction. The time dependencies of the microwave signal transmitted through the cavity are presented for different regimes in fig.11. The unnumbered curve belong to the periodic regime, the curves 1,2,3 to chaotic regimes at H=2.0 kOe, curve 1 represents the chaotic regime with the

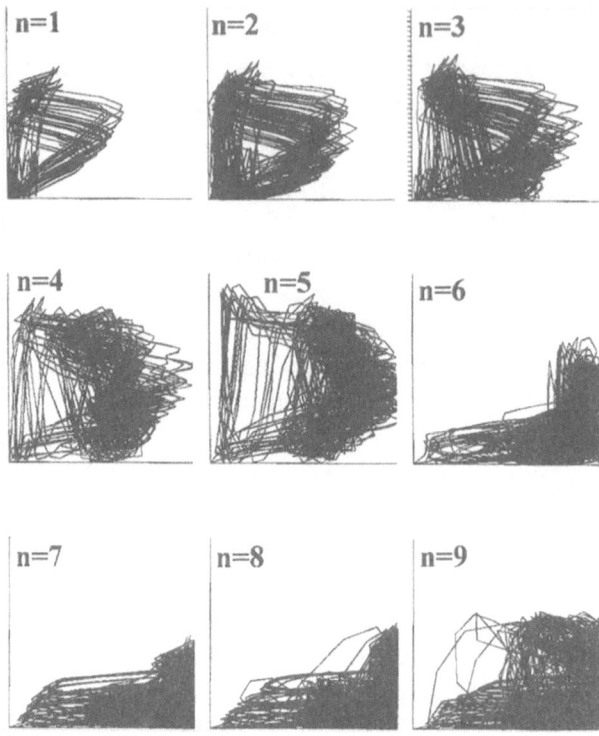

Figure 13. Portraits $a_{n+1}(t)$ *vs* $d_n(t)$ of the regime number 5 from fig. 11 for different n

pumping power below the cycle 3 island (see diagram fig.6), for the curve 2 the pumping power is above cycle 3 on the diagram, curve 3 corresponds to the further enlarging of power. Curves 4,5 represent the regimes from chaos "b" area on the fig.6, the pumping power is larger for curve 5.

The projections of the phase portraits of these regimes on the plane X_1, X_2 with $\tau = 2\mu s$ is shown in fig.12. The application of this method of the embedding dimension to the chaotic regime indicated as chaos "b" in the diagram fig.6 is shown in the fig.13. It is seen that not more then five coordinates is required for the exact description of the system state for that kind of motion. For the regimes from the area of chaos "a" the enough number of coordinates is 3.

5. Topology of strange attractors of spin wave turbulence

The regimes within the area of chaos "a" on the diagram fig.6 have the dimension of the reproductive space equal to 3. For this case we constructed the topological equivalents of the phase portraits.

Figure 14. Sections of the attractor of the chaotic regime 3 on the fig.11

To realize the position of the attractor in the 3-dimensional space we plotted numerous sections of the attractor with different planes. Several sections for the regime 3 from fig.11 are shown in the fig.14. The directions of corresponding sections are indicated in fig.12.

Analyzing such sections, the models of attractors embedded in the 3-dimensional space were constructed, the topology of this models is given in fig.15. The investigation of the sections shows that attractors consist from the pieces of the plane strip. Strange attractors of spin wave turbulence possess the important topological elements - foldings and branching off. The folding may be illustrated considering sections number 4 and 3. The comparison of these sections demonstrates that the lower points of the right wing of the section 4 are superimposed on the lower part of the left wing of the section 3 in the process of the attractors strip folding.

For the chaotic regime 1 the attractor consists from the plane strip, winded in two loops (trace of the period-doubling), with foldings. For the regime 2 the branching and superimposing of the plane strip is added. The more complicated structure has the attractor of the regime 3, where the two loops merge and the branch merges with the outer part of the large loop.

The last data on the topology of the spin wave turbulence attractors show, that this attractors are combined from the basic elements of the well known attractors of Rössler and Lorenz (fig.16). The Rössler attractor,

272

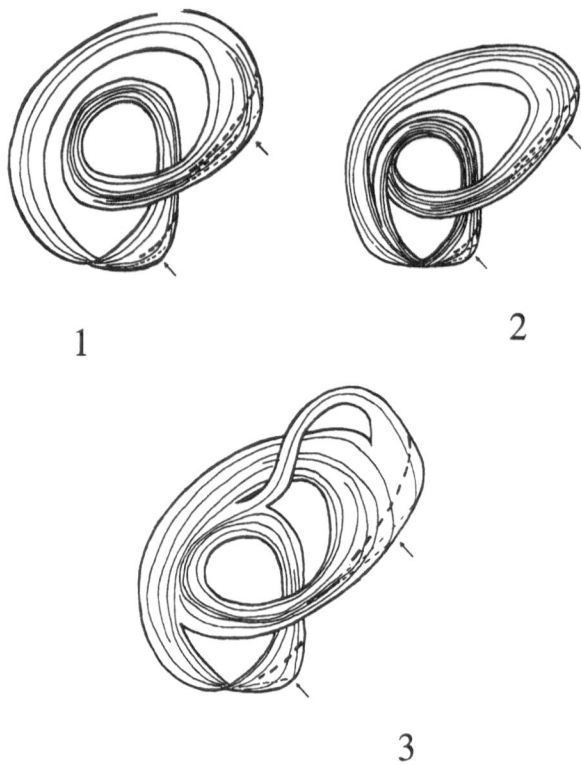

1

2

3

Figure 15. Topological equivalents of the strange attractors for the regimes 1,2,3 indicating on the fig.11. Foldings are indicated by arrows

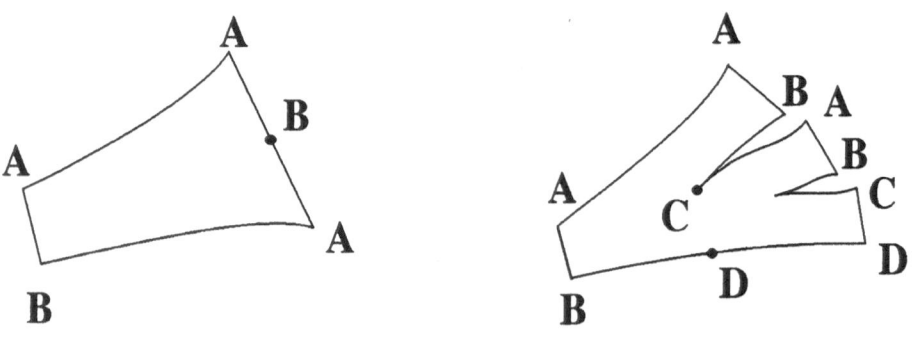

Figure 16. Topology of Rössler (left) and Lorenz (right) attractors. Points marked by identical letters are to be superimposed

originating from a nonlinear system of 3 differential equations, performs stretching and folding of the plane strip of the phase trajectories. Therefore the trajectories on different sides of the strip become close and indistinguishable with time, resulting in chaotization. In the Lorenz attractor which arisen from the Reylay-Bénard convection modeling the trajectories are indistinguishable mixed due to the branching of the basic strip followed by the superimposing of the branch on the basic strip.

Thus the attractors of spin wave turbulence for the 3-dimensional regimes are of the hybrid Rössler-Lorenz type, containing several foldings and branches.

6. Resume

The chaotic behavior of the spin wave system in an antiferromagnet realizes both in space and time, is driven by the nonlinear frequency shift of magnons, and shows many characteristic features of the deterministic chaos.

7. Acknowledgment

This publication is supported in part by the INTAS research project number 94 0968

References

1. F.Dyson Phys.Rev. (1956)
2. V.E.Zakharov, V.S.Lvov, S.S.Starobinets Usp.Fiz.Nauk 114, 609 (1974) [Sov.Phys.Usp. 17, 896 (1975)]
3. L.A.Prozorova, A.I.Smirnov, Zh.Eksp.Teor.Fiz. 74, 1554 (1978) [Sov.Phys.JETP 47, 812 (1978)]
4. I.V.Krutsenko, V.S.L'vov, G.A.Melkov. Zh.Eksp.Teor.Fiz. 75, 1114 (1978) [Sov.Phys.JETP 48, 561 (1978)]
5. A.I.Smirnov Zh.Eksp.Teor.Fiz. 88, 1369 (1985) [Sov.Phys.JETP 61, 815 (1985)]
6. L.A.Prozorova, A.I.Smirnov Zh.Eksp.Teor.Fiz. 67, 1952 (1974) [Sov.Phys.JETP 40, 970 (1975)]
7. B.Ja.Kotiuzhanski, L.A.Prozorova, Zh.Eksp.Teor.Fiz. 65, 2470 (1973) [Sov.Phys.JETP 38, 1233,(1974)]
8. B.Ja.Kotiuzhanski, L.A.Prozorova, Pisma Zh.Eksp.Teor.Fiz. 25, 412 (1977) [JETP Lett 25, 385,(1977)]
9. A.I.Smirnov, S.V.Petrov Zh.Eksp.Teor.Fiz. 80, 1628 (1981) [Sov.Phys.JETP 53, 838 (1981)]
10. A.J.Lichtenberg, M.A.Lieberman, Regular and Stochastic Motion, Springer-Verlag, New York, 1983
11. J.P.Eckmann, Rev.Mod.Phys.53, 643 (1981)
12. N.H.Packard, J.P.Crutchfield, J.D.Farmer, R.S.Show, Phys.Rev.Lett. 45, 712 (1980)
13. S.N.Lukaschuk, A.A.Predtechenski, G.E.Falkovich, A.I.Chernych. Calculation of attractor dimensionality from experimental data. Preprint 280, Institute of Automatic and Electrometry, Sib.Div. Acad. Sc.USSR, Novosibirsk,1985

Chapter III
Solitons and Chaos

MICROWAVE SOLITONS IN MAGNETIC MEDIA : A REVIEW OF FUNDAMENTAL PROPERTIES

A D BOARDMAN, R C J PUTMAN, K XIE, S A NIKITOV AND
H M MEHTA
Photonics and Nonlinear Science Group
Joule Laboratory
Department of Physics
University of Salford
Salford, M5 4WT
United Kingdom

1. Introduction

It is now accepted that nonlinear effects are important in science and engineering. The simplest view is that nonlinearity shows itself whenever the square of a displacement, or a higher-order than that, has to be accounted for. Nonlinear effects can be used to transmit and control pulses in various waveguides and the frequency sweep across pulse envelopes can be exploited. In fact, it is the change in phase, frequency and amplitude, associated with the nonlinearity (e.g. [magnetic field]²) that is being looked at, globally, as a possible route to new and exotic devices. Just as photonics, through [electric field]² effects, is expected to replace electronic switching, nonlinearity in magnetic materials, [magnetic field]² effects, will enable all-microwave devices to emerge. In other words, 'magnonics' may well be a valid replacement for the type of switching electronics, hitherto done at microwave frequencies. Thus, one of the principal objectives will be to generate an all-microwave switch that exploits nonlinearity, showing up as a power-dependence of the frequency. Nonlinear effects, like self-phase modulation, soliton creation, pulse compression, cross-phase modulation and a host of parametric processes, will soon attract a lot of attention [1].

Before a detailed discussion of nonlinear magnetic films is given however, some of the more general points about the interplay between dispersion and nonlinearity will be made. The ideas to be exposed lead, directly, to the concept of solitons and, because spin waves in magnetic materials are involved, these will be pulses where centre the frequency is in the, very useful, microwave (2-10GHz) frequency range.

Long ago, in August 1834, John Scott Russell, a naval architect, was working out a problem that used a boat (called a barge) on the Edinburgh-Glasgow canal [2,3]. For some reason, the boat stopped, rather suddenly, and a

R. Marcelli and S.A. Nikitov (eds.), Nonlinear Microwave Signal Processing: Towards a New Range of Devices, 277–304.
© *1996 Kluwer Academic Publishers.*

wave of water was generated. This wave, in fact, a significant hump of water stretching across the rather narrow canal, rose up at the front of the boat and proceeded to travel down the canal. Scott-Russell, immediately, observed that the wave was something special [3]. It was 'alone', in the sense that it sat on the canal with no disturbance to the front or the rear; nor did it die away until he had followed it for quite a long way. The word 'alone' is synonymous with 'solitary' and, although the, words 'solitary wave' did not appear in the literature for about 130 years, it is routinely used. In addition, the word 'solitary' tends to be replaced by the more generic word 'soliton' [2,4]. Once the physics behind Scott-Russell's soliton was understood, solitons, of one kind or another, appeared to be everywhere. It is interesting, however, that the underlying causes of soliton generation were not understood by Scott-Russell, and only partially by his contemporaries. Even so, by then, pirates and smugglers had, for hundreds of years, been manipulating the shape of the hulls of ships to achieve extra speed. Even the yacht that won the first America Cup benefited from the work of Scott-Russell. This shape is that of a sech function, or approximates to it, and it was the work of Korteweg and de Vries [5] that finally sorted out the precise nonlinear mathematics of water waves, to yield the prediction that what Scott-Russell saw was, indeed, a sech-shaped, sizeable, hump of water. This is now known as a Korteweg-de Vries soliton. In broad, physical terms, nonlinear water waves can easily be appreciated, by glancing at figure 1, which is a sketch of what can happen to water waves approaching a shore. Under certain conditions, to be discussed, properly, later on, dispersion and nonlinearity will balance and then solitons on the water surface will appear.

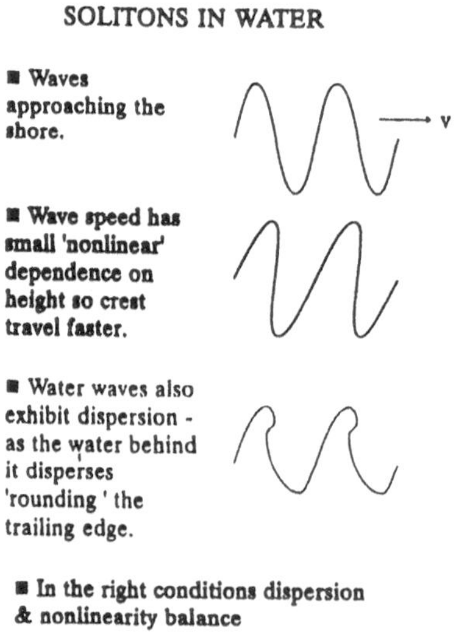

SOLITONS IN WATER

■ Waves
approaching the
shore.

■ Wave speed has
small 'nonlinear'
dependence on
height so crest
travel faster.

■ Water waves also
exhibit dispersion -
as the water behind
it disperses
'rounding ' the
trailing edge.

■ In the right conditions dispersion
& nonlinearity balance

Figure 1. Sketch to illustrate some water wave behaviour. Solitons are readily generated on the sea.

This balance can be thought of in a crude sort of way but it is more sophisticated than that, as will be shown below. In fact, for the solitons of interest here, it is the *chirps* that balance [4]. Chirp, although it is a concept that arouses interest, immediately, will not be defined, at this stage. It is better to introduce it later on.

Earlier on, the Scott-Russell observation was referred to as a *solitary wave* and the word *soliton* was, confidently, introduced. Some sort of qualification of this language is now necessary. Basically, it is quite simple [2,4]. If *solitary waves* pass through each other, without any change in amplitude, they are elevated to the status of *solitons*. Solitary waves, or solitons, also come in various forms, some of which are shown in figure 2. Some of the most famous are

- *Korteweg-de Vries*:   easily observed on shallow water and have a velocity that is proportional to their amplitude [2,5,6]
- *Envelope*:   easily observed on deep water and are solutions of the *nonlinear Schrödinger equation* and have a velocity independent of amplitude. Their most famous application is in optical fibres [4] but they are the ones that will be the centre of attention here.
- *Sine-Gordon*:   dislocations in solids are described by these solitons and are kinks or anti-kinks with velocities that are independent of amplitude.

Solitary Waves or Solitons?

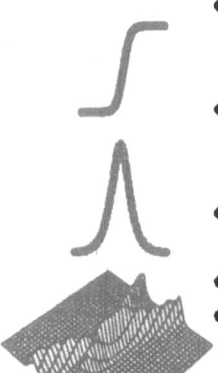

- Solitary waves can be curious entities like the kink on the left that seem to propagate without a change in shape
- A more familiar example is like the pulse on the left ... large amplitude water waves for example
- Why do we say solitary and why do we say soliton ?
- SOLITARY ... because they are alone
- SOLITON ... because they go through each other without being aware of this

Figure 2. Illustrations of solitary waves and some useful comments.

The point about solitons is that they are among Nature's generic entities, i.e. although their first observation was on water they turn up in optical fibres and should be looked for wherever excitations, or waves, can be created. In order to be selective, this chapter concentrates on *bright envelope solitons*. Dark solitons [7] (black holes!) are also possible but they will not be discussed here. A typical, undamped, evolution of what is called a fundamental *bright* envelope soliton is shown in figure 3. Figure 3, deliberately, for the benefit of this discussion, has no units attached to it. If a specific [magnetic] system is in mind, however, and that system is in the form of a thin film, then the units, i.e. the scales, will of course, matter [1]. First, however, it is necessary to find out if a magnetic material will respond to power changes [i.e. become nonlinear].

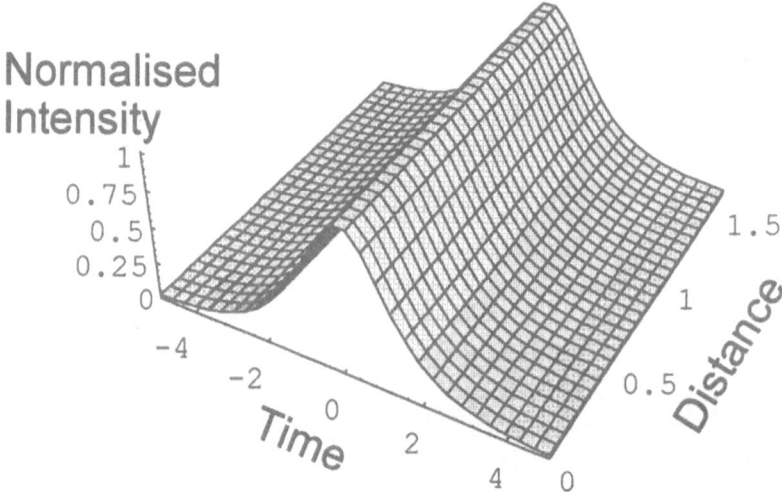

Figure 3.   The fundamental [lowest-order] bright envelope soliton, in the absence of damping.

This is, relatively, easy to determine. For a given magnetic material, an applied magnetic field $\mathbf{H}_0$ is applied. This saturates the magnetisation to a value $\mathbf{M}_0$ and the application of an alternating field $\mathbf{h}(t)$ causes the resultant magnetisation $\mathbf{M}$ to precess about $\mathbf{H}_0$. Figure 4 summarises this situation in which it is seen that increasing the power, introduced through $\mathbf{h}(t)$, causes M_z to decrease. Such a decrease in M_z causes the band edges of the magnetostatic wave dispersion bands to shift [8] with power. Hence, the angular frequency of a magnetostatic wave (a wave going through the spin system that represents the state of magnetisation) depends on power [proportional to the square of the wave amplitude] and this means that the system becomes *nonlinear* and solitons can be looked for.

2. Dispersion and Solitons

Figure 5 contrasts microwave pulse behaviour in a vacuum and in a material. It can be seen, there, that, for a vacuum, a pulse propagates without any spreading (i.e. dispersion). The figure, actually, shows a pulse in which the energy distribution is plotted as a function of what is called *local time* and shows how it evolves as it propagates through a certain distance. Local time simply means a time that is measured in a frame of reference that moves with a speed equal to the pulse group velocity, evaluated at the, centre, carrier frequency. Here the pulse has a centre [carrier] frequency ω_0 i.e. ω_0 is the instantaneous frequency of the pulse maximum.

Nutation of Magnetisation About DC Field

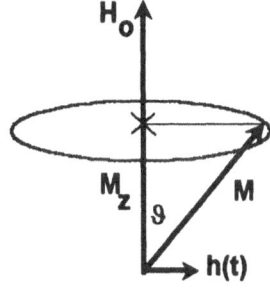

- magnetisation **M**
- angle ϑ is very small in linear systems
- power increase means nonlinearity appears
- component M_z decreases with power
- band edges will move up or down with power
- solitons can be looked for

Figure 4. An explanation of the appearance of nonlinearity in a magnetic medium, biased with an external field $\mathbf{H}_0$.

A pulse is a superposition of waves with different frequencies so that a spectrum of frequencies (its bandwidth) exists around ω_0. If this bandwidth is $\Delta\omega$, then the condition $\Delta\omega \ll \omega_0$ is necessary, if the concept of carrier frequency is to mean anything. In other words, although, in principle, a wave packet [pulse] can be decomposed into Fourier components, their amplitudes are small [negligible] outside $\Delta\omega$. Hence, as shown in figure 5, $\omega(k)$, where **k** is a wave vector, is represented rather well by a Taylor expansion, in the neighbourhood of ω_0. In fact, figure 5 shows that the linear dispersion for a vacuum contrasts rather nicely with the typical (ω, k) variation that can occur for a material, which, in this case, is a magnetic thin film biased with $\mathbf{H}_0$ in the plane of the film, and parallel to the propagation direction. It is indicated that it is the presence of a finite value of $\dfrac{\partial^2\omega}{\partial k^2}$ that causes pulse dispersion, i.e. spreading with local time as the evolution progresses. Figure 6 emphasises the, typical, spreading due to dispersion in a linear system and the fact that there is no spreading in the frequency domain. The pulse propagates along the x-axis and is sketched as a function of local time T. Its *spectrum* is the Fourier transform and the Fourier amplitude distribution is also shown. It is important to observe the following points [9]

- the group velocity, $V_g(\omega)$, depends upon the frequency
- the arrival times of the frequency components spread around ω_0, at a particular point z, are

$$\tau = \frac{z}{V_g(\omega)} = z\left(\frac{\partial k}{\partial \omega}\right)$$

- for a bandwidth $\Delta\omega$, the spread in arrival times of all the ω values is equal to the pulse width $\Delta\tau$, where

$$|\Delta\tau| = z\left[\frac{\partial}{\partial \omega}\left[\frac{\partial k}{\partial \omega}\right]\right]_{\omega_0} \Delta\omega = z\left(\frac{\partial^2 k}{\partial \omega^2}\right)_{\omega_0} \Delta\omega = \left[\frac{z}{|V_g|^3}\frac{\partial^2 \omega}{\partial k^2}\right]_{\omega_0} \Delta\omega$$

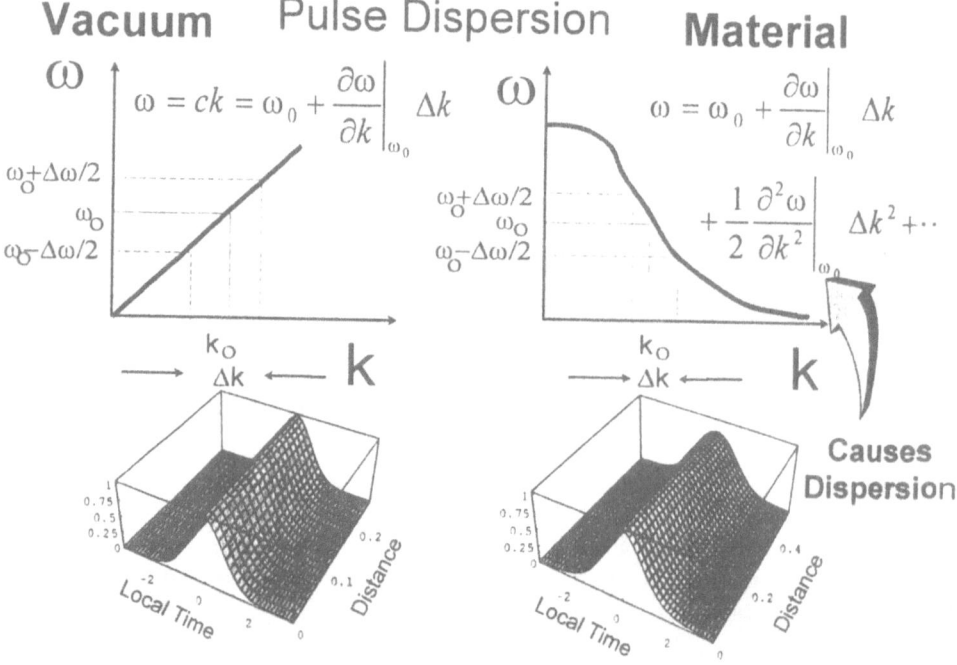

Figure 5. Comparison of pulse behaviour in a dispersionless (vacuum) medium and a medium with material dispersion.

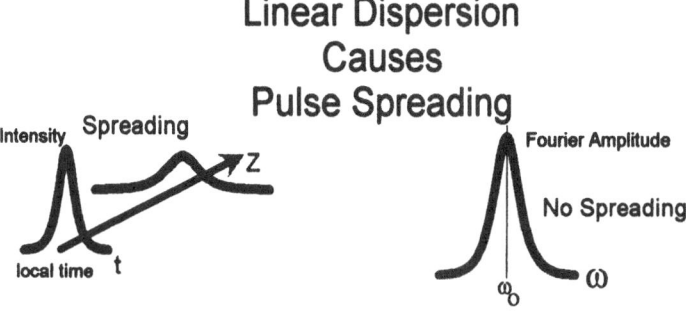

Linear Dispersion Causes Pulse Spreading

- carrier frequency ω_0
- envelope creates frequency distribution

- group velocity v_g depends on frequency
- arrival times of ω components at point Z

$$\tau = \frac{Z}{V_g(\omega)} = Z\left(\frac{\partial k}{\partial \omega}\right)$$

- spread in time=spread in pulse width

$$\Delta\tau = Z\left(\frac{\partial^2 k}{\partial \omega^2}\right)_{\omega_0} \Delta\omega$$

Figure 6. Pulse spreading in the time-domain due to material dispersion in a *linear* medium. There is not a corresponding spreading in the frequency-domain.

This proves, rather dramatically, that it is the existence of a finite value of $\left(\dfrac{\partial^2 k}{\partial \omega^2}\right)_{\omega_0}$, and hence $\dfrac{\partial^2 \omega}{\partial k^2}$, that *causes* dispersion. In optics, $\left(\dfrac{\partial^2 k}{\partial \omega^2}\right)_{\omega_0}$ is called the *group-velocity* dispersion. This is so-called because $V_g(\omega)$ is a function of ω and this alone causes arrival time spreading of signals. In magnetics, $\dfrac{\partial^2 \omega}{\partial k^2}$ is, currently, a favoured quantity and it will be referred to, here, as the *dispersion parameter* to avoid confusion with the adoption of *group-velocity dispersion* for $\left(\dfrac{\partial^2 k}{\partial \omega^2}\right)_{\omega_0}$.

The dependence of angular frequency ω on both wavenumber k and power, P, is sketched in figure 7. For all practical purposes, the dependence upon P is a linear function, so that any change $\Delta\omega$, due to power, is simply γP. Hence, pulse propagation can be described by the expansion [1]

$$\omega = \omega_0 + \left(\frac{\partial \omega}{\partial k}\right)_{\omega_0} \Delta k + \frac{1}{2}\left(\frac{\partial^2 \omega}{\partial k^2}\right)_{\omega_0} \Delta k^2 + \gamma P \qquad (2.1)$$

$$\underbrace{\hspace{5cm}}_{\text{LINEAR}} \qquad \underbrace{\hspace{2cm}}_{\text{NONLINEAR}}$$

284

where $\Delta k = k - k_0$ and $\gamma = \dfrac{\partial \omega}{\partial P}$ is the nonlinear coefficient. In operator language, $\omega - \omega_0 = i\dfrac{\partial}{\partial t}$, $\Delta k = -i\dfrac{\partial}{\partial z}$, where t is *global time* and z is the propagation direction. Equation (2.1) then shows that the pulse amplitude, defined as $\phi(z,t)$, slowly evolves, during propagation, according to the following *nonlinear Schrödinger equation* [1,5]

$$i\frac{\partial \phi}{\partial t} + i\left(\frac{\partial \omega}{\partial k}\right)_{\omega_0} \frac{\partial \phi}{\partial z} + \frac{1}{2}\left(\frac{\partial^2 \omega}{\partial k^2}\right)_{\omega_0} - \gamma |\phi|^2 \phi = 0 \qquad (2.2)$$

where the power has been normalised [4] to be $|\phi|^2$. Typical numerical solutions of (2.2) are shown in figure 8. The solutions include (with dispersion) or leave out (dispersionless) the dispersion parameter $\left(\dfrac{\partial^2 \omega}{\partial k^2}\right)_{\omega_0}$. The labelling 'linear' and 'nonlinear' leave, or include, the $\gamma |\phi|^2$ term. It is interesting that a pulse in a vacuum is, in fact, a solitary wave even though this is of no practical interest. Also the more cautious label 'solitary' has been used even though, when dispersion is included, 'soliton' solutions of (2.2) are possible, two of which are displayed in figure 8.

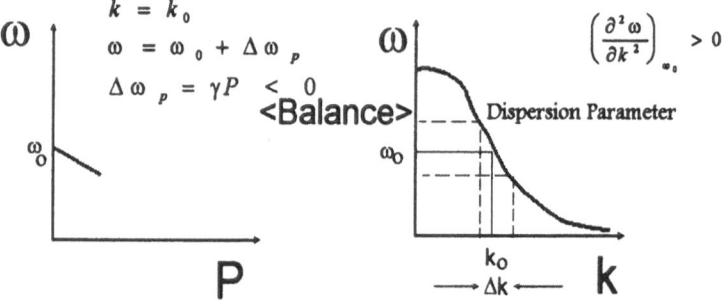

Variations of Angular Frequency [ω] with Power [P] and Wavenumber [k]

●[ω_0,k_0] operating point

Figure 7. Illustration showing that, in a dispersive medium, a change Δk drives the frequency away from ω_0, the carrier value. As far as *dispersion* goes, terms upto $\left(\dfrac{\partial^2 \omega}{\partial k^2}\right)$ are involved in the neighbourhood of ω_0.

The power level P can change ω_0 by an amount γP. The *combined* affect gives equation (2.1) but γP and $\left(\dfrac{\partial^2 \omega}{\partial k^2}\right)$ must be opposite sign

to result in a soliton.

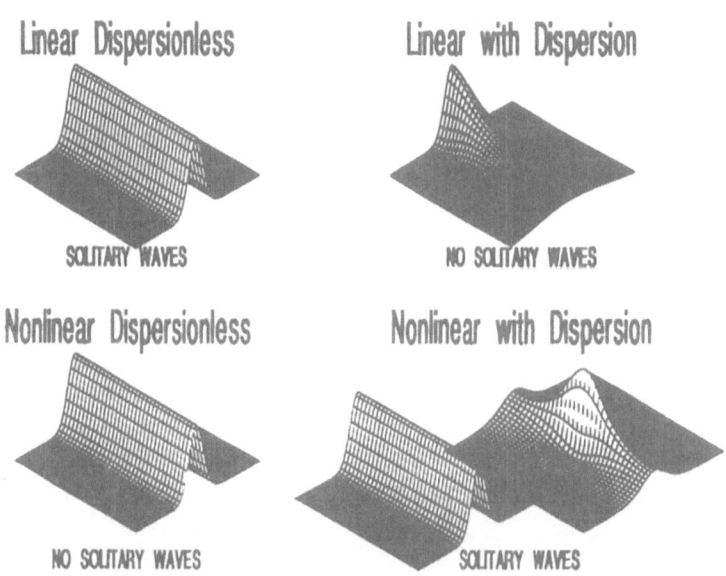

Figure 8. Numerical solutions of equation (2.2). The parameters are: *linear*

dispersionless $\left[\gamma = 0, \left(\dfrac{\partial^2 \omega}{\partial k^2}\right) = 0\right]$; *linear with dispersion*

$\left[\gamma = 0, \dfrac{\partial^2 \omega}{\partial k^2} > 0\right]$; *nonlinear dispersionless* $\left[\gamma > 0, \dfrac{\partial^2 \omega}{\partial k^2} = 0\right]$;

nonlinear with dispersion $\left[\gamma > 0, \dfrac{\partial^2 \omega}{\partial k^2} > 0\right]$.

3. Chirping of Pulses and Soliton Formation

At this stage of the discussion, it is useful to introduce the concept of chirp [4,10-12]. To begin with, figure 9 summarises the idea of phase.

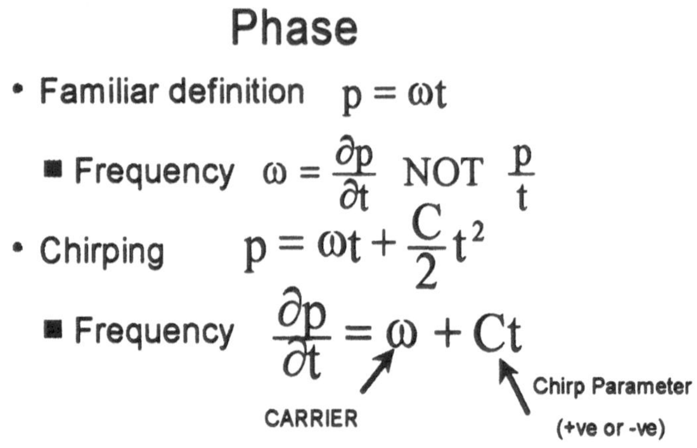

Phase

- Familiar definition $p = \omega t$

  - Frequency $\omega = \dfrac{\partial p}{\partial t}$ NOT $\dfrac{p}{t}$

- Chirping $\quad p = \omega t + \dfrac{C}{2} t^2$

  - Frequency $\dfrac{\partial p}{\partial t} = \underset{\text{CARRIER}}{\omega} + \underset{\substack{\text{Chirp Parameter} \\ \text{(+ve or -ve)}}}{Ct}$

Figure 9. Summary of the concepts of phase and chirp parameter.

The definition that phase is $p = \omega t$ is a very familiar one but the definition that $\omega = \dfrac{\partial p}{\partial t}$ seems, at first sight, to be unnecessary, for such a simple relationship. Why not just use p/t? Such a step leads, immediately, into error, however, if p is not as simple as ωt. The problem is that it is perfectly possible for the frequency to deviate from the, central, carrier frequency, as time is swept across the pulse i.e. the instantaneous frequency can be lower (higher) than the centre, carrier, frequency as the front, or back, end of the pulse is reached. It is perfectly possible for a smooth change to be established, through the existence of a so-called chirp parameter C, which can be positive or negative. The historical background of the concept of chirp is summarised in figure 10. In a Bell Laboratory report in 1951, B M Oliver [10], referring to work on frequency-modulated radar, introduced the use of chirping. The aim was to produce compressed radar signals and the remark attributed to him is "*not with a bang but a chirp*". The word chirp is, perhaps, more obviously attributable to birds, which emit chirped (frequency-modulated) pulses, as a matter of course. The ability to use what birds, and bats [12], do naturally, for the benefit of soliton creation can be said to be an example of Pasteur's view that "*fortune favours the prepared mind*" [10,11]. The original idea in the B M Oliver report is, first, to generate a square pulse envelope in contrast to a continuous wave which is uninterrupted and has a frequency ω_0, for example. A rectangular or square pulse envelope simply 'chops' off the wave, front and rear, encapsulating a carrier, which has a frequency ω_0, *all the way* across the pulse. In other words, in the *time-domain* $\omega = \omega_0$ is not a function of time across the pulse i.e. it is unchirped. This situation can be changed by making the frequency vary with time across the pulse. A linear 'frequency ramp' is shown in the second segment of figure 10 and the frequency variation (chirping) that is established across the pulse is shown in the last segment of figure 10.

Chirp: Historical Background

Bell Laboratory Report: B.M.Oliver (1951)

"Not with a bang, but with a chirp"

Frequency-modulated [chirped] radar signals

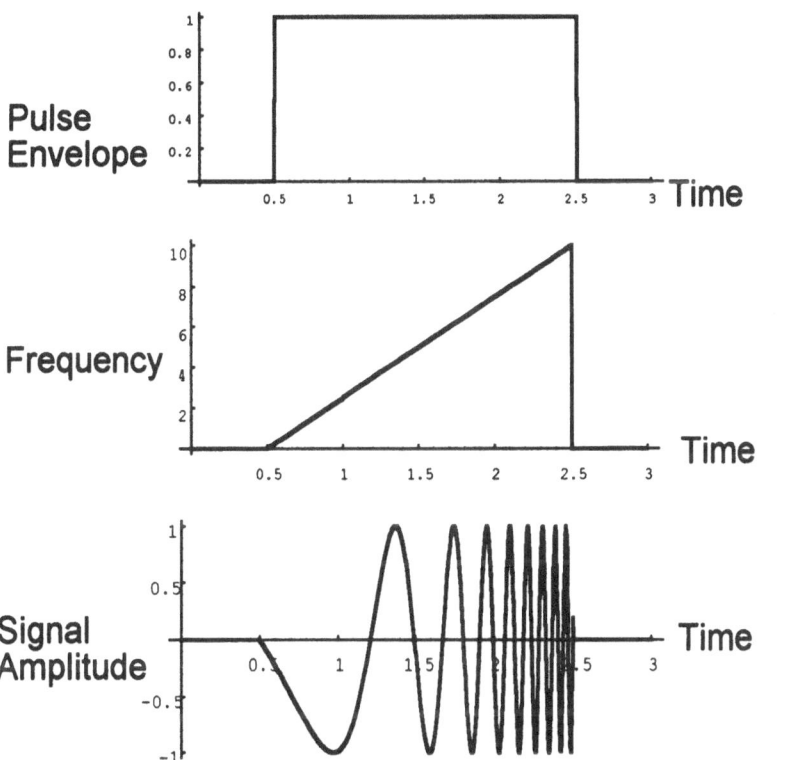

This is an example of Pasteur's opinion

"Fortune favours the prepared mind"

Figure 10. Basic idea on chirped radar put forward, long ago, by Oliver.

There are a number of ways, in a radar system, to add this feature to an outgoing rectangular pulse but it is not of interest to go in to that here. We are interested only in the fact that both material dispersion and the power level (nonlinearity) add a chirp to an evolving pulse, without external intervention. In the original radar example [10,11], the technique was used with a delay line to compress the pulses i.e. because different frequencies have different propagation times through the delay line, delaying the first arrivals causes 'a traffic jam' and, hence, a compression.

As an illustration of the behaviour of pulses in a material, figures 11-13 contain numerical simulations that show, in detail, what happens as a microwave pulse evolves.

Dispersion of a Sech Input in a Linear Medium

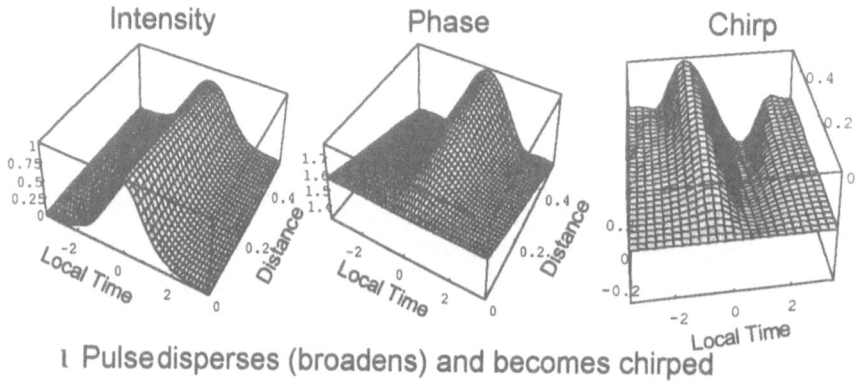

ι Pulse disperses (broadens) and becomes chirped

Figure 11. Dispersion of a pulse in a linear medium.

Figure 11 shows that, in the absence of any nonlinearity, i.e. there is no power dependence of the frequencies, a pulse will disperse. This is a well-known, and expected, feature of the propagation but now we come to the question of the chirp. As can be seen the phase p, which is a constant across the pulse at the beginning of the propagation develops a local time dependence as the evolution proceeds.

The derivative $\frac{dp}{dt}$, where t is now *local time*, is the chirp and across the centre of the pulse a very clear linear chirp develops. The slope is negative so the type of dispersion selected is called *anomalous* and the chirp parameter is negative. If the medium is said to be purely nonlinear then what is meant is that there is a significant power level but no dispersion at all. Of course, this is an artificial situation [dispersion is usually present] but it is being illustrated in figure 12 to make an important point.

Behaviour of Sech Input In A Purely Nonlinear Medium

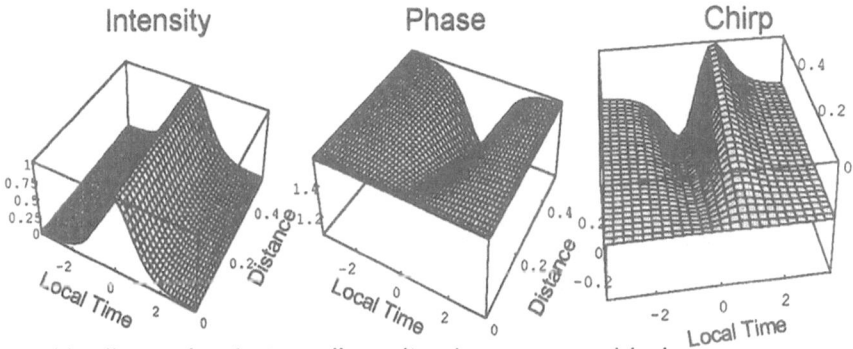

Intensity   Phase   Chirp

ι No dispersion but nonlinearity also causes chirping
ι Chirp is opposite to that caused by material dispersion

Figure 12. Behaviour of a pulse, in the time-domain, in a purely nonlinear medium.

This is that the action of nonlinearity alone does not compress the pulse in the time-domain. Indeed, in the time-domain the pulse progresses without change of shape. The phase behaviour, on the other hand, is very interesting. A positive chirp develops as the evolution progresses. What then happens in a nonlinear dispersive medium? The answer is it is possible for the chirps from dispersion and nonlinearity to cancel. When this happens, a sech(t)-shaped soliton can be created. The solitons are nonlinear states of the system and can exist in various forms, dependent upon the energy available. The lowest-order one is shown in figure 13. Clearly the chirps cannot cancel if hardly any energy enters the material so it is to be expected that some power threshold, for a sech(t) pulse, should be reached before a chirp cancellation, and hence a soliton, will appear. That this is, indeed, the case is shown in figure 14, where a 0.2 sech(t) input disperses, completely, and a 0.8 sech (t) input does not. The 0.8 sech(t) input sheds (disperses) energy it does not require but, because it is above the threshold energy for soliton formation, then goes on to form a stable, lowest-order soliton. If the threshold energy level is exceeded, therefore, the precise input shape does not matter. A dramatic example is given in figure 15 in which a rectangular input pulse, above the threshold energy, "gives birth to a soliton".

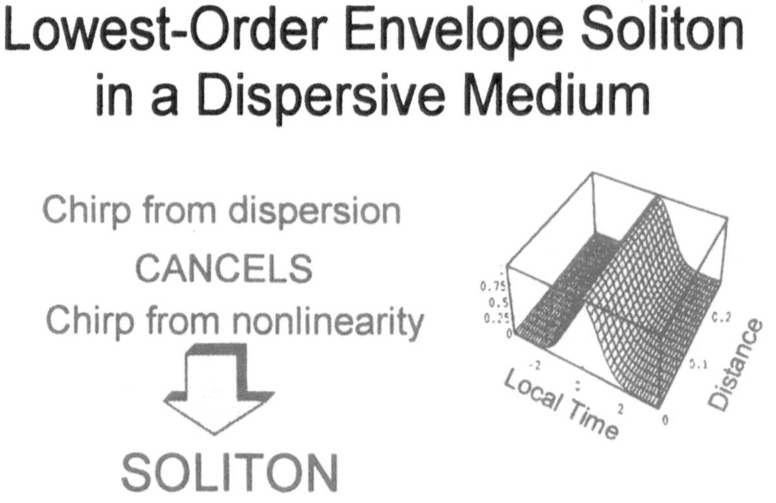

Lowest-Order Envelope Soliton
in a Dispersive Medium
Chirp from dispersion
CANCELS
Chirp from nonlinearity
SOLITON
Local Time
Distance

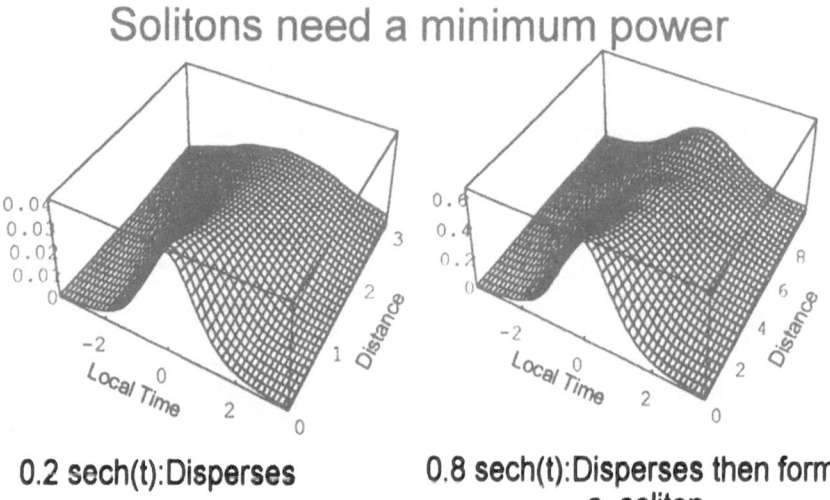

Figure 13. How the lowest-order, envelope, soliton arises.

Threshold Power
Solitons need a minimum power
Local Time
Distance
0.2 sech(t):Disperses
0.8 sech(t):Disperses then form
a soliton

Figure 14. A numerical check on the power needed to create the lowest-order, envelope soliton.

The emerging soliton has the sech shape it is required to have, as a solution of equation (2.2), the nonlinear Schrödinger equation. The equation has rather special properties because an Nsech(t) input generates a fundamental soliton if N = 1 and higher-order solitons if N > 1. Technically, fundamental solitons are of great interest because they do not change their shape as they propagate i.e. they

remain sech-shaped. Higher-order solitons $N = 2$, $N = 3$,... change shape [they breathe but, unfortunately, so does the chirp!] as they propagate but keep returning to Nsech(t), periodically. The period is known as the soliton period and is a useful length scale of the system. Apart from that, higher-order solitons are not useful in a switching device, for example.

Birth Of A Soliton

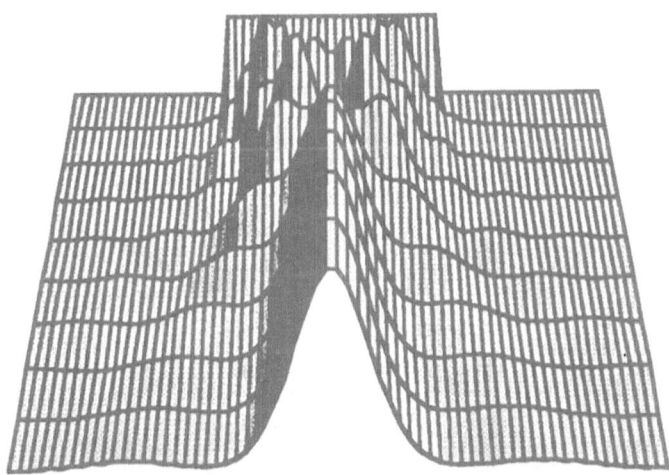

Figure 15. A square input pulse gives birth to an envelope soliton.

This is because even a higher-order, bright, envelope soliton is a bound state of $N = 1$, sech-shaped, fundamental solitons and a higher-order soliton is vulnerable to perturbations. Hence an $N = 2$ soliton, for example, could degenerate quickly to two $N = 1$ solitons. This means that N is the *soliton content* or *soliton number* for a given input pulse. Figures 16(a) and (b) contain a diagramatic explanation of what happens when either a Bsech(t) input pulse is used for a nonlinear system described by the nonlinear Schrödinger equation, or a rectangular pulse Brect(t). The soliton "content" is shown as a function of B. For example, if a microwave pulse Bsech(t) is entered into a dispersive material then, provided $0.5 < B < 1.5$, a fundamental soliton is created. On the other hand, if B rect(t) is entered, a fundamental soliton is created only if $0.5\pi < B < 1.5\pi$. These conclusions come from an exact mathematical treatment of equation (2.2), using inverse-scattering theory [IST]. They are, of course, borne out by numerical simulations.

Rules For Soliton Numbers
Sech Input Condition B sech(t)

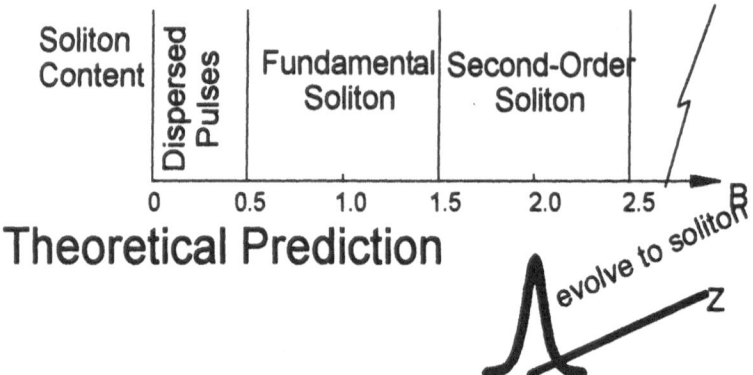

Rules For Soliton Numbers
Square Input Condition B rect(t)

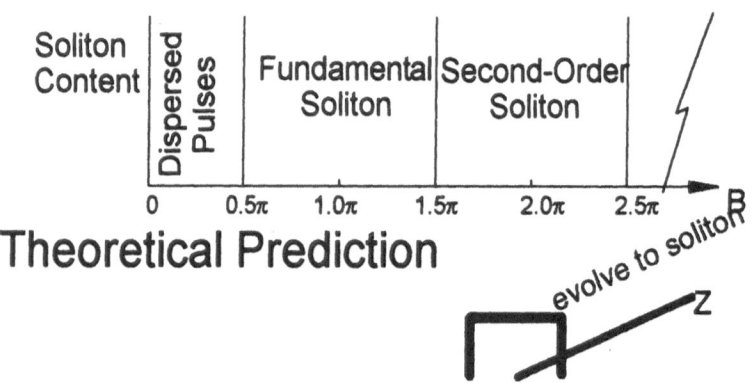

Figure 16. (a) how solitons are created from Bsech(t) input pulses (b) how solitons are created from Brect(t) input pulses.

4. Bright Envelope Solitons in Magnetic Films

In this section the evidence for solitons in magnetic thin films will be reviewed and some ideas on why solitons should be studied and what can be done with them will be addressed. First, there is the question of scales. It is easy to construct both time and length scales for nonlinear pulse propagation in a dispersive medium, yet they are of immense value. The appropriate quantities are [1,13-16]

$$L_D = \frac{T_0^2}{\left|\dfrac{\partial^2 k}{\partial \omega^2}\right|} = T_0^2 \frac{|V_g|^3}{\left|\dfrac{\partial^2 \omega}{\partial k^2}\right|}, \quad L_{NL} = \frac{|V_g|}{\gamma P_0} \tag{4.1a}$$

$$t_D = \frac{D_0^2}{\left|\dfrac{\partial^2 \omega}{\partial k^2}\right|}, \quad t_{NL} = \frac{1}{\gamma P_0} \tag{4.1b}$$

where L_D is called the *dispersion length*, t_D is called a *dispersion time*, L_{NL} is called a *nonlinear length* and t_{NL} is called a *nonlinear time*. T_0 is the half-width of a pulse, P_0 is the peak power and D_0 is the spatial half-width of the pulse. L_D is the distance over which a pulse in a linear, dispersive medium will double its width. L_{NL} is the distance over which the phase of the *centre of the pulse* increases, entirely due to the nonlinearity. Now nonlinearity, alone, changes the phase of the pulse to $e^{i\phi}$, where the maximum value of ϕ is $\gamma P_0 z$, z being the propagation distance when $\phi = 1$ then $z = L_{NL}$. Typical values and scales, in comparison to optical phenomena, are given in the table below.

QUANTITY	MAGNETICS	OPTICS
pulse width dispersion	100ns	1ps
observable within	1cm	50m
group velocity	$1.8\text{cm.}\mu\text{s}^{-1}$	$2 \times 10^8 \text{ms}^{-1}$
damping coefficient	$6\mu\text{s}^{-1}$	negligible

Envelope solitons have been observed using the typical magnetostatic wave (MSW) delay line shown in figure 17. It consists of an yttrium iron garnet (YIG) thin film, magnetised to saturation with an externally applied magnetic field, sitting upon a gadolinium gallium garnet (GGG) substrate. Excitation of microwaves [magnetostatic waves in the spin system of the magnetised film] takes place through metal striplines, positioned as above. The idea is to excite one stripline with a pulse and then hope that a microwave soliton is formed on the film. The initial pulse is expected to undergo a rapid evolution to a soliton state [or multiple soliton state] which is then received by the second antenna. The basic design, shown in figure 17, has been used by all the experimental groups so what is the state-of-the-art?

The pioneering experimental work is referred to in figure 18, which shows that work began long ago under the leadership of Kalinikos and Slavin. It has been followed up vigorously by the Rome group and, in recent years, consolidated, extended and elegantly presented under the leadership of Patton in Colorado.

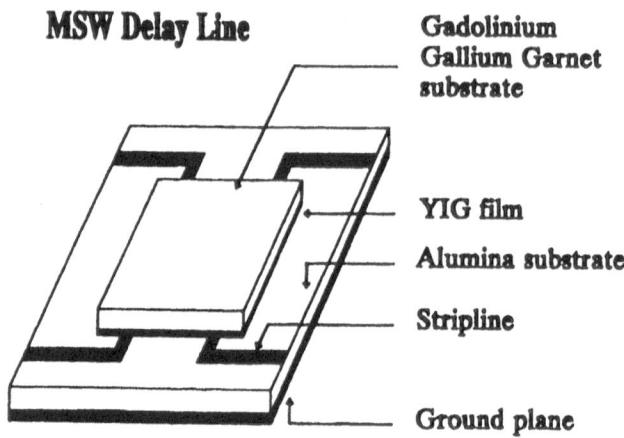

Figure 17. Typical magnetostatic (MSW) delay line type of device.

State of the Art Experimental

• Pioneering experiments by

- Kalinikos, Slavin, Kovshikov 1983, 1984--1990 (St.Petersburg)
- de Gasperis, Marcelli, Miccoli, 1987,1988 (Rome)
- Patton, Chen, Tsankov, Nash, Kabos 1994 (Colorado State University)

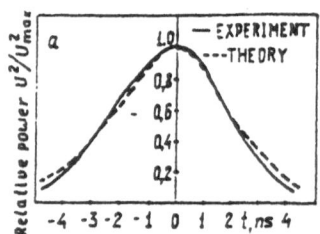

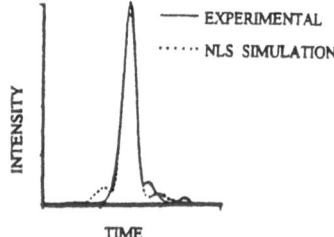

Kalinikos et al (Fit to sech shape)   Patton et al (Fit to numerics)

Figure 18. Brief description of important microwave soliton experiments, using magnetic thin films.

Some of the most convincing results are included in figure 18, in which attempts to fit theory to experiment have been rather successful.

Generally, experimentalists use square or rectangular pulses and the evolved pulses have been observed to have asymmetric shapes, quite close to the input antennae. In fact, nonlinear pulses, during their evolutionary stage, may be

- narrow
- multi-peaked
- split
- symmetric   - antisymmetric

Can all these outcomes be claimed as solitons? The answer is no, but solitonic behaviour can always be expected, provided that the evolution distance can be long enough. Damping may well intervene before solitons can be formed, however. The experimental evidence for solitons is as follows [16-23]

- observed threshold of nonlinearity $\propto \dfrac{1}{(\text{pulse width})^2}$

- sech-shaped pulses have been observed
- collision experiments preserve the pulse shapes [this is rather convincing evidence]
- the measured phase is constant across the centre of the nonlinear pulses [this is also convincing evidence of chirp cancellation]

5. Simulation of Nonlinear Pulse Evolution

The numerical simulations use equation (2.2) and the following data and notation is used: $|U(z,t)|$ is a normalised amplitude [1,24,25], which it is customary to define as $m(z,t)/(\sqrt{2}M_s)$, where $m(z,t)$ is the microwave magnetisation response [24] and M_s is the saturation magnetisation;

$$\omega_1 = \left(\frac{\partial \omega}{\partial k}\right)_{\omega_0} = 1.8\,\text{cm}/\mu s, \qquad \omega_2 = \left(\frac{\partial^2 \omega}{\partial k^2}\right)_{\omega_0} = -0.3\,\text{cm}^2/\mu s,$$

$\gamma = 3 \times 10^4\,\mu s^{-1}$. If damping is included in a simulation, then a term in $i\eta U$ is added to the left-hand side of equation (2.2), where $\eta = 6.0\mu s^{-1}$. All the calculations, unless stated to the contrary, are performed in the *laboratory frame* [1,4,24]. There are now some choices as to how to set the boundary conditions to generate a numerical solutions of equation (2.2). Basically, a pulse must be started up on the t or the z axis. Conventionally [4,24,26], the initial condition is set upon the z axis. Such a condition presents problems of physical interpretation in *real units*, however. In practice a pulse is 'switched on', allowing the initial pulse to be of a certain duration, so using the time axis makes physical sense [1,24,25]. Accordingly, all the simulations, from now on, have been

296

generated from this type of initial condition. Specifically the NLS has been solved in the laboratory frame and the method of solving it is called *forcing* [27]. An immediate problem is that inverse-scattering [28] does not strictly apply to this 'forced' problem. Yet it is inverse scattering results we usually turn to to *count* the number of solitons in a pulse or pulse train [28]. Hence *soliton counting* based upon inverse-scattering results, may now not be accurate enough to say how many solitons have been generated, for a given input. Instead of counting, using an inverse scattering result, we have attempted to fit, numerically, mathematically accurate forms for the shapes of $N = 1$ and $N = 2$ solitons to the computer-generated results. In this way, a very reasonable estimate is obtained for the order of the solitons emerging as propagation evolves. Figure 19, on a time-distance plane, shows what happens, under *forced boundary conditions*, to a 180ns rectangular input pulse with a (normalised) magnetostatic wave input amplitude $\phi_0 = 0.03$.

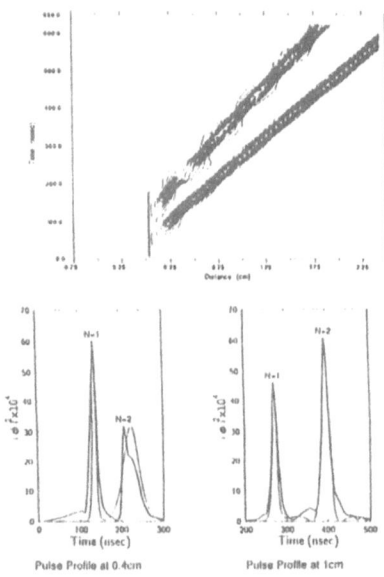

Figure 19. Evolution of an undamped, 180ns, rectangular, input pulse, shown as contours on a time-distance plane. Pulse cross-sections are also shown as they would appear at 0.4cm and 1cm from the input microstrip.

The evolution of the magnetostatic wave amplitude $\left[|\phi|^2 \times 10^4 \right]$ with time is also shown. The pulse profiles, computed at distances 0.4cm and 1cm from the input antenna show clearly the development of $N = 1$ and $N = 2$ solitons. The dotted lines are the theoretical soliton shapes with the overlap being smoothed out, without loss of significance. Linear damping is always severe in a magnetic film so this must always be borne in mind. An example is given in figure 20, which shows how dramatically a damping coefficient of $6.0 \mu s^{-1}$ affects a 40ns input pulse.

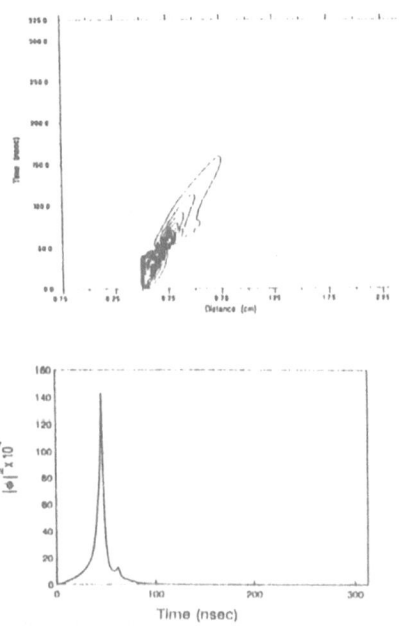

Figure 20. Evolution of a, damped, rectangular 40ns pulse in a YIG film with a damping coefficient equal to 6.0μs⁻¹. Note that the pulse shape at 0.17cm, from the input microstrip, shows strong asymmetry.

At even only a distance of 0.17cm from the input antenna, strong asymmetry is observable in the nonlinear pulse. This sort of asymmetry has been seen experimentally but it could not, at this stage, be used to say this is evidence of soliton formation. Damping of solitons is a complicated issue because both linear and nonlinear damping can occur. Linear damping is included through a term η U, added into the Schrödinger equation but nonlinear damping can be included through a term proportional to $i|U|^2 U$. There does not seem to be any information available, yet, on whether nonlinear damping is a real possibility in magnetic thin films, so no knowledge of the *nonlinear damping coefficient* is available. Accordingly, no further discussion of this type of damping will be included here. If a real material attenuates a soliton, through linear damping, it raises the question of how to introduce amplification and, having done so, does such a process destroy the soliton? First, it is not so easy to amplify microwave envelope solitons in magnetic thin films. In optical fibres, it is done by doping sections of the fibre with erbium [4]. The erbium is then pumped to provide the extra photons needed to 'amplify' the solitons. In magnetic thin films it is magnetostatic waves that are used so what can be used as an amplifier? A recent experiment [20] on magnetostatic forward volume waves used GaAs amplifiers. In the beginning, an amplified pulse was fed into a magnetostatic wave device. The narrowed output, soliton, was then amplified again and fed into a second device. The outcome was further pulse compression and soliton generation. The schematic flow of the experiment is sketched in figure 21. Pulses as short as 12ns were observed and it should be expected that this type of cascaded YIG film device could be very useful in soliton switches and couplers.

Amplification

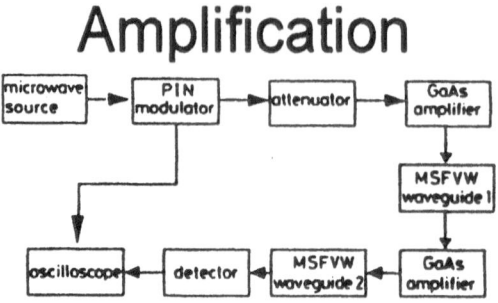

- Experiment performed by Priye & Tsutsumi
- High power pulse is fed into a magnetostatic wave device
- Narrow output pulse is amplified and fed into a second device
- Outcome- further pulse compression
- Application - use first device to generate soliton - allow soliton interactions in the second waveguide

Figure 21. Amplification of nonlinear pulses leading to soliton formation in the magnetostatic forward volume mode configuration.

In the presence of linear damping, a term $i\eta U$ is added into the nonlinear Schrödinger equation. Furthermore [4] a soliton, under these conditions will increase its width by a factor $\exp(2\eta z)$, where z is the propagation distance [4], and its amplitude will fall by $\exp(-2\eta z)$. Using L_D, as defined in equation (4.1a), the damping coefficient can be written as $\Gamma = \eta L_D$ and z can be made dimensionless, by measuring distance in units of L_D. Adopting these units, figure 22 shows the concept of *average soliton* [4]. It shows that the damped soliton decays as $\exp(-2\Gamma z)$, but, provided that the, dimensionless, amplitude is maintained between 1.5 and 0.5, no shape loss occurs and the soliton can be restored through amplification. The knowledge that the range 0.5 to 1.5 matters arises from inverse scattering theory [25,28] and it is assumed, in figure 22 that the soliton regime has been reached by propagating far enough away from the input plane of a magnetostatic. It is not difficult to appreciate, from figure 22, that a spacing of $z_a \approx 0.55/\Gamma$ between the amplifiers is enough to sustain, *on average*, a perfectly good soliton.

6. Conclusions

This brief review is intended to bring out some of the fundamental properties of the kind of solitons that can be generated, at microwave frequencies, on magnetic thin films. The emphasis here is on physical properties, rather than on mathematical detail. Accordingly, the text is liberally illustrated with illustrations and numerical simulations. The overall conclusion is that envelope solitons are not only possible on magnetic thin films but that they have been observed on a number of occasions. The field looks set, therefore, for an expansion into some prototype applications. It is appropriate in this concluding section, therefore, to comment on some of the possibilities for future experiments.

Average Soliton

ι amplitude of a damped soliton decays as $\exp[-2\Gamma]$
ι Γ is the damping coefficient
ι if amplitude is 1.5 ... 0.5 no shape loss occurs
ι soliton can then be amplified

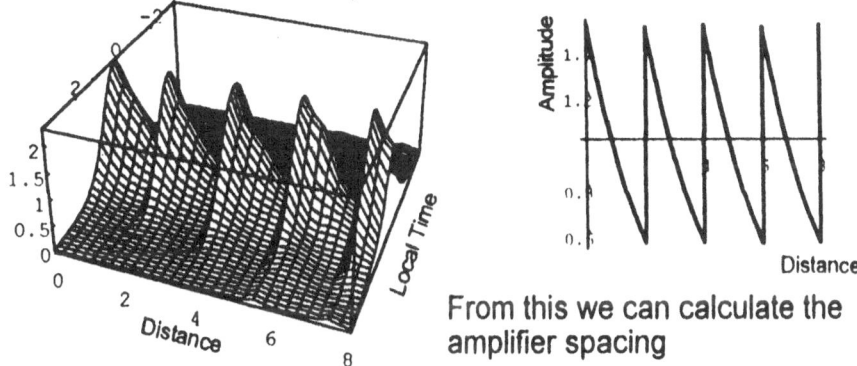

From this we can calculate the amplifier spacing

Figure 22. Average soliton concept.

The first point to make, concerning applications, is that in a device that is nonlinear an attempt to control its operation by adjusting the power level can be made. Figure 23 gives a simplified overview [29] of a possible directional coupler switch that uses this idea. It suggests the use of microwaves in signal, or control, beam form to effect a transmission change in two waveguides, coupled together, by proximity to each other. If a pulse of energy is launched into the top guide then the outcome is emission from port (1), or port (2). The interesting thing is that increasing the input power changes the outcome. As the power increases, the energy tends, more and more, to go straight through to port (1). At a certain level of power, almost all the energy comes out of (1) and so, in reality, *a switch* has been made. If solitons are used in the guides then the response, shown in figure 23 for the simpler, CW (continuous wave) case, becomes much sharper [30].

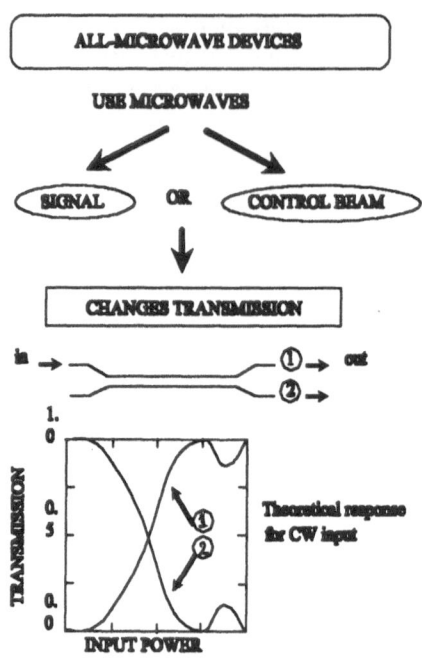

Figure 23. Introduction to the idea of an all-microwave directional coupled switch.

Although the basic idea of a switch is shown in figure 23 what devices are of interest in nonlinear microwave research? A short answer is provided in figure 24. Here we sketch a simple microwave coupler, a very simple directional coupler and a device known as a Mach-Zehnder interferometer. All of these are of potential interest.

It is interesting that although the preceding discussions concern temporal (pulse) solitons the work also can be used to describe spatial (beam) solitons. In the latter, it is diffraction that is being eliminated by the nonlinearity, to create stationary beam states. Such beams are ideally suited for manipulation in planar waveguides. Figure 25 shows one of the possibilities [31], adapted to show that this situation is also possible for (microwave) magnetostatic waves. Note that the scale in the propagation direction is precisely what is to be expected for magnetostatic waves on magnetic thin films and that the beams are rather narrow. The 'device' operates by permitting nonlinear beams (spatial solitons) to propagate in the *plane* of thin film, so there is guiding in the other dimension. As a result these solitons are stable because one dimension is *frozen out* by the guiding i.e. the diffraction/nonlinearity interplay is in the plane of the film along the x-axis.

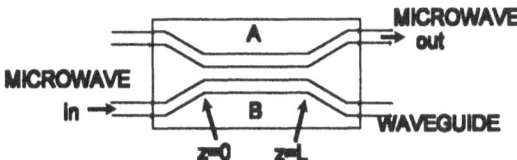

● SIMPLE INTEGRATED MICROWAVE COUPLER

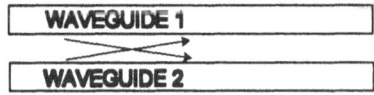

● EVEN SIMPLER DIRECTIONAL COUPLER

● MACH-ZEHNDER

Figure 24. Devices of interest in the search for all-microwave processing.

The general idea [31] is that A,C,D and H are to be *control* guides and B,E,F and G are to be *signal* guides. Although the original work [31] designated *photonic switching*, it is, nevertheless, generic and can just as well be *magnonic switching*, at microwave frequencies, as indicated in figure 25. The operation is as follows. In the absence of a control beam, a signal beam, excited on the input plane, propagates for a little while and then crosses the nonlinear segment. How this is to be achieved is a problem for the experimentalists and need not trouble us here. The signal goes into the centre guide on leaving the middle nonlinear section but is allowed to couple to its two neighbouring guides. Suppose now that, at the same time as the signal is initiated, a control beam comes towards it, from the left. These two beams *collide* in the nonlinear medium but, *because they are solitons*, experience only a transverse position shift. This is a well-known and beautiful feature of a two-soliton interaction. For this scenario, the control beam exits to the right, but the signal 'jumps' into the left-hand signal guide. Nevertheless, because of proximity coupling, this signal ends up exiting from the *right-hand* signal guide. Of course, the opposite will happen, if the control beam comes from the other side. So we have 1 signal beam ending up in 3 possible positions [31], depending upon what is done. Using more collisions can increase the potential of this device.

302

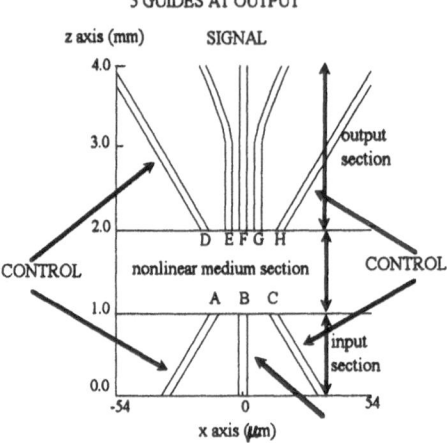

Figure 25. Proposal to use nonlinear beams (spatial solitons) for all-microwave processing.

Finally, the prospects for all-microwave (magnonic) switching look very bright, indeed, with the possibility that magnetic films and/or layered structures will be the perfect testbeds for soliton ideas, many of which arise in photonics.

References

1. Boardman, A.D., Nikitov, S.A., Xie, K. and Mehta, H. (1995) Bright magnetostatic spin-wave envelope solitons in ferromagnetic films, *J. Magn. and Mag. Mat.* **145**, 357-378.
2. Drazin, P.G. (1983) *Solitons*, Cambridge University Press, Cambridge.
3. Scott-Russell, J. (1844) Report on waves, *Proc. Roy. Soc. Edinburgh*, 319-320.
4. Agrawal, G.P. (1995) *Nonlinear Fibre Optics*, Academic Press, New York.
5. Korteweg, D.J. and de Vries, G. (1895) On the change of form of long waves advancing in a rectangular canal and on a new type of long stationary waves, *Phil. Mag.* **39**, 422-443.
6. Scott, A.C., Chu, F.Y.F. and McLoughlin, D.W. (1973) The soliton: a new concept in applied science, *Proc. IEEE* **b1**, 1443-1483.
7. Chcn, M., Tsankov, M.A., Nash, J.M. and Patton, C.E. (1993) Microwave magnetic-envelope dark solitons in yttrium iron garnet thin films, *Phys Rev Lett* **70**, 1707-1710.
8. Zvezdin, A.K. and Popkov, A.F. (1983) Contribution to the nonlinear theory of magnetostatic spin waves, *Sov. Phys. JETP* **57**, 350-355.
9. Lee, D.L. (1986) *Electromagnetic Properties of Integrated Optics*, John Wiley & Sons, New York.
10. Oliver, B.M., Bell Telephone Laboratories Technical Memorandum MM-51-150-10, Case 33089, March 8, (1951)
11. Klauder, J.R., Price, A.C., Darlington, S. and Albershein, W.J. (1960) The Theory and Design of Chirp Radars, *Bell Syst. Tech. J.* **39**, 745-808.
12. Dawkins, R. (1986) *The Blind Watchmaker*, Penguin, London.
13. Boardman, A.D., Wang, Q., Nikitov, S.A., Shen, J., Chen, W. and Mills, D. (1994) Nonlinear waves in ferromagnetic films, *IEEE Trans. Mag.* **30**, 1-30.
14. Boardman, A.D., Nikitov, S.A. and Waby, N.A. (1993) Existence of spin-wave solitons in an antiferromagnetic film, *Phys. Rev. B* **48** 13602-13606.
15. Boardman, A.D., Gulyaev, Yu-V., Nikitov, S.A. and Qi, Wang (1990) Nonlinear waves in ferromagnetic films in *Nonlinear Waves in Solid State Physics*, NATO ASI Series B : **247**, A.D. Boardman, M. Bertolotti and T. Twardowski (Eds), Kluwer Academic Publishers, New York and London 235-258.
16. Tsankov, M.A., Chen, M. and Patton, C.E. (1994) Forward volume wave microwave envelope solitons in yttrium iron garnet : propagation, decay and collision, *J. App. Phys.* **76**, 4274-4289.
17. Kalinkos, B.A., Kovshikov, N.G. and Slavin, A.N. (1983) Observation of spin-wave solitons in ferromagnetic films, *JETP Lett.* **38**, 13-47.
18. Gasperis, P. de, Marcelli, R. and Miccoli, G. (1987) Magnetostatic soliton propagation at microwave frequency in magnetic garnet films, *Phys. Rev. Lett.* **59**, 481-484.
19. Kalinikos, B.A., Kovshikov, N.G. and Slavin, A.N. (1990) Experimental observation of magnetostatic wave envelope solitons in yttrium iron garnet films, *Phys Rev B* **42**, 8658-8660.

304

20. Priye, V. and Tsutsumi, M. (1995) Observation of short pulse in cascaded magnetostatic soliton waveguide, *Elec Lett* **31**, 464-465.

21. Chen, M., Tsankov, M.A., Nash, J.M. and Patton, C.E. (1994) Backward-volume wave microwave-envelope solitons in yttrium iron garnet films, *Phys. Rev. B* **49**, 12773-12790.

22. Kovshikov, N.G., Kalinikos, B.A., Nash, J.M. and Patton, C.E. (1994) Reflection and collision of backward volume wave microwave envelope solitons in yttrium iron garnet films, Unpublished conference report

23. Nash, J.M., Kabos, P. and Patton, C.E. (1994) Phase profiles for microwave magnetic envelope solitons. Unpublished conference report

24. Chen, M., Nash, J.M. and Patton, C.E. (1993) A numerical study of nonlinear Schrödinger equation solutions for microwave solitons in magnetic thin films, *J App Phys* **73**, 3906-3909.

25. Slavin, A.N. and Dudko, G.M. (1990) Numerical modelling of spin wave soliton propagation in ferromagnetic films, *J Magn and Mag Mat* **86**, 15-123.

26. Satsuma, J. amd Yajima, N. (1974) Initial value problems of one-dimensional self-modulation of nonlinear waves in dispersive media, *Prog. Theor. Phys. Suppl.* **55**, 284-306.

27. Kaup, D.J. and Hansen, P.J. (1986) The forced nonlinear Schrödinger equation, *Physica* **18D**, 77-84.

28. Zakharov, V.E. and Shabat, A.B. (1972) Exact theory of two-dimensional self-focusing and one-dimensional self-modulation of waves in nonlinear media, *Soviet Physics JETP* **34**, 62-69.

29. Smith, P.W. (1993) All-optical devices: materials requirements, *SPIE* **1852**, 2-9.

30. Pare, C. and Florjanczyk, M. (1990) Approximate model of soliton dynamics in all-optical couplers, *Phys Rev A* **41**, 6287-6295.

31. Shi, T. and Chi, S. (1990) Nonlinear photonic switching by using the spatial soliton collision, *Optics Letters* **15**, 1123-1125.

SOLITON-LIKE PACKETS OF PARAMETRICALLY COUPLED SPIN-WAVES

A.F. POPKOV
Zelenograd Research Institute of Physical Problems
103460 Moscow, Russia

N.V. OSTROVSKAYA
Moscow Institute of Electronic Technology
103482 Moscow, Russia

AND

L.L. SAVCHENKO
Department of Physics, Moscow State University,
Moscow, 119899, Russia

1. Introduction

Dielectric magnetic films of high quality such as YIG-films are promising materials for superhigh-frequency spinwave devices: filters, delay lines, resonators, spectrum analyzers and others [1, 2]. Nonlinear three- and four-magnon processes govern dynamical range of linear magnetostatic wave (MSW) devices, but also may be used for signal to noise enhancement and other nonlinear microwave signal processing [3–6]. It is attractive also to put into practice of creation and control of nanosecond and sub-nanosecond powerful pulses by using nonlinear dispersion and self-action properties of spin waves. It may be interesting for communications, experimental practice and other fields of microwave applications. One of the way in the creation and control of soliton-like pulses, which has been proposed earrlier in nonlinear optics, was the using of distributed feedback coupling of nonlinear waves. A number of phenomena such as gap soliton formation, self-pulsating and others connected with parametric coupling of nonlinear optic waves has been considered [7–9]. Similar phenomena in nonlinear spin wave dynamics have not been analyzed yet in detail. Some hysteretic effects connected with nonlinear MSW rejection by grating have been calculated in [10].

R. Marcelli and S.A. Nikitov (eds.), Nonlinear Microwave Signal Processing: Towards a New Range of Devices, 305–323.

Nonlinear soliton-like transmission and multisoliton decay of MSW pulses have been demonstrated at first in YIG-films, where MSW spectrum was complicated due to strong spin pinning at the surface of magnetic films [11]. Soliton-like phenomena of MSW propagation in such films have some analogy with the effects of self-induced transparency in nonlinear nuclear-acoustic resonanse, which has been observed in KMnF$_3$ [12]. After that, soliton-like evolution and self-compression of MSW pulses in perfect YIG-films without surface spin pinning have been discovered [13] and investigated in detail [14–17]. The treatment of these nonlinear phenomena is based usually on Witham-Lighthill-Karpman theory of wave instability [18–20] as well as on nonlinear equations of envelope evolution, deduced by reduction of magnetodynamic and magnetostatic equations in approximation of nonlinear geometric optics [21–23]. The parabolic nonlinear equation modified with the dissipation and high dispersion of MSW short pulses taken into account seems to be rather satisfactory for their evolution description. As to the films with complicated MSW spectrum, caused by interaction of MSW with other kind of waves, excited in magnetic film (spin wave resonance modes, acoustic waves) the existed theory is insufficient for their description. Witham-Lighthill approximate theory applicable to the hybridised waves gives only guiding ideas for interpretation of experiment. It is not clear for example how solitons should behave, when pulse frequency lies in nontransmission band of interacting waves. It is unclear how the gap influences the soliton stability et.c. On the other hand, the intersection of spectrum branches arises also in multilayer film structures [24], and it is typical for processes of resonant MSW scattering by periodical gratings [25]. In all these cases, there arises stable parametrical interaction of MSW harmonies near their phase synchronization point due to the creation of distributed feedback coupling. Only nonlinear waves can exist in the formed nontransmission gap. We shall consider soliton-like properties of hybridised nonlinear waves in nontransmission gap, soliton stability, evolution under local electromagnetic pumping as well as scattering of nonlinear MSW by confined periodical grating. In conclusion we shall discuss the limitations of considered effects connected with four-magnon MSW decay at high power.

2. Nonlinear Coupling Equations

Let us consider monochromatic one-dimensional spin waves of carrying frequency w and wave number k propagating in y-direction of magnetic film with thickness d. It is convenient to introduce an amplitude of spin-wave envelope as a small angle averaged through the period of fast time $T = 2\pi/w$ of magnetization vector $\vec{M}(y,t)$ deviation from its equilibrium state by

$$\phi(y,t) = \langle |\vec{M}(y,t) - \vec{M}_0| \rangle / M_0.$$

An interaction of two nonlinear spin-waves near the point of intersection of their spectrum branches $w_\pm(k_\pm)$, when

$$w(k_+^0) = w(k_-^0), \qquad k_+^0 + k_-^0 = 0, \tag{1}$$

may be described by nonlinear coupled equations [26]:

$$
\begin{cases}
(i\partial_t + iv_g^+ \partial_y + \beta_+ \partial_y^2 - A_+|\phi_+|^2 - \\
\qquad - B_+|\phi_-|^2)\phi_+ + w_B^+ \phi_- = -i\delta w_+ \phi_+ + h_+(y,t) \\
(i\partial_t + iv_g^- \partial_y + \beta_- \partial_y^2 - A_-|\phi_-|^2 - \\
\qquad - B_-|\phi_+|^2)\phi_- + w_B^- \phi_+ = -i\delta w_- \phi_- + h_-(y,t)
\end{cases}, \tag{2}
$$

where $v_g^\pm = \frac{\partial w_\pm}{\partial k_\pm}$ are group velocities of interacting waves, $\beta_\pm = \frac{1}{2}\frac{\partial^2 w_\pm}{\partial k_\pm^2}$ are dispersion parameters, $A_\pm$, $B_\pm$ are nonlinear parameters, $w_B^\pm$ are linear coupling coefficients, $\delta w_\pm$ are MSW line widths, $h_\pm(y,t)$ describe electromagnetic linear pumping.

Here, for the simplicity, we shall consider the same kind of interacting waves propagating in opposite directions so that $v_g^+ = -v_g^- = v_g$, $\beta_\pm = \beta$, $A^\pm = B^\pm/2 = \alpha$, where $\alpha = \partial w/\partial|\phi|^2$ is nonlinear frequency shift, $w_B^\pm = w_B$ is half-width of nontransmission gap. In general case of $v_g^+ \neq v_g^-$ one can reduce initial system to the symmetrical form by the replacement $\tilde{y} = y - (v_g^+ + v_g^-)/2$ after which $v_g = (v_g^+ - v_g^-)/2$.

The parameters of equations (2) are mainly defined in long wave approximation ($kd \ll 1$) [22]. Particularly, in the case of surface kind of MSW one has $v_g = w_M^2 d/4w_0$, $\beta = -(w_M^4 d^2/16w_0^3)(1 + 8w_M^2/w_0^2)$, $\alpha = w_H w_M/4w_0$, $\delta w = \gamma \Delta H$, where $w_M = \gamma 4\pi M_0$, $w_H = \gamma H$, $w_0 = (w_H w_M + w_H^2)^{1/2}$, γ is magnetomechanical ratio, ΔH is field line-width of MSW.

The parameter w_B, which characterizes the strength of MSW coupling and the value of rejection gap, is proportional to the amplitude of modulated magnetic field $h_\sim$ creating feedback coupling of MSW, i.e.

$$w_B \sim \gamma h_\sim F(w),$$

where $F(w)$ is nondimensional coefficient determined by magnetodynamical properties of magnetic film (see for example [26]), and its value lies approximately in the range $F \sim 1 \div 10$.

The value of the gap frequency w_B widely varies from less than line width ($w_B \ll \delta w$) up to the carrying frequency of MSW, i.e. $w_B \sim w_H$, w_0, w_M. It is because of that the field of periodical modulation $h_\sim$ may be created either by induced currents in metallic strips of grating [27], magnetoelastic fields of surface acoustic wave [28], or by alternating currents in serpentine conductors as well as currrents in the coil [29, 30] creating magnetic field modulated in time.

Parameters w_B and v_g determine the dispersion of hybridized linear wave in absence of dissipation ($\delta w = 0$) and waveguide dispersion ($\beta = 0$) in accordance with well known relation

$$\Delta w^2 = w_B^2 + v_g^2(\Delta k)^2 \qquad (3)$$

where $\Delta w = w - w^0$, $\Delta k = k - k^0$ are detuning of the frequency and wave number from phase synchronization point w^0, k^0. This dispersion prevail wave guide MSW dispersion characterized by parameter β for not too strongly localized wavepackets, i.e. when $\Delta k \ll (w_B/v_g)\{(v_g^2/w_B\beta)^{2/3} - 1\}^{1/2}$. In the last case one can neglect wave guide dispersion and put $\beta = 0$ in coupled equations (2).

Let us discuss now right-hand side terms of equations (2). In absence of coupling, when $w_B = 0$, each equation of the system (2) describes envelope evolution of appropriate MSW and the first term in right-hand side describes MSW damping. It must be noted that envelope evolution equations are valid only for narrow wave packets, when $\Delta k/k_0$, $\Delta w/w_0 \ll 1$. Therefore spatial dimension of packet Δy is supposed to be much greater than the wave length, i.e. $\Delta y k \gg 1$. It means that microstrip transducer, which dimension satisfy opposite condition $\Delta y k \lesssim 1$, may be described as the point source of travelling wave formated by transducer in the far wave zone. One can use the Dirac δ-function for point source description by $h_\pm(y,t) = \eta_\pm \delta(y)j(t)$, where $j(t)$ is time dependent amplitude of exciting current at MSW carrying frequency. Parameters $\eta_\pm$ are easily related to the radiative resistances $R_\pm$ of transducer. Really, the steady-state solution of equations (2) for $j_w = j_w e^{i\Delta wt}$ in the absence of coupling ($w_B = 0$), waveguide dispersion ($\beta = 0$) and dissipation ($\delta w = 0$) is

$$\phi_\pm = \eta_\pm v_g^{-1} j_w e^{i\Delta w(t \mp y/v_g)}. \qquad (4)$$

MSW power in this case is equal to

$$P = æ(|\phi_+|^2 + |\phi_-|^2) = \frac{æ}{v_g^2}(\eta_+^2 + \eta_-^2)|j_w|^2, \qquad (5)$$

where parameter æ is defined for main kinds of MSW in [22]. Particularly for the surface MSW, it is equal to

$$æ = \pi L d^2 M^2 w_0, \qquad (6)$$

where L is film width. This power on the other hand may be expressed by [32]

$$P = (R_+ + R_-)|j_w|^2. \qquad (7)$$

Therefore, it follows that

$$\eta_\pm = v_g(R_\pm/\text{æ})^{1/2}. \tag{8}$$

One can relate the exciting currents $j_w(t)$ (if necessary) to the input electromagnetic power. However, we shall stop at it and return to the MSW soliton problem.

3. Gap soliton

The initial system with $\beta = 0$ in normalized variables $[t] = w_B t$, $[y] = y w_B/v_g$, $[\varphi] = \phi[\alpha/w_B]^{1/2}$ takes the form

$$\begin{cases} (i\partial_t + i\partial_y + sH_-)\varphi_+ + \varphi_- = -i\Gamma\varphi_+ + f_+ \\ (i\partial_t - i\partial_y + sH_+)\varphi_- + \varphi_+ = -i\Gamma\varphi_- + f_- \end{cases} \tag{9}$$

where $H_\pm = Q + |\varphi_\pm|^2$, $Q = |\varphi_+|^2 + |\varphi_-|^2$, $s = \pm 1$ is the sign of non-linearity, $\Gamma = \delta w/w_B$ is damping parameter, $f = h(\alpha w_B)^{1/2}$ is normalized generation source.

One should note that this system is described by the lagrangian

$$\mathcal{L} = \bar{\varphi}(i\hat{\partial} - 1)\varphi + G_{\mu\nu}J^\mu J^\nu, \tag{10}$$

and dissipative function

$$\mathcal{R} = i\Gamma\bar{\varphi}\gamma^0\partial_0\varphi, \tag{11}$$

were

$$\varphi = \begin{pmatrix} \varphi_+ \\ \varphi_- \end{pmatrix}, \qquad \bar{\varphi} = \begin{pmatrix} \varphi_+^* & \varphi_-^* \end{pmatrix}\gamma^0,$$

$$\gamma^0 = \begin{pmatrix} 0 & -1 \\ -1 & 0 \end{pmatrix}, \qquad \gamma^1 = \begin{pmatrix} 0 & 1 \\ -1 & 0 \end{pmatrix},$$

$$G_{\mu\nu} = \frac{s}{2}\begin{pmatrix} 3 & 0 \\ 0 & -1 \end{pmatrix}, \qquad J^\mu = \bar{\varphi}\gamma^\mu\varphi, \quad \mu, \nu = 0, 1.$$

When the gap value w_B is sufficiently large compared with δw so that $\Gamma \ll 1$, one can put $\Gamma = 0$ in zero approximation. Let the pumping be also absent: $f = 0$. The system (9) will be then hamiltonian and have three first integrals:

$$\partial_\mu J^\mu = 0, \qquad \partial_\mu T^{\nu\mu} = 0, \qquad \nu = 0, 1, \tag{12}$$

where

$$
\begin{aligned}
Q &= J^0 = |\varphi_+|^2 + |\varphi_-|^2 \\
I &= J^1 = |\varphi_+|^2 - |\varphi_-|^2 \\
W &= T^{00} = -\tfrac{i}{2}(\varphi_+^* \partial_y \varphi_+ - \varphi_-^* \partial_y \varphi_- - \partial_y \varphi_+^* \varphi_+ + \partial_y \varphi_-^* \varphi_-) \\
&\quad - (\varphi_+^* \varphi_- + \varphi_-^* \varphi_+) \\
&\quad - \tfrac{s}{2}(|\varphi_+|^4 + |\varphi_-|^4 + 4|\varphi_+|^2|\varphi_-|^2) \\
P &= T^{10} = -\tfrac{i}{2}(\varphi_+^* \partial_y \varphi_+ - \varphi_-^* \partial_y \varphi_- + \partial_y \varphi_+^* \varphi_+ - \partial_y \varphi_-^* \varphi_-) \\
T^{01} &= \tfrac{i}{2}(\varphi_+^* \partial_t \varphi_+ - \varphi_-^* \partial_t \varphi_- - \partial_t \varphi_+^* \varphi_+ + \partial_t \varphi_-^* \varphi_-) \\
T^{11} &= \tfrac{i}{2}(\varphi_+^* \partial_t \varphi_+ + \varphi_-^* \partial_t \varphi_- - \partial_t \varphi_+^* \varphi_+ - \partial_t \varphi_-^* \varphi_-) \\
&\quad + (\varphi_+^* \varphi_- + \varphi_-^* \varphi_+) \\
&\quad + \tfrac{s}{2}(|\varphi_+|^4 + |\varphi_-|^4 + 4|\varphi_+|^2|\varphi_-|^2)
\end{aligned}
\tag{13}
$$

It is supposed here also that $\partial_t = \partial_0 = \partial^0$, $\partial_y = \partial_1 = -\partial^1$. It should be noted that in such notation parameters W, P, Q, and I play the role of energy, pulse, charge and current densities, as it is customary in the theory of spinor fields [31].

The nondissipative system (9) with lagrangian (10) has localized soliton-like travelling solution in self-similar form $\varphi_\pm = R_\pm(y-vt) \exp[i(ky - wt + \chi_\pm)]$, where v is soliton velocity, k and w are wave number and frequency detuning from phase synchronization point $k = 0$, $w = 0$, $\chi_\pm$ is phase shift of corresponding wave. In this case, the amplitudes $R_\pm$ of soliton depend on space variable according to formula

$$
R_\pm = \left[\frac{1 \pm v}{2} Q\right]^{1/2},
\tag{14}
$$

where

$$
Q = \frac{4[1 - v^2 - (w - kv)^2]/(3 - v^2)}{s(w - kv) + \sqrt{1 - v^2} \cosh\left[\frac{2(y-vt)\sqrt{1-v^2(w-kv)^2}}{\sqrt{1-v^2}}\right]}.
\tag{14a}
$$

The soliton phase shift is equal to

$$
\chi_\pm = \chi_0 + \frac{wv - k}{1 - v^2}(y - vt) + \frac{2v\chi}{3 - v^2} \pm \frac{\chi}{2}
\tag{15}
$$

where

$$
\chi = \pi - \arccos\left[\left(w - kv + \frac{3 - v^2}{4} sH\right)\Big/\sqrt{1 - v^2}\right].
\tag{15a}
$$

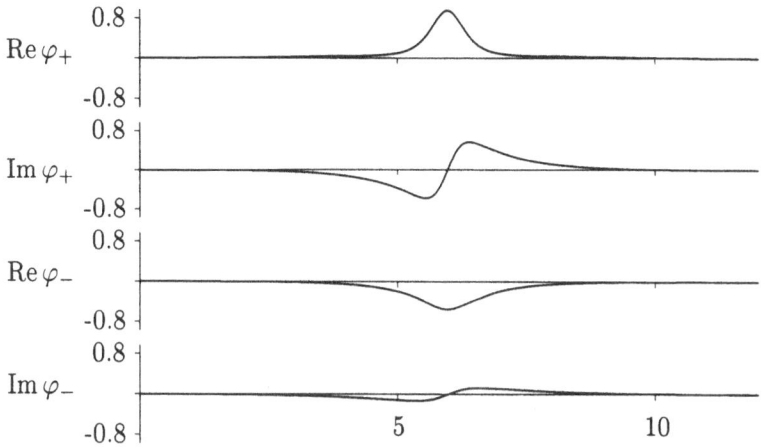

Figure 1. Gap soliton structure at $v = 0.5$, $w = 0.4472$, $k = 0$

The soliton structure discribed by (14)–(15) is shown in Fig.1.

The condition for existence of the solution obtained is not violated provided the soliton velocity in the range $v_+ > v > v_-$, where $v_\pm = (wk \pm \sqrt{1 + k^2 - w^2})/\sqrt{1 + k^2}$. Hence it follows also that the soliton frequency w must lie between branches of the dispersion curves of the linear MSW, i.e., $|w| < (1 + k^2)^{1/2}$. Thus the soliton can not move with a velocity exceeding the group velocity of the coupled waves which are its components, and its frequency for a given wave number can not exceed the frequency of the MSW. For $k = 0$ solitons of coupled MSW exist only inside of the nontransmission band ($|w| < 1$) and they can be called conventionally "gap" solitons. The amplitude of a "gap" soliton at rest ($v = 0$) tends to zero as $sw \to +1$, i.e. close to one of the edges, near which the Lighthill-Witham criterion for the formation of small amplitude solitons is satisfied, and reaches a maximum ($R = \sqrt{\frac{8}{3}}$) near the opposite edge, when $sw \to -1$, where this criterion is violated. In terms of dimensional variables the maximum soliton amplitude depends on nonlinear coefficient $\partial w/\partial|\phi|^2$ and coupling strength which is characterized by w_B according to

$$\phi_{\max} = \left[\frac{8w_B}{3\partial w/\partial|\phi|^2} \right]^{1/2}. \tag{16}$$

As it will be seen further the soliton amplitude limitation is the cause of pulse autogeneration in the presence of linear electromagnetic pumping and in processes of nonlinear MSW scattering by confined grating.

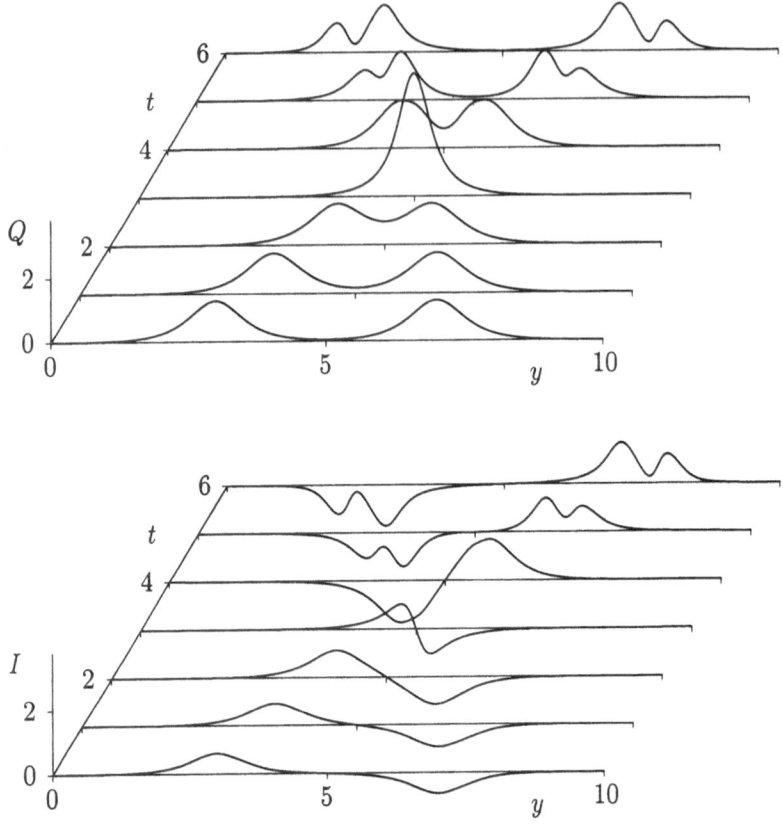

Figure 2. Collision of gap solitons with initial velosities $|v_\pm| = 0.5$, at $w = 0$. Here $Q = |\varphi_+|^2 + |\varphi_-|^2$, $I = |\varphi_+|^2 - |\varphi_-|^2$

3.1. SOLITON STABILITY

Found soliton-like packets are stable only without external action. The results of numerical simulation of the collision process of oppositely moving soliton-like waves are shown in Fig. 2. After solitons collision they decay into fast soliton-like wave packets scattered with different velocities. Time duration of decay process depends on initial packets velocities. It decreases if initial soliton velocities increase.

In dissipative media solitons decay due to their amplitude damping. In this case, not only amplitude goes to zero, but also soliton frequency approaches the edge of the nontransmission gap at which the soliton disappears. This may be easily seen from equation for soliton energy evolution averaged over space [26], solving of which gives $\langle Q \rangle = \langle Q(0) \rangle \exp{(-2\Gamma t)}$, where $\langle Q \rangle = \int\limits_{-\infty}^{+\infty} (|\varphi_+|^2 + |\varphi_-|^2)\, dy$. Hence, for the soliton at rest ($v = 0$),

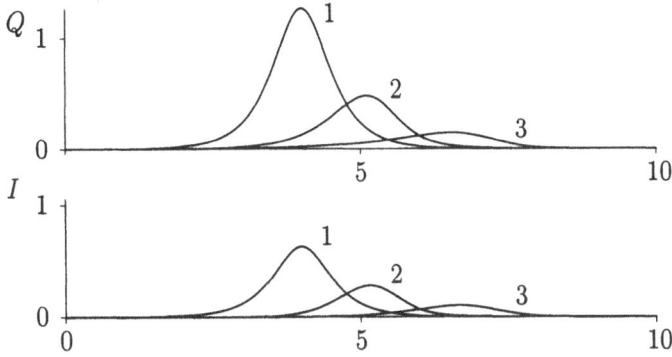

Figure 3. Gap soliton decay in a lossy media at $\Gamma = 0.2$:
$1 - t = 0$; $2 - t = 2$; $3 - t = 4$.

one can get $w = s \cos [a \exp (-2\Gamma t)]$, where $a = \arccos [sw(0)]$, using (14a). The time of decay is determined by MSW line width: $[t] \sim (\delta w)^{-1}$. This process is demonstrated in Fig. 3.

3.2. LOCAL ELECTROMAGNETIC PUMPING

Thus, the distributed feed back acts as wave resonator for coupled MSW. It leads to soliton creation in nonlinear media, which decays due to dissipation. The damping of the resting soliton, however, should be compensated by local electromagnetic pumping [26].

Numerical simulation showed that the steady-state localized solution exists in the infinite nonlinear resonator, if the pumping is not zero, namely $f(y) = f_0 \delta(y - y_0)$, where f_0 is an intensity of generation, y_0 is the source location, until the intensity exceeds the threshold value ($f_0 \simeq 1.4$ at $\Gamma = 0.1$). The structure of localized excitation is shown in Fig. 4a. Then it becomes unstable and, at a higher pumping level ($f_0 \geq 1.4$ at $\Gamma = 0.1$), the process of pulse autogeneration evolves. Fast soliton-like packets of scattered waves are generated after that periodically with the period depending on pumping excess of the threshold value. The initial soliton changes its amplitude and shape also periodically, as it is shown in Fig. 4b. The autogeneration frequency increases with pumping (Fig.5). The process of self-pulsation becomes chaotic at a very high level of electromagnetic pumping.

4. Nonlinear Bragg scattering of MSW

Now we shall consider the dynamics of nonlinear scattering by confined Bragg grating. This process may be also described by coupled equations (2).

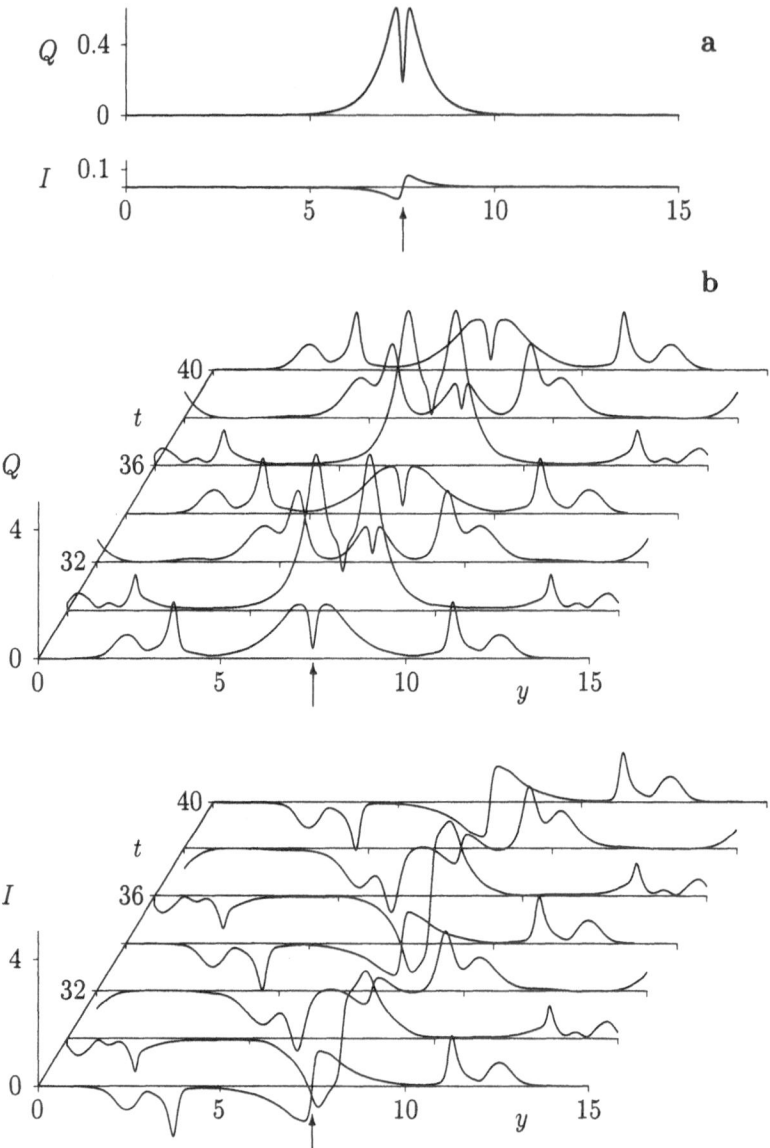

Figure 4. Evolution of soliton-like parametrically coupled wave packets in the presence of local linear pumping in nonlinear lossy media at $\Gamma = 0.1$:
a — stationary regime, $f_0 = 1.18$; b — pulse autogeneration, $f_0 = 1.6$.
Source location is shown by arrow.

In this case, one should put the coupling coefficients ω_B equal to zero outside the reflecting region $0 < y - y_0 < L$, where L is a length of grating. Let the point source of generation be located just at the edge of scattering zone so that $f_+ = f_0 \delta(y - y_0)$, but $f_- = 0$, because the generated back

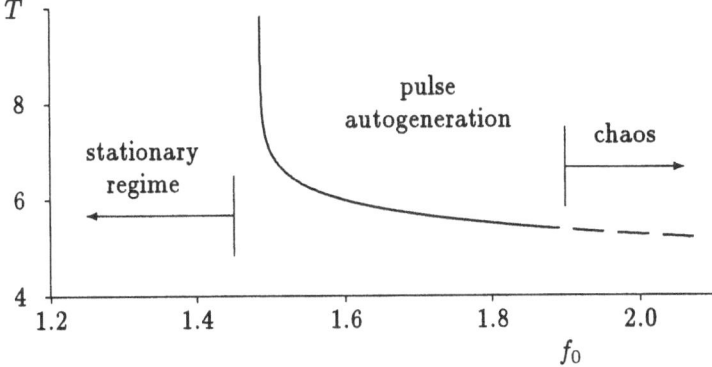

Figure 5. Dependence of self-pulsing time period on pamping amplitude at $\Gamma = 0.1$

wave is not essential for considered effects. Numerical solution of equations (9) shows that there are two kinds of scattering regimes, one of which is stationary and the other is unstable in time. The option depends on the microwave input power $P \sim \varkappa a^2$, where $a = |\varphi_+(y = 0, t = 0)|$ is an amplitude of the incident signal wave. At a small power, when $a \ll 1$, the scattering is of linear character with the exponential wave amplitude distribution along the grating, which is established after the wave front crossed the scattering region (see Fig. 6). Due to nonlinearity of the medium, increasing of input power changes the wave distribution along the grating qualitatively. Instead of exponential decreasing at the edge of grating, the soliton-like formation appears. When the input MSW power exceeds the threshold level, the process of pulse autogeneration evolves like that in the case of two coupling waves in the unconfined grating. This effect is shown in Figs. 7, 8. The threshold power depends on the grating length L, damping parameter Γ, and frequency detuning $\Delta\omega$ from the phase synchronization point ($\Delta\omega = 0$).

The threshold of autogeneration is immediately connected with the hysteresis in dependence of reflectivity $r = |\frac{\varphi_-(0)}{\varphi_+(0)}|^2$ on the input power $P \sim |\varphi_+(0)|^2 = a^2$. It may be calculated by means of solving of (9) in the region $0 < y < L$ seeking the stationary solution $\varphi_\pm(y, t) = R_\pm(y) \exp(i\Delta\omega t)$ with the boundary conditions $R_-(L) = 0$, $|R_+(0)| = a$.

In the case of $\Gamma = \Delta\omega = 0$, the problem has the known analitical solution (see, for example, [33]):

$$r^2 = \frac{1 - \mathrm{dn}\,(2L/k, k)}{1 + \mathrm{dn}\,(2L/k, k)}, \tag{17}$$

$$a^2 = \frac{\sqrt{1 - k^2}}{k}\frac{1 + \mathrm{dn}\,(2L/k, k)}{\mathrm{dn}\,(2L/k, k)}, \tag{18}$$

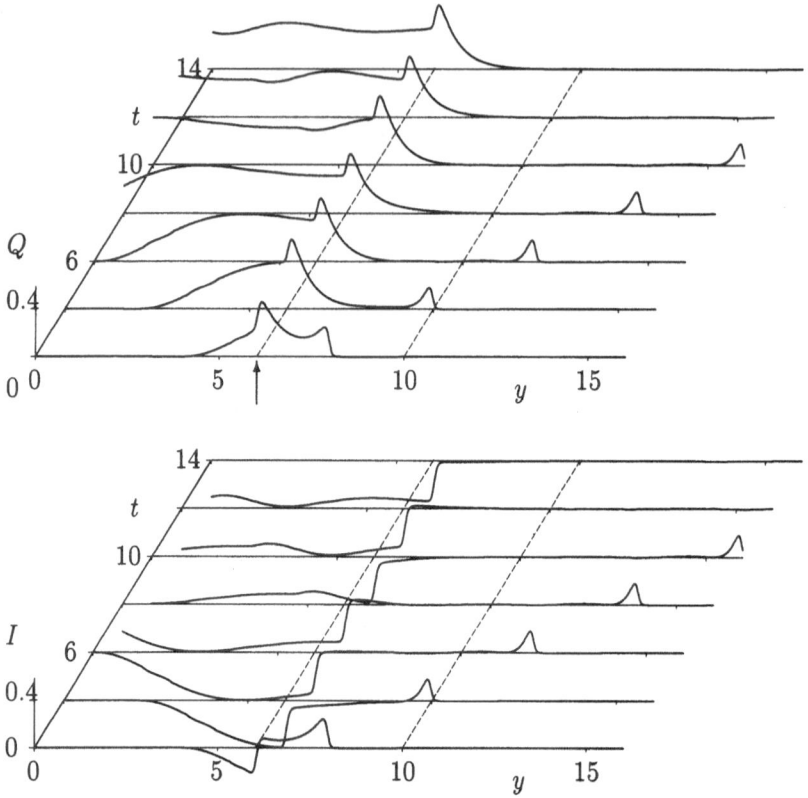

Figure 6. Dynamics of wave scattering in linear regime, $L = 4$, $\Gamma = 0$. Source location is shown by arrow. Dashed lines confine scattering region.

where $\mathrm{dn}\,(u, k)$ is Jacoby function with modulus k $(k < 1)$.

In the general case of $\Gamma, \Delta\omega \neq 0$ it is easy to obtain the required dependences of reflectivity by numerical integration of (9). Some of them are shown in Fig. 9, from where it is seen that for a long grating $(L \gg 1)$ and low damping $(\Gamma \ll 1)$, the hysteretic fenomena may be observed at $a_p^2 \approx \frac{2}{3} \simeq 0.66$ that corresponds to the amplitude of the gap soliton at $\Delta\omega = 0$. The finiteness of the grating leads to the threshold level increase and to the hysteresis dissappearence at $L \simeq 1.45$. The damping has a strong influence also on the reflectivity, as it is shown in Fig. 10. There is a critical value $\Gamma \approx 0.3$, at which self-pulsation becomes impossible before the gap is closed at $\Gamma = 1$.

In Fig. 11, the detuning dependences of the reflectivity $r(\Delta\omega)$ are presented for the linear and nonlinear regimes. At the high input power $(a = 0.5)$, they are like those of the nonlinear resonance responce of unharmonic oscillator. Some similarities of the amplitude and detuning dependences in

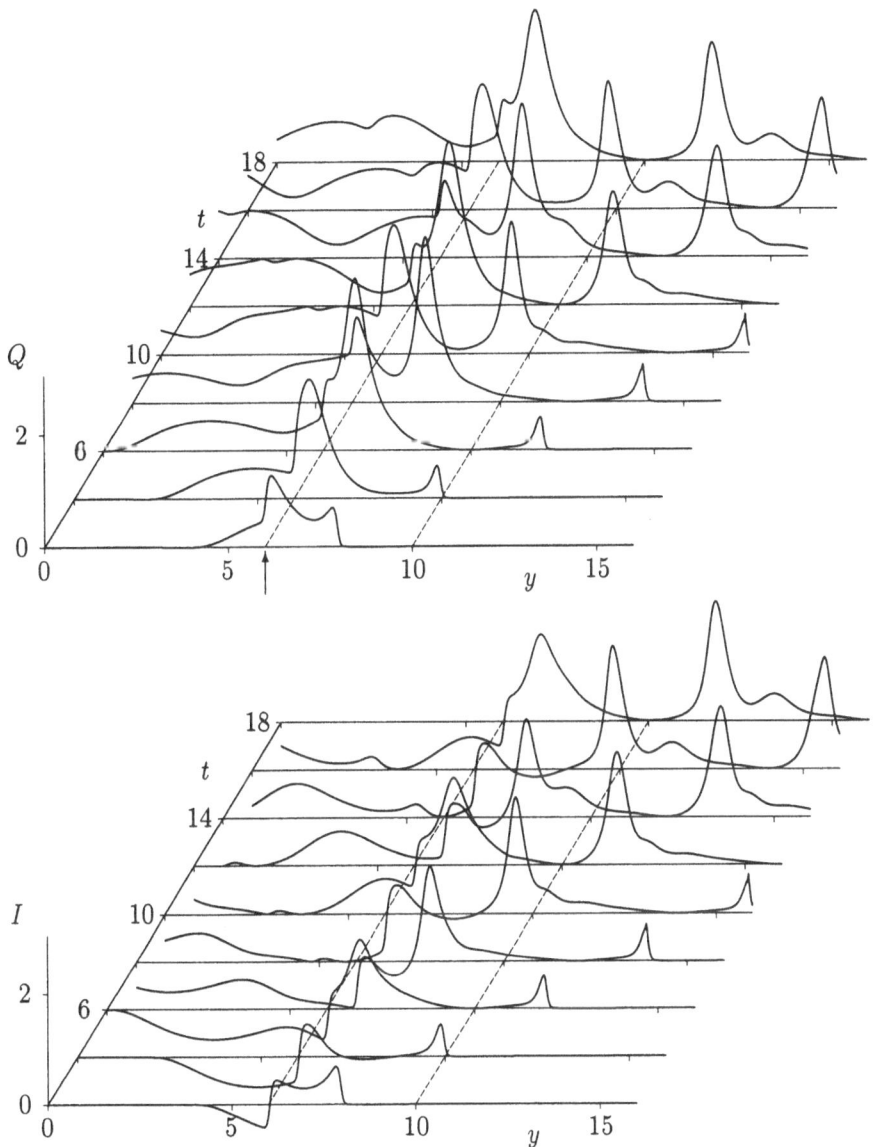

Figure 7. Pulse autogeneration in nonlossy media ($\Gamma = 0$, $L = 4$).

Fig. 9 and Fig. 11 are connected with the frequency detuning caused by nonlinear frequency shift $\Delta\omega \sim a^2$.

5. Discussion and conclusions

Thus, the restriction of nonlinear frequency shift of parametrically coupled waves imposed by the gap directly connects effects considered nonlinear.

318

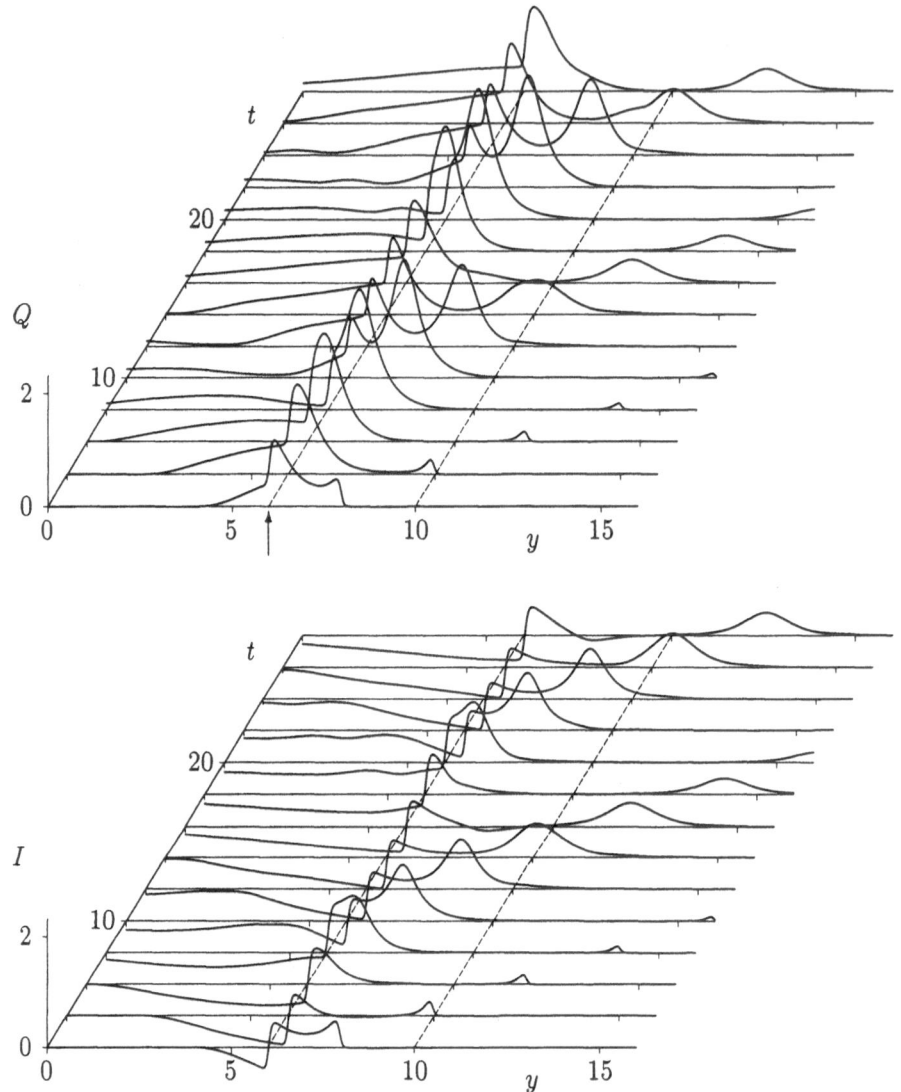

Figure 8. Pulse autogeneration in lossy media ($\Gamma = 0.1$, $L = 4$).

Strong dependence of these effects on damping is very important for their observation in magnetic films. Besides that, many-magnon decay processes lead to the power limitations of the nonlinear MSW. Therefore, it should be necessary to discuss the influence of such limitations on the considered nonlinear fenomena.

At first, as we have already seen, nonlinear hysteretic effects are possible only if $\Gamma \geq 0.3$, that means the gap value ω_B must exceed the line-width $\delta\omega$:

$$\omega_B \geq 3\delta\omega. \tag{19}$$

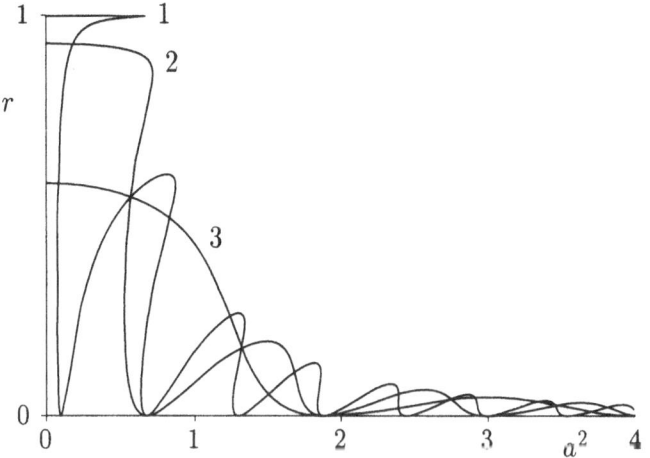

Figure 9. Nonlinear reflectivity dependence on grating length at $\Gamma = 0$
$1 - L = 4$; $2 - L = 2$; $3 - L = 1$.

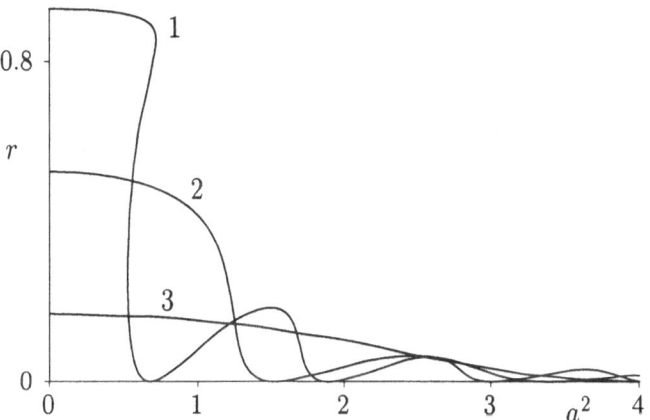

Figure 10. Nonlinear reflectivity dependence on input power at $L = 2$:
$1 - \Gamma = 0$; $2 - \Gamma = 0.3$; $3 - \Gamma = 1$.

Secondly, the threshold amplitude of autogeneration is about $a^2 \approx 0.5$.
That means

$$|\phi|^2 \sim \left(\frac{m_\perp}{M_0}\right)^2 \sim 0.5\omega_B \left/ \frac{\partial\omega}{\partial|\phi|^2}\right. .$$

On the other hand, the amplitude of nonlinear MSW is limited by parametric decay instabilities due to three- or four-magnon processes. It is necessarry to choose the carrying MSW frequency ω and external magnetic field H in the region, where four-magnon decay processes prevail because of very low value of resonant three-magnon power limitations [34, 22].

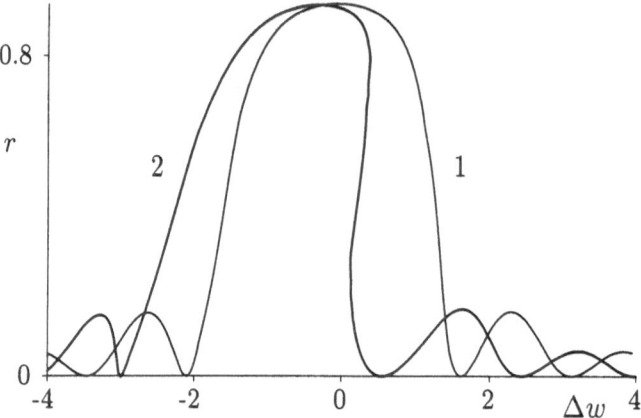

Figure 11. Reflectivity dependence on frequency detuning at $\Gamma = 0$, $L = 2$.
$1 - a^2 = 0.1$, $2 - a^2 = 0.5$

Four-magnon limitation of the MSW amplitude is approximately equal to $(\frac{m_\perp}{M_0})^2 \sim \frac{\delta\omega}{\omega_M}$ [34]. Thus, we have to satisfy the following condition:

$$|\phi|^2 \sim \frac{\delta\omega}{\omega_0} > \omega_B \left/ \frac{\partial\omega}{\partial|\phi|^2} \right. . \tag{20}$$

One can see that conditions (19) and (20) are in some contradiction and rather difficult for realization. They require the nonlinearity of MSW to be as large as possible and the carrying frequency to be near the bottom of magnon spectrum, where four-magnon processes $2\omega = 2\omega_k$ are depressed. From this point of view, a normally magnetized magnetic film and forward volume magnetostatic waves are favourable for observation of autogeneration effects because of their frequency being close to the bottom of magnon spectrum in long-wave limit.

Thus, the maximum frequency gap, which determines the autogeneration periodicity, is limited by the maximum nonlinear shift in the spin-wave packet

$$\omega_B^{max} \sim \frac{\partial\omega}{\partial|\phi|^2}|\phi|_{max}^2 \sim \frac{\partial\omega}{\partial|\phi|^2}\frac{\delta\omega}{\omega_M}. \tag{21}$$

It cannot exceed significantly the line-width of MSW, and for YIG films, we have $\omega_B \geq 3\delta\omega_k \simeq 3\gamma\Delta H \sim 3\cdot 10^7 s^{-1}$.

On the other hand, the low MSW dispersion $d_k^2\omega \sim v_g^2/\omega_B$ caused by distributed backward coupling can be achieved at large gap values $\omega_B \sim \omega_M$ (especially, in the case of compensation of wave guide dispersion). It may be used for creation of ultra-short powerfull pulses with the frequency being near the gap edge, where small-amplitude solitons would be excited. Really, using formulas for threshold power of soliton creation

$$|\phi|^2 \sim \frac{d^2\omega}{d\,k^2} \left(\frac{\partial\omega}{\partial|\phi|^2} \right)^{-1} (v_g\tau)^{-2}, \tag{22}$$

where τ is pulse duration, and the estimation of four-magnon limitation $|\phi| \approx \delta\omega_k/\omega_M$, one can obtain

$$\tau \leq \left[\frac{\omega_M}{\omega_B\, \delta\omega\, \partial\omega/\partial|\phi|^2} \right]^{1/2} \sim (\delta\omega\ \omega_B)^{-1/2}. \tag{23}$$

If we suppose that $\omega_B > \omega_M$, $\partial\omega/\partial|\phi|^2 \sim \omega_M$, the rough estimation for YIG gives $\tau \leq 2$ ns.

To sum up, we can make following conclusions.

The soliton-like packets may be excited inside the nontransmission spectrum gap of two parametrically interacting nonlinear spin waves. They are "softly" produced (with zero amplitude) near the edge of the gap, where the Lighthill criterion is fulfilled, and "rigidly" produced (with finite amplitude) near the opposite edge. Soliton-like packets are unstable in collision with each other. Local linear pumping may be used for creation of localized spin-wave packets of parametrically coupled spin waves in the periodical scattering structure. This creation is realized by the effect of pulse autogeneration, if the gap value is greater than the width of MSW and the signal power exceeds the threshold value. Similar effect may be observed at MSW scattering by confined grating. Time periodicity of pulse autogeneration is determined by the gap frequency and input power level. Because of four-magnon processes determinating spin-wave power limitations hysteretic and self-pulsation effects can be observed if the gap value does not too exceed MSW line width. FVMSW are favoured type of spin waves for pulse autogeneration observation in thin magnetic films. Wide gap enables to ultra-short soliton-like microwave pulses be created at the edge of the gap due to low dispersion of hybridized MSW in distributed coupling structure.

References

1. Adam, J.D., Daniel, M.R. (1981) The status of magnetostatic devices, *IEEE Trans. Magn.* **MAG-17**, 2951–2956.
2. Ishak, W.S. (1988) Magnetostatic wave technology: A Review, *Proc. IEEE* **76**, No 2, 171–184.
3. Gurevich, A.G., Melkov, G.A. (1994) *Magnetic Oscillations and Waves*, Nauka, Moscow.
4. Mednikov, A.M. (1981) Nonlinear effects of surface spin waves propagation in YIG films, *Fiz. Tverdogo Tela* **23**, no.1, 242–246.
5. Adam, J.D., Stitzer, S.N. (1980) A magnetostatic wave signal-to-noise enhancer, *Appl. Phys. Lett.* **36**, no.6, 485–487.

6. Emtage, P.R., and Stitzer, S.N. (1977) Interaction of signals in ferromagnetic microwaves limiters, *IEEE Transaction on microwave theory and techniques* **MTT–25**, no.3, 210–213.

7. Winful, H.G., Marburger, J.H., Garmire, E. (1979) Theory of bistability in nonlinear distributed feedback structure, *Appl. Phys. Lett.* **35**, no.5, 379 –381.

8. Winful, H.G., Cooperman, G.D. (1982) Self-pulsing and chaos in distributed feedback bistable optical devices, *Appl. Phys. Lett.* **40**, 298–300.

9. Voloshchenko, Yu., Ryzhov, Yu.N., Sotin, V.E. (1981) Stationary waves in nonlinear modulated media with great group delay, *Zhurn. Tekhn. Fiz.* **51**, no.5, 902–907.

10. Boardman, A.D., Nikitov, S.A., and Wang, Q. (1994) Theory of multistable magnetostatic waves, *IEEE Trans. Magn* **MAG–30**, 1–13.

11. Kalinikos, B.A., Kovshikov, N.G., Slavin, A.N. (1983) Spin-wave soliton observation in ferromagnetic films, *Pis'ma v JETF* **38**, no.7, 343–377.

12. Bogdanova, Kh.G., Golenishchev-Kutuzov, V.A., Monakhov, A.A. et al. (1980) Magneto-acoustic soliton observation in monocrystals of $KMnF_3$ near NMR, *Pis'ma v JETF* **32**, no.7, 476–479.

13. De Gasperis, S.P., Marchelli, R., and Micolli, G. (1987) Magnetostatic soliton soliton propagation at microwave frequency in magnetic garnet films, *Phys. Rev. Lett.* **59**, no.4, 481–484.

14. Kalinikos, B.A., Kovshikov, N.G., and Slavin, A.N. (1990) Experimental observation of magnetoelastic wave envelope solitons in yttrium-iron-garnet films, *Phys. Rev.* **B42**, p.8058–8666.

15. Chen, M., Tsankov, M.A., Nash, J.M., and Patton, C.E. (1993) Microwave magnetic envelope dark solitons in Yttrium Iron Garnet thin films, *Phys. Rev. Lett.* **70**, no.11, 1707–1710.

16. Chen, M., Tsankov, M.A., Nash, J.M., and Patton, C.E. (1994) Backward volume wave microwave envelope solitons in Yttrium Iron Garnet films, *Phys. Rev.* **B49**, p.12773.

17. Tsankov, M.A., Chen, M., and Patton, C.E. (1994) Forward volume wave microwave envelope solitons in Yttrium Iron Garnet films — propagation, decay, and collision, *J. Appl. Phys.* **73**, 3906.

18. Witham, G.B. (1965) Nonlinear dispersive waves, *Proc. of the Royal Soc.* **A283**, *Math. and Phys. Sc.*, no.1393, 238–261.

19. Lighthill, M.J. (1965) Contributions to the theory of waves in non-linear dispersive systems *J. Inst. Math. Appl.* **1**, p. 269–306.

20. Karpman, V.I. (1968) *Nonlinear waves in dispersive media*, Izd. NGU, Novosibirsk.

21. Lukomskii, V.P. (1978) Nonlinear magnetostatic waves in ferromagnetic plates, *Ukr. Fiz. Zh.* **23**, no.1, 134–139.

22. Zvezdin, A.K., and Popkov, A.F. (1983) Contribution to the nonlinear theory of magnetostatic spin waves, *Sov. Phys. JETP* **57**, no.2, 350–354.

23. Boardman, A.D., Nikitov, S.A., Xie, K., Mehta, H. (1995) Bright magnetostatic spin-wave envelope solitons in ferromagnetic films, *JMMM* **45**, 357–378.

24. Grünberg, P. (1980) Magnetostatic spinwave modes of a ferromagnetic double layer, *J. Appl. Phys.* **51**, 4338–4341.

25. Seshadry, S.R. (1979) Magnetic wave interactions in a periodically corrugated YIG film, *IEEE Trans. on microwave theory and techniques* **27**, no.2, 199–204.

26. Popkov, A.F. (1993) Spin-wave solitons in a sound "lattice", *JETP* **77**, no.3, 486–491.

27. Brinlee, W.R., Owens, J.M., Smith, C.V., Jr., and Cater, R.L. "Two-port" magnetostaticwave rezonators utilizing periodic metal reflective arrays, *J. Appl. Phys.* **52**, 2276–2278.

28. Mednikov, A.M., Popkov, A.F., Anisimkin, V.I. et al. (1981) Nonelastic scattering of surface spin waves in thin YIG films by surrface acoustic wave, *Pis'ma Zh. Exp. Teor. Fiz.* **33**, 646–649.

29. Myasoedov, A.N., Fetisov, Yu.K. (1989) Magnetostatic volume wave scattering by

the dynamic magnetic lattice, *Zhurnal Techniceskoi Fiziki* **59**, no.6, 133–136.

30. Danaev, N. and Fetisov, Yu.K. (1992) Parametric interaction of magnetostatic waves in uniform nonstationary-magnetized garnet films, *Electronics Letters* **28**, no.21, 1998–2000.

31. Bogolyubov, N.N., Shirkov, D.V. (1980) *Quantum fields* Nauka, Moscow.

32. Ganguly, A.K. and Webb, D.C. (1975) Microstrip Excitation of Magnetostatic Surface Waves: Theory and Experiment, *IEEE Trans. on Microwave Theory and Techniques* **MTT-23**,, no.12, 998–1006.

33. Karpov, S.Yu., Stolyarov, S.N. (1993) Wave propagation and transformation in in one-dimensional periodical media, *Uspekhi Fizicheskikh Nauk* **163**, no.1, 63-89.

34. Suhl, H. (1957) The theory of ferromagnetic resonance at high signal power, *J. Phys. Chem. Solids*, Pergamon Press **1**, 209–227.

MACROSCOPIC QUANTUM TUNNELLING OF SOLITONS IN ULTRATHIN FILMS

A.K.ZVEZDIN
Institiute of General Physics
Vavilova st., 38, Moscow 117942, Russia

AND

V.V.DOBROVITSKI
Moscow State University, Physics Department
Vorobievy gory, Moscow 119899, Russia

Abstract. Processes of macroscopic tunnelling of magnetic solitons are considered. The significance of such processes for fundamental physics (for connection between macro- and microscopic descriptions of the world) and for applications (in magnetic memory, in quantum computer creation) is briefly discussed. The application of instanton approach to theoretical consideration of such phenomena is described for two cases: for tunnelling of magnetic soliton through the external pinning potential and for tunnelling change of topological charge of a soliton. An influence of dissipative environment is briefly discussed. The review of corresponding experimental results is given; it is demonstrated that theoretical ideas about macroscopic tunnelling phenomena are supported by experiments.

1. Introduction. Motivations of an interest in MQT and MQC

In recent years magnetic nanostructures attract a great attention. It is caused by very unusual properties specific for these systems; these properties make the magnetic nanostructures not only extremely promising materials for various practical applications but also cause them to be important for fundamental science.

Significance of these structures in modern physics results from their unusual position among other physical objects, the position between microscopic and macroscopic objects. Thus, investigation of such systems allows

R. Marcelli and S.A. Nikitov (eds.), Nonlinear Microwave Signal Processing: Towards a New Range of Devices, 325–353.
© *1996 Kluwer Academic Publishers.*

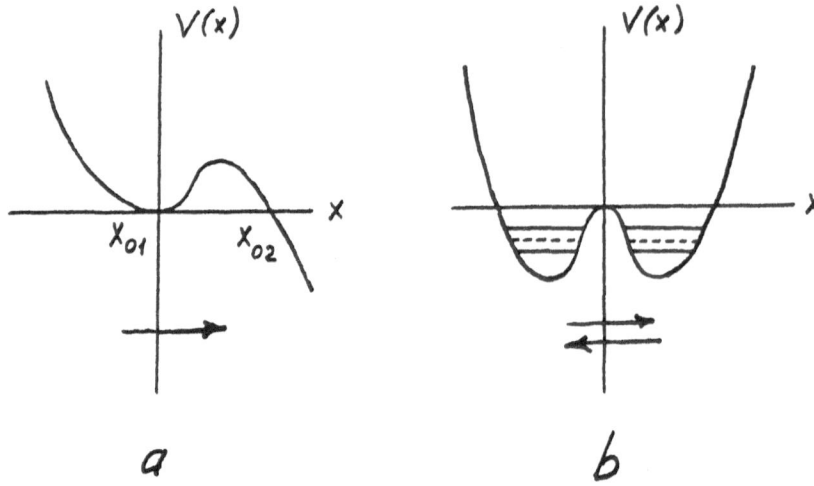

Figure 1. Two most attractive at present macroscopic quantum phenomena: (a) – MQT; (b) – MQC. $V(x)$ is the potential energy versus some macroscopic coordinate x.
In the first case (a) the particle being put into left (metastable) minimum has the finite probability to escape to the right and never returns to the left minimum; thus, the probability to stay at the metastable state decreases exponentially with time. In the second case (b) the particle being initially put into the left minimum gradually "leak" to the right minimum, then gradually returns to the left etc.

us to span the bridge between microscopic and macroscopic description of the world. Moreover, for the same reasons such structures are important for creation of new devices based on new physical principles.

Probably, one of the most prominent subsequences of intermediate position of nanostructures in physics is the possibility of manifestation of specifically quantum properties at mesoscopic (or even at macroscopic) scale. At present, the greatest attention among other such phenomena attract two effects, usually referred in literature as macroscopic quantum tunnelling (MQT) and macroscopic quantum coherence (MQC). The first one corresponds to the situation when the metastable state decays into continuum due to the tunnelling through the potential barrier (which separates this metastable state from the continuum) (see Fig.1a). In the second case two degenerate energetic levels separated by potential barrier split due to the tunnelling from one minimum to the other one so that the degeneracy is lifted (Fig.1b). It is worth to note that in both cases we deal with the coherent (or almost coherent) tunnelling of a macroscopically large number of elementary (microscopic) objects.

So why these specific effects occur to be important for modern physics?

1.1. FUDAMENTAL ASPECTS

At present, a physicist considering a nonrelativistic problem can use two general approaches: classical and quantum-mechanical, applying the first one to "large enough", or macroscopic, objects and the second to the "small enough", or microscopic, objects. But there is no principal limitations for possibility of quantum mechanics to describe macroscopic bodies such as, for example, Schrödinger's Cat; it is well known that the Schrödinger's paradox appears in this way: it seems strange that the macroscopic body can exist in the superposition of two macroscopically distinct states, such as a superposition of living and dead cat in Schrödinger's example. There is an objection that macroscopic bodies are coupled with a large number of microscopic degrees of freedom so that we deal not with the wave function of a pure state but with a statistical density matrix corresponding to a mixed state. But, obviously, it is not the answer to the questions of interest: we just put the question "how large is the matrix element" instead of "how large is the object under consideration". Eventually, there is no doubt that an electron in the crystal can tunnel though it is coupled with phonons, photons etc. Thus, the real question is how should we account for the interaction of a macroscopic object with microscopic degrees of freedom, to what extent should we do it, how weak the interaction should be to admit us to neglect it etc.

This brief discussion shows that the problem of relation between micro- and macrodescription requires a comprehensive study. Macroscopic tunnelling phenomena, MQT and MQC, seem to be quite useful in such studies since, for example in MQT case, initial and final states can be relatively easy made macroscopically distinguishable, or, in MQC case, we can measure the two-time correlation function or the levels splitting. Besides, it seems difficult to find some quantum-mechanical effect which is more "specifically quantum" than tunnelling.

Moreover, problems of quantum tunnelling and quantum coherence are important for new concepts of mesoscopic physics, such as a wave function collapse [1]. This concept considers the quantum measurement process as a decoherence between different quantum states leading to the collapse of the system into one of these states, with subsequent separation of one of these states by measuring device. In this context we can consider the tunnelling as an example of such collapse. In addition, in MQC case the decohering interaction of the system with its dissipative environment can be considered as a "measurement" of the system, when the decoherence between system's quantum states leads to the collapse of the system into the left or right well

(Fig.1b).

Thus, we see that the study of macroscopic tunnelling phenomena is important for understanding of the interplay between microscopic and macroscopic descriptions of the world. But the realm of importance of these phenomena is not restricted by only fundamental science.

1.2. POSSIBLE DIRECTIONS OF PRACTICAL USE

Now, we shall discuss two cases when macroscopic tunnelling phenomena are crucial for applications.

The miniaturization of a computer memory leads to the miniaturization of a piece of magnetic medium used to store a bit of information, but it can not continue as long as we wish. Eventually, this piece of magnetic medium will be so small that it will be unstable against remagnetization by means of motion of domain walls or by means of rotation of magnetization. If it occurs at room temperature, we can lower the temperature and decrease the scale of thermal fluctuations. But, even at zero temperature we still have the finite quantum fluctuations, and, therefore, we have the principal, quantum-mechanical limitation on the size of an elementary piece of magnetic media.

However, the tunnelling phenomena play not only negative role in computer technology.

As we noted above, the miniaturization of computer hardware leads to the necessity to account for the laws of quantum mechanics, and the idea of Benioff [2] that the computer operating according to quantum-mechanical laws can be constructed is extremely interesting and attractive.

This idea is closely connected with another serious problem occuring in the course of miniaturization: the problem of a heat dissipation. The heat dissipation restricts the size of logical devices, but as was pointed by Landauer [3], the heat production during execution of logical operations can be made, in principle, as small as we need, since the only irreversible operations (and, subsequently, the only operations which are always accompanied by the heat production) are operations of creation and removing of information. Moreover, as was shown by Bennett [4], we do not need to remove results of intermediate operations since the computation programme can be reversed so that after complete execution of the reversed programme we obtain the initial situation, without any intermediate result. Thus, the only really irreversible operations are the input and output of information. The absence of dissipation can be provided by using of quantum-mechanical system; in this case the temporal evolution will be governed by the Schrödinger equation which is reversible in time. It was shown (see, for example [5]) that it is possible to construct the Hamiltonian which represents the necessary sequence of logical operation.

In such a computer (usually referred as quantum computer) every bit is represented by the bistable quantum-mechanical system and the bistable magnetic moment is quite natural candidate to play this role. As a simple example, the ferromagnetic particle possessing easy-plane anisotropy can be considered. The external magnetic field initiates the precession of magnetic moment of the particle. In a time of half-period of the precession the magnetic moment occurs to be directed oppositely to its initial orientation, i.e. the logical operation NOT is performed. If the magnetic field is modulated by the magnetic moment of another particle, the controlled NOT can be realized. Moreover, a magnetic soliton propagating in a chain of coupled magnetic particles can serve as a synchronization pulse.

Although here we can not discuss the questions concerning a problem of quantum computer, it is seen that macroscopic tunnelling phenomena are extremely important for this promising application.

2. General approach to the problem of a metastable state

The problem of a decay of metastable state is a very general problem and it was considered in various contexts during a long time (see, for example [6]). But, as far as we know, the first conceptually complete study of this problem was performed by Langer [7], [8]. In these papers he dealt with properties of a metastable phase in the vicinity of the point of first-order phase transition. In this case the cluster expansion for a free energy of metastable phase diverges and, strictly speaking, looses its meaning. The main feature of Langer's approach was the conjecture that the free energy of metastable phase can be nevertheless meaningful if we consider it as an analytic continuation into the complex plane. In this case the real part of resulting complex quantity is the free energy of metastable phase and the imaginary part is the decay rate of this phase. The latter conjecture about the connection between the imaginary part of free energy and the decay rate was proved in the work [8].

Though Langer considered a decay via thermal fluctuations, his idea was successfully applied by Coleman and Callan [9], [10] to the problem of a quantum decay by means of tunnelling through the potential barrier.

In their work [10], Callan and Coleman used the Feynman's path integrals formalism. They started from Euclidean, or imaginary-time, version of a path integral, i.e. they were interested in the quantity

$$\rho(x_1, x_2) = \langle x_2 | e^{-H\tau/\hbar} | x_1 \rangle = \int Dx \exp(-S[x]/\hbar), \qquad (1)$$

which can be considered as a transition amplitude from the state $|x_1\rangle$ to the state $|x_2\rangle$ in the imaginary time $\tau = -it$; the functional $S[x]$ is the

330

action in imaginary time on the trajectory $x(\tau)$:

$$S[x] = \int_{-T/2}^{T/2} L(x, \dot{x}) d\tau, \tag{2}$$

where L is the imaginary-time Lagrangian. Here we assume that the motion starts at the moment $\tau = -T/2$ at $x = x_1$ and finishes at $\tau = T/2$ at the point $x = x_2$.

It is known (see [11]) that the quantity $\rho(x_1, x_2)$ is an element of the density matrix corresponding to the temperature $\Theta = \hbar/k_B T$. In the work [10] the diagonal element of a density matrix was considered. Such an element $\rho(x, x)$ also can be rewritten as

$$\rho(x, x) = \sum_n \psi_n^*(x) \psi_n(x) \exp(-E_n T/\hbar), \tag{3}$$

where $\psi_n(x)$ and E_n are the eigenfunctions and eigenvalues of the Schrödinger's equation correspondingly.

If a decay of a metastable state into continuum is considered, the situation shown in Fig.1a is relevant. It is convenient to assume that the minimum of potential $V(x)$ is placed at $x_{01} = 0$, $V(0) = 0$ and the point $x = x_{02}$ has the same energy $V(x_{02}) = 0$.

Now, Langer's ideas can be used. The system in the decaying state is nonequilibrium and nonstationary. Nevertheless, we can expect that if a temporal evolution (decay) of metastable state is slow enough, the equilibrium stationary density matrix (1) is a meaningful quantity for description of this state. Also, we expect that this quantity is a complex number, and its imaginary part is connected with a decay rate. If the system is in the lowest metastable state, it is natural to examine the element $\rho(0, 0)$ since in the equlibrium stationary case this is a probability to find the system at the point $x = 0$. It is worth to emphasize that the time T is not a real time and the quantity that we are interested in is *not* a probability to find the system at the position $x = 0$ at the moment T.

We study the system at zero temperature, i.e. at $T \to \infty$. Using (3) we see that in this limit

$$\rho(0, 0) \propto \exp(-E_0 T/\hbar), \tag{4}$$

where E_0 is the energy of the lowest metastable level, since all the other terms (with higher E_n) are exponentially small. From Langer's analysis it is expected that E_0 is a complex number.

Dropping details (reader can find them in the original paper [10]) we now discuss only results of the work of Callan and Coleman.

The quantity $\rho(0,0)$ in WKB-approximation is determined by the contributions from classical trajectories (where $\delta S/\delta x = 0$) and their neighbourhood. There are two such trajectories: the trivial one $x \equiv 0$, and the instanton x_{inst}, which starts at $\tau \to -\infty$ in $x = 0$, at $\tau = 0$ passes through the turning point $x = x_{02}$ and goes back to $x = 0$ when $\tau \to +\infty$. Contribution to E_0 from the trivial trajectory is real and equal to $\hbar\omega/2$, where ω is the frequency of small oscillations near the minimum $x = 0$ (often referred as an "attempt frequency"): $\omega^2 = V''(0)/m$.

Contribution to E_0 from the instanton's neighbourhood is, strictly speaking, divergent, but, being considered as an analytic continuation into the complex plane, remains finite complex number with small imaginary part, that corresponds to decay rate Γ. For this quantity Callan and Coleman obtained

$$\Gamma = A \exp(-B), \quad B = S_{inst}/\hbar, \tag{5}$$

where S_{inst} is the action in imaginary time calculated along the instanton trajectory; the quantity B is often referred as Gamov's constant. The pre-exponential factor A is

$$A = \sqrt{\frac{m\nu}{2\pi\hbar}} \left| \frac{\det'[-\partial_\tau^2 + V''_{inst}/m]}{\det[-\partial_\tau^2 + \omega^2]} \right|^{-1/2}, \tag{6}$$

$$\nu = \int_{-\infty}^{+\infty} [\dot{x}_{inst}(\tau)]^2 \, d\tau, \quad V''_{inst} \equiv \left. \frac{\partial^2 V(x)}{\partial x^2} \right|_{x \equiv x_{inst}},$$

where determinant of an operator is defined as a product of its eigenvalues. The notion "$\det'$" means that the zero eigenvalue should be excluded from the determinant; this operator possesses a zero mode due to translational invariance of the instanton solution (the correct integration over this mode produces the factor ν in Eq.(6)).

The work [10] requires some notes. First, the same result can be obtained using the conventional WKB-method (which is more rigorous), but in this case we can not consider so easily many-dimensional problems and systems with more than one degrees of freedom (such as fields), whereas the path integrals approach can be easily extended to such kind of problems. Second, as a basic quantity for consideration the element $\rho(0,0)$ was chosen; consideration of other quantities (such as a density matrix trace $\text{Tr}\,\rho(x,x)$) gives the same result. Third, the case of zero temperature was considered; but, the generalization to non-zero temperatures can be made (see below). And, last but not least, we emphasize that the instanton is sufficiently nonlinear object, appearing in nonlinear problems; it can not be treated using perturbation theory (it is well known that WKB-method gives sufficiently nonperturbative results).

In conclusion of this section, we shall make some general remarks.

The fact that instantons are extremely important for various tunnelling problems was realized first in quantum field theory. It is the field of physics where the close connection between solitons and instantons was the subject of a large number of papers (see, for review, the book [12]). What can be seen immediately is the sufficiently nonlinear, nonperturbative nature of both objects. In general, the instanton in N dimensions is the static soliton in $N + 1$ dimensions, the energy of soliton is the action on the instanton field configuration. Though we have no space to discuss these questions more thoroughly, these facts are interesting in their own right.

The instanton trajectories (or field configurations) appear quite naturally within the framework of path integral approach and it allows to extend instanton approach to the case of finite temperatures. It was shown (as a most simple example see [13]) that in the case of tunnelling at finite temperature the tunnelling rate is

$$\Gamma = \frac{2}{\hbar} \operatorname{Im} \mathcal{F}, \tag{7}$$

where $\mathcal{F}$ is the free energy of metastable state continued into the complex plane. It is remarkable that for a thermoactivated decay rate Langer [8] obtained

$$\Gamma = \frac{|\tilde{\omega}|}{\pi k \Theta} \operatorname{Im} \mathcal{F}, \tag{8}$$

where k is the Boltzman's constant, Θ is the temperature and $\tilde{\omega}$ is the increment of the unstable mode corresponding to the direction of a decay (in functional space).

Furthermore, the path integrals formalism allows to account for an interaction of the macroscopic system with microscopic degrees of freedom in the processes of MQT and MQC [14], [15]. This account is extremely important since the decohering effect of dissipative environment is the most serious problem for fundamental researches in the field of macroscopic tunnelling phenomena as well as for possible practical use. Actually, the problem of decoherence is the most serious barrier for a quantum computer creation: the decoherence time, obviously, determines the number of steps in the programme for such computer.

3. Tunnelling of a magnetic soliton through the external potential barrier

Although this section is devoted to the tunnelling of magnetic solitons, as a preliminary, let us briefly discuss the tunnelling of magnetization in small magnetic particles. Actually, the problem of consideration of a magnetization tunnelling appeared first in experiments performed by L.Weil on small superparamagnetic nickel particles in 1955.

All theoretical papers concerning this question (see, for example, [16], [17]) treat the magnetization of the small particle as a large spin, though it comprises a large number of elementary spins. Thus, the individual motion of spins is completely removed from the generally used approach. Nevertheless, this model is far from oversimplification; such treatment can be justified most simply by means of path integral approach. The trajectories in phase space which correspond to individual spins deflecting noticeably from each other have very large action due to strong exchange interaction between the spins, and, according to Eq.(1), contribute almost nothing to the total tunnelling probability because of exponential cutoff. Hence, we can restrict ourselves only by collective motion of spins and describe it by means of relatively small number of coordinates which are usually referred as collective coordinates.

The description by means of collective coordinates is crucial for consideration of the tunnelling of magnetic solitons. We shall see below how it can be applied to the case we are interested in.

3.1. THEORY

As far as we know, the first theoretical study of tunnelling of a magnetic soliton, domain wall (DW), was performed by Egami [19]. He considered a ferromagnetic uniaxial material where DW can tunnel through a barrier of the Peierls potential. These barriers are important for the case of magnetically hard material where a domain wall is thin enough. Due to the discreteness of the crystal lattice the wall motion over one lattice constant is accompanied by noticeable rotation of spins; it leads to the increase in wall energy in the intermediate positions due to the deflection of spins from the easy axis.

In 1991 Stamp [18] considered more usual case of DW tunnelling through the potential created by defect (nonmagnetic inclusion) in the uniaxial ferromagnet. He calculated the tunnelling rate using the instanton approach and discussed possible influence of dissipative environment on the tunnelling process. In this work he pointed out the feature of extremal importance: the domain wall due to its solitonic nature is significantly decoupled from the dissipative environment.

Later, in [20], the possible sources of dissipation (phonons, magnons, photons produced by the process of magnetization rotation and conducting electrons) were examined more thoroughly. The Stamp's conclusion was confirmed for the case of insulating material and moderate DW "velocities in imagninary time" (i.e., for not too high potential barriers). The DW tunnelling in conducting ferromagnet was considered by Tatara et al. [21] using the Hubbard's model. In this case conducting electrons are not decoupled enough from the DW and can have a noticeable impact on the tunnelling rate. Nevertheless, under certain conditions the rate of tunnelling can be high enough to be detectable in experiments.

The central idea of modern theoretical investigations of DW tunnelling is the application of instanton approach to the wall Lagrangian obtained by the method of collective coordinate. Reasoning for use of this idea is similar to the arguments mentioned above, when we discussed the magnetization tunnelling in small particles. There are various ways in imaginary time connecting the magnetization configurations representing the DW ahead and behind the defect, but only the ways which correspond to the motion of the wall as whole have the least action. All the other ways, which deflect the magnetization configuration from the structure of classically moving (in imaginary time) domain wall have much greater action (due to the exchange energy increase) and, therefore, much smaller probability. Thus, only collective coordinates (such as wall center position) survive.

Now, we shall deal with the most simple example of DW tunnelling. It is the case of uniaxial ferromagnet with strong easy-axis anisotropy considered in [20]. In this case Slonczewski's equations [22] for description of DW motion can be applied. In Slonczewski's approximation the wall is represented by two-dimensional membrane which is described using two collective coordinates: q – local position of DW and ψ – local azimuthal angle of magnetization vector. Let us assume that z-axis coincides with the easy axis, y-axis is directed along the domain wall normal and the angle ψ is measured from the x-axis. The Slonczewski's approximation is applicable if $\nabla q, \nabla \psi$ are small; actually, in the problem under consideration the curvature radius of the wall is small in comparison with the wall thickness $\Delta = (A/K)^{1/2}$ (A is the nonuniform exchange stiffness constant and K is the anisotropy constant), so we can treat the DW surface during the motion as a plane. Thus, we obtain the one-dimensional problem and in this case the Lagrangian of the wall (more exactly, the Lagrangian density per unit area of DW) in imaginary time is

$$L = \frac{\sigma_0 \dot{q}^2}{2c^2} + U(q), \tag{9}$$

where $c = 2\pi\gamma M\Delta$ is the Walker velocity, M is the magnetization, $\sigma_0 = 4(AK)^{1/2}$ is the energy density of the wall and $U(q)$ is the total wall potential energy (including the potential produced by the defect as well as the potential of externally applied drive field). The total effective magnetic field acting on the wall is

$$H_{eff} = -\frac{1}{2M}\frac{\partial U}{\partial q}. \tag{10}$$

The DW Lagrangian (9) has the same form as the Lagrangian used by Callan and Coleman [10], so that results of their work can be directly applied.

But still it is necessary to make some notes. To obtain an observable tunnelling rate it is necessary to have a reasonable value of the instanton action. Eq.(9) represents the Lagrangian density per unit area, hence, the total area of tunnelling section of DW should be small; it implies the small size of the defect. Thus, the tunnelling section of the wall is almost plane (the noticeable curvature at small distance corresponding to small size of defect would significantly increase the instanton action). However, there is another possibility: if the defect area is large enough, the tunnelling can proceed by nucleation process; in this case the critical portion of the wall tunnels through the defect and then expands classically, depinning the entire wall. We do not consider this possibility here: exact results concerning this tunnelling scenario [23] and the qualitative analysis in [20] and [19] show that the tunnelling rate do not differs considerably from the case discussed here.

Furthermore, to have a reasonable instanton action we should have quite small barrier height. It can be reached if we apply the external drive field H_0 quite close to the defect coercive field H_c. In this case the total effective magnetic field acting on the DW can be approximated by quadratic function as follows:

$$H(q) = H_c\left(1 - \frac{(q-b)^2}{2a^2}\right) - H_0, \tag{11}$$

where b is the position of the defect center and a is the effective width of the barrier (for small defect it is of order of DW thickness Δ). Accounting for the fact that for correct normalization of the potential we should have the metastable minimum at $q = 0$ and $U(0) = 0$, it is easy to obtain the standard "qudratic-plus-cubic" form of the wall potential:

$$U(x) = -\frac{MH_c}{3a^2}q^3 + \frac{MH_c}{a}q^2\sqrt{2(1-H_0/H_c)}. \tag{12}$$

The instanton action can be easily found:

$$S_{inst} \simeq 16A_w a\sqrt{mH_c Ma}(1-H_0/H_c)^{5/4}, \tag{13}$$

where A_w is the area of the tunnelling section of the wall and $m = \sigma_0/c^2$ – effective mass of the wall. The calculation of the pre-exponential factor is more complicated so that we do not present it here; in addition, the tunnelling rate depends on the exponent much more strongly than on the pre-exponential factor.

It is worth to note that the effective mass of the wall in the ferromagnetic material is governed by the magnetostatic interaction: the increase in M leads to the decrease in m. But in weakly ferromagnetic (WFM) materials the effective mass of DW is governed by the exchange coupling, which is much stronger than the anisotropy interaction. It is due to the fact that the weak ferromagnet consists of two equivalent aniferromagnetically coupled magnetic sublattices which are slightly canted with respect to one another. The limiting velocity of DW (which is equal to the velocity of magnons in the linear region of their dispersion curve) is determined by the frequency of small oscillations of these sublattices with respect to one another; obviously, this frequency is governed by the interaction between sublattices. Thus, the weak ferromagnets can be more promising materials for observation of the wall tunnelling.

The DW tunnelling in WFM was very briefly discussed by Chudnovski [24]. Independently, more comprehensive investigation was undertaken by present authors [25]. Our investigation was initiated by experiments performed by Zhang et al. [26] on the sample of terbium orthoferrite $TbFeO_3$.

We considered the weak ferromagnet (such as $TbFeO_3$) within two-sublattices approximation, using the antiferromagnetic vector $\vec{l}$. Corresponding Lagrange density is (e.g., [27]):

$$\mathcal{L} = \frac{\chi_\perp}{2\gamma^2}(\dot{\vec{l}})^2 - \frac{\chi_\perp}{\gamma}\vec{H}\,[\vec{l}\dot{\vec{l}}] - F, \tag{14}$$

$$F = A(\nabla\vec{l})^2 - \frac{\chi_\perp}{2}\left(H^2 - (\vec{H}\vec{l})^2\right) - M_z^0 H_z l_x + M_x^0 H_x l_z + K_{ac}l_z^2 - K_{ab}l_x^2,$$

here $\vec{H}$ is the total external field acting on the wall (we assume $\vec{H} = H\vec{e}_z$), M_z^0 and M_x^0 are the values of magnetization in the phases $\Gamma_4(\vec{l} \parallel \vec{x})$ and $\Gamma_2(\vec{l} \parallel \vec{z})$ respectively, K_{ac} and K_{ab} are the anisotropy constants corresponding to crystal planes ac and ab. The z-axis is directed along the crystal c-axis. Without loss of generality, we consider the case $K_{ac} < K_{ab}$, so magnetization inside the wall rotates in the ac-plane. Let us introduce spherical coordinates with polar axis directed along the c-axis and azimuthal one — along the a-axis. Taking into account that azimuthal angle $\phi < 0.1°$ if $H \leq 200\ Oe$ [27], we ignore it.

Our first goal is to obtain the reduced Lagrangian for domain wall containing only collective coordinates. For the freely moving DW we can obtain:

$$\theta_0 = -\pi/2 + 2\arctan e^\xi, \quad \xi = \frac{x - x_0}{\Delta}, \tag{15}$$

$$\Delta = \Delta_0\sqrt{1 - v^2/c^2},$$

here $v = \dot{x}_0$ is DW velocity, $\Delta_0 = \sqrt{A/K}$ is the static wall width, $c = \gamma\sqrt{A/\chi_\perp}$ is the spin-wave velocity. The influence of external field is weak in comparison with anisotropy and exchange, so we can apply the perturbation theory for soliton motion [27, 28]. Substituting in Eq.(14) the structure of moving DW, described by $x_0(t)$, $\Delta(t)$ varying with time, and integrating the Lagrange density over x, we obtain Lagrange function per unit area of domain wall depending on the only collective coordinate, wall center position x_0:

$$L = -mc^2\sqrt{1 - v^2/c^2} - U(x_0),$$

where $mc^2 = 4\sqrt{AK}$ and $U(x_0) = -\int 2M_z^0 H(x_0)\,dx_0$, $H(x_0)$ is the total external field, including the effective field of defect and drive field.

Now, we turn to the problem of the wall tunnelling. Let us consider a domain wall, pinned by defect at the position x_{01}, which is under influence of externally applied drive field (see Fig.1a). The tunnelling rate in this case can be calculated in WKB-approximation using the path integral technique in imaginary time τ. But we can not directly apply usual Feynman's formulation because Lagrange function is not quadratic in DW velocity. We begin with Hamilton formulation, with canonically conjugated variables p and q:

$$\rho(x, y) = \int\limits_{(q=x)}^{(q=y)} Dq \int\limits_{(-\infty)}^{(+\infty)} \frac{Dp}{2\pi\hbar} \exp\left[\frac{1}{\hbar}\int_0^T d\tau\left(ip\dot{q} - H(p,q)\right)\right]. \tag{16}$$

Within WKB-approximation we expand the effective action in the neighbourhood of classical trajectories $\{p_0(\tau), x_0(\tau)\}$, obtaining

$$\rho(x, y) = e^{-S_{cl}(x,y)/\hbar}\int\int\frac{D\xi D\eta}{2\pi\hbar}\exp\left(-\frac{1}{2\hbar}\delta^2 S\right), \tag{17}$$

$$\xi = q - x_0, \quad \eta = p - p_0.$$

Here $S_{cl}(x, y)$ is an action along the classical trajectory, which at $\tau = 0$ starts at $x_0 = x$, and at $\tau = T$ finishes at $x_0 = y$:

$$S_{cl} = \int_0^T d\tau\left(mc^2\sqrt{1 + \dot{x}_0^2/c^2} + U(x_0)\right),$$

338

and

$$\tfrac{1}{2}\delta^2 S \;=\; \int_0^T \left(\frac{1}{2m}\eta^2\bar{u}^2 + \frac{1}{2}\xi^2\alpha^2 - i\eta\dot{\xi}\right) d\tau,$$

$$\alpha^2 = \left.\frac{\partial^2 U}{\partial q^2}\right|_{q=x_0(\tau)}, \qquad \bar{u}^2 = \frac{m^3 c^3}{(p_0^2 + m^2 c^2)^{3/2}}.$$

In spite of the fact that the integral in (17) is gaussian, we can not simply perform an integration over η since the corresponding path integral measure would have rather complex form:

$$\prod_{k=1}^N \sqrt{\frac{m}{2\pi\hbar\varepsilon}} \frac{1}{(1+\dot{x}_k^2/c^2)^{3/4}}, \qquad \varepsilon = T/N.$$

This difficulty can be easily overcome by the following substitution

$$\theta = \int_0^\tau [\bar{u}(\tau')]^2 \, d\tau'.$$

Then Eq.(17) will be rewritten

$$\rho(x,y) \;=\; e^{-S_{cl}/\hbar} \iint \frac{D\xi\,D\eta}{2\pi\hbar} \exp\left[-\frac{1}{\hbar}\int_0^\Theta \left(\frac{\eta^2}{2m} + \frac{\xi^2\alpha^2}{2\bar{u}^2} - i\eta\xi'_\theta\right) d\theta\right],$$

$$\Theta = \int_0^T [\bar{u}(\tau)]^2 \, d\tau,$$

and an integration over η gives the path integral with usual Feynman's normalization:

$$\rho = e^{-S_{cl}/\hbar} \int_{(\xi=0)}^{(\xi=0)} D\xi \, \exp\left[-\frac{1}{\hbar}\int_0^\Theta \left(\frac{m}{2}(\xi'_\theta)^2 + \frac{\alpha^2}{2\bar{u}^2}\xi^2\right) d\theta\right] \qquad (18)$$

The tunnelling rate can be obtained by method, developed in [10]:

$$\Gamma = C\exp{(-B)}, \qquad \frac{B}{A_w} = S_{inst}/\hbar,$$

A_w is the area of tunnelling section of the wall. The pre-exponential factor is (cf. [20], [10])

$$C = \sqrt{\frac{m\nu}{2\pi\hbar}}D, \qquad (19)$$

$$D = \left| \frac{\det'\left[-\partial_\tau\{(1/\bar{u}^2)\partial_\tau\} + (\alpha_{inst})^2/m^2\right]}{\det[-\partial_\tau^2 + \omega^2]} \right|^{-1/2},$$

$$\nu = \int_{-\infty}^{+\infty} [\dot{x}_{inst}(\tau)]^2 \, d\tau$$

where

$$\omega^2 = \frac{1}{m}\frac{\partial^2 U}{\partial x^2}\bigg|_{x\equiv x_{01}}, \quad (\alpha_{inst})^2 = \frac{\partial^2 U}{\partial x^2}\bigg|_{x=x_{inst}}.$$

Equations of motion for instanton $x_{inst}(\tau)$ can be readily integrated, yielding

$$\tau = \int_{x_{02}}^{x_{inst}} \frac{dx}{\pm c\sqrt{(mc^2/U)^2 - 1}}, \tag{20}$$

$$S_{inst} = 2\int_{x_{02}}^{x_{01}} \sqrt{2mU(x) - [U(x)]^2/c^2} \, dx,$$

where x_{02} is the turning point of instanton trajectory.

For comparison of our results with experiments [26], we assume that the barrier can be approximated by "quadratic-plus-cubic" potential which was discussed above:

$$U(x) = -\frac{M_z^0 H_c}{3a^2}x^3 + \frac{M_z^0 H_c}{a}x^2\sqrt{2(1 - H_0/H_c)}, \tag{21}$$

where a is the width of barrier, H_0 is drive field, H_c — coercive field of defect. Preliminary estimates show that the condition $U \ll mc^2$ is satisfied here. Thus, using that

$$\frac{B}{A_w} \simeq (16/\hbar)a\sqrt{mH_c M_z^0 a}(1 - H_0/H_c)^{5/4}, \tag{22}$$

$H_c \simeq 600$ Oe is obtained from $\ln\Gamma(H)$ dependence reported in [26]. Assuming that the barrier is created by nonmagnetic inclusion of length d, we obtain for barrier height at $H_0 = 0$:

$$U_{max} = mc^2(d/2\Delta_0). \tag{23}$$

With typical values of TbFeO$_3$: $A = 1 \cdot 10^{-7}$ erg/cm, $M_z^0 = 10$ G, $c = 2 \cdot 10^6$ cm/s, the comparison of Eq.(21) with Eq.(23) gives $d \simeq 10$ Å. With evaluated in [26] value of B it gives $A_w \simeq 10^3$ Å^2. On the other hand, U_{max} at $H = 0$ is U/A_w, so $A_w \simeq 4 \cdot 10^3$ Å^2. Furthermore, the tunnelling volume

$$V = \frac{U}{2M_z^0 H_c} \simeq 7 \cdot 10^5 \text{Å}^3,$$

So, with $a \simeq \Delta_0 \simeq 100$ Å, we get $A_w \simeq 7 \cdot 10^3$ Å^2. We see that the model of nonmagnetic inclusion gives reasonable values of d and A_w and that three different characteristics of tunnelling give the values of A_w reasonably close to one another.

Above we discussed the case of MQT of a magnetic soliton. The possibility of experimental observation of domain wall MQC was discussed by Gaitan [29], [30]. He considered the symmetric double well potential (see Fig.1b) created by two identical voids in the magnetic material (of course, these voids should be placed quite closely to one another). As a result of the study, he concluded that the coherence between two energetically degenerate states of the wall is destroyed too easily due to small deviations from the exact symmetry between the voids. Thus, perspectives of observation of domain wall MQC at macro- or mesoscopic scales are rather gloomy.

Among other MQT problems, the case of DW depinning via tunnelling is, probably, most close to the practice: it is the process restricting the information storage density in magnetic devices. The review of theoretical results given above shows that the significant progress was achieved in theory. Below, we briefly describe experiments performed in this field.

3.2. EXPERIMENTS

The technique usually used in modern experiments on MQT phenomena, known as "magnetic after-effect" measurements, is more than 50 years old. When the strong enough external magnetic field pulse is applied, the magnetization reversal in a sample starts, but this process is retarded by the potential barriers created, for example, by defects. After-effect technique is based on the measurement of the rate of a change of sample's total magnetization. Specifically, the quantity $S = -dM/d(\ln t)$, referred as magnetic viscosity, is measured (since for uniform height distribution of barriers $n(E) \equiv$ const the magnetization M varies with time as $\ln t$, for review see [31]). Also, the mean relaxation time τ, which is defined as

$$1/\tau = (1/2M_s)(dM/dt)_{(M=0)}$$

is often used for description of magnetization relaxation rate. At thermoactivated (TA) regime S decreases linearly with temperature (τ increases as $\exp(E/kT)$, E is the "mean" barrier height, k is the Boltzman's constant and T is the temerature). When the temperature decreases, thermoactivated regime is replaced by MQT at certain temperature T_c (crossover temperature). At MQT regime both quantities S and τ do not depend on the temperature (although a certain distribution of potential barriers, $n(E) \propto (1/E)$, also produces the constant S, [31]).

The classic case of MQT of domain walls through potential barriers is the relaxation in highly anisotropic rare-earth-based systems [32]. But there is a problem of comparison of these experiments with theory: these systems are disordered magnets and the problem of quantitative comparison is hard. Nevertheless, the qualitative agreement presents: for example, experiments on bulk samples gave $T_c = 5$ K for Dy$_3$Al$_5$ system, $T_c = 10$ K for SmCo$_{3.5}$Cu$_{1.5}$ [33], whereas from the theory value of order of few kelvins is expected. Analogous experiments were performed on small particles and multilayered random magnets [33], [34]. In both cases features specific for DW tunelling were detected (as an example, data corresponding to experiments on small particles are shown at Fig.2): theory developed in [32] for this case predicts for τ:

$$\tau(H,T) = \tau_0 \exp\left[p(1/H - 1/H_0)/kT^*\right],$$

where p and τ_0 are some constants specific for given sample, H is the applied field, H_0 is the maximum corcive field, T – temperature, T^* is the effective temperature of magnetization reversal, $T^* = T$ for TA regime (high T) and $T^* = T_c$ in MQT region (low T); experimental data correspond to the scenario of two-dimensional nucleation of the critical portion of a wall.

To check the present theories of domain wall MQT, it is necessary to create a system with better defined barrier parameters. Now, two main ways are used: (i) the study of artificially created "DW junction", the idea proposed by Barbara et al. [35] and (ii) the investigation of relaxation via DW tunnelling in nanowires [36], [37].

In the work [36] the array of nickel wires was used; each wire was 1 μm in width, 3 mm in length and 500 Å in thickness. The interval between adjacent wires was 6.5 μm. The magnetization change with time was logarithmic with a good accuracy (see inset in Fig.3), nevertheless, more thorough analysis performed in [36] shows Gaussian distribution. The curve $S(T)$ (see Fig.3) shows approximate saturation below 20 K; but this value is too high for MQT crossover temperature. This discrepancy can be caused by interaction between adjacent wires. Also, Ni wires (though prepared by another method) were studied in [37], but in this work changes in the resistance of the wire were used for determination of changes in distribution of magnetization.

The domain wall junctions (DWJ) also have been used for MQT experiments. The first example is w-DWJ where the artificial defect with

342

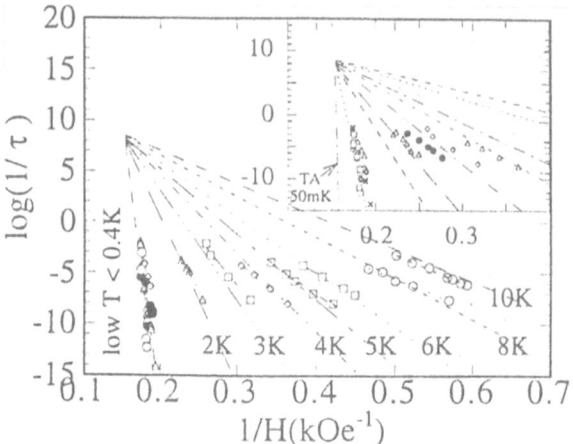

Figure 2. Evidence for the DW tunnelling in small particles $Tb_{0.5}Ce_{0.5}Fe_2$. The experimental curves $\log(1/\tau)$ versus $1/H$ are shown. The linear dependence shows that the magnetization reversal proceeds by motion of the domain wall. The constant value of τ at low temperatures ($T < 0.4$ K) shows the crossover from TA to MQT regime ($T_c = 0.6$ K). Picture is taken from [33].

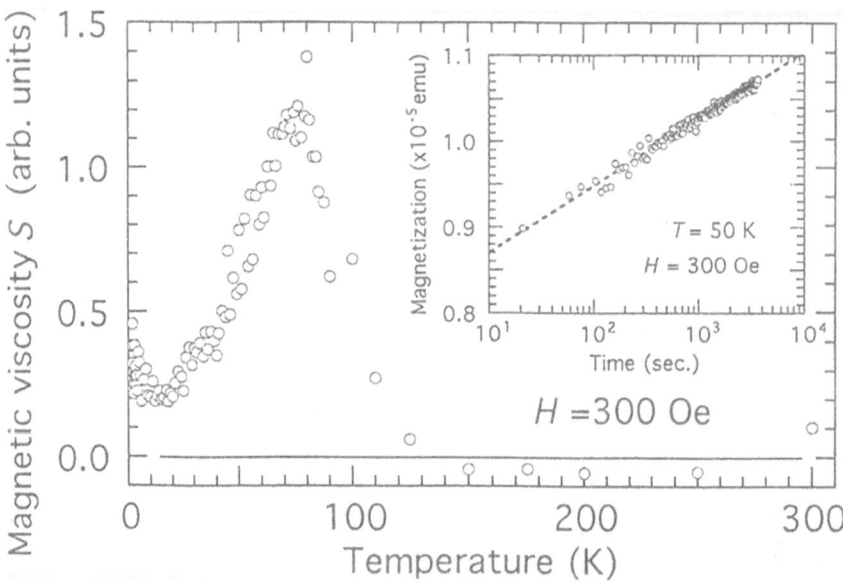

Figure 3. DW tunnelling in Ni nanowires. Temperature dependence of magnetic viscosity $S(T)$ at $H = 300$ Oe. The plateau below 20 K can be caused by tunnelling. Inset: typical dependence of the total magnetization of sample versus time; logarithmic dependence is clearly seen. Picture is taken from [36].

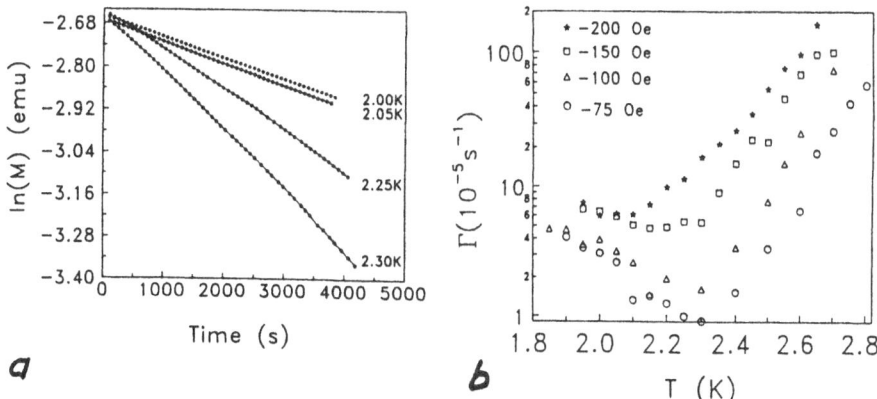

Figure 4. DW tunnelling in $TbFeO_3$ bulk sample. (a) – The dependence of magnetization versus time. The exponential relaxation is seen. (b) – Temperature dependence of the tunnelling rate Γ. Plateau at low temperatures is clearly seen. Pictures are taken from [26].

reduced exchange energy was created. This junction was produced in multilayer system Co/CoCu/Co [33], and the thin central layer served as a defect. Hysteresis curves showed that the wall is pinned by "defect" layer. The crossover temperature of such system is about 5 K. The second case is *b*-DWJ, where artificially created defect has higher energy. In this case the system GdCo/SmCo/GdCo was used with SmCo layer serving as a defect; the increase in defect energy is due to increase in anisotropy energy. The crossover temperature for this case is also about 5 K.

In conclusion of this subsection we remind the paper [26]. This work was performed on the bulk sample of $TbFeO_3$, nevertheless, the extremal purity of the sample led to the single relaxation time, that is, to the situation when all barriers have the same parameters (height, width etc.); it was concluded from the exponential (not the logarithmic) relaxation of magnetization. Parameters of the defects can be extracted from high-temperature region (TA relaxation). In addition, the crossover temperature is not very high (see Fig.4) Thus, even bulk materials can be very useful for quantitative investigations of MQT process.

Thus, the brief review given above shows that ideas about MQT of domain walls are supported by a solid experimental base.

4. The change of topological charge of a magnetic soliton via the

macroscopic tunnelling

In this subsection we discuss another very interesting phenomenon concerning the tunnelling of a magnetic soliton: the change of its topological charge. This problem is not totally academic: magnetization reversal in small ferromagnetic particles (having ultrasmall thickness, of order of a few tens nanometers) [38] and in two-dimensional systems (e.g., in multilayers) [33] can proceed via nucleation of the vortex structures, which is, in our terms, the switching of the topological charge.

The topological charge is one of most basic parameters of a soliton. In classical theory of solitons the fact can be proved (see [12]) that the topological charge is conserved in the process of soliton motion. Nevertheless, if we account for a possible singularity in the configuration corresponding to the transition from one topological charge to the other, such a process can take place.

To be more specific let us discuss the vortex in a two-dimensional easy-plane antiferromagnet (AFM); the change of topological charge of such vortex via macroscopic tunnelling was considered by Galkina et al. [39]. This model is very similar to the nonlinear $O(3)$-model widely used in the field theory.

Magnetic structures in AFM can be described by normalized antiferromagnetic vector $\vec{l}$, $l^2 = 1$. It is convenient to introduce spherical angles θ and ϕ so that $l_z = \cos\theta$, $l_x + il_y = \sin\theta\exp(i\phi)$. The vortex structures can be described as follows:

$$\theta = \theta_0^\pm(r), \quad \phi = q\chi + \phi_0, \quad \theta_0^\pm(\infty) = \pi/2, \quad \theta^+(0) = 0, \quad \theta_0^- = \pi, \quad (24)$$

where r and χ are polar coordinates in the plane of antiferromagnet; $\theta^\pm$ describes two possible states of the vortex (differing by the direction of vector $\vec{l}$ at the center of the vortex). Functions $\theta^\pm$ differ from $\pi/2$ considerably only inside the vortex's core which has the size of order of $\Delta_0 = (A/K)^{1/2}$, where A is the exchange constant and K is the anisotropy constant of the material (K is not large in the case under consideration, so that the vortex with $l_z \equiv 0$ is unstable). In the work [39] the case of macroscopic quantum coherence between the states θ^+ and θ^- was considered, when energy levels corresponding to θ^+ and θ^- configurations are splitted due to the process of macroscopic tunnelling between these two states.

According to the general theory of MQC (which is somewhat similar to MQT) [15], the tunnelling splitting of levels $\tilde{\Gamma}$ in WKB-approximation is given by the formula

$$\tilde{\Gamma} = C\exp(-S_{inst}/\hbar), \quad (25)$$

where the instanton in this case is the trajectory in the imaginary time (in the reversed potential $-V$, see Fig.1b) starting at $\tau \to +\infty$ from one minimum x_1 and approaching at $\tau \to -\infty$ to the other one, x_2.

The Lagrangian density of two-dimensional easy-plane AFM in imaginary time τ has the form

$$L = A\left\{\frac{1}{c^2}\left(\frac{\partial \vec{l}}{\partial \tau}\right)^2 + \left[(\nabla_2\vec{l})^2 - (\nabla_2\vec{l}^0)^2\right] + \frac{1}{\Delta_0^2}\left[l_z^2 - (l_z^0)^2\right]\right\}, \quad (26)$$

where $\nabla_2 = \vec{e}_x \partial/\partial x + \vec{e}_y \partial/\partial y$ – gradient in the plane (x,y), c is the spin-wave velocity; $\vec{l}^0$ describes the structure of static vortex (see Eq.(24)). Terms including $\vec{l}^0$ do not change equations of motion and are included in the Lagrangian for correct normalization (the energy in potential minima should be zero).

Equations of motion for the angles θ and ϕ are

$$\nabla_E^2\theta + \sin\theta\cos\theta[\Delta_0^2 - (\nabla_E\phi)^2] = 0, \quad \nabla_E(\sin^2\theta\nabla_E\phi) = 0, \quad (27)$$

where ∇_E is the gradient in the Euclidean space $(c\tau, x, y)$. The instanton solution of the system (27) has the form of "hedgehog" (see Fig.5) and it is easy to see that in the center of this configuration the singularity exists, where $\vec{l}$ is not defined. It is the very singularity which allows the tunnelling change of topological charge.

In the neighbourhood of the center of hedgehog (inside the core of instanton configuration) the spherically-symmetric analytical solution has the form:

$$\cos\theta = c\tau/\sqrt{x^2 + y^2}, \quad \tan\phi = y/x. \quad (28)$$

The contribution from the core of hedgehog is

$$S_{inst}^0 = 4\pi(A/c)R, \quad (29)$$

where R is the size of the core of instanton; this value, of course, is not exactly defined, it is clear only that $R \ll \Delta_0$. It is remarkable that the contribution to the action from the singularity at the center of hedgehog is finite.

Since the exact analytical instanton solution outside of the hedgehog core is unknown, to obtain the total action on the instanton configuration the variational approach was used in [39]. It was suggested that $\theta(c\tau, x, y)$ at the distances $r \geq \Delta_0$ has the form

$$\theta = \pi/2 + F(\tau)[\pi/2 - \theta_0^+(r)], \quad r = \sqrt{x^2 + y^2}, \quad (30)$$

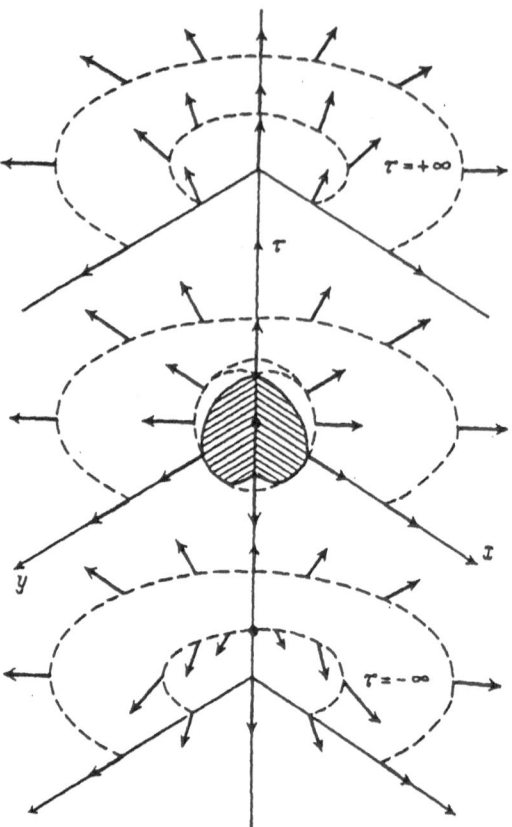

Figure 5. Instanton "hedgehog" configuration $\vec{l}(x, y, \tau)$ corresponding to the tunnelling switching of the vortex's topological charge. The picture displays a gradual change of the distribution of vector $\vec{l}$. Three moments of imaginary time τ are shown: $\tau \to -\infty$, $\tau = 0$ and $\tau \to +\infty$. Pictures at $\tau \to \pm\infty$ correspond to static vortices with different topological charges. Note the singularity at $\tau = 0$. Picture is taken from [39].

where $F(\tau)$ is a fitting function: $F(\tau) \to +1$ at $\tau \to +\infty$ and $F(\tau) \to -1$ at $\tau \to -\infty$ respectively. This variational procedure yields for the total action

$$S_{inst} \simeq 2\pi\xi A\Delta_0/c, \tag{31}$$

where ξ is the dimensionless quantity of order of unity.

Another case of tunnelling change of a soliton's topological charge was considered in [41]. In this paper we dealt with a switching of vortex structure in ferromagnetic (FM) material.

We considered the vortex structure, vertical Bloch line (VBL) [40], in the domain wall. Dealing with the case of easy-axis ferromagnet, for description of the wall we use Slonczewski's equations, chracterizing the structure by its center position q and the azimuthal angle ψ of magnetization vector at the center of DW. The VBL corresponds to the intermediate region between wall subdomains $\psi = 0$ and $\psi = \pm\pi$.

We consider the ultrathin film so the distribution of polar angle ψ is independent on z-coordinate (we assume that z-axis is directed along the film normal and y-axis coincides with the normal to the wall plane). The topological charge of VBL under consideration is defined by the condition

$$\psi(x \to -\infty) = 0, \quad \psi(x \to +\infty) = -\pi. \tag{32}$$

This configuration has the same energy as

$$\psi(x \to -\infty) = 0, \quad \psi(x \to +\infty) = \pi, \tag{33}$$

so the tunnelling can lift the degeneracy and the value of splitting is given by above-mentioned WKB expression. The instanton configuration corresponding to the tunnelling event is "hedgehog"-like structure in $\tau - x$ plane (see Fig.6), reminiscent of the vortex in AFM considered above or Bloch point (BP) structure [40] in thick films, but there is essential difference: BP (as well as a vortex in AFM) is "pseudorelativistic hedgehog", variables x and y appear in an equivalent manner in Laplace operator, whereas our hedgehog is nonrelativistic, with essentially distinct x and τ: as will be seen later, the spatial scale X of our hedgehog is of order of $\Lambda\sqrt{4\pi\gamma MT}$ where T is the scale in imaginary time.

To provide the DW stability against the bending, it should be held by external gradient field H'; we assume $H'\Delta \ll 4\pi M$ (it is true for the most of experimental situations). For such system Slonczewski's equations have the form

$$2M\gamma^{-1}\dot{\psi} = \sigma q_{xx} - 2MH'q, \tag{34}$$

$$-(\gamma\Delta)^{-1}\dot{q} = 2AM^{-1}\psi_{xx} - 2\pi M \sin 2\psi. \tag{35}$$

It is convenient to introduce dimensionless variables: $q \to q/\Delta$ ($\Delta = \sqrt{A/K}$ is the DW width), $x \to x/\Lambda$ ($\Lambda = \sqrt{A/2\pi M^2}$ is the Bloch line width), $t \to 4\pi\gamma Mt$; this replacement implies dimensionless Lagrangian (more exactly, its density per unit length) of the system $L \to L/8\pi M^2 \Delta d$ (d is the film thickness), which has the form:

$$L = -q\dot{\psi} - \frac{1}{2}(q_x)^2 - \frac{1}{2}(\psi_x)^2 - \frac{1}{2}\sin^2\psi - \frac{\kappa^2}{2}q^2, \tag{36}$$

where $\kappa^2 = H'\Delta/4\pi M$, $\kappa \ll 1$. In imaginary (dimensionless) time τ it should be rewritten as:

$$L = iq\dot{\psi} + \frac{1}{2}(q_x)^2 + \frac{1}{2}(\psi_x)^2 + \frac{1}{2}\sin^2\psi + \frac{\kappa^2}{2}q^2, \tag{37}$$

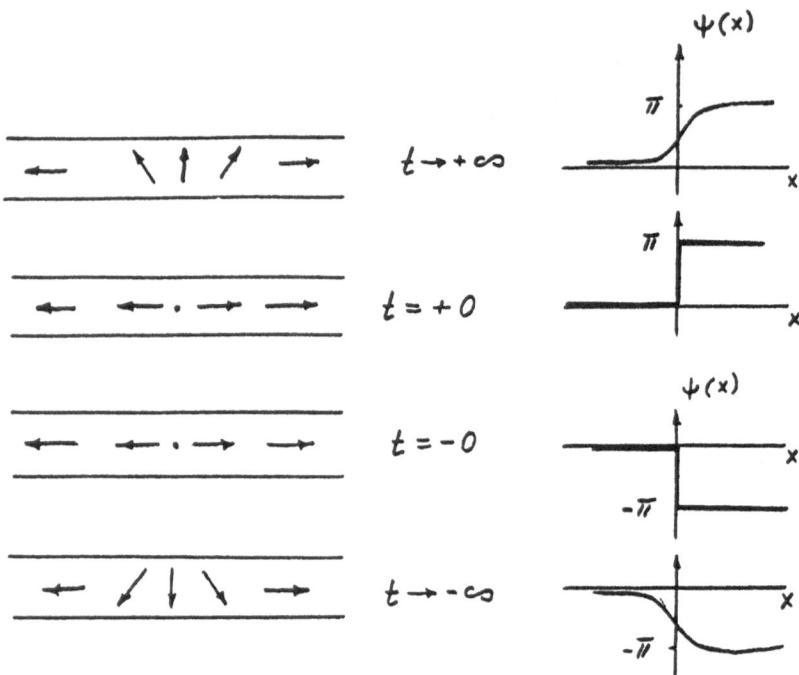

Figure 6. Instanton "hedgehog" configuration $\psi(x,\tau)$ corresponding to the tunnelling switching of VBL's topological charge. The left picture displays a gradual change of the distribution of magnetization vector, the right picture shows the correspondent graphs $\psi(x)$. Four moments of imaginary time τ are shown: $\tau \to -\infty$, $\tau \to -0$, $\tau \to +0$ and $\tau \to +\infty$. Pictures at $\tau \to \pm\infty$ correspond to static vertical Bloch lines with different topological charges. Note the singularity point: the transition from $\tau \to -0$ to $\tau \to +0$ changes nothing in the real structure of magnetization and changes crucially $\psi(x)$ graphs.

and Eqs. (34), (35) become:

$$i\dot\psi = q_{xx} - \kappa^2 q, \tag{38}$$

$$-i\dot q = \psi_{xx} - \frac{1}{2}\sin 2\psi. \tag{39}$$

The main contribution into instanton action comes from the neighbourhood of the center (core) of "hedgehog", where we can neglect second terms in Eqs. (38) and (39). Then, in imaginary time these equations give

$$\dot\psi = \pm\psi_{xx}, \tag{40}$$

The sign "+" in (40) corresponds to $\tau > 0$, the sign "−" corresponds to $\tau < 0$. Such a choice is equivalent to above-mentioned boundary conditions: (32) at $\tau < 0$, and (33) at $\tau > 0$.

We can solve (40) in both directions: from $\tau = 0$ to $\tau \to +\infty$ (forward) and from $\tau = 0$ to $\tau \to -\infty$ (backward). In this case Eqs. (32) and (33) give the initial condition for either direction: $\psi(\tau = 0) = \pm\pi 1(x)$, $1(x)$ is the standard step function. Then, solution of Eq. (40) is known very well:

$$\psi(x,\tau) = \pm\frac{1}{2\sqrt{\pi|\tau|}}\int_{-\infty}^{+\infty}\pi 1(\xi)\exp[-(x-\xi)^2/4|\tau|]d\xi \qquad (41)$$

$$= \pm\frac{\pi}{2}\text{erfc}(-x/2\sqrt{|\tau|}).$$

To find an instanton action, we need to know the $q(x)$ profile. But now, unless $\kappa \ll 1$, we can not drop the second term in (38), because it provides the fulfillment of boundary condition $q(x \to \pm\infty) \to 0$. So, we write

$$q = -i\int\frac{\exp(-\kappa|x-\xi|)}{2\kappa}\dot{\psi}(\xi,\tau)d\xi. \qquad (42)$$

and go further keeping κ where it is necessary for convergence (working only with lowest order in κ). Nevertheless, it is easy to see that (37) does not depend on κ if κ is small: $\dot{\psi}(x,\tau)$ is antisimmetric with respect to x, so (42) does not contain the term $O(1/\kappa)$ and the lowest order in q is $O(1)$. Thus, the term $\kappa^2 q^2/2$ is of order of κ^2 and can be neglected. Futhermore, the others "dangerous" terms are: (i) $q_x{}^2/2$, which can be obtained from (42) (when $\kappa = 0$):

$$q_x = -\pi\frac{\exp\left(-x^2/4|\tau|\right)}{2\sqrt{\pi|\tau|}}.$$

and converges even in this case due to convergence of $\dot{\psi}(x,\tau)$; (ii) $iq\dot{\psi}$, which converges too (even if $\kappa = 0$), since $\dot{\psi}(x,t)$ cuts off this term at large x. Thus, computing the instanton action we can take $\kappa = 0$ from the very beginning.

Computation itself is quite simple. We drop in (37) fourth and fifth terms, playing minor role in the center of instanton and integrate over x from $-\infty$ to $+\infty$ and over τ from $-T$ to $+T$ ($2T$ is the temporal "size" of instanton core). All integrals over x are different versions of erfc-like integrals and can be exactly found. The only difficulty is the integral of $iq\dot{\psi}$, which is the integral of error function. But it has the form

$$\int\Phi(x)\Phi''(x)dx, \qquad \Phi(x) = \int_x^{+\infty}\exp{-\xi^2/4|\tau|}d\xi$$

and can be reduced to erfc-like integral as follows:

$$\int\Phi(x)\Phi''(x)dx = \Phi\Phi' - \int[\Phi'(x)]^2dx.$$

Thus, performing all computations, we obtain the action of the core of hedgehog:

$$S^0_{inst} = \pi\sqrt{2\pi T} \sim 8\sqrt{T} \tag{43}$$

It is easy to check that the fifth term in (16) is $O(\kappa)$. The fourth term is $(4/3)T^{3/2}$ and therefore in the core of hedgehog (where $T \ll 1$) is much less than the S^0_{inst}.

So, we know solutions for internal core and external "vacuum" (static Bloch line solution). Unfortunately, we can not find the solution for instanton "shell" (intermediate region) due to nonlinearity of equations. But the very general feature of solitons and instantons is that the shell is quite thin, so we remove the shell at all and take the spatial distribution in the hedgehog as follows: if $T < 1$ we take our solution (41) and (42), if $T > 1$ we take the static Bloch line distribution. So, the total instanton action is (43) with $T \sim 1$, or, in dimensional units,

$$S_{inst} \sim 16Ad/\sqrt{2\pi\gamma^2 K}, \tag{44}$$

where d is the film thickness, and the levels splitting

$$\tilde{\Gamma} \propto \exp -S_{inst}/\hbar. \tag{45}$$

The interesting and very unusual feature of the general problem of tunnelling switching of topological charge is that the height of potential barrier is formally infinite, in our case it diverges as $1/\sqrt{\tau}$ at $\tau \to 0$, whereas the action is finite. Thus, strictly speaking, in the approach proposed the concept of the crossover temperature looses its meaning. Also, the attempt frequency goes to infinity. Of course, our description, using Slonczewski equations is correct only up to the order of Δ/Λ, but it is easy to see that above-mentioned quantities are still divergent in completely micromagnetic consideration. They arise because the conditions of applicability of micromagnetic approach are violated at distances of order of lattice constant a and energy barrier height for the hedgehog is of order of exchange integral J. Nevertheless, the convergence of action integral allows to assume that corrections to micromagnetic description will be small, of order of a/Δ.

Above we considered the case of MQC. But, the MQT of topological charge is, probably, more easy to detect experimentally. In this case it is necessary to apply the external field which forces the initially degenerate levels (corresponding to solitons with different topological charge) apart. Moreover, such a field decreases the instanton action and, therefore, increases the tunnelling rate. In experiments [38], [33] the switching of magnetization was driven by external field.

5. Conclusion

We observed modern state-of-art in the field of quantum tunnelling of magnetic solitons. Our attention was focused mainly on the tunnelling itself, effects of dissipation were only briefly described.

Nevertheless, it seems that the next step should be the thorough investigation of influence of dissipative effects, since their impact is not connected only with decoherence and reducing of the tunnelling rate. Sometimes the dissipation can even drive the tunnelling process. For example, the magnetic moment in exactly uniaxial system with easy-axis anisotropy can not tunnel since it would lead to the change of projection of mechanical moment J_z to the anisotropy axis, whereas this quantity is the integral of motion in uniaxial system: $J_z \equiv$ const. But the interaction with microscopic degrees of freedom can destroy the uniaxial symmetry and allow the magnetization to tunnel.

Probably, this situation occur in the case of molecular magnets, such as Mn_{12} clusters (see [42]). Experiments show that these clusters possesses the strong uniaxial anisotropy. Actually, crystals of this substance have tetragonal symmetry but the corresponding magnetic anisotropy is very weak and can not produce the observed in [42] tunnelling rate. One of possible explanations is the interaction with phonons in crystal lattice (it was proposed in [43]); but in general the situation is far from complete understanding.

References

1. Kadomtsev B.B. (1994) Dynamics and information, *Sov. Phys.-Usp.* **37**, 425-499.; *Usp. Fiz. Nauk* **164**, 449-529 (in Russian).
2. Benioff P. (1982) Quantum mechanical Hamiltonian models of Turing machines, *J. Stat. Phys.* **29**, 515-546.
3. Landauer R. (1961) Irreversibility and heat generation in the computing process, *IBM J. Res. Develop.* **5**, 243-245.
4. Bennett C.H. (1973) Logical reversibility of computation, *IBM J. Res. Develop.* **17**, 525-532.
5. Feynman R.P. (1986) Quantum mechanical computers *Foundations of Phys.* **16**, 507-531.
6. Kramers H.A. (1940) Brownian motion in a field of force and the diffusion model of chemical reactions, *Physica* **7**, 284-304.
7. Langer J.S. (1967) Theory of the condensation point, *Annals of Phys.* **41**, 108-157.
8. Langer J.S. (1969) Statistical theory of the decay of metastable states, *Annals of Phys.* **54**, 258-275.
9. Coleman S. (1977) Fate of the false vacuum.Semiclassical theory, *Phys. Rev.* **D15**, 2929-2936.
10. Callan C.G. Jr. and Coleman S. (1977) Fate of the false vacuum.II.First quantum corrections, *Phys. Rev.* **D16**, 1762-1768.
11. Feynman R.P. (1972) *Statistical mechanics*, Benjamin, Reading, Massachusetts.
12. Rajaraman R. (1987) *Solitons and Instantons*, North-Holland, Amsterdam.

13. Affleck I. (1981) Quantum-statistical metastability, *Phys. Rev. Lett.* **46**, 388-391.
14. Caldeira A.O. and Leggett A.J. (1983) Quantum tunnelling in a dissipative system, *Annals of Phys.* **149**, 374-456.
15. Leggett A.J., Chakravarty S., Dorsey A.T. et al. (1987) Dynamics of the dissipative two-state system, *Rev. Mod. Phys.* **59**, 1-85.
16. Chudnovski E.M. and Gunther L. (1988) Quantum tunneling of magnetization in small ferromagnetic particles, *Phys. Rev. Lett.* **60**, 661-664.
17. Scharf G., Wreszinski W.F. and van Hemmen J.L. (1987) Tunnelling of a large spin: mapping onto a particle problem, *J. Phys. A: Math. Gen.* **20**, 4309-4319.
18. Stamp, P.C.E. (1991) Quantum dynamics and tunneling of domain wall in ferromagnetic insulators, *Phys. Rev. Lett.* **66**, 2802-2805.
19. Egami, T. (1973) Theory of Bloch wall tunnelling, *phys. stat. sol. (b)* **57**, 211-224; Egami, T. (1973).Theory of intrinsic magnetic after-effect, *phys. stat. sol. (a)* **20**, 157-165.
20. Chudnovski, E.M., Iglesias, O. and Stamp, P.C.E. (1992) Quantum tunneling of domain walls in ferromagnets, *Phys. Rev.* **B 46**, 5392-5404.
21. Tatara, G. and Fukuyama, H. (1994) Macroscopic quantum tunneling of a domain wall in a ferromagnetic metal, *Phys. Rev. Lett.* **72**, 772-775.
22. Slonczewski, J.C. (1972) Dynamics of magnetic domain walls, *Intern. J. Magnetism* **2**, 85-97.
23. Chudnovski, E.M. and Gunther, L. (1988) Quantum theory of nucleation in ferromagnets, *Phys. Rev.* **B37**, 9455-9459.
24. Chudnovski E.M. (1995) Magnetic tunnelling *J. Magn. Magn. Mat.* **140-144**, 1821-1824.
25. Dobrovitski, V.V. and Zvezdin, A.K. (1996) Quantum tunnelling of domain wall in weak ferromagnet, *Sov. Phys. JETP*, in press.
26. Zhang X.X., Tejada J., Roig A., Nikolov O. and Molins E. (1994) Quantum exponential relaxation of antiferromagnetic domain walls in $FeTbO_3$ single crystals, *J. Magn. Magn. Mat.* **137**, L235-238.
27. Zvezdin A.K. and Mukhin A.A. (1992) Magnetoelastic solitary waves and domain wall supersound dynamics *Zh.Eksp.Teor.Fiz.* **102**, 577-599 (in Russian).
28. McLaughlin, D.W. and Scott, A.C. (1981) Many-solitons perturbation theory, in K.Lonngren and A.C.Scott (eds.), *Solitons in Action*, "Mir", Moscow, pp.210-268 (in Russian).
29. Gaitan F. (1994) On domain wall quantum coherence in magnetic insulators, *Physica* **B194-196**, 265-266.
30. Gaitan F. (1994) On the observability of mesoscopic or macroscopic quantum coherence of domain walls in magnetic insulators, *J. Phys.: Condens. Matter* **6**, 7565-7579.
31. Barbara B. and Gunther L. (1993) The 'barrier plot' as a tool in magnetic relaxation studies, *J. Magn. Magn. Mat.* **128**, 35-41.
32. Uehara M. and Barbara B. (1986) Noncoherent quantum effects in the magnetization reversal of a chemically disordered magnet: $SmCo_{3.5}Cu_{1.5}$, *J. Physique* **47**, 235-238.
33. Barbara B., Sampaio L.C., Wegrowe J.E., Ratnam B.A., Marchand A., Paulsen C., Novak M.A., Tholence J.L., Uehara M. and Fruchard D. (1993) Quantum tunnelling in magnetic systems of various sizes, *J. Appl. Phys.* **73**, 6703-6708.
34. Tejada J., Zhang X.X. and Balcells Ll. (1993) Nonthermal viscosity in magnets: quantum tunnelling of the magnetization, *J. Appl. Phys.* **73**, 6709-6714.
35. Gunther L. and Barbara B. (1994) Quantum tunneling across a domain-wall junction, *Phys. Rev.* **B49**, 3926-3933.
36. Yamazaki H., Tatara G., Katsumata K. and Aoyagi Y. (1995) Magnetic relaxation in Ni wires *report P3.62 at 2nd International Symposium on Metallic Multilayers, 11-14 September, Cambridge, UK*; to be published in (1996) *J. Magn. Magn. Mat.*
37. Kimin Hong and Giordano N. (1995) Approach to mesoscopic magnetic measure-

ments, *Phys. Rev.* **B51**, 9855-9862.

38. Barbara B., (1995) Novel magnetic structures and nanostructures, *report NS.1 at 2nd International Symposium on Metallic Multilayers, 11-14 September, Cambridge, UK*; to be published in (1996) *J. Magn. Magn. Mat.*

39. Galkina E.G. and Ivanov B.A. (1995) Quantum tunnelling in the magnetic vortex of two-dimensional easy-plane magnetic, *Zh. Eksp. Teor. Fiz.* **61**, 495-498 (in Russian).

40. Malozemoff A.P. and Slonczewski J.C. (1979) *Magnetic domain walls in bubble materials*, Academic Press, New-York.

41. Dobrovitski, V.V. and Zvezdin, A.K. (1995) Macroscopic quantum tunnelling of solitons in ultrathin films, *report P3.81 at 2nd International Symposium on Metallic Multilayers, 11-14 September, Cambridge, UK*; to be published in (1996) *J. Magn. Magn. Mat.*

42. Barbara B., Wernsdorfer W., Sampaio L.C. et al. (1995) Mesoscopic quantum tunneling of the magnetization, *J. Magn. Magn. Mat.* **140-144**, 1825-1828.

43. Politi P., Rettori A., Hartmann-Boutron F. and Villain J. (1995) Tunneling in mesoscopic magnetic molecules, *Phys.. Rev. Lett.* **75**, 537-540.

Controlling Chaos in Thin YIG Films at Microwave Frequencies

D.W. Peterman and P.E. Wigen
The Ohio State University
Columbus, Ohio USA

1 Introduction

From cream mixing in a hot cup of coffee to leaky faucets to beating hearts, chaotic phenomena is everywhere in our daily lives. The irregular, unpredictable dynamics of chaos has long been an academic puzzle, yielding many surprising and interesting discoveries. For practical applications, chaos has been avoided, since it would appear to resist any control or utility. However, many chaotic systems have been controlled experimentally [1, 2, 3], providing new information about chaotic systems and suggesting possible applications.

This chapter will present recent results in controlling chaos in thin magnetic films at ferromagnetic resonance (FMR). Through small perturbations to an experimental parameter, chaotic dynamics in a film sample at FMR has been controlled. The control has stabilized periodic orbits from chaotic dynamics, and has synchronized chaotic FMR signals.

2 Chaos in Ferromagnetic Resonance

For over a decade, chaotic behavior in FMR has been investigated. The first reported observations of chaos in FMR date from Gibson and Jeffries [4] investigations of a gallium-doped yttrium iron garnet (YIG) sphere. In this experiment,

R. Marcelli and S.A. Nikitov (eds.), Nonlinear Microwave Signal Processing: Towards a New Range of Devices, 355–380.
© 1996 Kluwer Academic Publishers.

auto-oscillations displayed doubling routes to chaos. This experiment was modelled qualitatively by Zhang and Suhl [5].

These results stimulated a great deal of activity in this field. Other researchers analyzed various routes to chaos in spheres at ferromagnetic resonance [6, 7, 8] and characterized the chaotic dynamics [9, 10]. Others analyzed chaotic transients in magnetic spheres [11]. While several models existed to describe the routes to chaos [12, 13, 14], they were successful only qualitatively in predicting the behavior of spheres in ferromagnetic resonance. This is due to the large number of spinwave modes participating in the chaotic dynamics at high resonant powers within magnetic spheres. To understand the difficulty in modelling the dynamics of spheres and other bulk materials in FMR, it is necessary to understand the spinwave dispersion relation, given by

$$(\frac{\omega}{\gamma})^2 = (H_0 - 4\pi N_z M_s + Dk^2)(H_0 - 4\pi N_z M_s + Dk^2 + 4\pi M_s \sin^2\theta_k). \quad (1)$$

The spinwave band developed by Holstein and Primakoff [15] is presented in Figure 1 and indicates the complexity of FMR dynamics in different material geometries. The uniform precession for bulk materials lies somewhere in the middle of this band. For spheres, the location is a third of the way down from the top of the band. Thus, the uniform precession in spheres degenerate with a large number of spinwaves. At large precession amplitudes, the degenerate spinwaves are parametrically excited through coupling to the uniform mode. Zakharov, L'vov, and Starobinets [16] approximated the Hamiltonian of the system to develop the S theory, which in part determines the coupling between the uniform modes and other spinwaves present in the system. The S theory indicates that the uniform precession parametrically excites many degenerate short wavelength spinwaves at the outer edge of the spinwave band, creating a highly complex dynamical system. These complicated interactions give rise to the observed periodic and chaotic auto-oscillations. Since accounting for all the participating spinwaves in a numerical simulation of FMR in a sphere would be highly impractical, severe truncations were performed in order to model the observed data, leading to numerical results that only qualitatively agreed with experiments [5, 12].

In contrast to magnetic spheres, only a few modes participate in the FMR dynamics of magnetic films. This is largely due to the fact that the demagnetization fields of the film shift the uniform precession mode to the bottom of the spinwave band where the uniform precession is not degenerate with any short wavelength spinwaves. Only the magnetostatic modes are exited which participate in the dynamics. The dynamics of films at resonance can be numerically modelled to accuracies as high at 10-15% [17]. In addition, period doubling routes to chaos have been observed in thin films at resonance [17, 18]. Since magnetic films at resonance provide an ideal system to test numerical results with experiment. The results presented in this chapter review effects in magnetic films, specifically circular films.

3   Ferromagnetic resonance in circular films

In circular films, the lowest energy modes have the spatial form of Bessell functions across the plane of the film, shown in Figure 2. The modes which have no net dipole moment do not couple to uniform excitation fields, and therefore are known as the hidden modes, since they do not appear in the FMR spectrum. These low energy modes are separated in energy due to dipole interactions, and constitute the magneto-static spectrum within each exchange branch. The experiments presented here were conducted in the lowest order exchange branch.

With the appropriate sample geometry, the magneto-static spectrum in a circular film consists of well defined, symmetric absorption peaks. At higher microwave powers, asymmetric broadening of the absorption peaks, known as foldover, is observed, as well as hysteretic behavior in the resonant frequencies of the modes. Figure 3 shows the typical foldover and hysteresis effects observed in circular films.

As the power is further increased, auto-oscillations are observed. A map of the occurrence of auto-oscillations in an experimental parameter space of applied field versus microwave power gives rise to finger-like regions of auto-oscillatory behavior [17]. Each finger generally corresponds to a magneto-static mode, as indicated in Figure 4. At the tips of the fingers, the auto-oscillations are simple periodic functions. As the microwave power is increased, period doubling routes to chaos are often observed. An example of a period doubling route to chaos is shown in Figure 5, with the observed oscillations plotted in delay coordinate space.

The nonlinear behavior observed in circular YIG films can be modelled by a

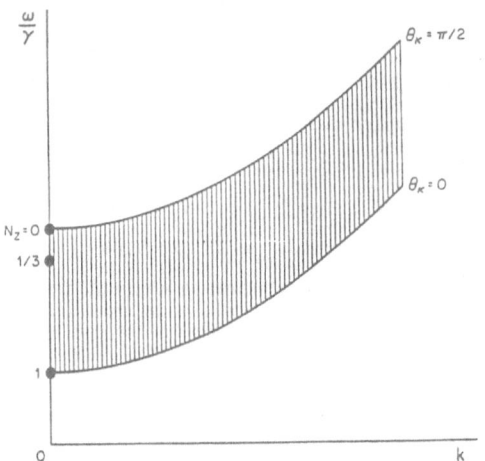

Figure 1: The spinwave band of a ferromagnet. The uniform precession of a sphere is indicated a third of the way down from the top of the band, while the uniform precession of a film is located at the bottom of the band.

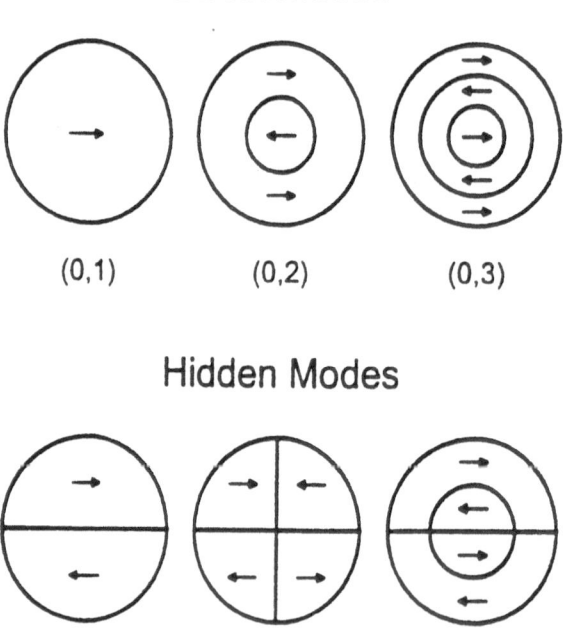

Figure 2: The lowest order direct and lowest order hidden magneto-static modes for a circular film.

Hamiltonian. The four terms included in the Hamiltonian $\mathcal{H}$ are the Zeeman energy, the demagnetization field energy, the dynamic dipole-dipole interaction energy, and the energy due to the excitation field. To express the Hamiltonian in terms of the sample magnetization $\mathbf{M}$, it is necessary to introduce the variables $M^+ = M_x + iM_y$ and $M^- = M_x - iM_y$, which are canonical provided $M_z \approx M_s$. The transformations

$$M^+ = a\sqrt{2\gamma M_s - \gamma^2 aa^*} \tag{2}$$

$$M_z = M_s - \gamma aa^* \tag{3}$$

are made to obtain time dependent equations from the expression

$$\dot{a} = -i\frac{\partial\mathcal{H}}{\partial a^*}. \tag{4}$$

The variable γ is the gyromagnetic ratio. The variable a is related to the dynamics of the sample magnetization $\mathbf{M}$ and therefore has a spatial dependence that can be expanded into orthogonal basis functions which represent the normal modes of the magnetostatic spectrum

$$a = \sum_i a_i m_i(\mathbf{r}) \tag{5}$$

such that

$$\frac{1}{V}\int_V m_i^* m_j(\mathbf{r}) = \delta_{ij} \tag{6}$$

where V is the sample volume. By evaluating the terms in the Hamiltonian using the transformations above, and including a Landau-Lifshitz phenomenological damping term, the following is obtained for the equations of motion for the complex mode c_j.

$$\frac{dc_j}{dt} = -i\gamma(H_\rangle - H_j^{res} - i\Gamma)c_j - 2\pi i\gamma M_s \sum_{klm} A_{jklm} c^* c_k c_l c_m - \frac{i}{\gamma}h_p I_j^* \tag{7}$$

where the nonlinear coupling parameter A_{ijkl} is expressed as

$$A_{ijkl} = \frac{1}{V}\int_V m_i^*(\mathbf{r})m_j^*(\mathbf{r})m_k(\mathbf{r})m_l(\mathbf{r}) \tag{8}$$

and the coupling to the excitation field h_p is

$$I_i = \frac{1}{V}\int_V m_i(\mathbf{r})dr. \tag{9}$$

The parameters H_j^{res} and Γ refer to the resonant fields and linewidths of the modes, and are determined from data taken at low microwave powers, where a linear response of the magnetostatic modes is observed. The equation parameters are determined from experimental data, material characteristics, and the spatial forms of the excited modes. There are no free parameters in this equation. It is also important to notice that the nonlinear term, the damping term, and the driving term make this a forced-damped nonlinear system. Forced-damped nonlinear systems are widely know to display chaotic behavior under certain experimental conditions.

4 Experimental apparatus

The experiments presented here were performed in the perpendicular resonance configuration. A field of 2000 Oe was applied perpendicular to the film plane. The sample was placed film side down on a microstrip waveguide structure [19]. Microwaves travelling through the structure at 1.0-1.5 GHz are coupled to the sample to excite the normal modes within it. By measuring the microwaves reflected from the film with a diode detector, the magnetostatic absorption peaks and auto-oscillations are observed in the diode voltage.

The film samples analyzed with this apparatus consisted of yttrium iron garnet (YIG). YIG is commonly used in nonlinear studies of ferromagnetic resonance due to its low loss characteristics and low crystalline anisotropy. The samples were grown on a gadolinium galium garnet substrate by liquid phase epitaxy [20]. The discs processed from the YIG wafer were 3mm in diameter and $1\mu m$ thick. This geometry created a well resolved magnetostatic mode spectrum in the main magneto-exchange branch, as well as widely separated higher order magneto-exchange branches.

To control the chaotic dynamics in the experiment, the magnetic field applied to the sample was chosen to be the dynamical variable perturbed. To achieve this, a coil 1 cm in diameter, consisting of 20 turns of wire was suspended next to the film to slightly perturb the applied field. Due to the inductive nature of the coil, a coil driver circuit was used to compensate for the distortion of broad band inputs to the coil.

The first control experiment performed on this apparatus involved the stabilization of periodic orbits from chaotic dynamics using time delayed control. In previous FMR experiments, periodic orbits have been stabilized by an external modulation of the applied magnetic field [21, 22] or through feedback perturbations to the microwave power [23]. In the results presented here, periodic orbits were stabilized by applying a perturbation to the applied magnetic field of the form

$$\delta H = K[V(t - \tau) - V_{d.c.}] \tag{10}$$

where V is the voltage measured by the diode detector, τ is the time the voltage signal is delayed before it is used to perturb the system. The goal of the control is to eliminate any deviations from the average d.c. component of the auto-oscillation voltage $V_{d.c.}$ measured by the diode. This is accomplished through experimentally determining the parameter K. Due to the inductive nature of the coil and time lags due to the experimental circuitry, there is a $0.1\mu s$ intrinsic delay time in the experiment. Since the broad band chaotic auto-oscillations lie in the 0.5-10 MHz range, this 0.1 μs delay can seriously degrade any attempt at control. Therefore, the voltage is delayed by some time τ until the perturbation applied to the system will be most appropriate. A time delay circuit was built which permitted delay times in 25 ns intervals ranging from 290-700 ns.

The second control experiment which involves the synchronization of chaotic FMR signals, the experimental technique following a scheme proposed by Pyragas [24] in which the output of a chaotic system is stored into a memory. This signal is referred to as the master signal. The signal is then played back using a LeCroy Arbitrary Function Generator. The real time output of the system, called the slave signal, is compared to the master signal, and a continuous perturbation to an experimental parameter proportional to the difference between the master and slave signals are generated. The perturbation induces the real time slave signal to follow the prerecorded master. In the case of thin films at FMR, the perturbation was made to the applied magnetic field in the form of

$$\delta H(t) = J[V_{master} - V_{slave}], \tag{11}$$

where J is experimentally determined. V_{master} and V_{slave} the master and slave voltages are measured by the diode.

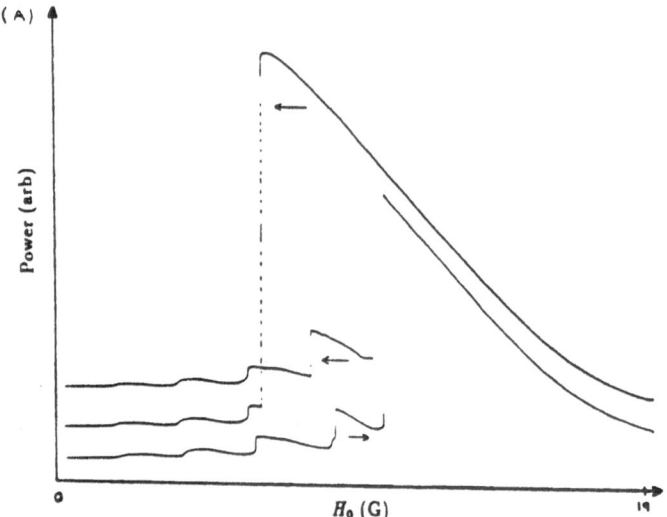

Figure 3: An example of foldover and hysteresis behavior in the ferromagnetic resonance spectrum of a circular film.

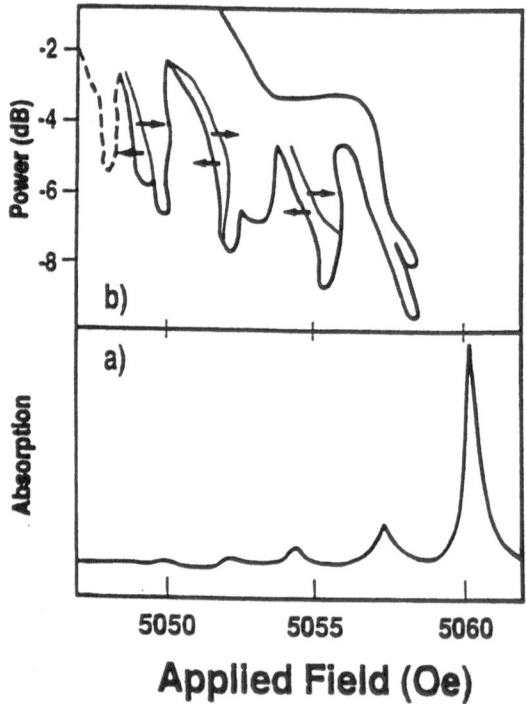

Figure 4: Fingers of auto-oscillation in the experimental parameter space of microwave power and applied field. The low power FMR spectrum of the sample in the bottom half of the figure indicates the positions of the magneto-static modes.

5 Experimental Results

5.1 Stabilization of Periodic Orbits

The time delayed control experiments in FMR were carried out with the static magnetic field at approximately 2000 Oe and the perpendicular microwave frequency oscillating at 1.0-1.4 GHz at powers of 1-20 milliwatts. The static field, microwave frequency and power were chosen so that the microwave absorption of the sample was chaotic. This generally occurred at the center of the "fingers of auto-oscillation", and towards the top of the auto-oscillation map. A typical condition is shown as point A in Figure 4.

By applying a control perturbation of the form

$$\delta H(t) = K[V(t - \tau) - V_{d.c.}], \tag{12}$$

periodic orbits of decreasing periodicity were stabilized as K was increased for a given delay time τ, essentially a reversal of the typical period doubling route to chaos presented in Figure 5. For example, a delay time τ of 440 ns stabilized period-4, period-2, and period-1 oscillations as the parameter K increased, as shown in Figure 6. In the unperturbed chaotic FMR signal Fourier spectrum, the peak frequency was found to be 1.2 MHz, corresponding to a period of 840 ns. The delay time of 440 ns is approximately half of this period. Delay times of 415 and 470 ns were also available and made it possible to stabilize periodic orbits. However, the period doubling reversal (debifurcation) all the way to period-1 could not be achieved by these delay times. Other delay times were able to stabilize periodic orbits for certain values of K as well. To a first approximation, delay times which are half-integer multiples of the period corresponding to the peak frequency in the unperturbed chaotic Fourier spectrum are found to be the optimal delay times for control and stabilization for the observed debifurcation routes from chaos.

In stabilizing the period-4 oscillations, the control gain K was such that the maximum value of $|\delta H|$ never exceeded 0.15 Oe. As K was increased to stabilize period-2 and period-1 oscillations, $|\delta H|$ never exceeded 0.30 and 0.50 Oe respectively. This compares with a width of the auto-oscillation finger where the oscillations were stabilized of 2-3 Oe.

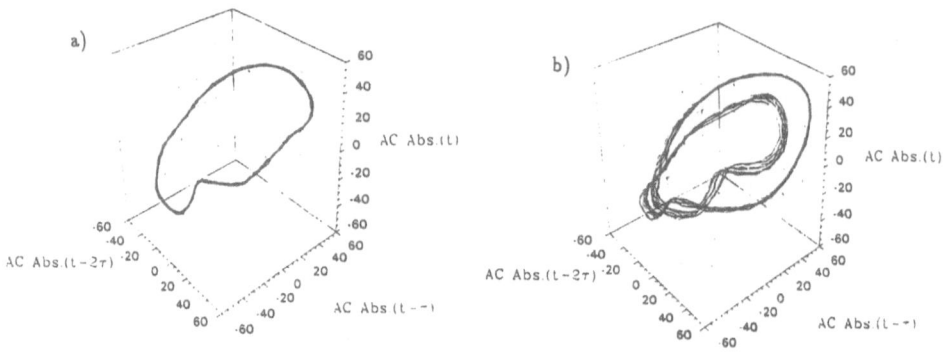

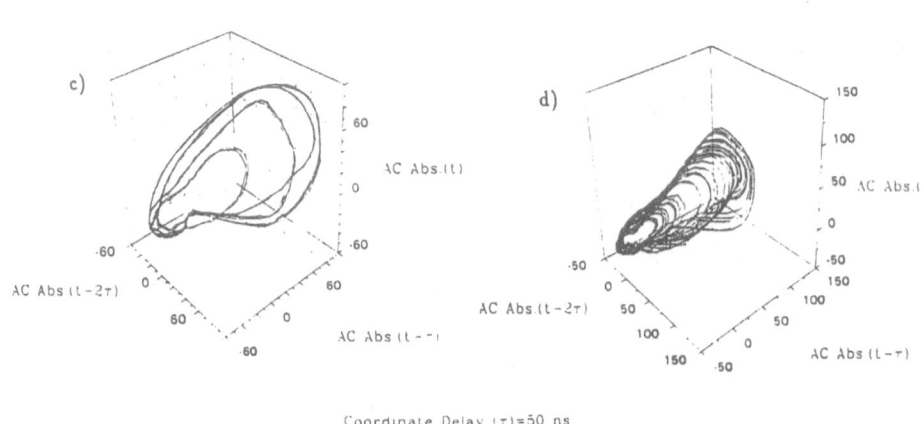

Coordinate Delay $(\tau) = 50$ ns

Figure 5: Three dimensional time delay plots of period-1 a), period-2 b), period-4 c), and chaotic d) auto-oscillations with increasing microwave power, indicating the period route to chaos observed in the sample at resonance. The coordinate delay, τ is 50 ns.

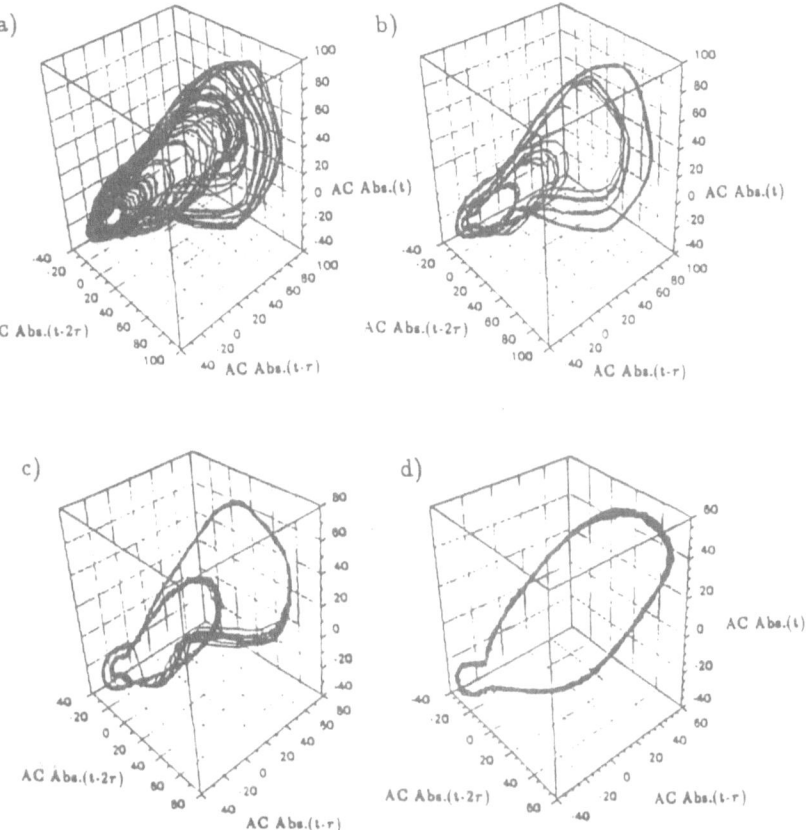

Figure 6: Three dimensional delay coordinate plots of the unperturbed chaotic signal a), the stabilized period-4 signal b), period-2 signal c), and period-1 signal d). The coordinate delay, τ is 50 ns.

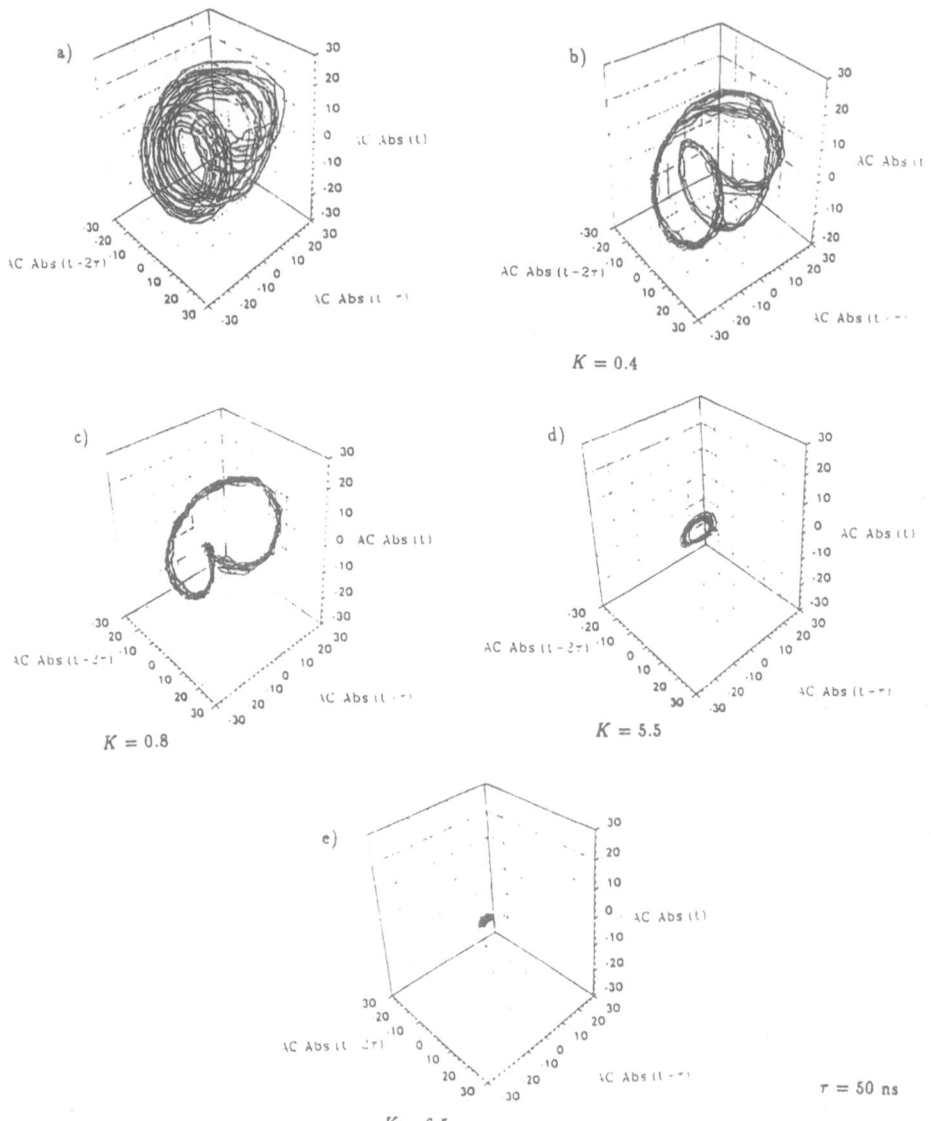

Figure 7: Three dimensional delay coordinate plots of an unperturbed chaotic signal a), the stabilized period-4 oscillation b), the stabilized period-2 oscillation c), the stabilized period-1 oscillation d), and the stabilized quiescent state. The coordinate delay, τ is 50 ns.

Another example of the stabilization of orbits in a chaotic system is presented in Figure 7. In Figure 7a the FMR signal of the film in the chaotic state is shown. The application of time delay control with a delay time of roughly 470 ns with a control gain of 0.4 arbitrary units (a.u.) produced a period-4 oscillation, shown in Figure 7b. Further increasing the gain to 0.8 a.u. resulted in a period-2 oscillation (see Figure 7c). Continuing to increase the gain, a period-1 oscillation was observed, as shown Figure 7d, and further increase of the gain eliminated the auto-oscillations entirely (Figure 7e). If the gain was increased to approximately 7.5 a.u., the resulting behavior of the YIG film reverted to a chaotic state, although visually more turbulent than observed in the system without control.

The size of the perturbations to the static field in this example rarely exceeded 0.1 Oe. Compared to the value of the static field, 2000 Oe, and the width of the finger, ~ 2 Oe, the fields used to stabilize periodic orbits are much smaller than the relevant fields in the experiment. In maintaining the quiescent state, perturbation fields no larger than 10 mOe were employed. If the polarity of the coil was reversed, effectively changing the sign of the gain, the system again became highly turbulent. When the control apparatus was switched off, the system abruptly returned to the original unperturbed chaotic state.

The delay time of 470 ns corresponded roughly to the period represented by the peak in the chaotic Fourier spectrum, 1.95 MHz. Other delay times in the vicinity of 470 ns were able to produce periodic oscillations, but not quiescence. Presumably, a delay time of 235 ns (470/2) with the opposite control gain polarity would also produce the observed effects, but this time delay was below the range of delays available in the experiment. Delay times of higher order $\frac{1}{2}$ integer multiples of 470 ns resulted in controlled periodic oscillations, but did not reduce the system to the quiescent state.

In general, a reduction of the periodicity was observed as the control gain was increased if the time delay used was approximately a half-integer multiple of the the period corresponding to a prominent peak in the FMR signal Fourier spectrum. It was observed that controlled auto-oscillations with other periodicities such as 6, 5, and 3 could be stabilized in many situations when the time delay was not half-integer multiple of the peak frequency period.

An important question in the analysis of stabilized periodic orbits is to ask how similar the stabilized orbits are to the unstable orbits of the unperturbed chaotic attractor. Determining the similarities between the unstable orbits of the unperturbed chaotic attractor and the stabilized periodic orbits provides a means to evaluate the time delayed control technique. Poincaré sections provide a way to measure these similarities. As the system trajectory evolves in time-delay space, the points at which the trajectory pierce a surface in this space comprise a Poincaré section. By examining the Poincaré sections of the unperturbed chaotic attractor with the maps of the stabilized periodic orbits, it can be determined to what extent the time delay control stabilizes unstable orbits of the unperturbed chaotic attractor.

Figure 8 shows the Poincaré sections of the chaotic and stabilized periodic oscillations from the data presented in Figure 6. The small dots indicate the Poincaré section of the unperturbed chaotic attractor and the controlled period-4 (circles), period-2 (triangles), and period-1 (square) sections are superimposed on top. The points corresponding to the periodic sections lie on or very near the region of the chaotic section. This suggests that the controlled periodic orbits are nearly identical to unstable orbits in the underlying unperturbed chaotic attractor. In the control sequence presented in Figure 7, the stabilized period-4 and period-2 sections are also found to overlap the unperturbed chaotic section.

5.2 Synchronizing Chaotic Signals

In addition to stabilizing periodic orbits with time delayed control, two chaotic FMR signals were synchronized by applying perturbations of the form

$$\delta H(t) = J[V_{master}(t) - V_{slave}(t)]. \tag{13}$$

The slave FMR signal was the real time output of the magnetic film sample in FMR. The master was a prerecorded time series segment of the sample FMR signal. The control perturbations altered the static magnetic field applied to the sample in real time. The experimental conditions that produced the slave and master signals were kept as identical as the experimental uncertainties would allow.

The results of the experiment are presented in Figure 9. As expected, when no perturbation is applied to the sample ($J = 0$), the prerecorded chaotic signal

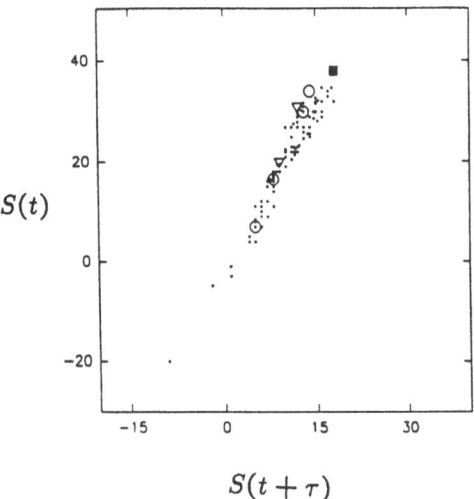

Figure 8: Poincaré section in three dimensional coordinate space of the experimental data presented in Figure 6. The piercings lie on the $S(t + 2\tau) = 72$ plane in delay coordinate space. The points corresponding to the controlled period-4 oscillation (open circles), period-2 oscillation (open triangles), and period-1 oscillation (solid square) are superimposed upon the section of the unperturbed chaotic section (dots).

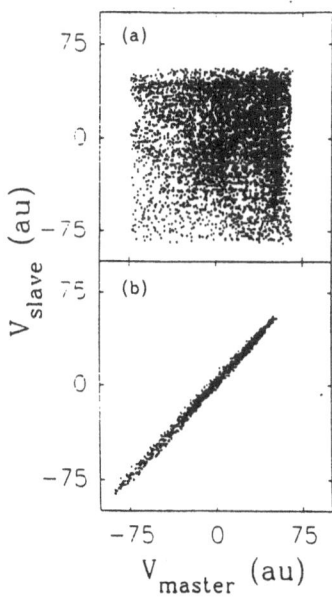

Figure 9: Correlation diagram of the slave and master signal showing the correlation in the absence of control perturbations a) and when the two signals were synchronized b).

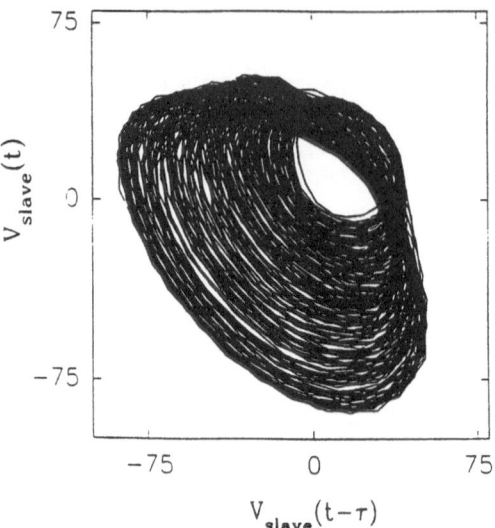

Figure 10: Two dimensional attractor of the slave signal in time delay space during synchronization, indicating the dynamics were chaotic during the synchronization process.

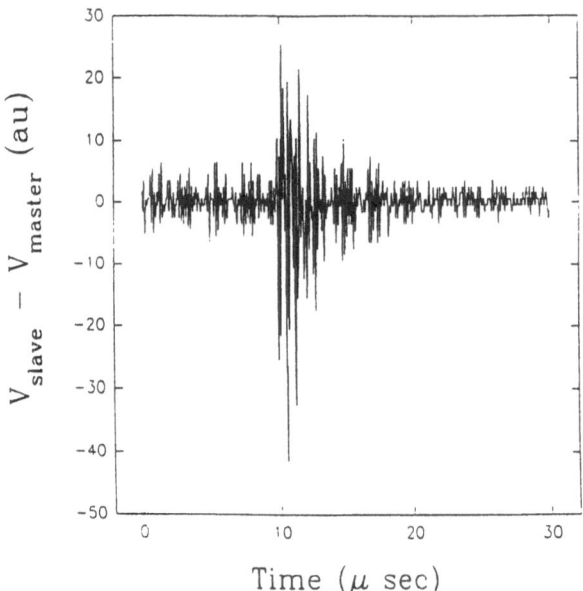

Figure 11: The difference between the master and slave signal at the discontinuity in the master signal, indicating the transient behavior of the synchronization control.

and the real time chaotic FMR signal have almost no correlation, seen in Figure 9a. However, the systems become correlated to almost the experimental noise level when the appropriate choice of K is made in the experiment, shown in Figure 9b. In addition, the fields required to achieve synchronization are no larger than 1 Oe in the initial transient stage, and fields of 0.1 Oe or less were required to preserve the synchronization. A two-dimensional time delay plot of the real time FMR signal attractor during synchronization is presented in Figure 10, which indicates the sample dynamics were chaotic during synchronization.

The transient behavior of this synchronization scheme is presented in Figure 11. The recorded FMR signal output segment was usually 100 μs long and played back in a continuous loop. This created a discontinuity in the master signal every 100 μs. This allowed for the transient behavior of the synchronization process to be observed as the signals would resynchronize after the discontinuity. The typical transient time, as indicated in the diagram, was about 10 μs or less. Once the signals were synchronized, they remained so to nearly the level of noise in the experiment.

6 Analysis

To run the model simulations requires parameters from the low power spectrum of the film sample. The linewidths and resonance positions of the magnetostatic modes in the spectrum determine the input parameters for the coupled nonlinear equations that describe the mode amplitude dynamics. In the model, period doubling routes to chaos are observed, as witnessed in the experiments. Figure 12 shows a period doubling route to chaos in the first finger of the model. As the power is increased in the model in increments of 0.2-0.5 dB, the resultant auto-oscillations range from period-1 to period-2, period-4, and finally chaotic oscillations.

When time delayed control was introduced into the model as the equations of motion were behaving chaotically, periodic orbits were stabilized in a reverse period doubling sequence similar to that observed in the experiments. For example, Figure 13a shows the uncontrolled chaotic FMR signal in the model. Time delayed control was applied to the static field in the model of the form

$$\delta H(t) = K[S(t - \tau) - S_{d.c.}], \qquad (14)$$

where $S_{d.c.}$ is the average d.c. component of the FMR signal and $S(t - \tau)$ was the value of the FMR signal at time τ earlier in experimental time. A delay time of 380 ns was able to stabilize a period-8 oscillation when $K = 2.75$ (Figure 13b). Further increases of the control gain parameter K stabilized period-4, period-2, period-1 oscillations, shown in Figure 13. Ultimately, the quiescent state was achieved, where all auto-oscillations in the model were eliminated. When K was further increased ($K \approx 40$), the quiescent state was over-corrected and the resulting auto-oscillations were highly turbulent.

The delay time of 380 ns was roughly half the period corresponding to the peak in the chaotic Fourier spectrum at 1.5 MHz. Other delay times in the vicinity of 380 ns allowed for periodic oscillations to be controlled, but not the quiescent state. It is not clear why 380 ns and not 330 ns (half of the period associated with 1.5 MHz) is the optimal delay time, but it may be that the particular combination of significant frequencies in the chaotic Fourier spectrum makes 380 ns the optimal delay time.

The perturbation fields used in the model to stabilize periodic oscillations were never greater than 0.1 Oe. In achieving the quiescent state, the perturbation field

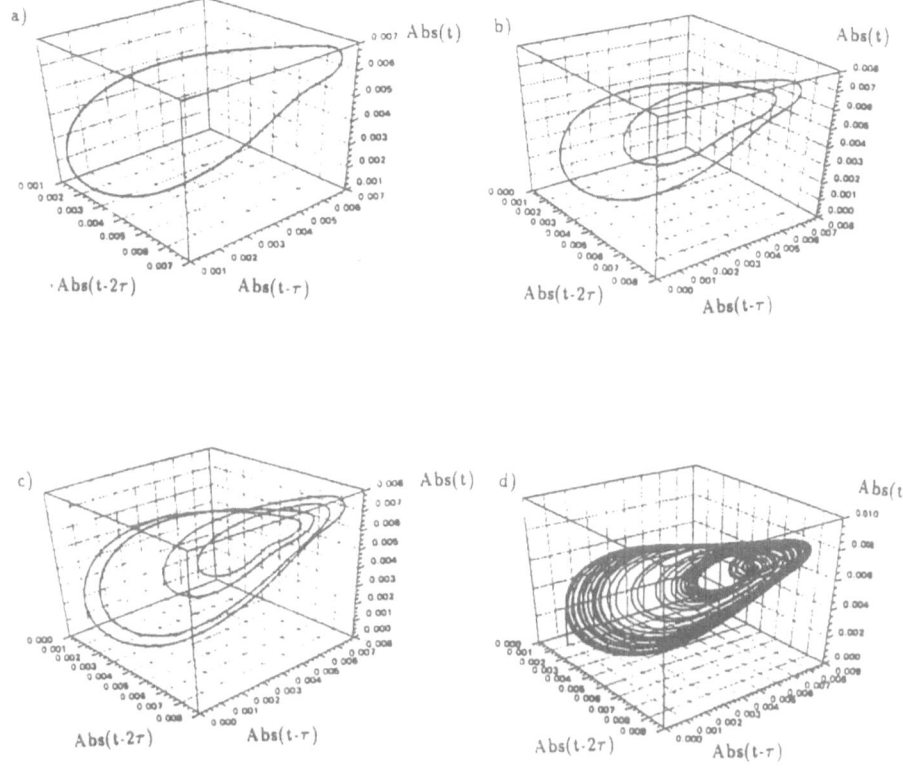

Figure 12: Period doubling route to chaos in the model in three dimensional delay coordinate space, showing the period-1 oscillation a), period-2 oscillation b), period-4 oscillation c), and chaotic oscillations d) produced as the microwave power was increased in the model. The coordinate delay time τ was 50 ns.

374

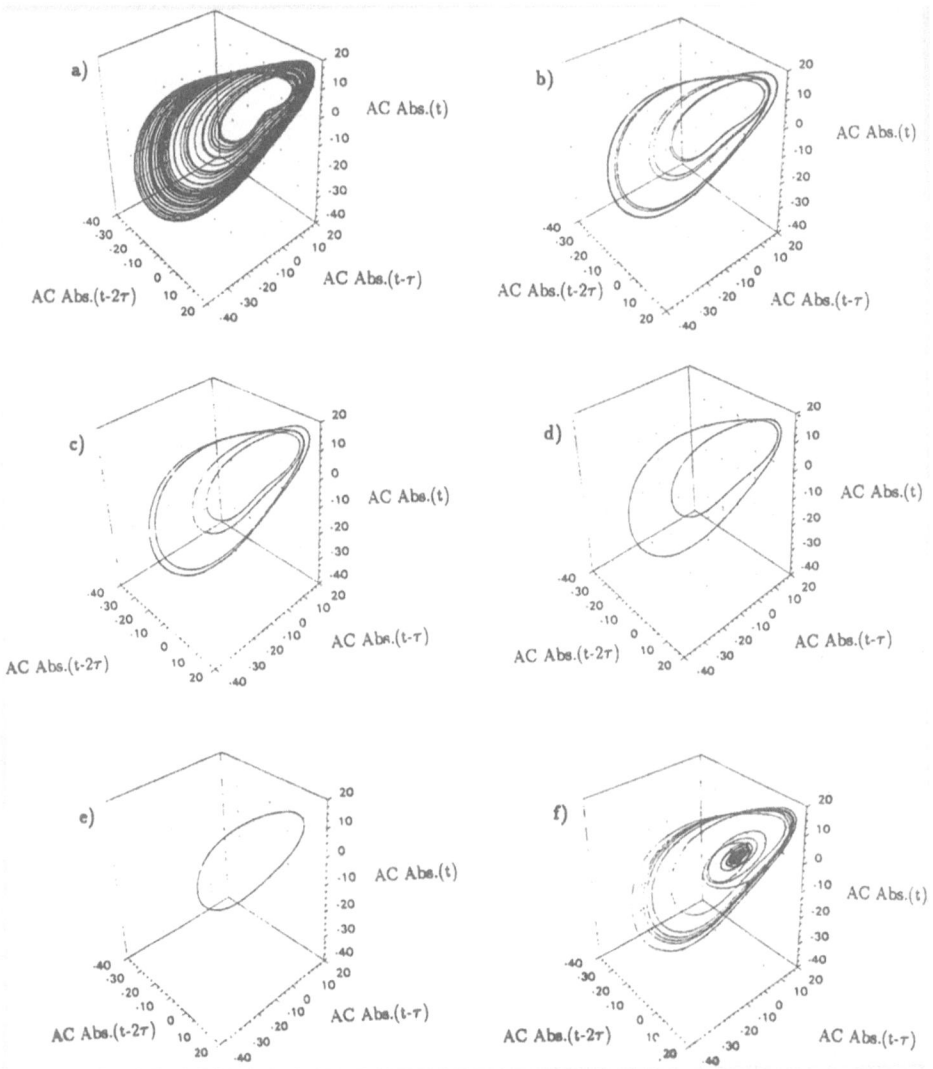

Figure 13: Three dimensional delay coordinate plots of the unperturbed chaotic oscillation in the model a), the stabilized period-8 oscillation ($K = 2.75$) b), the stabilized period-4 oscillation ($K = 3.5$) c), the stabilized period-2 oscillation ($K = 5.0$) d), the stabilized period-1 oscillation ($K = 15$), and the system spiraling into the quiescent state ($K = 30$). Two lines in the period-8 plot are barely resolvable in the reproduction. The coordinate delay time τ is 50 ns.

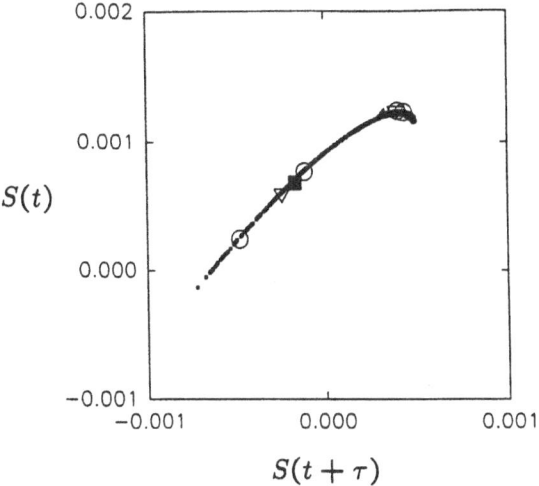

Figure 14: Poincaré section in three dimensional space of the numerical simulations presented in Figure 13. The piercings lie on the $S(t + 2\tau)$=-0.001 plane in delay coordinate space. The points corresponding to the stabilized period-4 oscillation (open circles), period-2 oscillation (open triangles), and period-1 oscillation (solid square) are superimposed upon the section of the unperturbed chaotic system (dots).

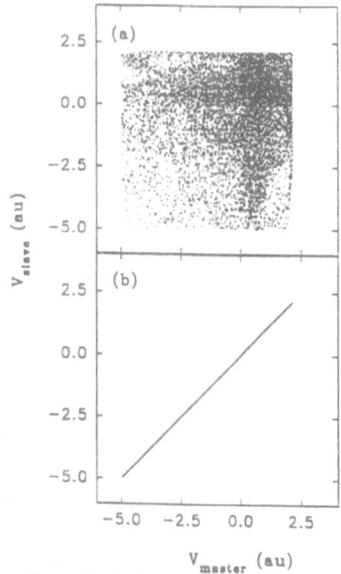

Figure 15: Correlation of the prerecorded model time series segment and the model output with $J = 0$ a) and with $J = 35$ b), indicating the synchronization effect in the model.

approached zero after the initial transients. Generally, a transient time of 5-10 μs was required for the control to convert a chaotic auto-oscillation into a periodic or quiescent state. However, transient times as long as 40 μs were observed in the numerical simulations. It would seem as if the current system trajectory at the time the control is first applied determines, to some degree, the transient time required before stabilization.

Figure 14 presents the Poincaré sections of the stabilized oscillations in the model superimposed upon the Poincaré section of the unperturbed chaotic attractor. The stabilized period-4, period-2, and period-1 oscillation points are seen to be nearly identical to points on the unperturbed chaotic attractor, in agreement with the experimental results. (The period-8 points are not presented here, but also lie on the chaotic attractor.) The proximity of the stabilized periodic points to the unperturbed chaotic attractor on the Poincaré sections suggests that the control perturbations act to stabilize periodic orbits of the underlying attractor.

In the quiescent state, the perturbation approaches zero since the goal of time delayed control is the quiescent state, where the precession angle of the moments are constant. While quiescence is not an unstable orbit of the periodic attractor, clearly the sample in the chaotic state is being controlled. As the moments precess about the static magnetic field, small corrections to the system keep the moments precessing at a constant angle, instead of oscillating chaotically.

The model was also able to predict the results of the synchronization experiments. To simulate the prerecording of the chaotic signal, input parameters were used in the model to describe the magnetic film sample in the chaotic state. The model was allowed to evolve for 400 μs of experimental time and every 10 ns, the output was written to a file. This file corresponded to the prerecorded master signal in the experiment. The model was then run again, only this time, perturbations to the static field from the equation

$$\delta H(t) = J[S_{master} - S_{slave}] \tag{15}$$

were applied. The master signal consisted of the data file of the initial simulation run. A random starting point from the master file was chosen and the current FMR signal in the model was compared every 10 ns of experimental time with the next entry in the saved data file. The difference between these two values was multiplied by K and added to the static magnetic field.

With the appropriate choice of J, the numerical model synchronized to the randomly chosen time series segment in the file. Figure 15 shows the correlation between the stored file and the model output with no control perturbations $(J = 0)$ and when $J = 35$, after the synchronization control had been applied for 10 μs to eliminate the initial transient behavior. From this figure, it can be seen that nearly perfect correlation between the two signals was achieved in the model. Once the control perturbations were applied in the model, synchronization was achieved after a transient period of about 5-10 μs. This is in reasonable agreement with the experimental results.

7   Conclusion

By applying small perturbations to an experimental parameter, chaotic auto-oscillations have been controlled in thin magnetic films at FMR. Control in FMR has both stabilized periodic orbits from chaotic dynamics and has synchronized chaotic FMR signals. Both experiments have been modelled by computer simulations.

In the experiments in which periodic orbits were stabilized from chaotic dynamics, time delayed control was used, where the FMR signal was delayed in time before being used to perturb the system. The time delay was determined from the peak frequency in the chaotic Fourier spectrum. The application of time delayed control applied to the static field often resulted in a reverse period doubling route from the chaotic state to a low periodicity oscillation. In some cases, the quiescent state could be achieved, in which all auto-oscillations were eliminated.

In the experiments involving chaotic synchronization, chaotic auto-oscillations were synchronized to a prerecorded chaotic auto-oscillation signal. The transient period before synchronization was achieved was typically 5-10 μs. The perturbation to the static field required to maintain synchronization was quite small, less than 0.1 Oe.

The proceeding results add to the growing body of knowledge that small perturbation to a chaotic system parameter can make the unpredictable dynamics of chaos predictable. We believe that this phenomena is more than just a nonlinear curiousity, and suggests many potential applications for chaotic systems that would otherwise be considered useless. Control of chaos has allowed lasers to be operated at significantly higher stable output powers than would normally be possible without control [25]. The potential for private communications through chaotic synchronization was demonstrated in a dramatic experiment by Cuomo and Oppenheim [26]. Since YIG films are used in many microwave applications, the stabilization of dynamics of YIG films suggest the potential for microwave devices to work at higher powers before failure. Chaotic synchronization in YIG films indicate the potential for private microwave communications. Chaotic phenomena, already present so much in our daily lives, may soon be found regularly in microwave applications.

References

[1] L.M. Pecora and T.L. Carroll, *Phys. Rev. Lett.* **64**, 821 (1990).

[2] W.L. Ditto, S.N. Rauseo, and M.L. Spano, *Phys. Rev. Lett.* **65**, 3211 (1990).

[3] R.Roy, T.W. Murphy, T.D. Maier, Z. Gills, and E.R. Hunt, *Phys. Rev. Lett.*, **68**, 1259 (1992).

[4] G. Gibson and C. Jeffries, *Phys. Rev. A*, **29**, 811 (1984).

[5] X.Y. Zhang and H. Suhl, *Phys. Rev. A*, **34**, 2530 (1985).

[6] F.Waldner, D.R. Barberis and H. Yamazaki, *Phys. Rev. A*, **31**, 420 (1985).

[7] F.M. deAguiar and S.M. Rezende, *Phys. Rev. Let.*, **56**, 1070 (1986).

[8] A.I. Smirnov, *Sov. Phys. JETP*, **63**, 222 (1986).

[9] H. Yamazaki, *J. Appl. Phys.* **64**, 5391 (1988).

[10] F.M. deAguiar, A. Azevedo, and S.M. Rezende, *Phys. Rev. B*, **39**, 9448 (1989).

[11] T.L. Carroll, L.M. Pecora, and F.J. Rachford, *Phys. Rev. Lett.* **59**, 2891 (1987).

[12] S.M. Rezende, O.F. de Alcantara Bonfim, and F.M. de Aguiar, *Phys. Rev. B*, **33**, 5153 (1986).

[13] X.Y. Zhang and H. Suhl, *Phys. Rev. B*, **38**, 4893 (1988).

[14] F. Waldner, *J. Phys. C*, **21**, 1243 (1988).

[15] T. Holstein and H.P. Primakoff, *Phys. Rev.*, **58**, 1098 (1940).

[16] V.E. Zakharov, V.S. L'vov, and S.S. Starobinets, *Sov. Phys.-Usp.*, **17**, 896 (1975).

[17] R.D. McMichael and P.E. Wigen, *Phys. Rev. Lett.*, **64**, 64 (1990).

[18] P.E. Wigen, H. Döetsch, Y. Ming, L. Baselgia, and F. Waldner, *J. Appl. Phys.*, **63**, 4157 (1988).

[19] B. Lührman, M. Ye, and H. Döetsch, and A. Gerspach, *J. Mag. Mat. Mat.*, **96**, 237-244 (1991).

[20] The YIG film was graciously provided by Dr. Roger Belt.

[21] A. Azevedo and S.M. Rezende, *Phys. Rev. Lett.*, **66**, 1342 (1991).

[22] M. Ye, D.E. Jones, and P.E. Wigen, *J. Appl. Phys.*, **73**, 6822 (1993).

[23] R. Henn, E. Rödelsperper, and H. Benner, *Proc. of Ampere Congress*, Athens, (1992).

[24] K. Pyragas, *Physics Letters A*, **201**, 203 (1993).

[25] Z. Gills, C. Iwata, R. Roy, I.B. Schwartz and I. Triandaf, *Phys. Rev. Lett.*, **69**, 3169 (1992).

[26] K.M. Cuomo and A.V. Oppenheim, *Phys. Rev. Lett.*, **71**, 65 (1993).

SUPPRESSING AND CONTROLLING CHAOS IN SPIN-WAVE INSTABILITIES

T. BERNARD, R. HENN, W. JUST, E. REIBOLD, F. RÖDELS-
PERGER, and H. BENNER

*Institut für Festkörperphysik and SFB 185, TH Darmstadt,
Hochschulstr. 6, D-64289 Darmstadt, Germany*

Suppression of chaos has recently received increased attention. We report on
magnetic resonance experiments in yttrium iron garnet (YIG) spheres probing
spin-wave instabilities above the first-order Suhl threshold. Different techniques
have been applied to change the observed irregular auto-oscillations of magnetization into regular behaviour: (i) The suppression of chaos by fast parametric
modulation can be attributed to an effective change of the modulated parameter.
(ii) Control is achieved by means of a modified OGY method considering the
specific requirements of fast experimental systems. (iii) The delayed feedback
technique by Pyragas is studied both in theory and experimental application.

1. Introduction

In spite of the various efforts to observe, to analyse and to understand the
chaotic behaviour of nonlinear systems, irregular and unpredictable motion is
generally not desired in practical applications. Thus, suppression and control of
chaos, as a first applied concept of nonlinear dynamics, has become subject of
intensified research recently. Different strategies have been proposed how to
change the irregular behaviour of a nonlinear system into regular motion:
Modulation methods [1-3] are either based on synchronization to an external
periodic force with a frequency close to an intrinsic system frequency, or on the
change of stability induced by fast modulation of some system parameter. These
simple non-feedback methods generally require rather large controlling power.
More sophisticated feedback control methods aim at the stabilization of existing
unstable periodic orbits. This can be achieved by a simple time-delayed feedback [4] or by calculated time-dependent corrections on one of the system
parameters [5]. Since the latter techniques make use of the intrinsic properties
of the underlying chaotic attractor they generally can be run with very small
controlling power, but require very detailed information on the system and are
rather intricate to perform [6-13]. In all these approaches we meet a competition

R. Marcelli and S.A. Nikitov (eds.), Nonlinear Microwave Signal Processing: Towards a New Range of Devices, 381–408.
© *1996 Kluwer Academic Publishers.*

between the complexity of the algorithm applied and the strength of external perturbations on the system required to achieve stabilization.

In this chapter we discuss the application of three different approaches to spin-wave chaos. Parametric excitation of spin waves is observed in high power FMR experiments when the amplitude of the pumping microwave field exceeds a certain instability threshold well-known as the first- or second-order Suhl instability [14]. Parametrically excited spin waves show a variety of nonlinear phenomena, such as low-frequency auto-oscillations, quasiperiodicity, mode-locking, intermittency and chaos, which result from nonlinear couplings between them and have been widely studied for more than a decade [15]. The specific challenge of controlling real spin systems is mainly related to the very fast time scale of regular and irregular auto-oscillations, which is in the order of a microsecond. Thus, sophisticated concepts which are based on real-time numerical analysis are generally too slow and can only be applied in a modified way.

We start with a short review on the experimental background and observed scenarios (sec. 2). In sec. 3 we demonstrate the suppression of chaos by fast parametric modulation. It is shown that the increase of modulation amplitude actually results in an effective variation of the modulated control parameter, and the obtained suppression of chaos is essentially related to a global change of the system. The opposite case represented by the method of Ott, Grebogy, and Yorke is based on the precise knowledge of existing unstable periodic orbits, eigenvalues, and eigendirections and allows the suppression of chaos with very slight external perturbations by making use of the intrinsic system dynamics in a rather sophisticated way (sec. 4). The time-delayed feedback method of Pyragas (sec. 5) may be considered as an efficient compromise combining simplicity in application with small controlling power.

2. High Power Ferromagnetic Resonance at the First-Order Suhl Instability

Suhl's first-order spin-wave instability [14] is characterized by the decay of the externally driven uniform mode into two spin waves of half the pumping frequency $\omega_k = \omega_p/2$ and opposite wave vectors ($k, -k$) according to the conservation of energy and quasi-momentum. This instability can either be observed off resonance (i.e. with the pumping frequency far away from the usual ferromagnetic resonance, $\omega_p \neq \omega_0$) as a *subsidiary absorption*, or directly on the FMR line ($\omega_p = \omega_0$) within the *coincidence regime*. Profiting by the resonance amplification of the FMR mode, experiments in the coincidence regime require much less microwave power to reach the threshold (typically some $10\,\mu$W for high-quality YIG samples).

2.1 EXPERIMENTAL SET-UP

In view of such small thresholds, high power FMR experiments, in principle, can be performed with a conventional ESR spectrometer. We have studied the subsidiary absorption, for example, at $9.3\,$GHz by means of a bimodal

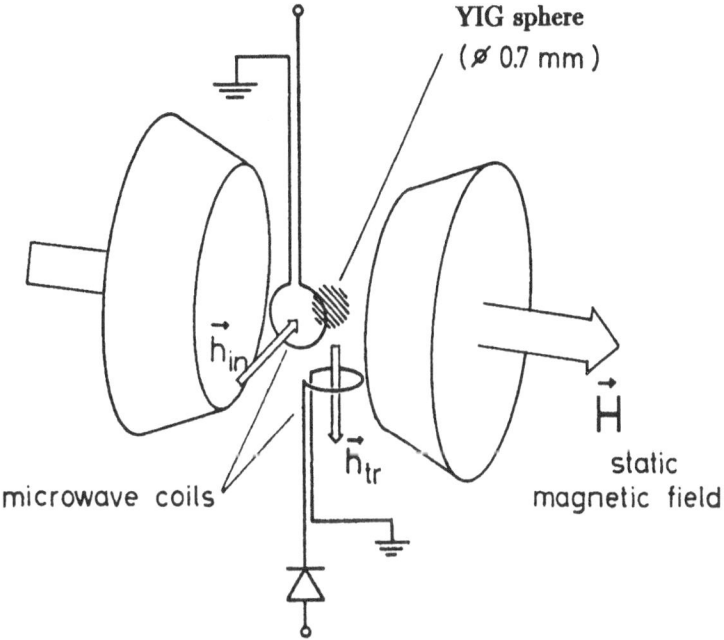

Figure 1. Experimental set-up. The driving coil (1 - 4 GHz) is directly fed by an extremely stable microwave generator. The signal transmitted to the pick-up coil is amplified, rectified, and recorded by a digital oscilloscope.

transmission-type cavity of quality factor 3000. For our experiments in the coincidence regime we preferred a broad-band (1 − 4 GHz) transmission-type set-up (Fig. 1). For the transverse excitation and detection of the uniform mode, instead of a microwave cavity, we used two microcoils with perpendicular orientation in order to minimize mutual disturbations by crosstalk. The signal transmitted to the pick-up coil was amplified and detected by a diode. In either setup the squared amplitude of the driving field h at sample position was proportional to the input power P_{in} supplied by the microwave source, and the rectified signal was proportional to the squared amplitude $|b_0|^2$ of the uniform mode, i.e. to the transmitted power P_{tr}. By means of a digital oscilloscope and an integrating voltmeter we recorded both the time dependence of $P_{tr}(t)$ and its average value $\overline{P_{tr}}$ with respect to the input power P_{in}. The presented data were obtained at room temperature on a highly polished sphere of pure YIG, 0.71 mm in diameter. We applied the magnetic field H either in $\langle 100 \rangle$ or in $\langle 111 \rangle$ direction.

2.2 THRESHOLD FOR RESONANT AND NONRESONANT PUMPING

In subsidiary absorption ($\omega_p \neq \omega_0$) Suhl's first-order instability shows up as an additional absorption structure in lower field, which is well separated from the main resonance and shows a drastic broadening with increasing microwave power, accompanied by auto-oscillations and sequences of bifurcations. We have systematically analysed [16] the dynamic behaviour of the subsidiary absorption signal at room temperature at a fixed pumping frequency of 9.26 GHz, as presented in Fig. 4. The lower line shows the dependence of the Suhl threshold on H (the so-called *butterfly curve*). The broad bumps at 1.6 and 1.9 kOe have been explained by the interaction with elastic or magnetostatic modes [17]. The bifurcation line above indicates the onset of auto-oscillations. The steep increase of the threshold at 2.2 kOe corresponds to the fact that the bottom of the spin-wave band becomes larger than $\omega_p/2$ and the parametric process is no longer resonant.

Resonant pumping of both the uniform FMR mode and a spin-wave pair ($\omega_p = \omega_0 = 2\omega_k$) is possible in the frequency range from 1.8 to 3.4 GHz ($680 - 1280$ Oe for $H \parallel \langle 100 \rangle$). For lower field the FMR vanishes due to the occurrence of magnetic domains. For higher field $\omega_0/2$ falls below the spin-wave band and, keeping the resonance condition $\omega_p = \omega_0$, a changeover to the second-order instability takes place. Within the coincidence regime the first-order threshold shows up as a sharp and asymmetric break at the top of the main resonance line accompanied by auto-oscillations. The break becomes broader with increasing input power and may be followed by further breaks resulting in a complex multistability, which is connected with a variety of auto-oscillations. Details have been presented in the literature [18-21] and have been explained in terms of a model including the specific properties of discrete magnetostatic modes in the parametric process, which result in a novel coupling mechanism [18]. According to this model the sudden jumps from one level to the other are induced by the nonlinear coupling or decoupling of certain spin-wave modes.

2.3 OBSERVED ROUTES TO CHAOS

To get a more systematic impression of the observed oscillations, a large number of time series of $P_{tr}(t)$ – up to 16 000 data points each – was recorded on variation of P_{in}, H, or some other control parameter. The corresponding power spectra were obtained by Fourier transformation, and their strongest spectral components were plotted versus the parameter under variation. This way, one obtains a number of 'maps' visualizing the dependence of the oscillation frequencies on various control parameters. Such maps are useful for classifying the observed routes to chaos.

As a general result, we found that a global correspondence to one of the well-known scenarios of Feigenbaum, Ruelle-Takens-Newhouse or Pomeau-Manneville [22] does not occur, but a variety of parts from all of them. This obviously corresponds to the fact that the nonlinearities of a real system are more complicated and based on a larger number of internal degrees of freedom than represented by the simple models where these standard routes have been derived from. The physical meaning of the degrees of freedom is probably that of specific eigenmodes or a collective motion of several of them.

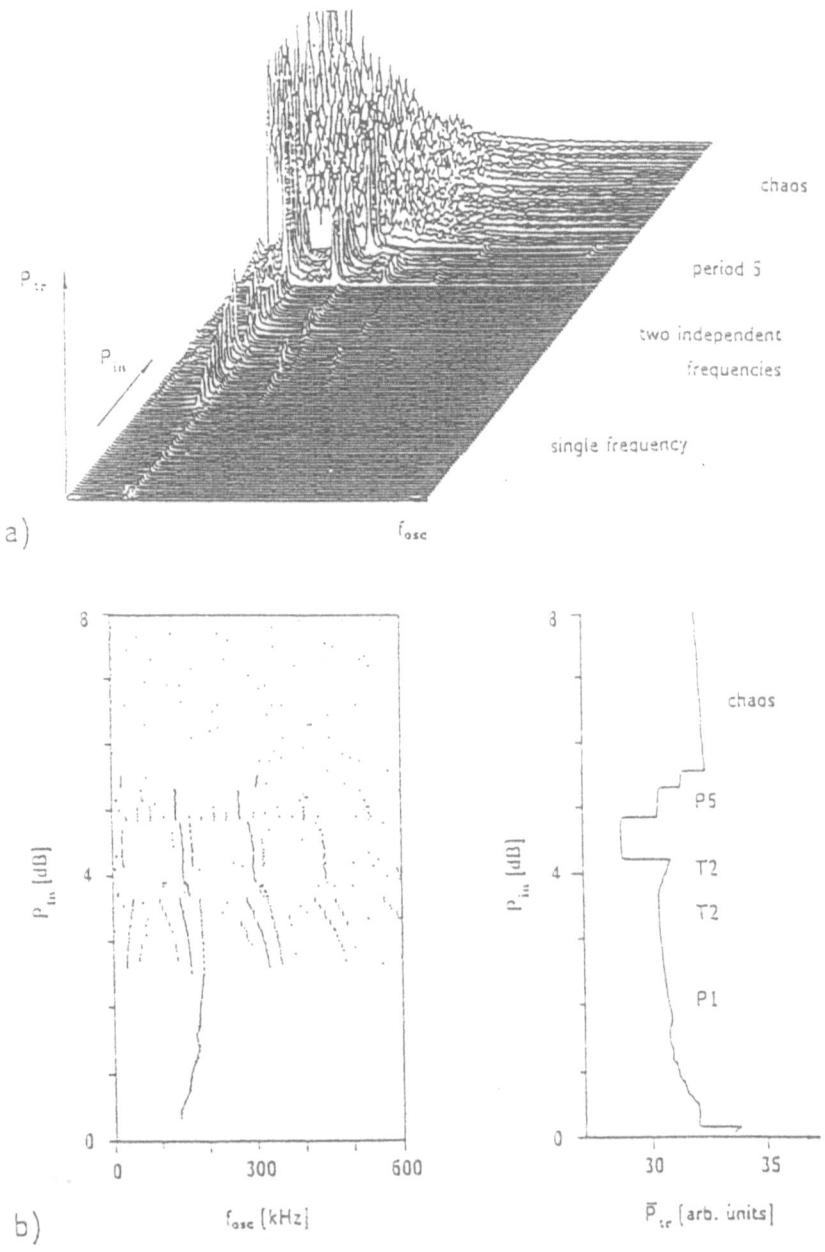

Figure 2. Power spectra of auto-oscillations at $\omega_p/2\pi = 2.375\,\mathrm{GHz}$ and $H = 838\,\mathrm{Oe}$ with respect to the input power P_{in} (normalized to the Suhl threshold). a) Three-dimensional 'landscape' of spectral components. b) 'Map' of oscillation frequencies; the corresponding level of $\overline{P_{tr}}$ is presented in the r.h.s. diagram. The system shows quasiperiodic behaviour and mode-locking.

Often *quasiperiodicity* was observed with up to 3 fundamental frequencies. A typical example, suggesting an interpretation in terms of the Ruelle-Takens-Newhouse scenario, is shown in Fig. 2. Very close above the threshold (denoted by '0 dB') the system starts oscillating at about 130 kHz, that means, a first Hopf bifurcation changes the *fixed point* into a *limit cycle*. At 2.5 dB a second fundamental frequency of 160 kHz occurs — corresponding to a second Hopf bifurcation — together with several sum and difference frequencies of harmonics. With increasing microwave power both oscillation frequencies vary independently, indicating that the attractor is a *2-torus*. This quasiperiodic oscillation remains stable for about 1 dB. Then, by a superficial inspection of the results of Newhouse, Ruelle and Takens [23] one would expect a third Hopf bifurcation and the immediate collapse of the resulting *3-torus* to chaos [1]. The different behaviour observed in our experiment can be interpreted by the spin system switching over to a coexisting stable attractor. Nevertheless, we also found experimental examples where a third fundamental frequency occurred for an extended parameter range. There are other levels where the different frequencies tend to lock. At 5 dB, for instance, instead of quasiperiodicity a period-5 (P 5) oscillation takes place. Low-period oscillations, such as P 2, P 3, P 4 or P 6 were observed rather often, but sometimes also higher periods of 11 or even 25. The changeover to chaos is generally accompanied by a jump of $\overline{P_{tr}}$. Since in most cases this changeover does not start from a 2-torus, it cannot be related to the third bifurcation of a Ruelle-Takens-Newhouse scenario. We rather suppose that the chaotic behaviour results from a sudden increase of the number of coupled modes, which is related to some global symmetry-breaking bifurcation [16], and does not follow one of the standard routes.

A *period-doubling route*, as reported previously from both transverse and parallel pumping experiments [25, 26], was observed up to period 8 but occurred rather seldom. More often, only a single period doubling occurred, remaining stable for an extended range of P_{in} and then changing directly over to chaos. Though the Feigenbaum route is known to be very sensitive to noise which might suppress the subsequent period doublings, we rather interpret the observed behaviour to represent an independent route. We also observed a sequence of period triplings (not to confuse with a period-3 window!) up to period 9.

Intermittency means the occurrence of a signal which randomly alternates between two (or more) different dynamic states, e.g. laminar phases and irregular bursts. Three universal types of intermittency (I – III) have been discussed by Pomeau and Manneville [27] corresponding to the basically different ways how a fixed point of a 1D map can loose its stability through a local bifurcation. Often these types can already be distinguished from their characteristic time behaviour, but also from a reconstruction of the generating map and from the distribution and scaling behaviour of the laminar lengths [22]. So far, intermittency has been found rather seldom in magnetic systems [28]. Both in the coincidence regime and for subsidiary absorption we observed various kinds

---

[1] However, one cannot exclude a stable 3-torus also, since the measure of parameter values for which a chaotic attractor occurs may vanish [24].

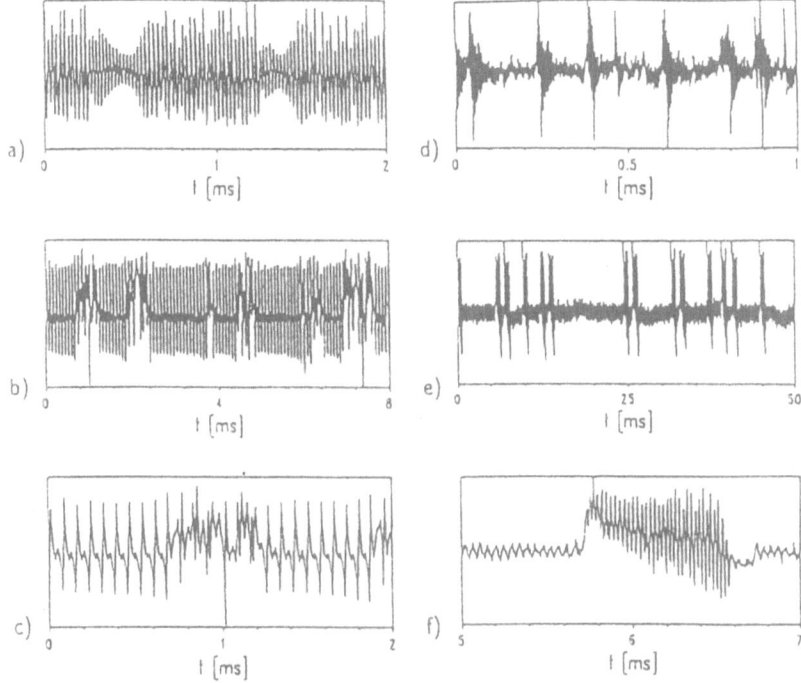

Figure 3. Different types of intermittency observed in the coincidence regime (a-c, e, f) and in subsidiary absorption (d). a) Pomeau-Manneville type I, b) type III, c) extended time scale of b. d) Chaos-chaos intermittency due to a homoclinic crisis, e) on-off intermittency, f) extended time scale of e.

of intermittency starting from a fixed point, a limit cycle, a 2-torus, or even alternating between different chaotic states (see Fig. 3). From analysing the distribution and scaling behaviour of the 'laminar' lengths, the observed signals could clearly be attributed to each of the universal types I, II or III or to crises [16, 29]. It is interesting to note that the Pomeau-Manneville types are generally observed in parameter regimes where the system remains *low-dimensional*, whereas chaos-chaos intermittency occurs at *higher dimensions*, especially in the coincidence regime. We found that in most cases the observed chaos-chaos intermittency scales like the Pomeau-Manneville type III [29] which is not consistent with the common interpretation of arising from a crisis [30]. Very recently this specific behaviour was interpreted in terms of a new type called 'on-off intermittency' which is based on a global symmetry-breaking bifurcation [31]. The underlying physical mechanism is probably that of the transitory excitation of an additional spin-wave mode through a 3-magnon process [16, 32].

The analysis of the observed chaotic behaviour shows a significant difference between resonant and nonresonant pumping: In the coincidence regime we generally observe higher-dimensional chaos ('hyperchaos') with correlation dimensions [33] $D_2 \simeq 7 \dots 15$, whereas in subsidiary absorption low-dimensional ('marginal') chaos of dimension $D_2 \simeq 2 \dots 3$ is prevailing. Accordingly, our attempts to suppress or control the irregular time behaviour were limited to the latter case. Details of the observed scenarios and the corresponding bifurcation map are presented in Fig. 4.

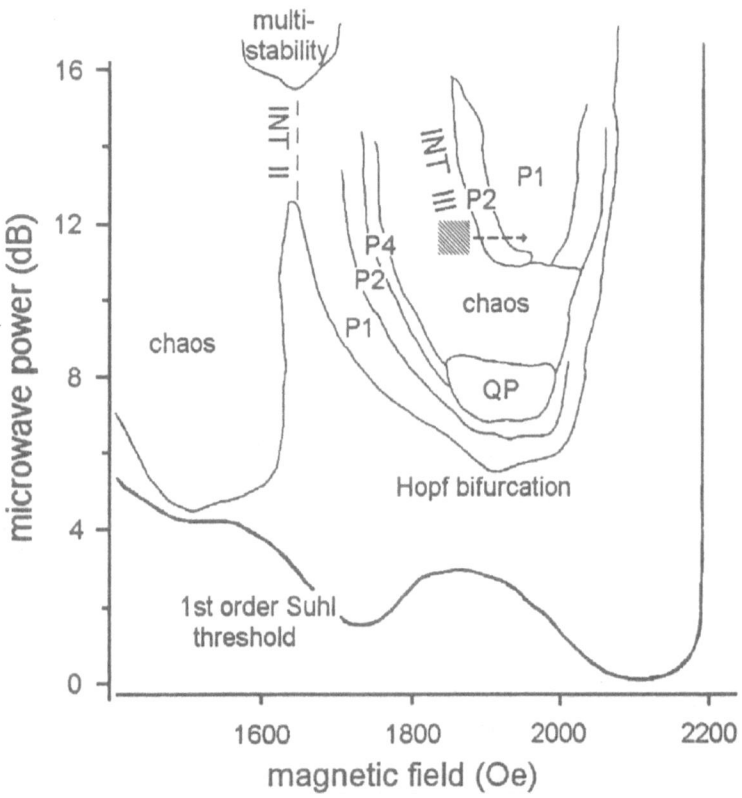

Figure 4. Bifurcation map for subsidiary absorption. The lower line indicates the Suhl threshold, the next upper line ('Hopf bifurcation') the onset of auto-oscillations. QP: quasiperiodic behaviour; P2, P4: period doubling bifurcations; INT II, III: intermittency of Pomeau-Manneville type II, III; dashed area: marginal chaos, application of control.

3. Suppression of Chaos by Nonresonant Parametric Modulation

We first focus on the method of nonresonant parametric modulation, which actually requires no knowledge about the system and can be applied to a wide class of systems. We demonstrate its simple applicability for spin-wave chaos observed at higher pumping power in subsidiary absorption. The large universality, however, can only be obtained by lifting the condition of low control amplitudes. Thus, parametric modulation actually results in an effective change of the modulated control parameter, and the obtained suppression of chaos is essentially related to a global change of the system.

3.1 ANALYTICAL APPROACH

A theoretical understanding of the underlying mechanism can be achieved by considering the arbitrarily chosen example of the Duffing oscillator [3]

$$\ddot{x} - \alpha\, x + \beta\, x^3 \; = \; -\gamma\, \dot{x} + F\cos(\omega\, t) \; , \tag{1}$$

with the control parameter α being modulated with a high frequency $\Omega \gg \omega$:

$$\alpha(t) \; = \; \alpha\,[\,1 + \epsilon\cos(\Omega\, t)\,] \; . \tag{2}$$

By means of an analytical method based on the separation of slow and fast time scales of $x(t)$ and on the comparison of independent harmonics of the rapidly oscillating part we found that the dynamics of the slowly varying part $\bar{x}(t)$ (averaged over an oscillation period $2\pi/\Omega$) approximately follows the same equation of motion (1) as the original Duffing oscillator without modulation, but with renormalized coefficients $\tilde{\alpha}$, $\tilde{\beta}$ and $\tilde{\gamma}$, which up to the order of α^2/Ω^4 are given by

$$\tilde{\alpha} \; = \; \alpha\left\{1 - \frac{\alpha}{2}\left(\frac{\epsilon}{\Omega}\right)^2\right\}, \quad \tilde{\beta} \; = \; \beta, \quad \tilde{\gamma} \; = \; \gamma \; . \tag{3}$$

Thus, for the slow timescale the continuous increase of the modulation amplitude ϵ acts in the same way as a reduction of the effective parameter $\tilde{\alpha}$. This means, fast parametric modulation should result in the same scenario as obtained when directly decreasing the respective control parameter α. By means of the well-known Melnikov criterion [34] we were able to calculate the values of $\tilde{\alpha}$ (or ϵ) where the chaotic behaviour is suppressed and the signal becomes regular. For the case of small modulation amplitude the resulting condition for the suppression of chaos then reads

$$\epsilon \; \geq \; \sqrt{2/\alpha}\;\, \Omega \; . \tag{4}$$

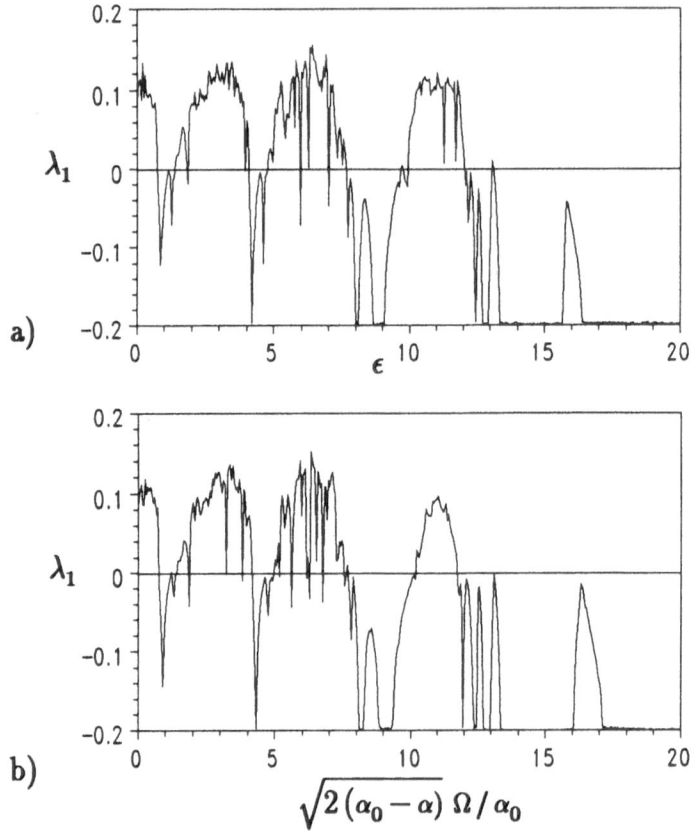

Figure 5. Numerical simulation of chaos suppression by fast parametric modulation. a) Decrease of the largest Lyapunov exponent with increasing modulation amplitude ϵ for constant system parameter $\alpha = \alpha_0$. b) For comparison: Scenario without modulation, but on direct variation of α, using a nonlinear scale related to eq. (3).

3.2 NUMERICAL INVESTIGATIONS

We checked these analytical results by numerical simulations of eqs. $(1, 2)$, calculating the time-dependent signal, the power spectrum, and the relevant Lyapunov exponent. For certain initial conditions ($\alpha_0 = \beta = \omega = 1$, $F = 0.35$ and $\gamma = 0.4$) we found a symmetric chaotic attractor. Then, switching on the fast ($\Omega = 10$) parametric modulation, and increasing its amplitude ϵ from 0 to 15, the system actually steps back through a scenario of different bifurcations, periodic windows, and finally ends up in periodic oscillations [3, 16]. For further illustration of the scenario, we extracted the characteristic Lyapunov exponent λ_1 from the simulated time series with respect to the modulation amplitude ϵ (Fig. 5 a). Note that there is only one relevant exponent λ_1 for the Duffing oscillator; the second exponent corresponds to the external forcing $\lambda_2 = 0$, while the third one is determined by dissipation: $\lambda_3 = -\gamma - \lambda_1$.

Two important results were obtained [3]: (i) The exponent λ_1 becomes negative for large values of ϵ, and a regular motion is actually recovered. (ii) Periodic windows with a negative exponent λ_1 are observed, and bifurcations occur as indicated by the vanishing of λ_1. The equivalence between the continuous increase of the modulation amplitude ϵ and the continuous decrease of the effective control parameter $\tilde{\alpha}$ in the unmodulated system was shown by calculating the Lyapunov exponent of the unmodulated system in dependence on α (Fig. 5 b). The equivalence was emphasized by rescaling the α-axis according to eq. (3), i.e. using the nonlinear scale $f(\alpha) = \Omega \sqrt{2 (\alpha_0 - \alpha)}/\alpha_0$. This direct comparison of numerical simulations yields striking evidence of the fact that the direct variation of α gives rise to exactly the same scenario as the high frequency parametric modulation, and that the basic mechanism of chaos suppression by fast parametric modulation can be well understood to arise from a renormalization of external control parameters.

The analytical approach formulated above can, in principle, be applied to any system of differential equations. The effect of modulation on one of the control parameters, as depicted by eq. (3) for the case of the Duffing oscillator, however, specifically depends on the system under consideration. Especially the sign of change of an effective parameter needs not be negative, which means that chaos suppression is not achieved in every attempt. In all cases of resonantly driven oscillators, however, the sign of effective parameter change is not relevant, since any shift of "eigenfrequencies" will move the system out of resonance and, thus, increase the threshold for chaotic excitation.

3.3 SUPPRESSION OF SPIN-WAVE CHAOS

The generality of the fast modulation approach could be demonstrated by applying it to the real experimental situation. We chose the chaotic regime around $H = 1900$ Oe and $P_{in} = 10$ dB in Fig. 4. The chaotic behaviour was analysed [16] to be of marginal type, i.e. characterized by only one positive Lyapunov exponent and a correlation dimension between 2 and 3. This chaotic regime is entered from the high field side through the following scenario (reverse to the dashed arrow in Fig. 4): Starting at 2200 Oe/10 dB and decreasing H, one first passes through a pitchfork bifurcation, which corresponds to the Suhl threshold, then through a Hopf bifurcation resulting in auto-oscillations, and finally through an inverse period doubling bifurcation (denoted by P2) to the chaotic state. Such a bifurcation results in intermittency of the classical Pomeau-Manneville type III [22] if, as in our case, a reinjection process exists which remits the system trajectory from the chaotic attractor back to the neighbourhood of the former stable periodic orbit. Right before the bifurcation point, we already observed weak period-2 oscillations which are probably induced by thermal fluctuations and noise and already anticipate characteristic properties of the dynamics beyond that point. Thus, the complete scenario into chaos reads:

period-1 oscillation $\rightarrow$ period-2 oscillation $\rightarrow$ intermittency type III $\rightarrow$ chaos .

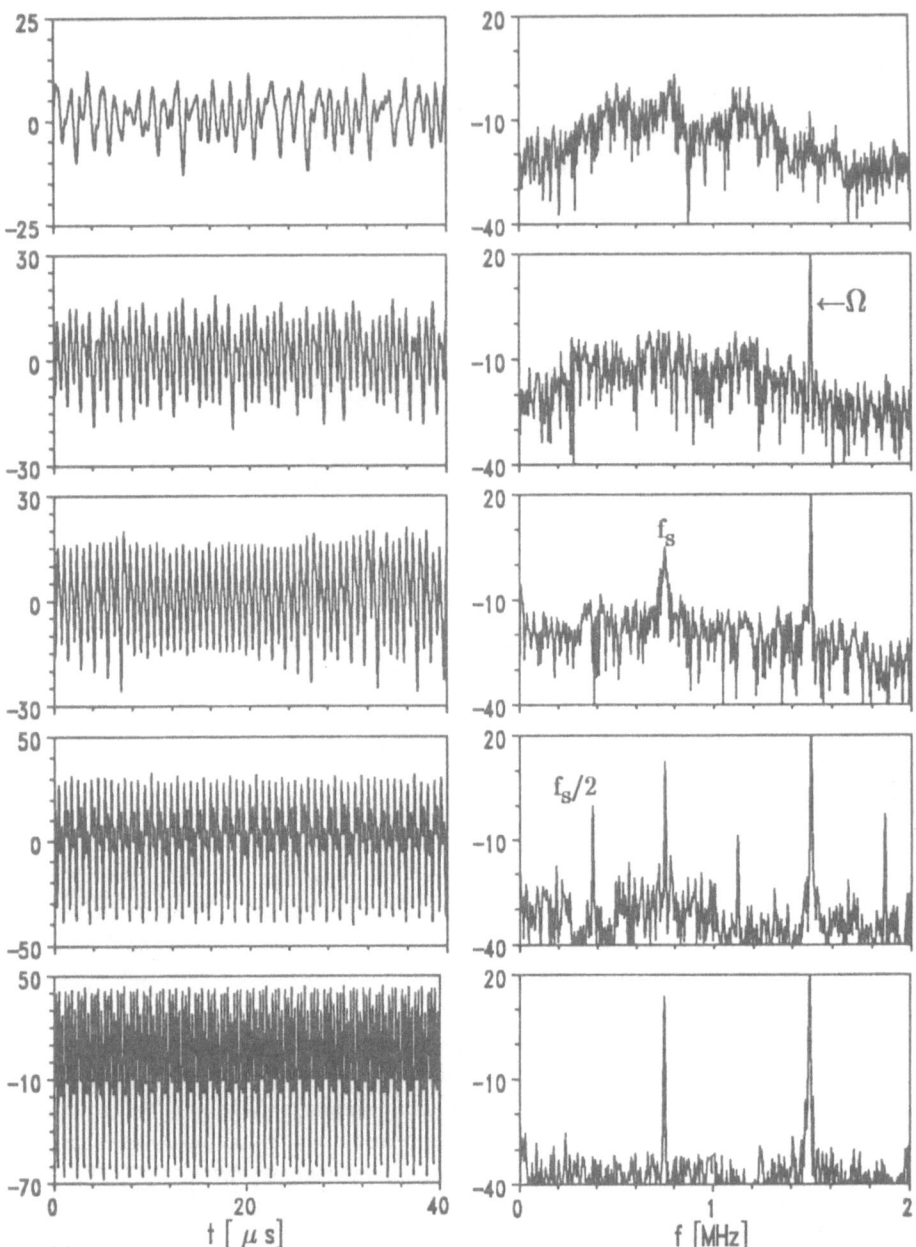

Figure 6. Scenario obtained for modulation of the microwave frequency in subsidiary absorption ($\omega_p/2\pi = 9.26\,\mathrm{GHz}$, $H = 1.90\,\mathrm{kOe}$, $P_{in} = 10\,\mathrm{dB}$). Time-dependent FMR signals [a.u.] and power spectra [dB] for increasing modulation depths $\Delta\omega/2\pi = 0$, 1.4, 2.0, 3.6, and 5.8 MHz. Note that the stabilized orbits are of larger amplitude and cannot have been embedded in the unmodulated chaotic attractor.

For our experiment on chaos suppression [35] we fixed the control parameters in the chaotic regime at $H = 1900$ Oe and $P_{in} = 10$ dB. A single time series of the unperturbed chaotic attractor is shown in Fig. 6. We then switched on a frequency modulation of the microwave field

$$\omega_p = \omega_{p,0} + \Delta\omega \cdot \sin(\Omega t) .$$ (5)

The modulation frequency was fixed at $\Omega/2\pi = 1.490$ MHz, while the modulation depth $\Delta\omega$ was varied continuously. The resulting time behaviour and the corresponding power spectrum of the observed single time series for increasing modulation depth are represented in Fig. 6. The upper line ($\Delta\omega = 0$) corresponds to the unmodulated chaotic system. The spectrum is broadband around some typical system frequency f_s. After switching on the modulation, its frequency Ω distinctly appears in the power spectrum. Increasing $\Delta\omega / 2\pi$ to 2.0 MHz we observe intermittency with the laminar phases showing a period-2 oscillation of diverging amplitudes. This is typical for intermittency type III. Simultaneously, the broad spectrum decreases in amplitude and a sharper peak at a system frequency $f_s = 750$ kHz develops from the noisy background [2], accompanied by a weak subharmonic at $f_s/2$. At $\Delta\omega = 3.6$ MHz the broad bump has moved down to the noise level, indicating the successful suppression of chaos. Simultaneously, the subharmonic at $f_s/2$ shows up very clear, but vanishes again at higher modulation depths. Thus, by increasing the modulation depth $\Delta\omega$ we observed the scenario:

chaos $\rightarrow$ intermittency type III $\rightarrow$ period-2 oscillation $\rightarrow$ period-1 oscillation .

This is just reverse to what we observed on direct variation of the magnetic field H and confirms that high-frequency parametric modulation results in a 'route out of chaos' as predicted by our analytical and numerical calculations.

The scheme of parametric modulation seems to be applicable to almost every system. In principle, it should be capable to suppress even more than only one positive Lyapunov exponent and, therefore, is basically not restricted to marginal chaos, as e.g. the OGY method. Moreover, parametric modulation is very easy to apply, since it requires no knowledge about the system under consideration, and even the sign of its influence is of no importance in many cases. In real experiments it allows a simple access to fast systems which cannot be controlled by other, more sophisticated methods. Nevertheless, it has to be classified as a brute force method which drastically changes the system. The indirect variation of a system parameter results in new regular orbits of different topology which becomes evident in our experiment when comparing the reconstructed attractors of the unperturbed and of the modulated system. In spite of the very small change of pumping frequency [3] by less than 10^{-3} the resulting regular orbit is far outside the original attractor and of different shape.

---

[2]   Since we could not sufficiently increase the modulation frequency for technical reasons, the system frequency was partly synchronized via its harmonics by the external modulation .

4. Controlling Unstable Periodic Orbits by a Modified OGY Technique

Controlling chaos means to change the irregular into regular behaviour without drastically affecting the system, just by making use of the internal system dynamics. Ott, Grebogi and Yorke (OGY) [5] have proposed a general way to achieve control by means of a feedback technique, where small time-dependent perturbations are made on one of the system parameters. Their algorithm is based on the idea that chaotic attractors are generally embedded in an infinite number of unstable periodic orbits. If the trajectory comes into the vicinity of such a hyperbolic orbit, it approaches the orbit as long as the distance vector is located close to the *stable manifold*, and leaves it again in the direction of the *unstable manifold*. Since these processes evolve exponentially in time, they are rather slow in the vicinity of the orbit and can be affected by weak external perturbations. The practical use of such a control would be twofold: to suppress undesired irregularity and to select specific regular oscillations among a large number of possibilities by applying very small controlling power.

4.1 METHOD OF OGY

Unstable periodic orbits can be detected by means of the recurrence time method [36]. In order to stabilize one of these orbits, one first has to calculate the evolution of a trajectory in its neighbourhood. Technically, this is achieved by reconstructing the trajectory (of e.g. a 3D flow) from a time series by means of time-delayed coordinates [37]. The problem is simplified by applying a Poincaré section perpendicular to the unstable periodic orbit, thus converting the 3D flow to a 2D discrete map (Fig. 7). This way, the unstable periodic orbit is mapped to a hyperbolic fixed point $\xi_F(\mu_0)$, the stability of which has to be analysed. From the evolution of previous intersection points the stable and unstable eigenvalues λ_s and λ_u and the respective eigendirections e_s and e_u are determined. Such analysis yields a linearized prediction of the system dynamics in the neighbourhood of $\xi_F(\mu_0)$ where the evolution matrix can be expressed in terms of the stable and unstable eigenvalues ($e_{s,u}$ and $e_{s,u}^*$ denote the covariant and contravariant eigenvectors, respectively):

$$\xi_{n+1}(\mu_0) - \xi_F(\mu_0) \simeq (e_s \lambda_s e_s^* + e_u \lambda_u e_u^*) \cdot [\xi_n(\mu_0) - \xi_F(\mu_0)] \qquad (6)$$

Next, one considers a small change of the control parameter μ and calculates the corresponding shift g of this fixed point:

$$g \equiv \frac{\partial \xi_F(\mu)}{\partial \mu} \Big|_{\mu_0} \simeq \frac{\xi_F(\mu) - \xi_F(\mu_0)}{\mu - \mu_0} \qquad (7)$$

---

[3] It is somewhat misleading to proof the smallness of perturbation by the relative change of the pumping frequency or the magnetic field, as has been done in the literature. In the case of YIG a relative change of 10^{-3} of these parameters is already sufficient to destroy any resonance condition. Thus, taking the FMR linewidth as a more adequate scale, "10^{-4}" means already a 10% perturbation!

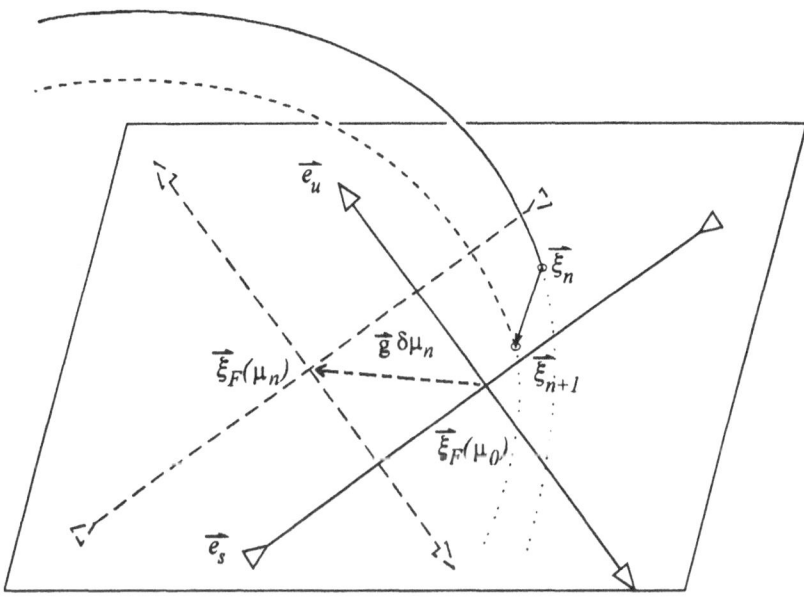

Figure 7. Controlling scheme of Ott, Grebogy and Yorke. The Poincaré section maps the unstable periodic orbit to the fixed point $\xi_F(\mu_0)$ with its stable and unstable directions e_s and e_u. The arrows indicate the effect of a small change of the control parameter μ (see text).

The basic idea of OGY is to shift the fixed point and the corresponding evolution for a short time in such a way that after one cycle the system — described by $\xi_{n+1}(\mu \neq \mu_0)$ — ends up on the stable manifold of the original fixed point $\xi_F(\mu_0)$, and then to switch the perturbation off again. This way the following intersection points $\xi_{n+2}(\mu_0)$, $\xi_{n+3}(\mu_0)$, ... exactly approach the fixed point without being repelled, i.e. after a well-targetted perturbation the intrinsic dynamics of the system is used for stabilizing the trajectory on the unstable periodic orbit. The value of this perturbation is obtained by combining eqs. (6), (7), and the condition that $\xi_{n+1} - \xi_F$ be orthogonal to the unstable manifold:

$$\Delta\mu = \frac{\lambda_u}{\lambda_u - 1} \cdot \frac{[\xi_n(\mu_0) - \xi_F(\mu_0)] \cdot e_u^*}{g \cdot e_u^*} \tag{8}$$

Once the system has approached the orbit, the still necessary corrections (due to linearization and noise) can be maintained by very small perturbations. Though originally developed for discrete maps, this concept can also be applied in low-dimensional continuous flows, but in practice is limited to only one unstable direction.

4.2 EXPERIMENTAL CONTROL BY ANALOG FEEDBACK DEVICE

Although the OGY method should apply to real experimental systems as well, in practice its application is restricted for the following reasons: (i) Experimental systems often show high-dimensional chaos, i.e. there is more than one unstable direction. (ii) The measured signal is disturbed by noise; this may either prevent the control to work at all, if in the case of strong noise the system is strongly pushed away from the neighbourhood of the fixed point, or at least reduces the sensitivity of the feedback in the case of weak noise. (iii) The characteristic time scale of real systems is often too fast. For the magnetic system investigated typical cycle times are in the order of μs (!), whereas the numerical calculation of the feedback signal requires at least some ms.

The first problem could be overcome by selecting chaotic signals of sufficiently low dimensionality. Grassberger-Procaccia analyses [33] show that this is nearly impossible for the coincidence regime but easy to obtain for subsidiary absorption. The dashed area in Fig. 4 marks the corresponding control parameter range where marginal chaos was observed. The chaotic signal to be controlled was of the same type as in the previous section, i.e. characterized by a correlation dimension $D_2 = 2.1 \pm 0.1$, only one positive Lyapunov exponent $\lambda_1 = 0.04 \, (\mu s)^{-1}$, and a mean cycle time of about 2 μs.

Since it was impossible to make numerical real-time predictions for a time scale of μs we had to modify the OGY algorithm in a way to be processed by an intelligent analog feedback device: For reconstructing the attractor we used analog time derivatives instead of time delay coordinates. The Poincaré plane and the location of intersection points were determined by analog window discriminators for the signal and its first and second time derivatives triggering a track-and-hold amplifier (Fig. 8 a). The amplitude of the control signal was not determined from a preceding stability analysis of the periodic orbit. Instead, we used a feedback signal which is proportional to the deviation of the momentary signal $U(t_n)$ from the set-point (the "U-coordinate" of the phase space window to be adjusted to the position of the unstable orbit)

$$U_{contr}(t_n) = A \cdot \left[U(t_n) - U_{ref} \right] \qquad (9)$$

and vary the phase space coordinates in order to meet the neighbourhood of some unstable periodic orbit where the given feedback results in a proper control. A similar technique [4] was applied by Hunt [7] for controlling a diode-capacity resonator who called it *occasional proportional feedback*. There is complete correspondence of eq. (9) to the original result of OGY, eq. (8), identifying the amplification factor A with $\lambda_u (\lambda_u - 1)^{-1} (g \cdot e_u{}^*)^{-1}$ and the deviation $U(t_n) - U_{ref}$ with $[\xi_n(\mu_0) - \xi_F(\mu_0)] \cdot e_u{}^*$.

---

[4] In contrast to our device Hunt is using a stroboscopic mapping for the Poincaré section and only one window, so his device is limited to non-autonomous, periodically driven systems.

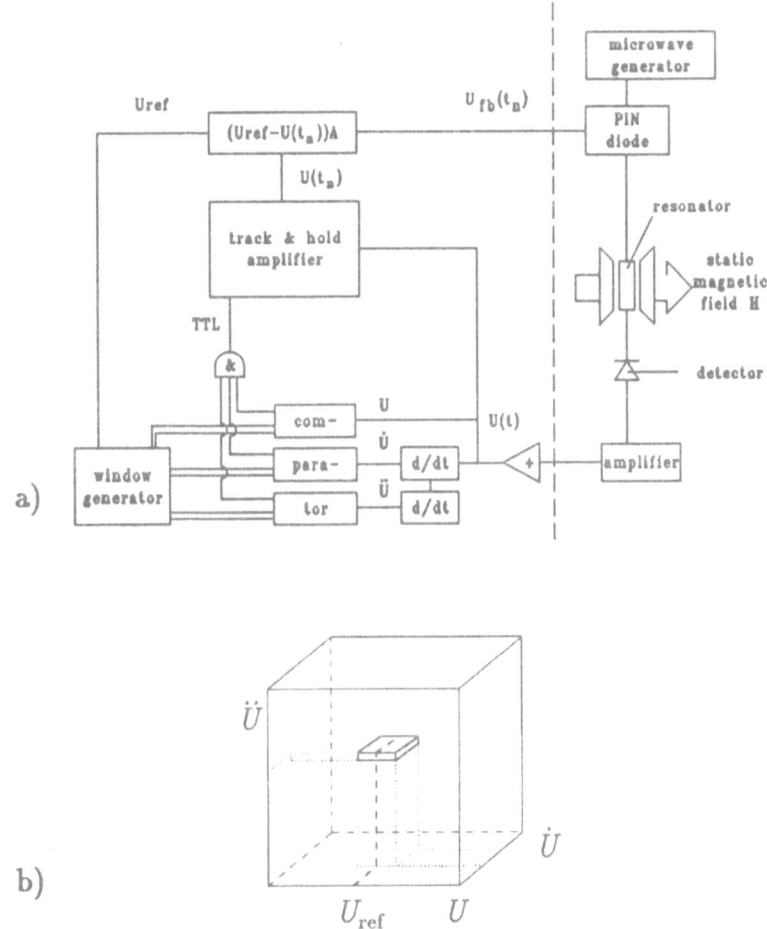

Figure 8. Analog feedback device for controlling the chaotic FMR signal. a) Experimental set-up. b) Reconstruction of the phase space by means of analog derivatives. The small probe volume can be moved around by varying the settings of the window generator.

In our experiment the position of the unstable periodic orbit was selected by setting windows for the observed signal $U(t)$ and its first and second time derivative by means of analog window comparators (Fig. 8 b). The corresponding signal was held by a fast track-and-hold amplifier. The deviation of this value from a given set-point is fed back to change the microwave pumping power. By careful adjustment of the windows and variation of the set-point we succeeded in stabilizing periodic orbits by means of a perturbation which is less than 10^{-3} of the actual pumping power P_{in}. The effect of control is illustrated in Fig. 9 , where we have compared the chaotic signal before and after switching on the

398

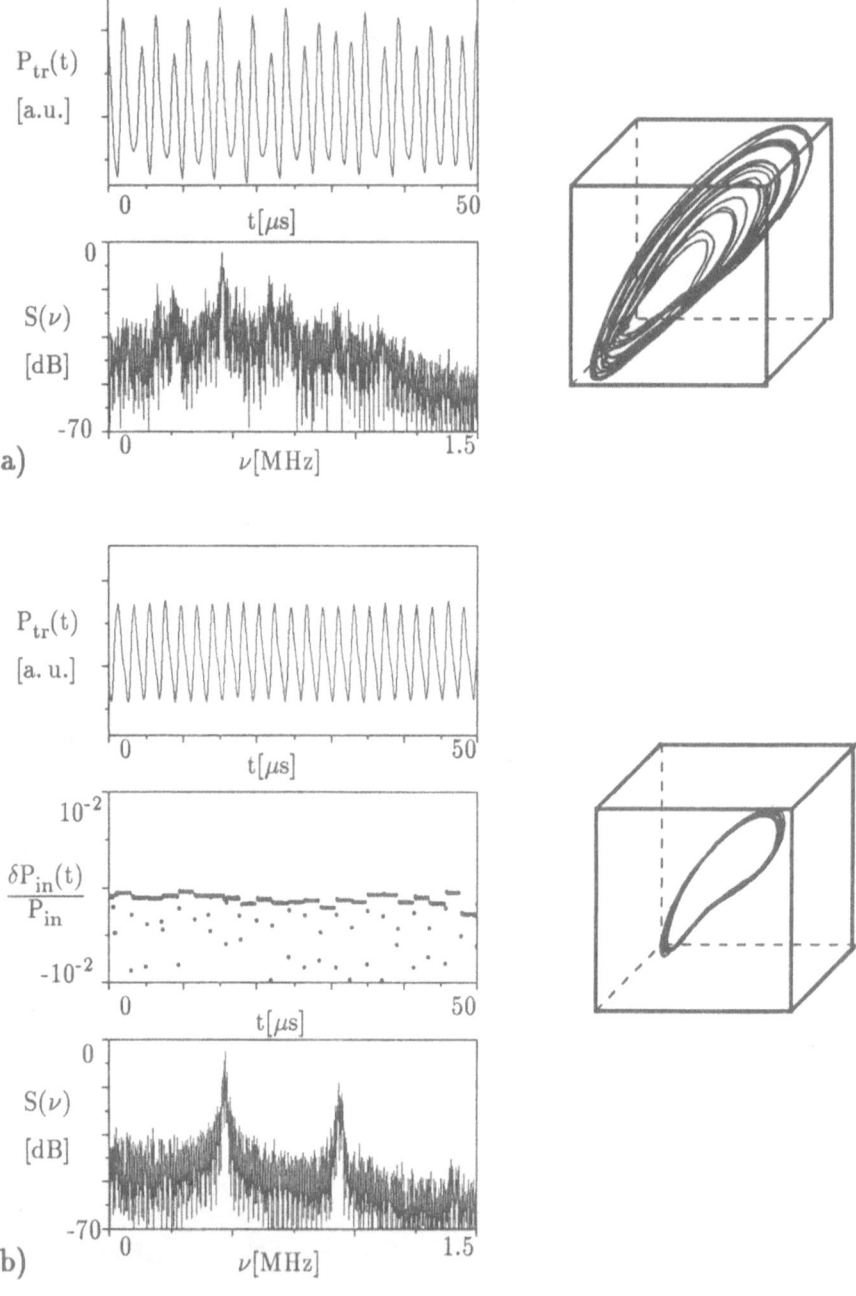

Figure 9. Result of the OGY-type control. a) Chaotic time signal, spectrum, and reconstructed attractor without control. b) Regular oscillations, control signal, spectrum, and stabilized orbit after switching on the control.

feedback. The controlled signal shows a very regular oscillation, and the corresponding phase space trajectory, in fact, consists of a single orbit slightly smeared out by noise. To our knowledge this is the first real feedback control of chaos in spin-wave turbulence which stabilizes the inherent periodic dynamics of a strange attractor.

5. Delayed Feedback Control

A different strategy which again aims at the stabilization of existing unstable periodic orbits (UPO) was proposed by Pyragas [4]. It is based on a continuous time-delayed feedback which in the neighbourhood of an UPO may directly act as a proper controlling force to stabilize the regular motion on the orbit. Because of its simplicity and robustness this method is especially appropriate for the control of fast experimental systems.

5.1 METHOD OF PYRAGAS

We consider a nonlinear system which is described by the following set of differential equations:

$$\dot{x} = P(x) , \tag{10}$$

where $x = (x_1, \ldots, x_N)$ are the system variables. At least one of them, x_1, should be accessible in experiment. The explicit knowledge of P is not required. The controlling scheme for an unstable periodic orbit ξ consists in adding a forcing

$$F(t) = -K[x(t) - \xi(t)] \tag{11}$$

which is directed towards the position of the UPO and just acts as a negative feedback. This simplest choice yields the evolution equation

$$\dot{x} = P(x) + F(t) . \tag{12}$$

Since the exact position of the UPO is *a priori* unknown, $\xi(t)$ is approximated by a part of the trajectory closer to it than the present position of $x(t)$. So, once the system has approached the UPO and is about to escape slowly, a proper correction is obtained by

$$F(t) = -K[x(t) - x(t - \tau)] , \tag{13}$$

where $x(t - \tau)$ denotes the position of the trajectory one cycle before (cf. Fig. 10). The vanishing of the control signal $F(t)$ can be taken as a criterion for successful control. The method is especially well suited for periodically driven systems, because τ is exactly known in this case, but can also be applied to autonomous systems. In this case the average cycle time of the chaotic signal can be used as a proper estimate. Pyragas [4] has only considered the forcing of one accessible variable, but the following ideas can also be extended to the case of general controlling, including a multi-component or parametric forcing [38].

One disadvantage of the delayed feedback control consist in the fact that the UPO to be stabilized is only selected by the choice of the delay time τ, so in the case of several UPOs having the same cycle time, one generally cannot predict which one will be stabilized. It is also open whether control can be achieved through every accessible system variable. On the other hand, because of the continuous control signal, the method is rather insensitive to noise. It has successfully been applied both in simulations and experiments [12, 13, 38-41].

5.2 THEORETICAL APPROACH AND SIMULATIONS

In contrast to the OGY method theoretical ideas concerning the efficiency of the control of real periodic orbits, its dependence on K and τ, the minimal K-value, or the dependence of the transient time on the previous instability of the orbit, are still missing. We have tried to fill this gap by some general considerations concerning the critical amplification K_c and the K-dependence of transient time. Moreover, numerical simulations were performed for the Toda oscillator and compared with experimental studies at a nonlinear diode resonator [38].

To begin with we remind the reader that linear stability analysis of the system (10) without control leads to a Floquet problem

$$\gamma \, v(t) + \dot{v} = D \, P(\boldsymbol{\xi}(t)) \, v(t), \quad v(t+\tau) = v(t) \tag{14}$$

where $\exp(\gamma \, t) \, v(t) = \boldsymbol{x}(t) - \boldsymbol{\xi}(t)$ denotes the deviation from the UPO and DP the Jacobian. The real part of the largest Floquet exponent $\lambda = \mathrm{Re}\,\gamma > 0$ determines the instability of the orbit whereas an imaginary part $\omega = \mathrm{Im}\,\gamma$ leads to a revolution around the UPO (cf. Fig. 10). If the delay of the control signal (13) coincides with the period of this orbit, then the driven system (12) admits the same periodic motion. Linear stability analysis according to $\boldsymbol{x}(t) = \boldsymbol{\xi}(t) + \delta \, \boldsymbol{x}(t)$ yields

$$\delta \dot{\boldsymbol{x}} = D \, P(\boldsymbol{\xi}(t)) \, \delta \, \boldsymbol{x}(t) - K[\delta \, \boldsymbol{x}(t) - \delta \, \boldsymbol{x}(t-\tau)] \ . \tag{15}$$

The (infinite dimensional) generalization of Floquet theory[5] tells us that the deviations obey $\delta \, \boldsymbol{x}(t) = \exp(\Gamma \, t) \, \boldsymbol{w}(t)$, so that eq. (15) reduces to

$$[\Gamma + K(1 - \exp(-\Gamma \tau))] \, \boldsymbol{w}(t) + \dot{\boldsymbol{w}} = D \, P(\boldsymbol{\xi}(t)) \, \boldsymbol{w}(t), \quad \boldsymbol{w}(t+\tau) = \boldsymbol{w}(t). \tag{16}$$

Since the eigenvalue equations (14) and (16) are identical, we obtain the desired relation between the Floquet exponent γ of the UPO, the Floquet exponent Γ of

---

[5] We claim that the subsequent analysis yields the isolated Floquet exponents. The investigation of a continuous part of the spectrum is more delicate, since in those situations the validity of linear stability analysis is not obvious.

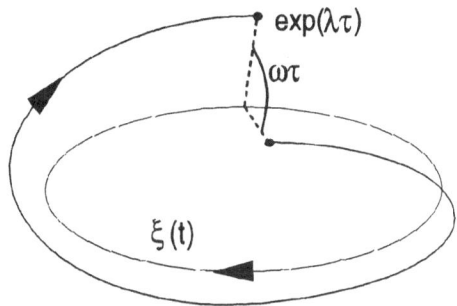

Figure 10. Uncontrolled trajectory in the neighbourhood of an unstable periodic orbit.

the controlled orbit, and the amplitude K of the control signal:

$$\lambda = \Lambda + K[1 - \exp(-\Lambda\tau)\cos(\Omega\tau)] \tag{17 a}$$

$$\omega = \Omega + K\exp(-\Lambda\tau)\sin(\Omega\tau) \ . \tag{17 b}$$

Again Λ and Ω denote the real and imaginary part of the largest exponent Γ, which govern the stability and the torsion of the controlled orbit. Stabilization is achieved when the real part Λ becomes negative, so that the critical amplitude K_c obeys

$$\lambda = K_c[1 - \cos(\Omega_c\tau)] \tag{18 a}$$

$$\omega = \Omega_c + K_c\sin(\Omega_c\tau) \ . \tag{18 b}$$

Equation (18 a) implies that only orbits with a finite torsion ($\Omega_c \neq 0$) can be stabilized. The critical amplitude is minimal if neighbouring trajectories flip during one turn ($\Omega_c = \pi/\tau$).

We illustrate our theoretical considerations with a simulation of the driven Toda oscillator

$$\dot{x} = y$$

$$\dot{y} = -\mu y - \alpha(e^x - 1) + A\sin\frac{2\pi t}{\tau} - K[y(t) - y(t-\tau)] \ . \tag{19}$$

At $\mu = 0.8$, $\alpha = 25$, $A = 105$, $\tau = 1$, and $K = 0$ the system has a chaotic attractor. A period-1 orbit which has become unstable in a period doubling bifurcation, can be stabilized at finite control $K > K_c \simeq 2.1$. Although the precise control force differs from that in the preceding paragraph the main features are not affected. The transient behaviour of the control signal $y(t) - y(t-\tau)$ is shown in Fig. 11 a. Its exponential decay is in accordance with the linear stability analysis mentioned above. The dependence of the decay rate Λ on the

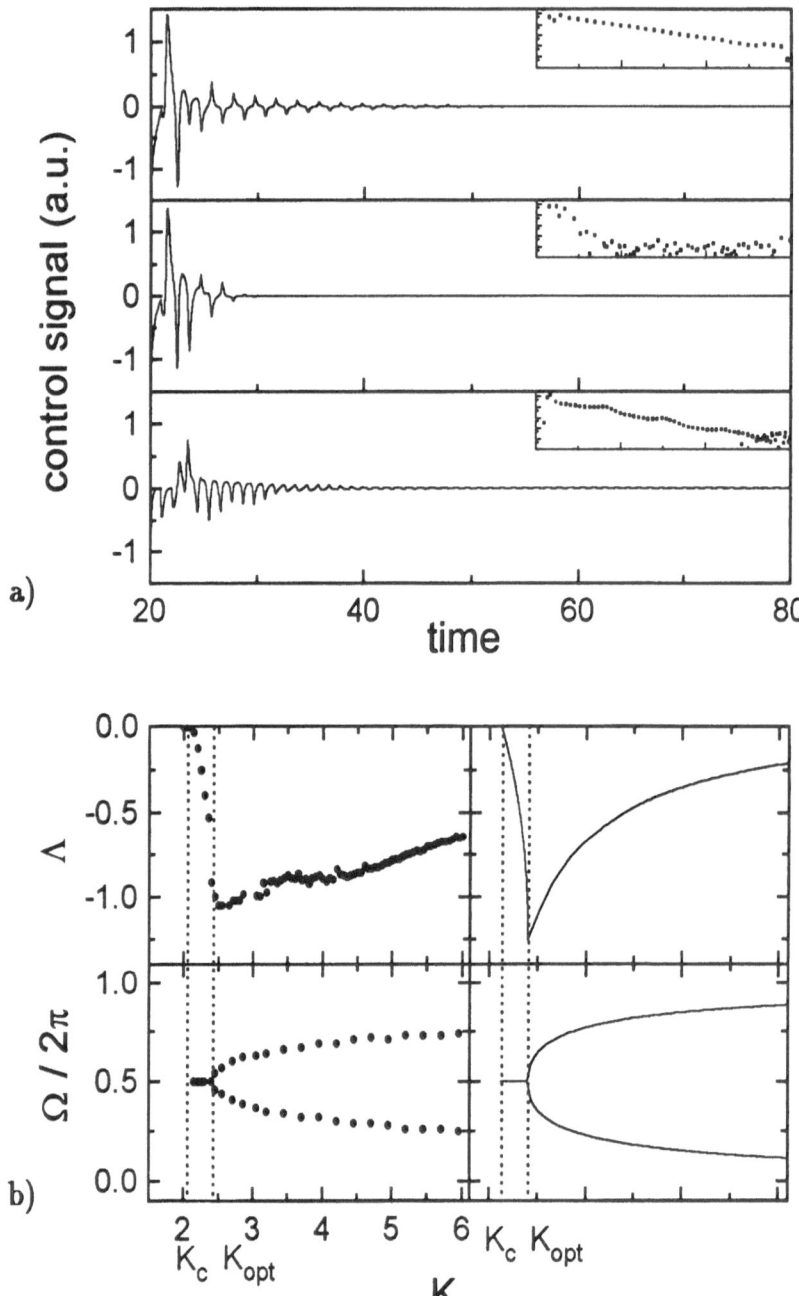

Figure 11. K-dependence of control signal for the Toda oscillator (simulation) a) Decay of control signal for K = 2.2, 2.5, and 6.0. Insets show amplitudes in logarithmic scale. The third picture shows an increase of the oscillation frequency accompanied by a modulation due to the second frequency ($K > K_{opt}$). b) Real and imaginary part of the Floquet exponent (simulation) and comparison with analytical result of eq. (17).

amplification factor K (Fig. 11 b) determines the transient time, i.e. the efficiency of the control. The minimum at K_{opt} indicates an optimal value for the control strength. In the region $K_c \leq K \leq K_{opt}$ the frequency of the control signal is given by $\Omega = \pi/\tau$, while a splitting into two frequencies and an increasing deviation from this optimal value occurs for $K > K_{opt}$.

These properties can be understood completely on the basis of our eqs. (17) (cf. Fig. 11 b, r.h.s.). Since the period-1 orbit has become unstable in a period doubling bifurcation, its frequency ω is given by π/τ. Then for moderate forcing ($K < K_{opt}$) the expression (17 b) gives the unique solution $\Omega = \pi/\tau$. This frequency allows for the maximal possible contribution to eq. (17 a) so that the decay rate increases dramatically with K. At $K = K_{opt}$ a deviation from this optimal frequency occurs and leads to a decrease of the decay rate. The complete qualitative agreement indicates that the proposed mechanism dominates the efficiency of the Pyragas control in typical situations.

5.3 APPLICATION TO SPIN-WAVE CHAOS

The application of delayed feedback control to spin-wave chaos in YIG spheres is more complicated for the following reasons: (i) Since we are dealing with a effectively autonomous system, the period of the orbit to be stabilized is not exactly known. For marginal chaos one can approximately use the averaged cycle time of the attractor as a delay parameter. This works well e.g. in simulations of the Lorenz system, but is much more critical in spin-wave chaos because of its higher sensitivity to external perturbations. So, already slight deviations from the exact period of the unstable orbit could make the control fail. (ii) In YIG spheres the spectrum of coupled modes is so complicated that we have no well-established model to work with. Therefore, we cannot predict which system variable would be the best for applying the feedback, and one is limited to 'trial and error'.

The experimental set-up which we used for our experiment (Fig. 12 a) was basically the same as in our OGY experiment. We chose the microwave input power as the feedback variable, which means that we applied a direct force on the uniform mode rather than a parametric feedback. The control device consisted of a cascade of electronic delay lines with a limiting frequency of about 2 MHz and several operation amplifiers acting as preamplifier, subtractor or inverter. The device allowed to apply a control signal of the form

$$F(t) = \pm K[x_1(t) - \epsilon \cdot x_1(t - \tau)] + F_0 \qquad (20)$$

with parameter ranges $K = 0 \ldots 100$, $\epsilon = 0 \ldots 2$, $\tau = 10 \ldots 7000$ ns, and $F_0 = -5 \ldots +5$ V. For conventional delayed feedback control ϵ has carefully to be adjusted to 1, the offset F_0 to zero, and K and τ are set according to the conditions of the system.

The attempt to stabilize orbits in the chaotic regime at $\omega_p/2\pi = 9.26$ GHz, $H = 1850$ Oe, and $P_{in} \geq 10$ dB was not successful, as long as our control signal had exactly the form of eq. (13). The reason could be that the torsion frequency

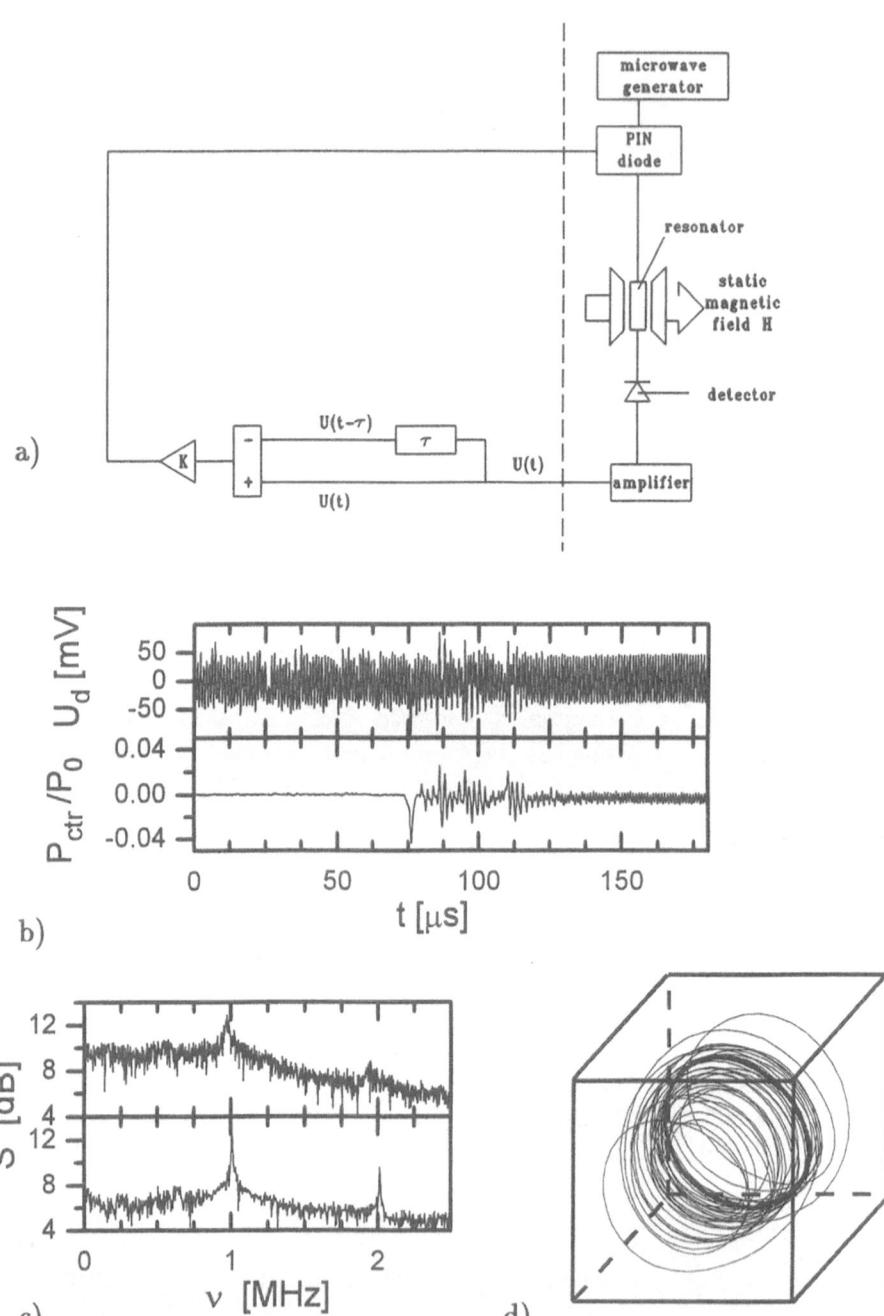

Figure 12. Time-delayed feedback control of YIG. a) Experimental set-up.
b) Signal and control signal. c) Spectrum before and after activation of control.
d) Phase space reconstruction of the uncontrolled and controlled attractors. The
new regular orbit is still embedded in the original attractor.

Ω_c was too small in comparison with Λ for the respective unstable orbit, or, more likely, the system too sensitive to a slight misadjustment of τ. However, by changing parameter ϵ from 1.0 to 0.8 (keeping $F_0 = 0$) the chaotic behaviour became regular, as presented in Fig. 12 b, c. In a strict sense, we should no longer call this control, since for $\epsilon \neq 1$ the corresponding control signal cannot vanish asymptotically. Instead, the feedback is included to change the system and to form a new regular state. The small frequency shift of the peak position in the corresponding power spectrum (Fig. 12 c) can be taken as a signature for that. In the present case, however, this new orbit is very close to the previous unstable one, as can also be seen from its phase space representation (Fig. 12 d).

When replacing the delayed feedback by some external periodic force the period of which was roughly in the order of the cycle time, we found that the chaotic dynamics could easily be suppressed even by perturbations of only a few percent magnitude. The effect was nearly independent of the special form of perturbation - sinusoidal, triangular, or spiked. This explains the number of papers in the literature claiming to report on 'feedback control experiments', but dealing, in fact, with non-feedback resonant modulation.

Another attempt of time-delayed feedback control in magnetic systems was made by Wigen and coworkers [13] on YIG films. They applied a parametric feedback by perturbing the magnetic field. Profiting from the very sharp resonance condition of YIG, they were able to stabilize regular orbits or fixed points by nominally extremely small field variations (see footnote 3 in sec. 3.3). Nevertheless, in their experiment, too, the system has obviously been changed by the feedback, which becomes evident from the reported reverse bifurcation route on variation of K.

6. Conclusions

A ferromagnetic sample externally driven by a strong microwave field represents an intriguing paradigm for studying general properties of nonlinear dynamics. The reasons are manifold: (i) Magnetic systems represent intrinsically nonlinear systems, whose nonlinearities originate from well-known interactions. (ii) Their nonlinearities give rise to spatio-temporal pattern formation and chaos. (iii) The chaotic state can be controlled with the help of very weak external time-dependent perturbations, as has been demonstrated in several experiments.

Nevertheless, in experiment one meets the problem that most of the interesting phenomena occur on rather inconvenient time and length scales. This makes the application of sophisticated control techniques often rather tedious or even impossible. For instance, present computers are still too slow for implementing a complex control algorithm like that of OGY for controlling on a timescale of microseconds. We could overcome these difficulties by means of an analog feedback device which models a part of the original analysis through fast electronic circuits.

In our FMR experiments on a YIG sphere suppression and control of chaos was achieved (i) by applying a fast parametric modulation to the microwave pumping field, (ii) by using a modified OGY method implemented by means of an analog feedback device, and (iii) by applying a time-delayed feedback signal to stabilize regular dynamic behaviour. Control in a strict sense could only be achieved by the second method, while in the first and last case the system was more or less affected by external forces.

For the parametric modulation method such a change of the system is the basic mechanism which makes this method work. By analytical as well as numerical calculations we were able to show that the increase of the modulation amplitude is equivalent to an effective variation of the parameter of the unmodulated system. Accordingly, with increasing modulation amplitude the system follows a bifurcation route 'out of chaos', which is exactly reverse to that observed on direct variation of the respective parameter.

The time-delayed feedback method, on the other hand, aims at the stabilization of existing unstable periodic orbits and should basically not affect the system in a global way. Because of its simplicity, universality, and robustness against noise this type of control represents a very promising method for practical use in fast technical systems. Nevertheless, its successful application strongly depends on the proper choice of the delay time, which may become a problem in autonomous systems. The fact that we were unable to achieve real control in YIG is probably related to an increased sensitivity of the system to a slight misadjustment of the delay time τ. Another criterion, which has not yet been considered in the literature, is related to the imaginary part of the respective Floquet exponent. In contrast to our results on YIG we found that in non-autonomous, periodically driven experimental systems, like the nonlinear diode resonator, the time-delayed control method works very well and also allows the stabilization of higher-periodic orbits (P2, P4, P5). By proper adjustment of the feedback parameters we even succeeded in 'tracking' the stabilized orbit far beyond the next bifurcations into parameter regimes of very different regular behaviour [38].

Though we have emphasized the distinction between chaos control and chaos suppression (which implies a change of the system), the limits are fluid and for practical purposes often not very important. Nevertheless, this distinction should be of interest when comparing the general efficiency of different control methods, as was the aim of this chapter. In any case one should distinguish between the specific properties of the system (e.g. the increased sensitivity due to resonance amplification) and general advantages or disadvantages of the underlying concepts, which have been mixed in previous investigations.

This project of SFB 185 *Nichtlineare Dynamik* has been supported by special funds of the *Deutsche Forschungsgemeinschaft*.

References

1. R. Lima and M. Pettini (1990) Suppression of chaos by resonant parametric excitation, *Phys. Rev. A* **41**, 726.

2. S. Parthasarathy (1992) Homoclinic bifurcation sets of the parametrically driven Duffing oscillator, *Phys. Rev. A* **46**, 2147.

3. Y.S. Kivshar, F. Rödelsperger, and H. Benner (1994) Suppression of chaos by nonresonant parametric perturbations, *Phys. Rev. E* **49**, 319.

4. K. Pyragas (1992) Continuous control of chaos by self-controlling feedback, *Phys. Lett. A* **170**, 421.

5. E. Ott, C. Grebogi, and Y.A. Yorke (1990) Controlling chaos, *Phys. Rev. Lett.* **64**, 1196.

6. W.L. Ditto, S.N. Rauseo, and M.L. Spano (1990) Experimental control of chaos, *Phys. Rev. Lett.* **65**, 3211.

7. E.R. Hunt (1991) Stabilizing high-periodic orbits in a chaotic sytem, *Phys. Rev. Lett.* **67**, 1953.

8. L.M. Pecora and T.L. Carroll (1991) Driving systems with chaotic signals, *Phys. Rev. A* **44**, 2374.

9. R. Henn, F. Rödelsperger, and H. Benner (1992) Controlling unstable periodic orbits and hyperbolic fixed points in spinwave turbulence, *Proc. XXVI Congress Ampère on Magnetic Resonance,* Athens, p. 371.

10. B. Hübinger, R. Doerner, and W. Martienssen (1993) Local control of chaotic motion, *Z. Phys. B* **90**, 103.

11. C. Reyl, L. Flepp, R. Badii, and E. Brun (1993) Control of NMR-laser chaos in high-dimensional embedding space, *Phys. Rev. E* **47**, 267.

12. A. Kittel, J. Parisi, K. Pyragas, and R. Richter (1994) Delayed feedback control of chaos in an electronic double-scroll oscillator, *Z. Naturforsch.* **49 a**, 843.

13. M. Ye, D.W. Peterman, and P.E. Wigen (1995) Controlling chaos in thin YIG films with a time-delayed method, *Phys. Lett. A* **203**, 23.

14. H. Suhl (1957) The theory of magnetic resonance at high signal powers, *J. Phys. Chem. Solids* **1**, 209.

15. see e.g. P.E. Wigen (ed.) (1994) *Nonlinear Phenomena and Chaos in Magnetic Materials,* World Scientific, Singapore.

16. F. Rödelsperger (1994) *Chaos und Spinwelleninstabilitäten,* Harri Deutsch, Frankfurt.

17. C.E. Patton and W. Jantz (1979) Anomalous subsidiary absorption in single-crystal YIG and evaluation of spin-wave linewidth, *J. Appl. Phys.* **50**, 7082.

18. G. Wiese and H. Benner (1990) Multistability and chaos by parametric excitation of longwave modes in a YIG sphere, *Z. Phys. B* **79**, 119.

19. G. Wiese, H.-A. Krug von Nidda, and H. Benner (1991) Temperature-induced nonlinearity at parametrically excited spin waves, *Europhys. Lett.* **15**, 585.

20. H. Benner, F. Rödelsperger, and G. Wiese (1992) Chaotic dynamics in spin-wave instabilities, in H. Thomas (ed), *Nonlinear Dynamics in Solids,* Springer, Berlin-Heidelberg, p. 129.

21. H.-A. Krug von Nidda, G. Wiese, and H. Benner (1994) Fine structure and critical modes at the first-order Suhl instability in YIG spheres, *Z. Phys. B* **95**, 55.

22. see e.g. H.G. Schuster (1988) *Deterministic Chaos (2nd rev. ed.)* VCH, Weinheim.

408

23. S. Newhouse, D. Ruelle, and F. Takens (1978) Occurrence of strange axiom-A attractors near quasiperiodic flow on T^m, $m \geq 3$, *Commun. Math. Phys.* **64**, 35.

24. C. Grebogi, E. Ott, and J.A. Yorke (1983) Are three-frequency quasiperiodic orbits to be expected in typical nonlinear dynamical systems, *Phys. Rev. Lett.* **51**, 339.

25. G. Gibson and C. Jeffries (1984) Observation of period doubling and chaos in spin-wave instabilities in yttrium iron garnet, *Phys. Rev. A* **29**, 811.

26. F.M. de Aguiar and S.M. Rezende (1986) Observation of subharmonic routes to chaos in parallel-pumped spin waves in yttrium iron garnet, *Phys. Rev. Lett.* **56**, 1070.

27. Y. Pomeau and P. Manneville (1980) Intermittent transition to turbulence in dissipative dynamical systems, *Commun. Math. Phys.* **74**, 189.

28. F.M. de Aguiar (1989) Intermittencies in the presence of symmetry in spin-wave experiments, *Phys. Rev. A* **40**, 7244.

29. F. Rödelsperger, T. Weyrauch, and H. Benner (1992) Different types of intermittency observed in transverse pumped spin-wave instabilities, *J. Magn. Magn. Mater.* **104-107**, 1072.

30. C. Grebogi, E. Ott, F. Romeiras, and J.A. Yorke (1987) Critical exponents for crisis-induced intermittency, *Phys. Rev. A* **36**, 5365.

31. F. Rödelsperger, A. Cenys, and H. Benner (1995) On-off intermittency in spin-wave instabilities, *Phys. Rev. Lett.* **75**, 2594.

32. A. Krawiecki and A. Sukiennicki (1995) On-off intermittency and peculiar properties of attractors in a simple model of chaos in ferromagnetic resonance, *Acta Phys. Pol.* **88**, 269.

33. P. Grassberger and I. Procaccia (1983) Measuring the strangeness of strange attractors, *Physica D* **9**, 189.

34. J. Guckenheimer and P. Holmes (1983) *Nonlinear Oscillations, Dynamical Systems, and Bifurcations of Vector Fields*, Springer, Berlin-Heidelberg.

35. F. Rödelsperger, Y.S. Kivshar, and H. Benner (1995) Route out of chaos by hf parametric perturbations in spin-wave instabilities, *J. Magn. Magn. Mater.* **140-144**, 1953.

36. D.P. Lathrop and E.J. Kostelich (1989) Characterization of an experimental strange attractor by periodic orbits, *Phys. Rev. A* **40**, 4028.

37. F. Takens (1981) *Lecture Notes in Mathematics* **898**, Springer, Heidelberg-New York.

38. T. Bernard, E. Reibold, W. Just, and H. Benner (1996) Basics of delayed feedback control and tracking in fast experimental systems, to be published.

39. D. Reznik and E. Schöll (1993) Oscillation modes, transient chaos, and its control in a modulation-doped semiconductor double-heterostructure, *Z. Phys. B* **91**, 309.

40. K. Pyragas and A. Tamasevicius (1993) Experimental control of chaos by delayed self-controlling feedback, *Phys. Lett. A* **180**, 99.

41. A. Kittel, J. Parisi, and K. Pyragas (1995) Delayed feedback control of chaos by self-adapted delay time, *Phys. Lett. A* **198**, 433.

Chapter IV
Magneto-Optic Interaction: Non-linear Effects and Devices

APPLICATIONS OF MAGNETIC GARNET FILMS IN OPTICAL COMMUNICATION

H. DÖTSCH, A. ERDMANN, M. FEHNDRICH, R. GERHARDT,
P. HERTEL, B. LÜHRMANN, M. SHAMONIN, H. P. WINKLER
AND M. WALLENHORST
University of Osnabrück, 49069 Osnabrück, Germany

Abstract. New types of optical waveguide isolators and circulators are proposed utilizing nonreciprocally coupled rib–waveguides. They rely on the nonreciprocal phase shift of TM modes which can be enhanced using double layers with opposite Faraday rotation. Detailed calculations yield the required geometry and material parameters. Furthermore, it is demonstrated that optical modulators can be realized using dynamical mode conversion induced by the ferrimagnetic resonance (FMR) and by the resonances of a stripe domain lattice: the domain wall resonance (DWR) and the two branches of the domain resonances (DR$^\pm$). For FMR the maximum precession angles depend on the saturation magnetization M_s. Precession cones of more than 80° can be obtained by local excitation. Preliminary conversion efficiencies of 6% are achieved. For DR$^\pm$ and DWR an improved version of the hybridization model is developed. It is in excellent agreement with experiments using [111] and [110] oriented films, if the quality factor exceeds 0.5. Dynamical conversion efficiencies of 18% at frequencies up to 3 GHz are measured. Finally, dynamic Bragg scattering by domain resonances is discussed.

R. Marcelli and S.A. Nikitov (eds.), Nonlinear Microwave Signal Processing: Towards a New Range of Devices, 411–465.
© 1996 *Kluwer Academic Publishers.*

Contents

1. Introduction

2. Material Properties

   2.1 Epitaxial films

   2.2 Magnetic anisotropy

   2.3 Optical anisotropy

   2.4 Film properties

3. Optical Isolators and Circulators

   3.1 Nonreciprocity

   3.2 Nonreciprocal effects in waveguides

   3.3 Isolator concepts

      3.3.1 Nonreciprocal mode conversion

      3.3.2 Nonreciprocal mode propagation

      3.3.3 Cutoff isolator

      3.3.4 Nonreciprocal interferometry

      3.3.5 Nonreciprocal coupling

   3.4 Optimizing the nonreciprocal phase shifter

      3.4.1 High refractive index top layer

      3.4.2 Double layer with opposite Faraday rotation

   3.5 Computational techniques

4. Modulators and Deflectors

   4.1 Principle of modulation

   4.2 Modulation by ferrimagnetic resonance

   4.3 Modulation by domain resonances

      4.3.1 Calculation of domain resonances

      4.3.2 Dynamic mode conversion

   4.4 Dynamic deflection by domain resonances

5. Conclusions

1. Introduction

Optical communication via glass fibers is developed in the near infrared region between 1.3 and 1.55 μm wavelength. Integrated optical circuits are required for all kinds of information processing. Lasers are developed as light sources on the base of III/V semiconductors. To protect the lasers from reflected light, optical isolators are required. Presently magnetooptical garnet films are the only materials studied to realize nonreciprocal integrated components like isolators and circulators. However, due to the large difference of refractive indices between garnets and III/V semiconductors, it is unlikely that both materials can be integrated on a common substrate.

On the other hand, magnetic garnets represent a class of materials perfectly suited to realize all–garnet integrated optical circuits. The garnets exhibit low optical damping combined with strong Faraday rotation. Thin garnet films produced by liquid phase or sputter epitaxy serve as dielectric waveguides. The material parameters can be adjusted in a wide range by controlling the chemical composition and preparation conditions [1]. Furthermore, modulators, deflectors, and switches can be made as well [2], and even the realization of lasers in magnetic garnet films seems feasible.

2. Material Properties

2.1. EPITAXIAL FILMS

Magnetic garnets discussed in this article are basically derived from yttrium-iron garnet (YIG):

$$\{Y\}_3 [Fe]_2 (Fe)_3 O_{12},$$

where $\{\}$, $[\ ]$, and $()$ denote the dodecahedral, octahedral and tetrahedral lattice sites, respectively. The iron and yttrium ions can be replaced by other rare earth or transition metal ions to optimize material properties. E. g. substituting yttrium by bismuth [3, 4] or cerium [5] results in a strong increase of the Faraday rotation, or substituting iron by diamagnetic gallium or aluminium reduces the effective saturation magnetization M_s [6]. At a substitution level of about 1.3 formula units (f.u.) a magnetic compensation of the tetrahedral and octahedral sublattices is obtained.

Thin films of these materials can be grown by liquid phase (LPE) [7] or sputter epitaxy [8] on substrates of gadolinium gallium garnet (GGG). The usual orientation of the film normal is the crystallographic [111] direction, but other orientations like [100], [110], [211] or [210] are used as well. For the two film orientations discussed in this article, the coordinate systems defined to describe material properties are sketched in Fig. 2.1. Two systems are used. The system (x_c, y_c, z_c) is parallel to crystallographic axes as

414

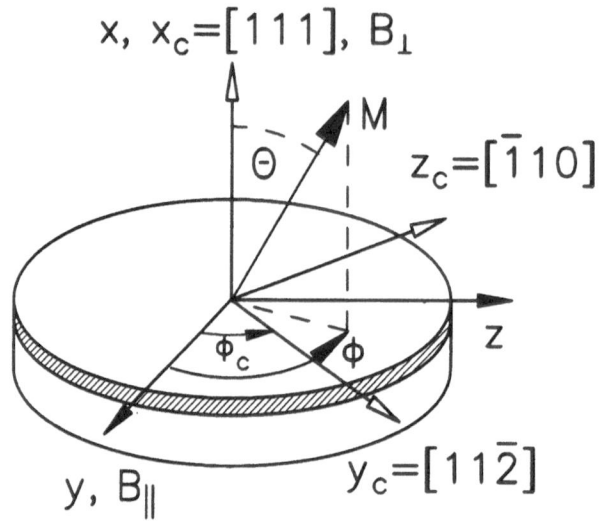

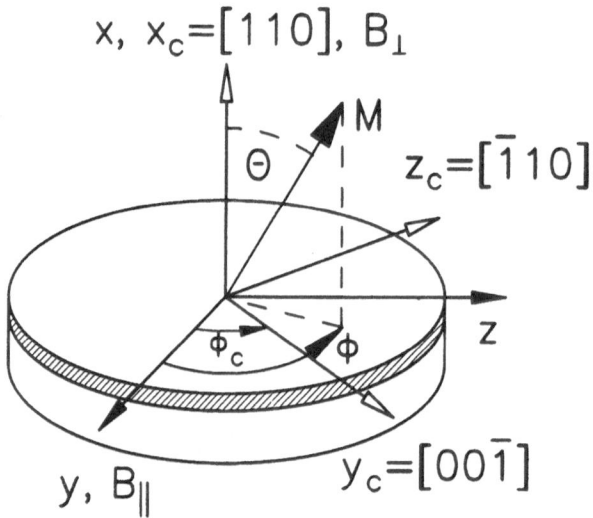

Figure 2.1. Coordinate systems used to describe [111] and [110] oriented films.

defined in Fig. 2.1; propagation of optical modes is always assumed to be parallel to $z_c = [\bar{1}10]$. The system (x, y, z) is used to describe the magnetic resonances and the domain structure. It is rotated about the film normal through the angle Φ_c with respect to the (x_c, y_c, z_c) system, where the film normal is parallel to the x– and x_c–directions. The static in–plane induction $B_\|$ and thus the stripe domains are parallel to the y–direction.

The mismatch between the lattice constants of film and substrate causes an elastic stress of the film. The lattice mismatch is denoted by:

$$\Delta a^\perp = a_S - a_F^\perp ,$$

where a_S is the lattice constant of the substrate and $a_F^\perp$ that of the film, measured along the film normal. The stress induces magnetic and optical anisotropies and thus reduces the symmetry of the garnet film which is originally cubic. Additional anisotropies are induced by the growth process. They have the same symmetry as the stress induced anisotropies and arise if more than one ion species is incorporated on the same class of lattice sites. In this case, ions distribute non-statistically which gives rise to an additional anisotropy [9]. The stress and growth induced contributions to the magnetic and optical anisotropy are denoted by superscripts S and G, respectively. They add up to the total magnetic or optical anisotropy.

2.2. MAGNETIC ANISOTROPY

The equilibrium position of the magnetization M is determined by the minimum of the free energy, consisting of Zeeman, demagnetizing, and anisotropy energies. The energy density of the original cubic anisotropy is given by:

$$F_c = K_1 \left(\alpha_1 \alpha_2 + \alpha_2 \alpha_3 + \alpha_3 \alpha_1 \right),$$

where the α_i are the components of the magnetization along the cubic crystallographic axes and K_1 is the cubic anisotropy constant, $-600 \ J/m^3$ for YIG at room temperature [10].

The induced anisotropy energies are:

$$F_{ind} = \left[K_u + K_o \sin^2 (\Phi - \Phi_c) \right] \sin^2 \Theta .$$

The angle $\Phi - \Phi_c$ denotes the azimuth of the magnetization M with respect to crystallographic axes as shown in Fig. 2.1 and Θ the polar angle of M with respect to the film normal; K_u is the uniaxial and K_o the in–plane, orthorhombic, anisotropy constant. For [111] oriented films K_o vanishes.

The stress induced contributions to the anisotropy constants are:

$$K_u^S = \frac{\Delta a^\perp}{a_S} B_2 \qquad \text{for } [111],$$

$$K_u^S = 0.5 \frac{\Delta a^\perp}{a_S} (B_2 + B_1) \qquad \text{for } [110],$$

$$K_o^S = 0.5 \frac{\Delta a^\perp}{a_S} (B_2 - B_1) \qquad \text{for } [110],$$

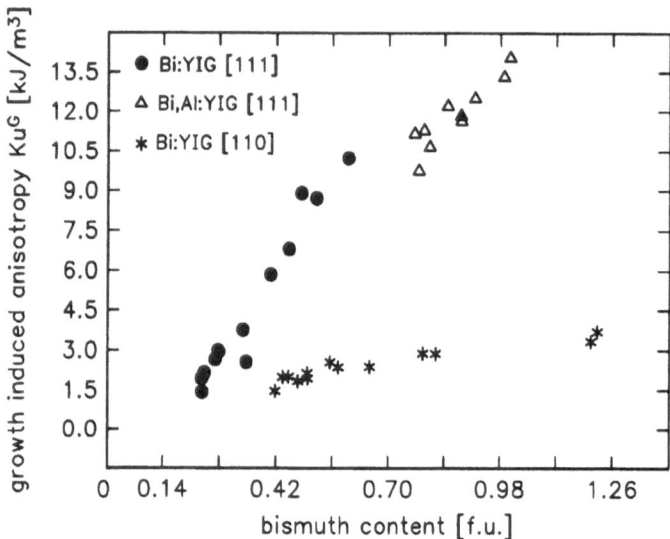

Figure 2.2. Growth induced anisotropy constant K_u^G vs bismuth content for bismuth and aluminum substituted films of [111] and [110] orientation.

where the B_i are the respective magnetoelastic constants. The growth induced contributions increase with the bismuth content of the films [11]; some further examples are shown in Fig. 2.2.

2.3. OPTICAL ANISOTROPY

The propagation of optical modes in the garnet waveguides is mainly determined by the permittivity tensor ϵ of the garnet film. As optical absorption is very small in the near infrared region, ϵ is assumed hermitean. It can be split into the following terms:

$$\epsilon = \epsilon^{IS} + \epsilon^S + \epsilon^G + \epsilon^{CB} + \epsilon^{LB} ,$$

where $\epsilon_{ij}^{IS} = n^2 \delta_{ij}$ is the isotropic contribution and n the refractive index of a bulk crystal of the same chemical composition as the film. ϵ^S and ϵ^G are the stress and growth induced optical anisotropies, respectively. For [111] oriented films they contain only diagonal components, where $\epsilon_{y_c y_c}^{S,G} = \epsilon_{z_c z_c}^{S,G}$. The difference $\epsilon_{x_c x_c}^{S,G} - \epsilon_{y_c y_c}^{S,G}$ is defined as the perpendicular anisotropy $\Delta\epsilon_\perp^{S,G}$. The stress induced contribution is related to the lattice mismatch by:

$$\Delta\epsilon_\perp^S = 2\,n^4\,P_{44}\,\frac{\Delta a^\perp}{a_S} .$$

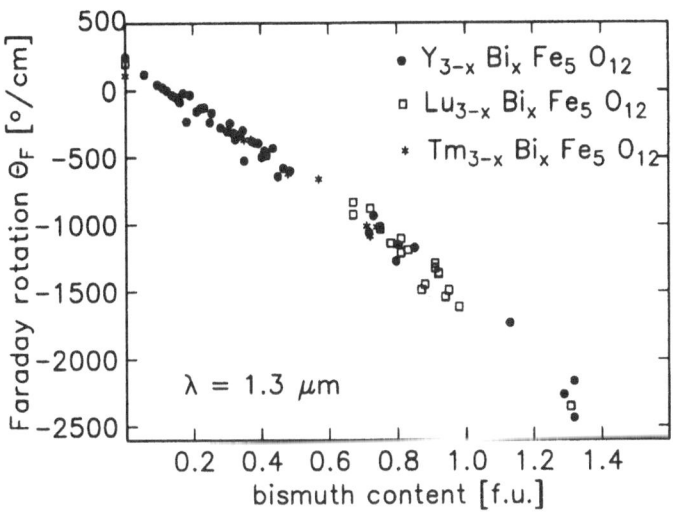

Figure 2.3. Specific Faraday rotation of yttrium, lutetium and thulium iron garnets vs bismuth content.

For [110] oriented films an additional in–plane anisotropy arises:

$$\Delta\epsilon_{\parallel}^{S} = \epsilon_{[00\bar{1}]}^{S} - \epsilon_{[\bar{1}10]}^{S} = 0.5\,n^4\,\Delta P\,\frac{\Delta a^{\perp}}{a_S}$$

with $\Delta P = P_{11} - P_{21} - 2\,P_{44}$, where the P_{ij} are the respective photoelastic constants. The growth induced optical anisotropies increase with the bismuth content [12, 13].

The term ϵ^{CB} is caused by the circular magnetic birefringence and gives rise to the Faraday rotation; it is described by an antisymmetric hermitean tensor

$$\epsilon^{CB} = i\,K\begin{pmatrix} 0 & M_z & -M_y \\ -M_z & 0 & M_x \\ M_y & -M_x & 0 \end{pmatrix}.$$

M_j denotes the components of the magnetization **M**. The constant K is approximately

$$K \approx \frac{n\,\lambda_0\,\Theta_F}{\pi\,M},$$

where λ_0 is the vacuum wavelength and Θ_F the specific Faraday rotation. Bismuth and cerium substitution cause a strong increase of the Faraday rotation [3, 4, 5]. This behaviour is demonstrated in Fig. 2.3 for bismuth substituted yttrium, lutetium and thulium iron garnets at the infrared wavelength of 1.3 μm.

The contribution ϵ^{LB} denotes the linear magnetic birefringence which leads to the Cotton–Mouton effect. It is a symmetric hermitean tensor bilinear in the magnetization:

$$\epsilon_{ij}^{LB} = \sum_{kl} G_{ijkl}\, M_k\, M_l\,.$$

The tensor G of rank four is determined by the three constants G_{11}, G_{12} and G_{44} with $\Delta G = G_{11} - G_{12} - 2\,G_{44}$. It has to be transformed into the film coordinate system $(x_c, y_c z_c)$. For [111] orientation the tensor elements read:

$$\epsilon_{x_c x_c}^{LB} = \frac{1}{6}\Delta G\, M^2 + 2\,G_{44}\, M_{x_c}^2\,,$$

$$\epsilon_{y_c y_c}^{LB} = 2\,G_{44}\, M_{y_c}^2 + \frac{1}{6}\Delta G\,(M_{x_c}^2 + 2\,M_{y_c}^2) - \frac{\sqrt{2}}{3}\Delta G\, M_{x_c}\, M_{y_c}\,,$$

$$\epsilon_{z_c z_c}^{LB} = 2\,G_{44}\, M_{z_c}^2 + \frac{1}{6}\Delta G\,(M_{x_c}^2 + 2\,M_{z_c}^2) + \frac{\sqrt{2}}{3}\Delta G\, M_{x_c}\, M_{z_c}\,,$$

$$\epsilon_{y_c z_c}^{LB} = (2\,G_{44} + \frac{1}{3}\Delta G)\, M_{y_c}\, M_{z_c} + \frac{\sqrt{2}}{3}\Delta G\, M_{x_c}\, M_{z_c}\,,$$

$$\epsilon_{x_c z_c}^{LB} = (2\,G_{44} + \frac{2}{3}\Delta G)\, M_{x_c}\, M_{z_c} + \frac{\sqrt{2}}{3}\Delta G\, M_{y_c}\, M_{z_c}\,,$$

$$\epsilon_{x_c y_c}^{LB} = (2\,G_{44} + \frac{2}{3}\Delta G)\, M_{x_c}\, M_{y_c} + \frac{\sqrt{2}}{6}\Delta G\,(M_{z_c}^2 - M_{y_c}^2)\,.$$

An additional isotropic term $(G_{12} + \Delta G/6)\, M^2$ has been omitted; it can be included in the isotropic refractive index n of the garnet film. The tensor components ϵ_{ij}^{LB} for [110] oriented films will not be reproduced here.

2.4. FILM PROPERTIES

The magnetic garnet films discussed in this article have the composition $Y_{3-x}Bi_x Fe_{5-y-z}Al_y Ga_z O_{12}$. They are grown by liquid phase epitaxy on [111] and [110] oriented substrates of gadolinium gallium garnet.

The compositions and material parameters are listed in Table 2.1. The film thickness is denoted by d, the period of the stripe lattice at zero induction by $2\,L_0$. The anisotropy constants K_u and K_o are derived from FMR measurements. The saturation magnetization M_s and the exchange constant A are determined by fitting these parameters to the period $2\,L$ of the stripe lattice as function of the in–plane induction $B_{\parallel}$. The optical absorption constant χ is derived from the exponential decay of the intensity of propagating modes.

TABLE 2.1. Material parameters of investigated garnet films
$Y_{3-x}Bi_xFe_{5-y-z}Al_yGa_zO_{12}$

film no.	1	2	3	4	5	6	7	8	unit
orient.	[111]	[111]	[111]	[111]	[110]	[111]	[111]	[111]	
x	0	0	0	0.96	0.70	0.45	0.34	0.31	
y	0	0	0	0	0.75	0.49	0.37	0.35	
z	1.1	1.25	0	0	0	0	0	0	
d	6.8	5.4	5.3	7.4	5.2	15.4	4.2	4.8	μm
$2 L_0$	27.0			3.2	$3.6^a/3.8^b$	5.1	2.5	2.6	μm
M_s	17	5	143	158	68	112	112	110	10^3 A/m
K_u	-0.1	0.03		14.8	4.1	5.5	3.5	3.1	kJ/m^3
K_o					1.65				kJ/m^3
A				9.2	3.6	9.3	11.3	10.1	10^{-12} J/m
Q	-0.58	1.8		0.94	1.42	0.70	0.45	0.41	
χ^*	0.8					1.5	1.7	2.1	cm^{-1}
$\Theta_F{}^*$	200					720	330	242	degree/cm
$M_s^2 G_{44}{}^*$	24					-6.4	-6.3	-5.9	10^{-5}
$M_s^2 \Delta G^*$	-7					2.5	2.5	2.5	10^{-5}
$\epsilon_{xx}{}^*$	4.538					5.064	4.956	4.954	
$\Delta\epsilon_\perp{}^*$	1.27					10.4	1.52	1.05	10^{-3}

*: all optical data are measured at a wavelength of 1.3 μm.
a: stripes parallel [00$\bar{1}$], b: stripes parallel [$\bar{1}$10].

3. Optical Isolators and Circulators

In this section we present the reciprocity theorem and elaborate on the privileged role of magnetooptic materials. Optical isolators or circulators are based either on nonreciprocal mode conversion or on nonreciprocal mode propagation. The isolator is a nonreciprocal device transmitting light in one direction while blocking it in reverse direction. A circulator does not block the reflected signal but redirects it into a different optical channel. Various concepts of such devices are discussed. We focus on the optimization of nonreciprocal phase shifters. A few remarks on numerical methods for simulating nonreciprocal devices conclude this section.

Here we will use a Cartesian coordinate system where the waveguide normal is the x-axis, modes propagate along z, and y runs in lateral direction.

3.1. NONRECIPROCITY

There is a domain of applications reserved for magnetooptic materials: they allow for nonreciprocal propagation of light. Magnetooptic materials defy the reciprocity theorem.

Consider a passive linear device with ports $1, 2, \ldots$. Denote by a_i the amplitude of an incoming wave at port i, by b_i the amplitude of an outgoing wave. The superposition principle demands that $b_i = \sum_k S_{ik} a_k$ hold, where S is the scattering matrix. If the permittivity tensor ϵ_{ab} is hermitean everywhere, the scattering matrix turns out to be unitary. If the permittivity tensor is also symmetric, the scattering matrix is symmetric likewise. The latter statement is one of many formulations of the reciprocity theorem [14].

The permittivity tensor depends on the angular frequency ω of the electromagnetic field. It also depends on all parameters which characterize the thermal equilibrium state of matter, such as temperature T, pressure p (or tension), and quasi-static electric and magnetic fields, $\bar{E}$ and $\bar{H}$, respectively. The Onsager relations, as applied to the permittivity tensor, require

$$\epsilon_{ab}(\omega, T, p, \bar{E}, \bar{H}) = \epsilon_{ba}(\omega, T, p, \bar{E}, -\bar{H}) \quad .$$

If the permittivity tensor does not depend on the applied quasi-static magnetic field strength (or magnetization), it will be symmetric, and the reciprocity theorem is a consequence.

In order to allow for nonreciprocal effects, the assumptions of the reciprocity theorem (passive, linear, symmetric permittivity) have to be renounced.

Nonreciprocal behaviour can be achieved either with active devices, with nonlinear media (e.g. employing the photorefractive effect), or with magnetooptic materials. Today, the first two possibilities are ruled out for reasons of speed, or because the nonlinearities are too small. An optical isolator or circulator requires a material the optical properties of which depend on the magnetization in odd order.

For cubic crystals, such as garnets, the first order magnetooptic (Faraday) effect is described by the following contribution to the permittivity tensor:

$$\delta\epsilon_{ab} = \epsilon_{ab}^{CB} = iK \sum_c \epsilon_{abc} M_c \quad , \tag{3.1}$$

where ϵ_{abc} is the totally antisymmetric Levi-Cività tensor and M_c are the components of the magnetization. In this section we shall assume that the circular magnetic birefringence is the only deviation from optical isotropy.

Unfortunately, the Faraday effect is rather small. $\xi = KM = 10^{-4}$ is a realistic, 10^{-3} a good, and 10^{-2} an optimistic value. Therefore, nonreciprocal effects have to be optimized by clever design and multiplied by the principle of repetition, in particular interferometry.

With this remark we hint at the difficulties of realizing practical devices: not only the nonreciprocal effect, but also the inaccuracies are amplified. Most concepts for nonreciprocal devices suffer from unacceptably high demands on fabrication tolerances.

3.2. NONRECIPROCAL EFFECTS IN WAVEGUIDES

Let us discuss slab waveguides first. With a diagonal permittivity tensor, modes are either TE or TM polarized. The electrical field for TE modes is $(0, E_y, 0)$, and $(E_x, 0, E_z)$ for TM modes, with $|E_z|$ much smaller than $|E_x|$ for ordinary waveguides.

With lateral magnetization $(0, M, 0)$ the off-diagonal components $\epsilon_{xz} = -i\xi$ and $\epsilon_{zx} = i\xi$ will not vanish. Therefore, TE modes are insensitive to this magnetooptic effect while TM modes are influenced. The eigenvalue equation for the propagation constant β of a guided mode now contains, besides the β^2 term, a $\beta\xi$ contribution. Hence, for fixed magnetization, $\beta \to -\beta$ is no longer a symmetry [15]. Instead, to every positive propagation constant β_+, there is a negative propagation constant $\beta_- = -(\beta_+ + \delta\beta)$. Propagation in forward and backward direction differ slightly.

With longitudinal magnetization $(0, 0, M)$ the off-diagonal components $\epsilon_{xy} = i\xi$ and $\epsilon_{yx} = -i\xi$ will not vanish. Consequently, TE and TM modes couple, although weakly.

Both effects, nonreciprocal propagation of TM modes and nonreciprocal coupling, occur simultaneously if the magnetization is neither lateral nor longitudinal, but tilted. An interesting isolator concept relies on this possibility [16].

The guided modes of channel waveguides are neither TE nor TM polarized, but hybrid. It turns out, however, that putting either $E_x = 0$ (quasi-TE) or $H_x = 0$ (quasi-TM) is a very good approximation in most cases. In channel waveguides with lateral magnetization, quasi-TM modes have different propagation constants in forward and backward direction, while quasi-TE modes are blind to the Faraday effect. Longitudinally polarized channel waveguides couple quasi-TE and quasi-TM modes. The quasi-TE/TM approximation fails if the propagation constants of differently polarized modes coincide.

3.3. ISOLATOR CONCEPTS

The standard microoptical (bulk) isolator contains two polarizers the principal axes of which are rotated by 45°, and a Faraday rotator between them. This optical element rotates the polarization vector by 45° clockwise with respect to the direction of light propagation. Light in forward direction passes the first polarizer, is rotated by 45°, and passes the second polarizer without losses. Reflected light which passes the second polarizer is rotated also clockwise by 45° and will be blocked by the first polarizer.

3.3.1. *Nonreciprocal mode conversion*
The waveguide analagon of nonreciprocal polarization rotation is nonreciprocal TE/TM conversion in a longitudinally magnetized waveguide.

The nonreciprocal conversion efficiency is characterized by

$$\eta = \frac{\Theta_F^2}{\Theta_F^2 + (\Delta\beta/2)^2} \sin^2\left\{ L\sqrt{\Theta_F^2 + (\Delta\beta/2)^2} \right\} \quad . \tag{3.2}$$

Here $\Theta_F = k_0\xi/2n$ is the specific Faraday rotation of a medium with refractive index n and off-diagonal permittivity $\xi = KM$, as defined above. $\Delta\beta = \beta^{TE} - \beta^{TM}$ denotes the difference of TE and TM propagation constants, or phase mismatch. The phase mismatch, if compared with the specific Faraday rotation, must be very small in order to guarantee good nonreciprocal conversion.

Unfortunately, the propagation constants of a TE and its corresponding TM mode differ, even if the waveguide is optically isotropic. The so called form birefringence, caused by different equations for TE and TM modes, must be compensated. Several techniques have been proposed: application of stress [17], growth induced birefringence [13], multilayered structure [18], anisotropic top layer [19], periodic reversal of magnetization [20], Dammann's geometry [21], to mention just a few. As of today, the precision requirement $|\Delta\beta| \ll \Theta_F$ could not be met.

Phase mismatch is not the only weak point of nonreciprocal TE/TM conversion isolators. Most designs demand a 50% nonreciprocal mode conversion followed by a 50% reciprocal mode conversion, which poses another accuracy problem.

Moreover, the counterpart of polarizers, i.e. TM or TE mode absorbers, have to be realized. In planar or broad channel waveguides TM modes spread out wider than TE modes. A properly positioned absorber may well catch all the energy of the TM mode while ignoring the TE mode. After all, fields decrease exponentially outside the waveguide proper. Again, the absorber must be positioned very precisely. More complicated TE/TM discriminators will certainly introduce additional complications.

3.3.2. *Nonreciprocal mode propagation*
As explained above, TE modes do not sense a lateral magnetization. TM modes, however, propagate differently in forward and backward direction.

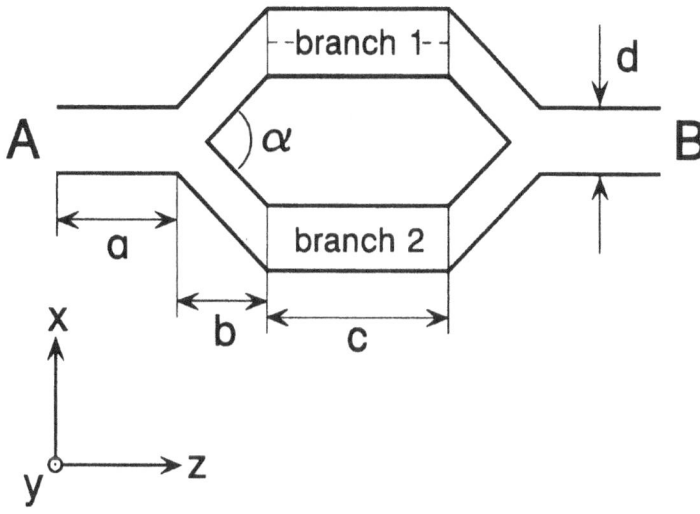

Figure 3.1. Basic design of the nonreciprocal Mach-Zehnder interferometer. The nonreciprocal branch 1 consists of two films with the same refractive index n_{b1} and thickness $d/2$, but opposite sign of the specific Faraday rotation Θ_F. The reciprocal branch is made of a nonmagnetic material with refractive index n_{b2}. Other nonmagnetic guiding and nonguiding regions of the structure have refractive indices n_f and n_s, respectively.

$\delta\beta$, the difference between forward and backward propagation constant of a TM-mode, is proportional to the specific Faraday rotation Θ_F.

3.3.3. *Cut-off isolator*

A magnetooptic waveguide with lateral magnetization may just guide a TM mode in forward and none in backward direction. This cut-off isolator [22] would be the simplest possible if the specific Faraday rotation were much larger. However, with todays garnets, a reflected wave must travel much too long a distance until it is completely radiated off.

3.3.4. *Nonreciprocal interferometry*

Already 20 years ago an isolator was proposed [23] which relied on nonreciprocal interferometry. Two Y-junctions are connected by a normal and an magnetooptic waveguide which acts as a nonreciprocal phase shifter, as depicted in Fig. 3.1.

If the device is well tuned, then light entering at A will leave the device at B, while a reflected signal entering at B will be radiated off at Y-junction A. Figure 3.2 illustrates the effect. Note the unrealistically large specific Faraday rotation constant.

Although such an isolator works [24, 25, 26], the accuracy requirements for a practical device are too high. The isolator effect is based on construc-

424

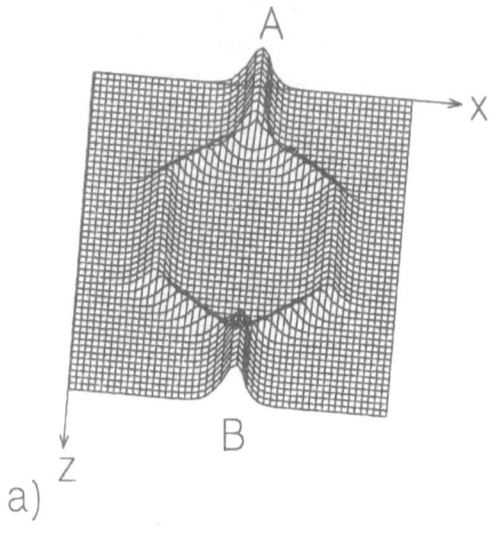

a)

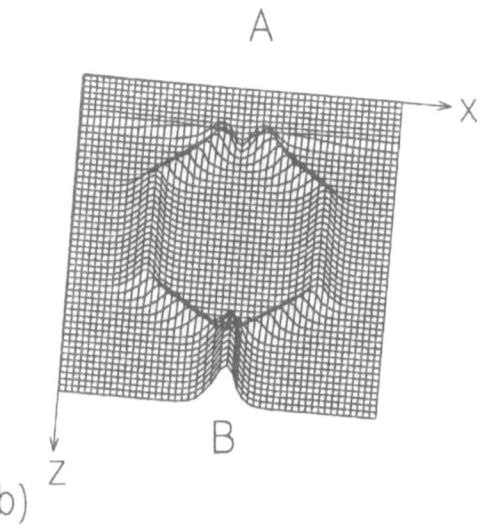

b)

Figure 3.2. Simulation of the nonreciprocal Mach-Zehnder interferometer. a) forward direction - input at A, b) backward direction - input at B. Parameters of the structure: $\lambda = 1.3\,\mu\text{m}$, $a = b = 50\,\mu\text{m}$, $c = 100\,\mu\text{m}$, $d = 0.5\,\mu\text{m}$, $\alpha = 3.0°$, $n_s = 1.95, n_{b1} = n_f = 2.30, n_{b2} = 2.29574, \Theta_F = 32000°/\text{cm}$. The large specific Faraday rotation was assumed for better presentation.

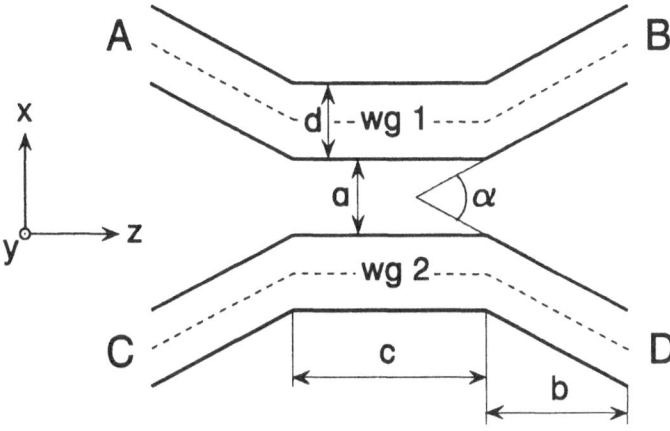

Figure 3.3. Basic design of the nonreciprocal coupler. Two coupled waveguides consist of two magnetooptic films with the same refractive index n_f and thickness $d/2$, but opposite sign of the specific Faraday rotation Θ_F. The refractive indices of the nonmagnetic layers between and outside the waveguides are n_s and n_c, respectively.

tive and destructive interference. The degree of isolation depends critically on the symmetry of the Y-junctions and on the form of the wave fronts (symmetry of the waveguides).

3.3.5. *Nonreciprocal coupling*

Recently we suggested a magnetooptic coupler with laterally magnetized waveguides [27]. The operation principle is well known for the microwave counterpart [28]. The coupling length of such a device is different for forward and backward propagation. The device should be designed such that there is an even number of couplings in forward and an odd number in backward direction. Figures 3.3 and 3.4 illustrate the principle. Again, note the unrealistically large specific Faraday rotation constant.

Coupling between two adjacent stripes is achieved by an overlap of modes of the individual waveguides. Since modes fall off exponentially outside the guiding structure, the coupling coefficient depends exponentially on the distance between the two stripes. Exactness and constancy of the distance is a strong requirement.

The number of beats required—such that it is even in forward and odd in backward direction—does not depend on the coupling coefficient, but on a geometrical factor multiplied by $|\xi| = |\epsilon_{xz}|$. Typical values, with todays garnets, are 50 to 100. Hence, the length of the coupling region is another critical parameter.

We propose a new concept for an isolator based on nonreciproal radia-

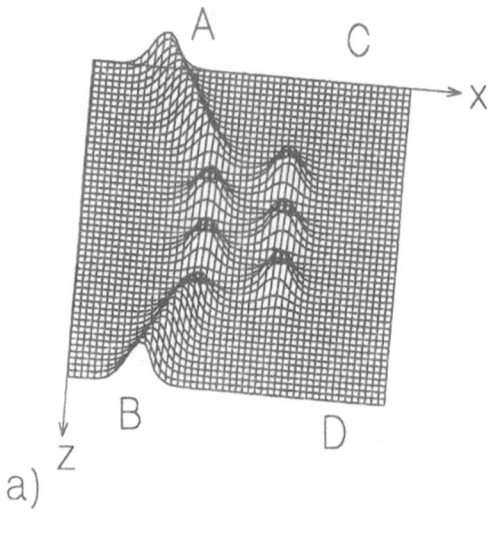

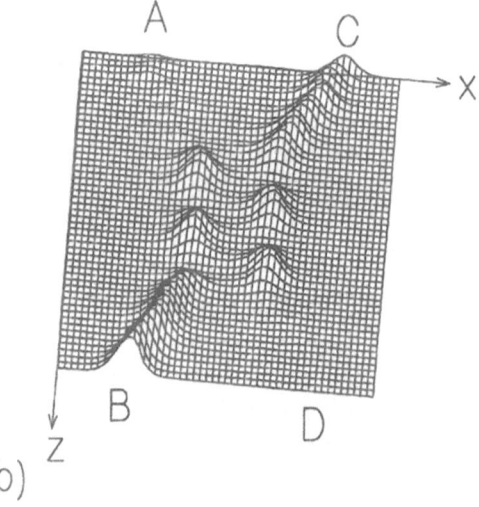

Figure 3.4. Simulation of the nonreciprocal coupler. a) forward direction - input at A, b) backward direction - input at B. Parameters of the structure are $\lambda = 1.3\,\mu\mathrm{m}$, $a = 0.6\,\mu\mathrm{m}$, $b = 335\,\mu\mathrm{m}$, $c = 500\,\mu\mathrm{m}$, $d = 0.6\,\mu\mathrm{m}$, $\alpha = 0.4°$, $n_f = 2.30$, $\Theta_F = 60200°/\mathrm{cm}$, $n_s = 1.98$, $n_c = 1.95$. The large specific Faraday rotation was assumed for better presentation.

tive coupling [29]. There is a medium of high refractive index between the waveguides, higher than the base mode effective index. Thereby, coupling is achieved by radiation modes the intensity of which is independent on the distance between the guides. Preliminary calculations show that this parameter is much less critical now.

3.4. OPTIMIZING THE NONRECIPROCAL PHASE SHIFTER

Isolators based on nonreciprocal coupling or nonreciprocal interferometry rely on a nonreciprocal phase shifter, a transversely magnetized waveguide. The nonreciprocal differential phase shift can be calculated by perturbation theory to be

$$\delta\beta = \frac{\int dx dy \, |H_y|^2 \partial(\xi/\epsilon^2)/\partial x}{\int dx dy \, |H_y|^2/\epsilon} \quad . \tag{3.3}$$

The magnetic field should be peaked where the Faraday constant changes. Note that the contributions of a magnetized film tend to cancel unless the magnetic field strength is different at the upper and lower interface. There are two strategies to optimize $\delta\beta$.

3.4.1. *High refractive index top layer*
Okamura et al. [30] suggested a four layer structure as shown in figure 3.5. It consists of a substrate, a magnetized film, and a high refractive index rib waveguide on top of it. The device is dimensioned such that the magnetic field is maximal at the ZnTe/YIG interface and already very small at the YIG/GGG interface. In this way, by concentrating $|H_y|^2$ at one surface, the differential phase shift could be enhanced by a factor of 2.5 with respect to an optimized YIG rib on top a GGG substrate [31].

3.4.2. *Double layer with opposite Faraday rotation*
The Faraday effect in garnets is the result of two oppositely magnetized sublattices. By modifying the composition, a negative or positive factor K in (3.1) may be achieved [32]. Hence it is possible to produce double layers with negative ξ in the upper and positive ξ in the lower layer [33]. In this way the contributions to (3.3) add up at the interface of the two layers. Compared with a single layer planar waveguide, an optimized double layer rib waveguide [15] can provide an improvement in $\delta\beta$ by a factor of 2.5 as shown in Fig. 3.6.

3.5. COMPUTATIONAL TECHNIQUES

The magnetooptic contribution to the permittivity tensor is small, unfortunately. It may be treated as a perturbation [35]. Therefore, any technique for

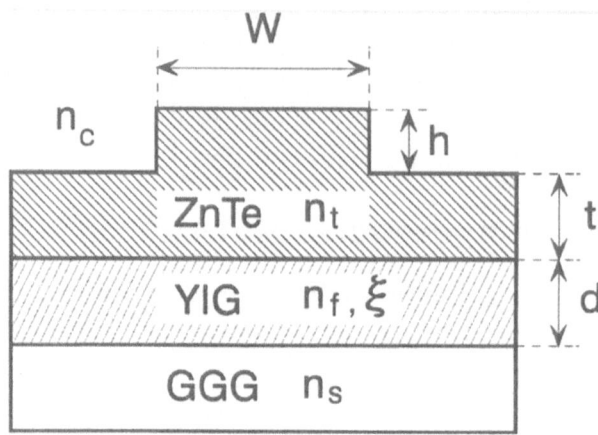

Figure 3.5. Basic geometry of a nonreciprocal phase shifter with a high-refractive-index top layer. Note the high value of dielectric constant, $n_t^2 = 7.64$, for ZnTe.

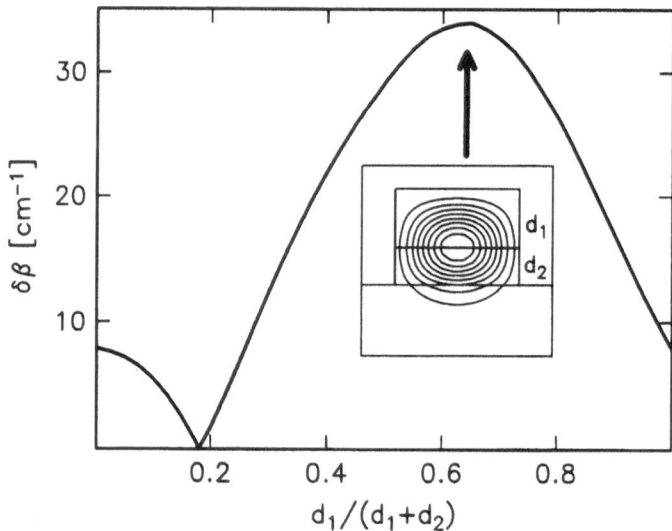

Figure 3.6. Dependence of the nonreciprocal phase shift $\delta\beta$ on the position of the interface between two gyrotropic layers with opposite sign of Faraday rotation. The rib, 0.61 μm high and 0.87 μm wide, supports only one quasi-TM mode. It is plotted in the actual computational window. Parameters: $\lambda = 1.3$ μm, permittivities $\epsilon_{cov} = 1$, $\epsilon_{rib} = 5.3$, $\epsilon_{sub} = 3.8$, $\xi = \pm 0.005$.

computing guided modes (fields and propagation constant) of an isotropic waveguide will serve the purpose.

We have employed and adapted some methods for directly computing magnetooptic waveguides, particularly with lateral magnetization. The methods of finite differences, effective index, spectral index, Galerkin and finite elements are described, applied, and compared in [15], [34] - [39]. To simulate the light propagation in real nonreciprocal integrated optical devices, a beam propagation method (BPM) is required. In general, the BPM calculations of 3D magnetooptic waveguide structures are involved and demand a large amount of computer resources [34]. However, slab structures can be treated in a relatively simple manner [34, 40].

1. Modulators and Deflectors

4.1. PRINCIPLE OF MODULATION

The modulators dicussed in this section rely on dynamic conversion of optical modes propagating in the waveguiding garnet film. This principle is sketched in Fig. 4.1. An incoming TE mode is partially converted to a TM mode. The modulated conversion depends on the dynamic components of the magnetization M excited by an rf magnetic field.

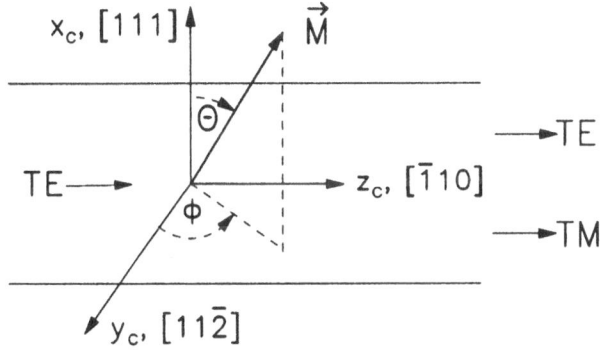

Figure 4.1. Basic geometry of dynamical mode conversion.

To calculate mode conversion, perturbation theory is used as described by Yamamoto et al. [41]. This procedure is justified, because all off–diagonal components of the permittivity tensor are small compared to the square of the isotropic refractive index n. The normal TE and TM modes of the planar waveguide are assumed to propagate along the z_c or [$\bar{1}$10] direction in the film; they are determined by the diagonal components of the permittivity tensor. The normalized electric fields of the optical modes read [41]:

$$\mathbf{e}^{TE} \;=\; \left\{0,\, e_{y_c}^{TE}(x_c),\, 0\right\} \exp\left[i\left(\omega t - \beta^{TE} z_c\right)\right]$$

$$\mathbf{e}^{TM} \;=\; \left\{e_{x_c}^{TM}(x_c),\, 0,\, e_{z_c}^{TM}(x_c)\right\} \exp\left[i\left(\omega t - \beta^{TM} z_c\right)\right],$$

where β^{TE} and β^{TM} are the propagation constants, respectively, and $\Delta\beta = \beta^{TE} - \beta^{TM}$ denotes the phase mismatch.

The off–diagonal components $\epsilon_{x_c y_c}$ and $\epsilon_{y_c z_c}$ cause coupling of the TE and TM modes, while $\epsilon_{x_c z_c}$ yields a phase shift of TM modes. Assuming a TE mode as input mode of intensity I^{TE}, the conversion efficiency η to a TM mode of intensity I^{TM} is [42]:

$$\eta \;=\; \frac{I^{TM}(z_c = z_{cl})}{I^{TE}(z_c = 0)} \;=\; \frac{|\kappa|^2}{|\kappa|^2 + \Delta\beta^2/4} \cdot \sin^2\left(z_{cl}\sqrt{|\kappa|^2 + \Delta\beta^2/4}\right) \quad (4.1)$$

where z_{cl} is the coupling length; the coupling constant κ is determined by the overlap of the fields:

$$\kappa = \omega\,\epsilon_0 \int_{-\infty}^{\infty} \left(\epsilon_{x_c y_c}^{*}\, e_{y_c}^{TE}\, e_{x_c}^{TM} + \epsilon_{y_c z_c}^{*}\, e_{y_c}^{TE}\, e_{z_c}^{TM}\right)\, dx_c\,.$$

* denotes complex conjugate. The maximum of the conversion is obtained at a coupling length

$$z_{cl,max} = \pi / \left(2\sqrt{|\kappa|^2 + \Delta\beta^2/4}\right).$$

To achieve a large conversion efficiency the following conditions must be satisfied.

i) The phase mismatch $\Delta\beta$ has to be as small as possible. This can be accomplished by adjusting the stress and growth induced contributions to the optical anisotropy $\epsilon^{S,G}$ or by adjusting the thickness of the waveguide.

ii) The coupling constant κ must be large. As in most cases the coupling is caused by the Faraday rotation, this parameter must be increased e. g. by bismuth substitution. Furthermore, coupling is stronger by the tensor element $\epsilon_{x_c y_c}$, which couples the strong transverse field components, than by $\epsilon_{y_c z_c}$ which couples a transverse field component to a small longitudinal one.

To determine the mode conversion, all elements of the permittivity tensor must be known. Since they depend on the static and dynamic components of the magnetization, these components have to be calculated first.

Various rf excitations of the magnetization $\mathbf{M}$ of garnet films have been used to obtain dynamic TE–TM mode coupling. Furthermore, if the dynamic magnetization has a spatially periodic structure, concurrently to the modulation a dynamic deflection by Bragg scattering can be achieved.

Dynamic TE–TM mode conversion by magnetostatic volume waves has been demonstrated by Fisher et al. [43], by Tamada et al. [44], by Talisa [45] and by Tsai et al. [46, 47, 48]. Also magnetostatic surface waves can be applied as shown by Tsai et al. [49]. These methods are applicable to cause deflection of the converted mode by Bragg scattering. Neite et al. [50] used the ferrimagnetic resonance (FMR) to obtain dynamic conversion, but no deflection is possible in this case.

Magnetic garnet films support periodic lattices of parallel stripe domains. The static magnetization inside the domains is oriented perpendicular to the film plane if a strong uniaxial anisotropy is present. The lattice constant can be adjusted by external magnetic fields.

Such domain lattices are excited by rf magnetic fields. Two resonances are observed:

i) The domain wall resonance (DWR), where the domain walls oscillate back and forth (breathing mode) [51]. Resonance frequencies up to 2 GHz are measured [52].

ii) The domain resonance (DR), where the magnetization precesses with different directions and phases in neighbouring domains. Two branches, DR^+ and DR^-, of this resonance exist [53, 54, 55]. Resonance frequencies of more than 7 GHz are observed [56].

Excitations of the domain lattice generate dynamic components of the precessing magnetization which couple TE and TM modes by the Faraday and the Cotton–Mouton effects. Thus the excited stripe domain lattice acts as a dynamic phase grating so that modulation and deflection of optical modes can be achieved. This has already been demonstrated for both cases: light traversing the garnet film perpendicular to the film plane [57] and dynamic TE–TM mode conversion [58].

4.2. MODULATION BY FERRIMAGNETIC RESONANCE

In this case the magnetization is precessing at a spatially uniform phase in the garnet film as sketched in Fig. 4.2. A static induction $B_\perp$ is applied along the film normal, being parallel to the [111] crystallographic direction. To achieve a strong TE–TM mode coupling a large precession cone θ_p is required.

To calculate the motion of the magnetization M, polar coordinates Θ and Φ are used. The Gilbert form of the equation of motion reads [59]:

$$\dot{\Theta} = -\frac{\gamma}{M_s \sin \Theta_0} \frac{\partial F}{\partial \Phi} - \alpha \dot{\Phi} \sin \Theta ; \qquad (4.2)$$

$$\dot{\Phi} = \frac{\gamma}{M_s \sin \Theta_0} \frac{\partial F}{\partial \Theta} + \frac{\alpha}{\sin \Theta} \dot{\Theta} ,$$

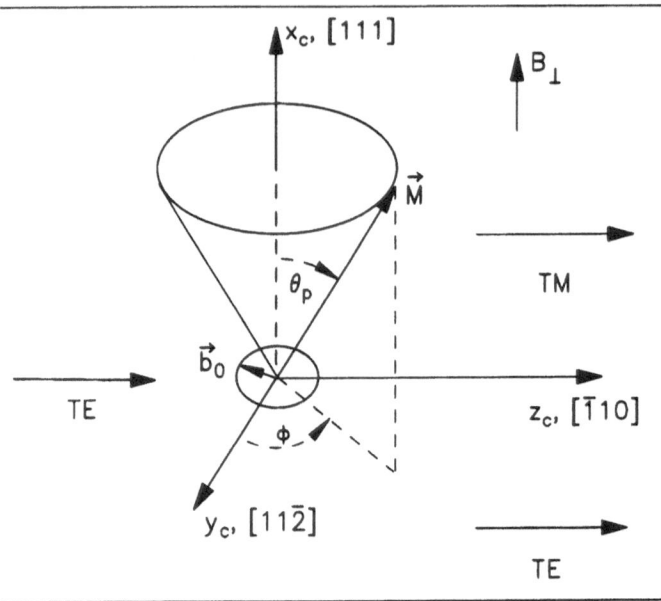

Figure 4.2. Basic geometry for dynamic mode conversion by ferrimagnetic resonance.

where

$$\mathbf{M} = M\left(\cos\Theta,\, \sin\Theta\cos\Phi,\, \sin\Theta\sin\Phi\right).$$

γ is the gyromagnetic ratio, α the damping constant and the density F of the free energy contains the Zeeman, demagnetizing and uniaxial anisotropy energies; the cubic anisotropy is neglected:

$$F \;=\; -B_\perp\, M\,\cos\Theta - b_0\, M\,\cos(\Phi - \omega t) + \left(K_u - \frac{\mu_0}{2}\, M^2\right)\sin^2\Theta\,.$$

b_0 is the amplitude of the circularly polarized rf induction of frequency ω which excites the ferrimagnetic resonance. At steady state the precession of $\mathbf{M}$ about the direction of $\mathbf{B}_\perp$ is circular, i. e. $\dot\Theta = 0$ and $\Phi = \omega t + \Phi_0$. One obtains for the steady state precession cone θ_p [60]:

$$\tan^2\theta_p \;=\; \gamma^2 b_0^2 \cdot \left\{\left[\omega - \gamma\, B_\perp - \gamma\left(\frac{2\,K_u}{M} - \mu_0\, M\right)\cos\theta_p\right]^2 \right. \tag{4.3}$$
$$\left. +\; \alpha^2\,\omega^2\,\cos^2\theta_p\right\}^{-1}.$$

This equation describes the foldover effect which is shown in Fig. 4.3 for the two cases $\mathbf{B}_{ueff} > 0$ and $\mathbf{B}_{ueff} < 0$ where the effective uniaxial

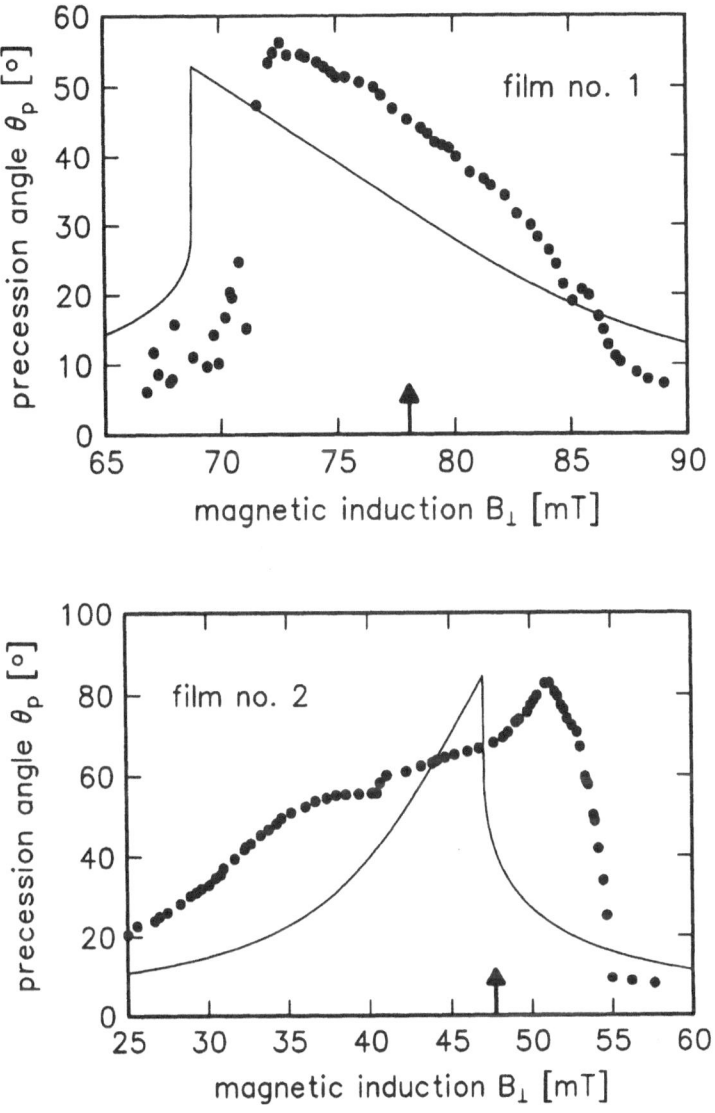

Figure 4.3. Measured precession angles of films no. 1 and 2 vs static induction $B_\perp$. The solid lines are calculated.

anisotropy is $B_{ueff} = 2\,K_u/M - \mu_0\,M$. The precession angle is measured by optical techniques and plotted vs the static induction. The microwave power and the excitation frequency of 1.42 GHz are kept constant. Precession angles up to 80° are achieved. The solid curves in Fig. 4.3 are calculated using equ. (4.3) and the parameters listed in Table 2.1; an amplitude of $b_0 = 3.5$ mT of the rf induction is chosen to fit the calculated maximum

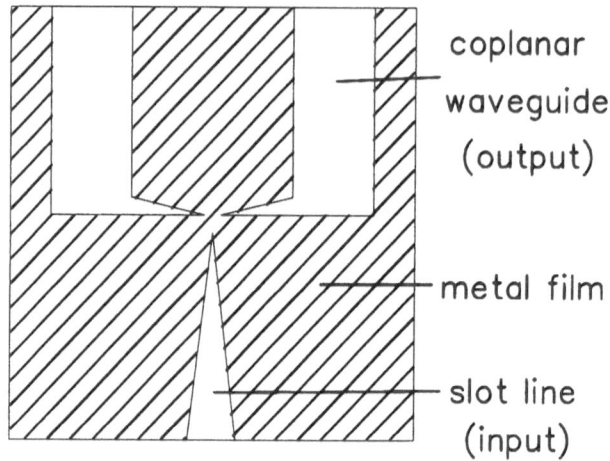

Figure 4.4. The planar microwave antenna used to excite and detect magnetic resonances.

precession angle to the measured one. The vertical arrows indicate the position of the FMR at low excitation power.

To excite the FMR a planar antenna structure is used as sketched in Fig. 4.4. A short circuited slot line serves as rf input while a short circuited coplanar waveguide is used as monitor output. The FMR is excited locally in a region of about 50 μm diameter at the center of the antenna. Within this region very large precesion cones can be obtained, if the saturation magnetization of the film is small. This behaviour is demonstrated in Fig. 4.5 where the spatial variation of the precession cone is shown for two films having small and large saturation magnetizations M_s. In the case of large M_s (pure YIG), precession cones of about 20° are observed, but the region of the precession extends to a large distance from the excitation area. On the other hand, for small M_s large precession cones are achieved only inside the excitation area. The measured dependence of the maximum precession cone on the saturation magnetization is shown in Fig. 4.6 for films of various orientations.

Preliminary measurements of dynamic TE–TM mode conversion using the FMR are shown in Fig. 4.7. Modulation is observed at both, the fundamental and the first harmonic frequency of the FMR. The modulation frequency is limited to about 3.5 GHz because of the finite bandwidth of the optical detection system used. A maximum dynamic conversion ratio of 6% is observed. The slopes of the straight lines agree within limits of error with γ and 2γ, where $\gamma=28.0$ MHz/mT is the gyromagnetic ratio of film no. 1.

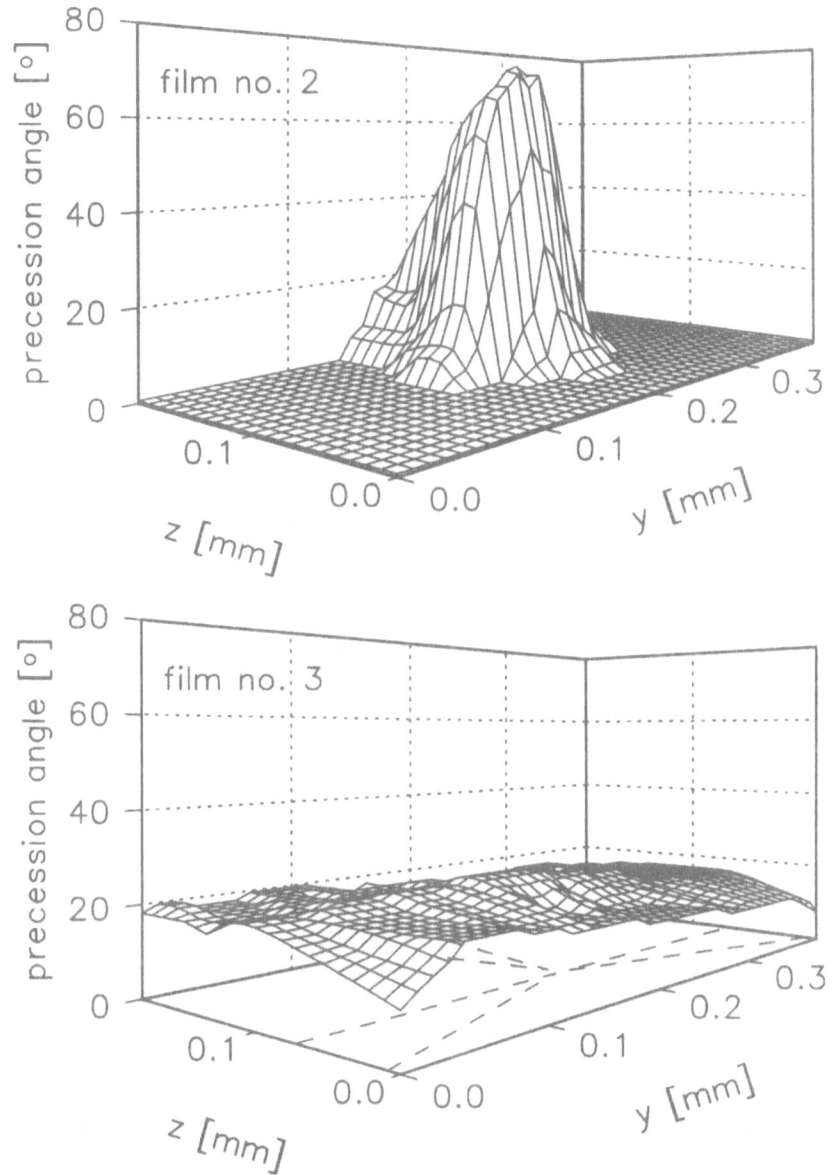

Figure 4.5. Measured spatial variation of the precession cone for two garnet films having small and large saturation magnetization. The planar microwave antenna is shown by dashed lines in the lower figure.

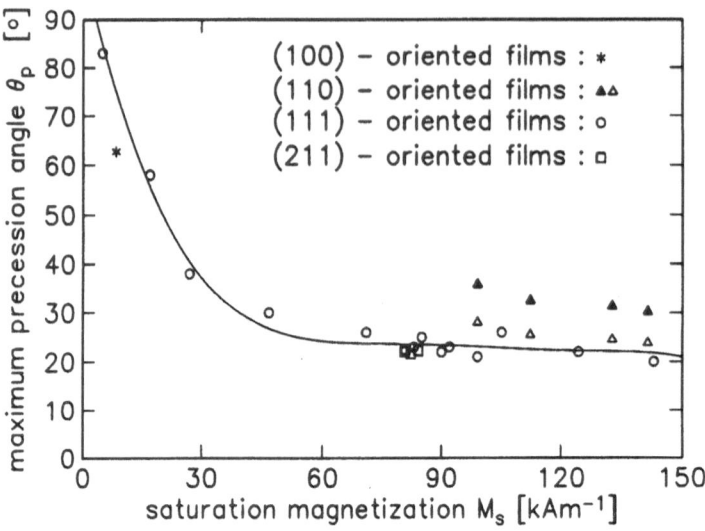

Figure 4.6. Measured maximum precession angles vs saturation magnetization for films of different orientations.

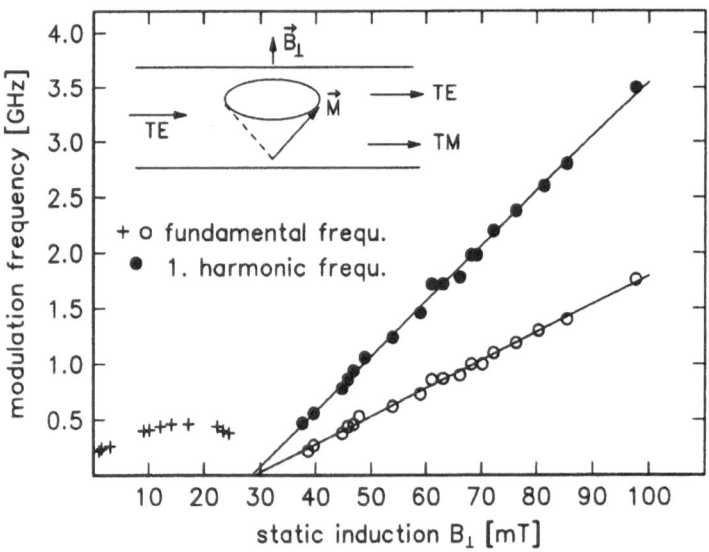

Figure 4.7. Measured dynamic conversion frequencies induced by ferrimagnetic resonance vs static induction $B_\perp$ for film no. 1.

4.3. MODULATION BY DOMAIN RESONANCES

The basic geometry is sketched in Fig. 4.8. The stripe domains are aligned along the direction of the static induction $\mathbf{B}_{\|}$ applied in the film plane. Optical modes propagate along the $[\bar{1}10]$ direction at an angle $\tau = \pi/2 - \Phi_c$ with respect to the stripe domains.

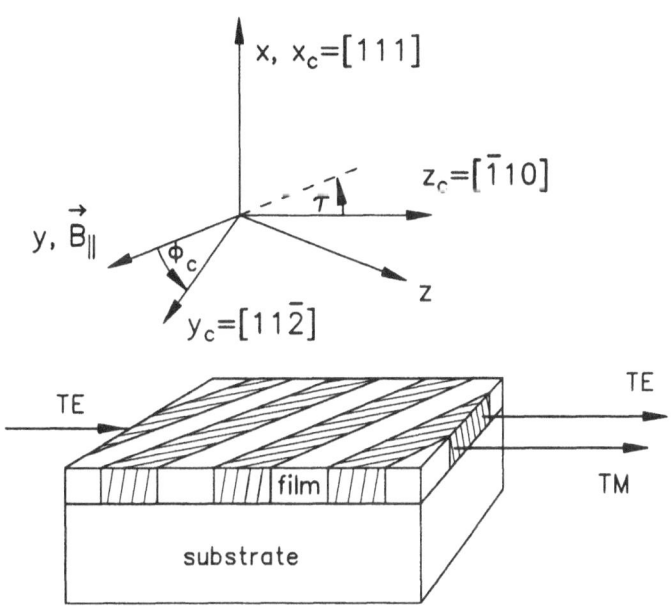

Figure 4.8. Basic geometry for dynamic mode conversion by domain resonances.

By improving material parameters, we could achieve frequencies of more than 7 GHz for the domain resonance. Thus these resonances are attractive for light modulation in optical communication systems. Due to the strong uniaxial anisotropy only very low bias fields are required to obtain such high resonance frequencies. Furthermore, the spatially periodic domain structure can be used to achieve dynamic Bragg scattering. Before studying dynamic TE–TM mode conversion, frequencies and eigenvectors of the domain resonances have to be calculated.

4.3.1. *Calculation of domain resonances*
To describe the domain resonances and the domain wall resonance, hybridization of all three resonance branches has to be taken into account as shown by Ramesh et al. [61, 62, 63]. In the following an improved version of the hybridization model is introduced which does not use any fitting parameters; it is in excellent agreement with measurements if the quality

factor Q of the film

$$Q = \frac{2\,K_u}{\mu_0\,M_s^2}$$

is larger than about 0.5. These results are obtained for [111] as well as for [110] oriented films.

Furthermore, it is demonstrated that the coupling of the domain resonance branches is caused by the cubic and the orthorhombic anisotropy. If the domains are aligned along certain symmetry directions, no hybridization occurs. In addition, the phase relation between the precessing magnetizations of neighbouring domains is derived, which is essential for optical applications.

The geometry of the stripe lattice and the motion of the magnetization and the domain walls during the domain resonance (DR) and domain wall resonance (DWR) are sketched in Fig. 4.9. In the following the up and down domains are denoted by the numbers 1 and 2, respectively. The widths of the domains are $L_1 = L + 2\,q$ and $L_2 = L - 2\,q$, where q is the amplitude of the oscillating domain walls. Inside the domains the direction of the magnetization is determined by the polar angles Θ_i and the azimuths Φ_i; $i = 1, 2$.

In the domain resonance (Fig. 4.9b) the magnetization precesses in neighbouring domains in opposite directions and at different phases. As there are two domains per unit cell of the domain lattice one obtains two resonance branches, denoted by DR^+ and DR^-. They are excited by linearly polarized rf magnetic fields applied in the film plane. In the domain wall resonance (Fig. 4.9c), the domain walls oscillate back and forth. This motion can be characterized as breathing mode; it is excited by a linearly polarized rf magnetic field applied perpendicular to the film plane.

The coupled equations of motion are [56]:

$$\dot{\Theta}_i = -\frac{\gamma}{M_s\,[1/2 - (-1)^i q/L]\,\sin\Theta_i}\,\frac{\partial F}{\partial\,\Phi_i} - \alpha\dot{\Phi}_i\sin\Theta_i \qquad (4.4)$$

$$\dot{\Phi}_i = \frac{\gamma}{M_s\,[1/2 - (-1)^i q/L]\,\sin\Theta_i}\,\frac{\partial F}{\partial\,\Theta_i} + \frac{\alpha}{\sin\Theta_i}\,\dot{\Theta}_i$$

$$\dot{q} = L\frac{\partial F}{\partial\,p} + \dot{p}\,\frac{\gamma\,\alpha\,\Delta}{M_s\,(\cos\Theta_1 - \cos\Theta_2)}$$

$$\dot{p} = -L\frac{\partial F}{\partial\,q} - \dot{q}\,\frac{M_s\,(\cos\Theta_1 - \cos\Theta_2)}{\gamma}\,\frac{\alpha}{\Delta}.$$

Special care is necessary to determine the contributions to the density F of the free energy. The form used in this article is a further improvement

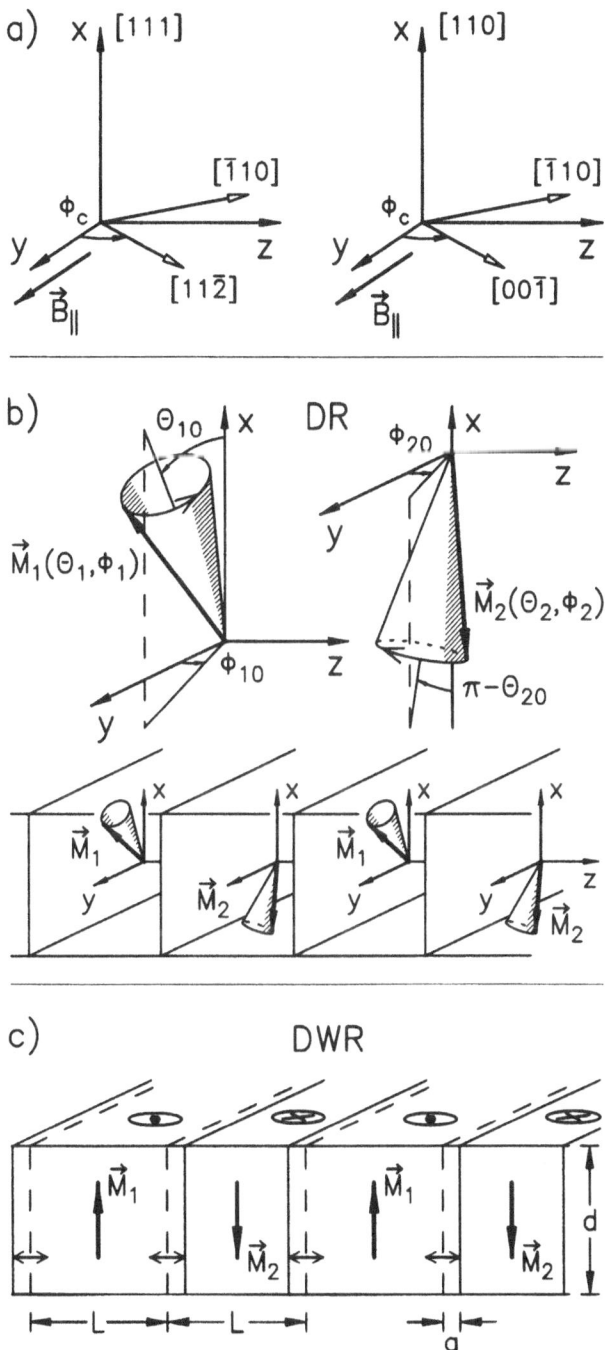

Figure 4.9. Film geometry and schematic sketch of the domain resonances (DR) and the domain wall resonance (DWR).

compared to [56]:

$$F = \left(\frac{1}{2} + \frac{q}{L}\right) \left(F_c^{(1)} + F_u^{(1)} + F_o^{(1)} + F_z^{(1)}\right)$$

$$+ \left(\frac{1}{2} - \frac{q}{L}\right) \left(F_c^{(2)} + F_u^{(2)} + F_o^{(2)} + F_z^{(2)}\right) + F_d + F_w .$$

F_c, F_u and F_o denote the cubic, the uniaxial and the orthorhombic aniso-tropy energies, respectively, F_z is the Zeeman energy, F_d the demagnetizing and F_w the wall energy. The superscripts (1) and (2) denote the two do-mains and $\Delta = \sqrt{A/K_u}$ is the wall width parameter. Whereas the energies F_c, F_u, F_o, and F_z differ in domains (1) and (2), the demagnetizing energy F_d and the wall energy F_w are common for both domains. The minimum of F determines the static equilibrium angles Θ_{i0} and Φ_{i0} of the magneti-zation. The wall momentum p is given by [64]:

$$p = \frac{\psi}{\gamma} M_s(\cos\Theta_1 - \cos\Theta_2),$$

where ψ is the angle the magnetization at the wall center forms with the wall plane.

The Zeeman energy density is:

$$F_z^{(i)} = -B_{\|} M_s \sin\Theta_i \cos\Phi_i . \qquad (4.5)$$

The cubic anisotropy energy reads for [111] oriented films:

$$F_c^{(i)} = K_1 \left[\frac{1}{4}\sin^4\Theta_i + \frac{1}{3}\cos^4\Theta_i \right. \qquad (4.6)$$

$$\left. + \frac{\sqrt{2}}{3}\sin^3\Theta_i \cos\Theta_i \cos 3(\Phi_i - \Phi_c)\right]$$

and for [110] oriented films:

$$F_c^{(i)} = \frac{K_1}{4}\sin^2\Theta_i \cdot \left[2\cos^2(\Phi_i - \Phi_c) - 3\sin^2\Theta_i\right.$$

$$\times \left. \cos^4(\Phi_i - \Phi_c) - 4\cos^2\Theta_i \sin^2(\Phi_i - \Phi_c)\right] .$$

For the uniaxial anisotropy energy one obtains:

$$F_u^{(i)} = K_u \sin^2\Theta_i .$$

While the orthorhombic anisotropy vanishes for [111] oriented films, it reads for [110] oriented ones:

$$F_o^{(i)} = K_o \sin^2 \Theta_i \sin^2(\Phi_i - \Phi_c).$$

Two contributions add to the demagnetizing energy $F_d = F_{d,x} + F_{d,z}$, where the first term is due to the magnetic charge distribution on the two surfaces of the film at $x = 0$ and $x = d$ and the second one is caused by charged domain walls. To calculate $F_{d,x}$ the method of Kooy and Enz [65] is followed; the angles Θ_1 and $\pi - \Theta_2$ need not be equal. One obtains the relation:

$$
\begin{aligned}
F_{d,\perp} \quad &- \frac{\mu_0 M_s^2}{8} \left\{ \left[\cos \Theta_1 + \cos \Theta_2 + \frac{2q}{L}(\cos \Theta_1 - \cos \Theta_2) \right]^2 \right. \\
&+ \frac{8L(\cos \Theta_1 - \cos \Theta_2)^2}{\pi^3 d} \times \sum_{n=1}^{\infty} \frac{\sin^2[n\pi(1 + 2q/L)/2]}{n^3} \\
&\left. \times \left[1 - \exp\left(-\frac{n\pi d}{L}\right) \right] \right\}.
\end{aligned}
$$

The second contribution $F_{d,z}$ is determined in the following way, see also Fig. 4.13 below. Inside each domain the z–component M_{iz} of the magnetization is assumed homogeneous yielding a charge density of the domain walls of $\Delta M = |M_{1z} - M_{2z}|$ with opposite signs of neighbouring walls. The charged walls cause a demagnetizing field $\mathbf{H}_d$. This field can be calculated at the position $\mathbf{r}$ by integrating over all domain walls which are counted by the index j:

$$
\mathbf{H}_d(\mathbf{r}) = \frac{\Delta M}{4\pi} \sum_{j=-\infty}^{+\infty} (-1)^j \int_0^d dx' \int_{-\infty}^{+\infty} dy' \frac{\mathbf{r} - \mathbf{r'}_j}{|\mathbf{r} - \mathbf{r'}_j|^3}. \tag{4.7}
$$

The demagnetizing energy $F_{d,z}$ is then obtained by averaging over one period, yielding:

$$
\begin{aligned}
F_{d,z} = {} &\frac{\mu_0 M_s^2}{\pi(L_1 + L_2)d} (\sin \Theta_2 \sin \Phi_2 - \sin \Theta_1 \sin \Phi_1) \sum_{i=1}^{2} (-1)^i \sin \Theta_i \sin \Phi_i \\
&\times \sum_{j=-\infty}^{+\infty} (-1)^j \int_0^d dx' \int_{-0.5L_i}^{+0.5L_i} dz' \arctan\left(\frac{x'}{z' + (j+0.5)L + (-1)^i q} \right).
\end{aligned}
$$

The wall energy F_w is composed of a static term F_w^{stat} and a dynamic term F_w^{dyn}. To calculate the static term, it is assumed that the azimuths

$\Phi_{i0} = \Phi_0$ are equal and constant across the domain wall and that $\Theta_{10} = \pi - \Theta_{20}$. Using Euler–Lagrange equations, the static wall energy density is given by:

$$
\begin{aligned}
F_w^{stat} &= \frac{2}{L} \sqrt{A K_u} \int_{\Theta_{10}}^{\Theta_{20}} \left\{ \left[1 + \frac{K_o}{K_u} \sin^2(\Phi_0 - \Phi_c) \right] \right. \\
&\quad \times \ (\sin^2 \Theta' - \sin^2 \Theta_{10}) + \frac{\mu_0 M_s^2}{2 K_u} \sin^2 \Phi_0 \sin^2 \Theta' \\
&\quad \left. - \frac{B_\| M_s}{K_u} \cos \Phi_0 \left(\sin \Theta' - \sin \Theta_{10} \right) \right\}^{0.5} d\Theta' .
\end{aligned}
$$

In this equation the demagnetizing, exchange, Zeeman, uniaxial and orthorhombic anisotropy energies are taken into account, while the contribution of the cubic anisotropy is neglected.

The dynamic wall energy is:

$$
F_w^{dyn} = \frac{p^2}{2 m L} ,
$$

where m is the wall mass, given by [66]:

$$
m = m_0 \int_{\Theta_{10}}^{\pi/2} \sin \Theta_1 \frac{\sin \Theta_1 - \sin \Theta_{10}}{\sin \Theta_1 + Q \sin \Theta_{10}} d\Theta_1
$$

with the Döring mass

$$
m_0 = \frac{2}{\mu_0 \gamma^2 \sqrt{A/K_u}} .
$$

In a first step the minimum of the free energy density F is calculated using the assumptions $q = p = 0$, $\Phi_{10} = \Phi_{20}$ and $\Theta_{10} = \pi - \Theta_{20}$. This procedure yields the equilibrium angles Θ_{i0} and Φ_{i0} and the strip width L as function of the induction $B_\|$. The six coupled equations of motion (4.4) are then solved numerically. In this way one obtains the eigenfrequencies and eigenvectors of the domain resonances and the domain wall resonance.

If $B_\|$ exceeds the collapse induction $B_\|^{coll}$ the magnetic film is homogeneously magnetized. In this case the resonances DR$^+$, DR$^-$ and DWR disappear and only the ferrimagnetic resonance (FMR) can be excited. For [111] oriented films the FMR–frequency is:

$$
\omega_{[111]}^2 = \gamma^2 B_\| \left[B_\| - \left(\frac{2 K_u}{M_s} - \mu_0 M_s \right) - \frac{K_1}{M_s} \right] + \frac{2 K_1^2}{M_s^2} \sin^2(3 \Phi_c)
$$

and for [110] oriented films:

$$\omega^2_{[110]} = \gamma^2 \left[B_{\|} - \left(\frac{2 K_u}{M_s} - \mu_0 M_s \right) - \frac{2 K_o}{M_s} \sin^2 \Phi_c \right.$$
$$+ \frac{3 K_1}{M_s} \cos^4 \Phi_c - \frac{K_1}{M_s} \sin^2 \Phi_c - \frac{K_1}{M_s} \right]$$
$$\times \left[B_{\|} + \frac{2 K_o}{M_s} \cos(2 \Phi_c) - \frac{K_1}{M_s} \cos(2 \Phi_c) \right.$$
$$\left. - \frac{3 K_1}{M_s} (4 \sin^2 \Phi_c - 1) \cos^2 \Phi_c \right] .$$

In Fig. 4.10 the period $2L$ of stripe domain lattices is plotted versus in–plane induction $B_{\|}$ for a [111] and a [110] oriented film. In the [110] oriented films an anisotropic behaviour of the domains is observed. This is demonstrated for the two extreme cases where the stripes are parallel to the [00$\bar{1}$] and the [$\bar{1}$10] direction. The error bars include the effect of hysteresis for increasing and decreasing induction. The solid curves are calculated from the minimum of the free energy; they are fit to the measurements at $B_{\|} = 0$ by adjusting the exchange constant A.

In Figs. 4.11 and 4.12 the resonance frequencies of DR$^+$, DR$^-$, DWR, and FMR are presented as function of $B_{\|}$ for a [111] and a [110] oriented film, respectively. The solid curves are calculated using the theory discussed above and the experimentally determined material parameters (Table 2.1); for all resonances the same damping constant α is used, which is derived from the FMR linewidth.

In Fig. 4.11a the stripe domains are parallel to the [11$\bar{2}$] direction ($\Phi_c = 0$). In this case DR$^+$ and DR$^-$ are coupled by cubic anisotropy and hybridization is observed. However, if the stripes are parallel to [$\bar{1}$10], the domain resonances are not coupled: There is no hybridization and the resonance branches DR$^+$ and DR$^-$ cross as shown in Fig. 4.11b.

For a [110] oriented film (film no. 5) the dependence of the resonance frequencies of DR$^+$, DR$^-$, DWR, and FMR on the in–plane induction are presented in Fig. 4.12 for three different alignments of the stripes. If the domains are parallel to [00$\bar{1}$] or [$\bar{1}$10] no hybridization of DR$^+$ and DR$^-$ takes place as shown in Fig. 4.12a and Fig. 4.12c, respectively. However, at a direction of 45° between [00$\bar{1}$] and [$\bar{1}$10] the resonances DR$^+$ and DR$^-$ are coupled yielding hybridization and thus repelling of the resonance branches is observed as shown in Fig. 4.12b.

For other samples having Q factors larger than 0.5 also an excellent agreement between theory and experiment is observed. However, if $Q < 0.5$, a large discrepancy between measured and calculated resonance frequencies occurs.

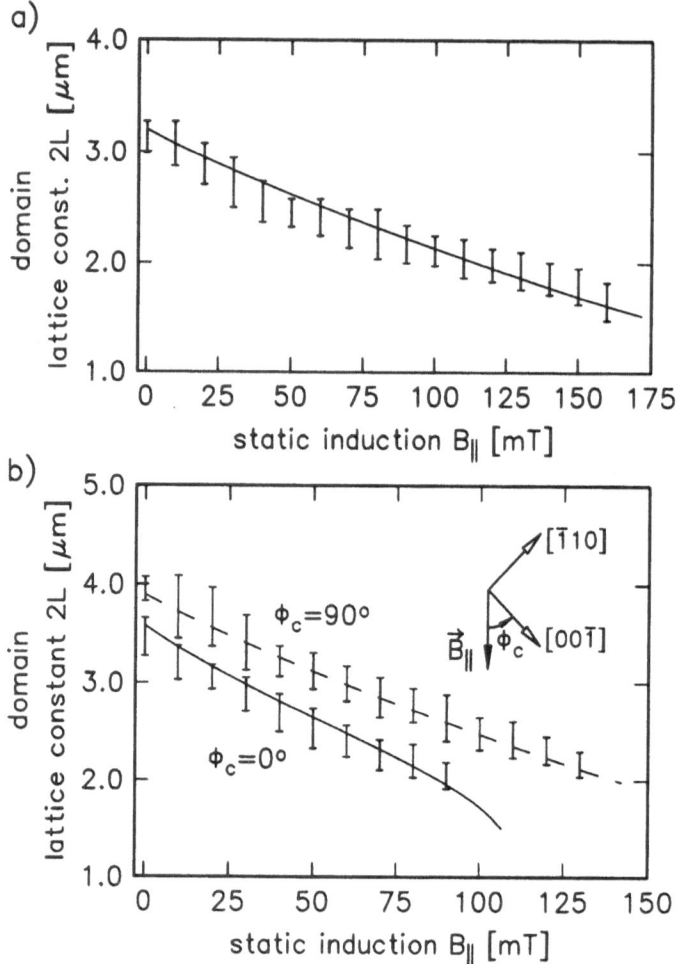

Figure 4.10. The period 2L of the stripe domain lattice vs in–plane induction $B_{||}$; a) film no. 4 ([111]), b) film no. 5 ([110]).

For the domain resonances very high frequencies are observed. The maximum is located at $B_{||} = 0$ and lies above 7 GHz. Such high resonance frequencies are caused by the large uniaxial anisotropy, which is induced by bismuth substitution and by the large saturation magnetization M_s. This behaviour can be explained in the following way for [111] oriented films. In Fig. 4.13 the motion of the magnetization in the two domains is sketched for both resonances DR^+ and DR^- at $B_{||} = 0$. The resonance DR^- can be regarded as uniform precession in the anisotropy field $2 K_u/\mu_0 M_s$. The resonance DR^+, however, has a different phase relation between the precessing magnetizations of neighbouring domains yielding charged domain

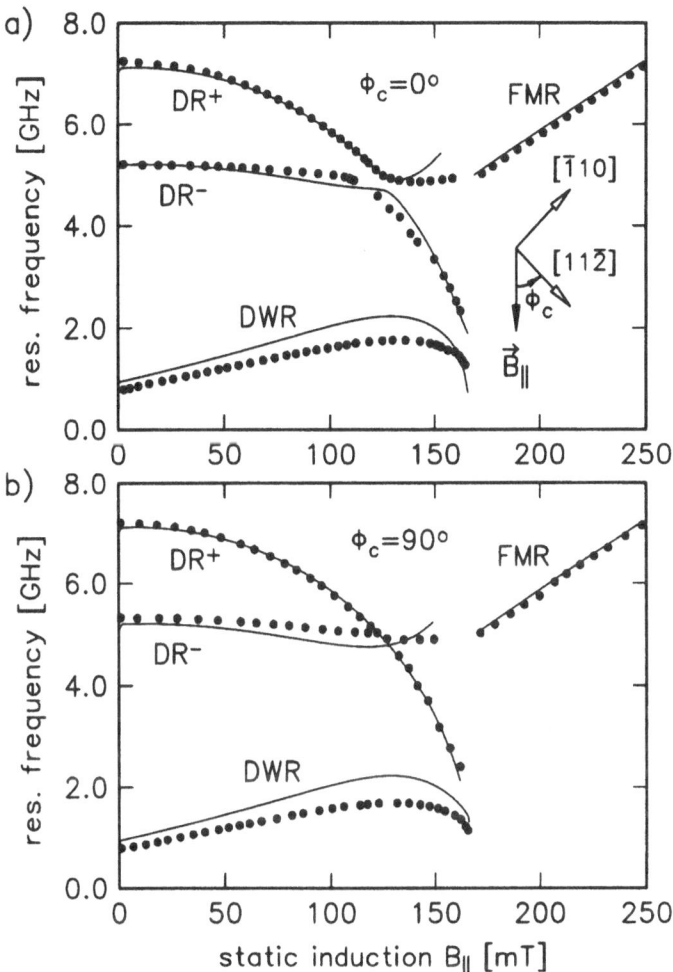

Figure 4.11. Measured and calculated frequencies of the domain resonances DR$^+$ and DR$^-$, of the domain wall resonance DWR and of the ferrimagnetic resonance FMR vs in-plane induction B$_{||}$ for film no. 4 ([111]) and two different orientations of the stripe domains.

walls. The induced demagnetizing fields $H_{d,iz}$ influence the motion of the magnetization, so that DR$^+$ resembles the precession in a thin film biased in the film plane. Thus one obtains the rough approximations for the resonance frequencies of DR$^-$ and DR$^+$ at B$_{||}$ = 0:

$$\omega_{DR-} \approx \gamma \frac{2 K_u}{M_s},$$

446

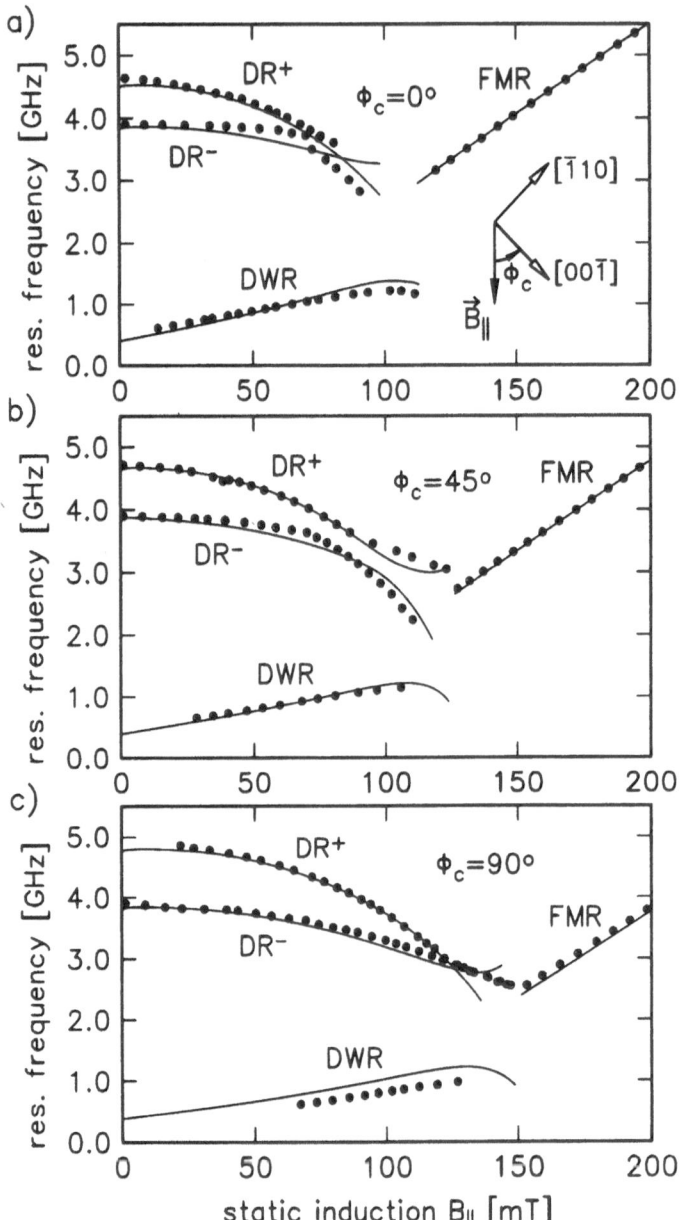

Figure 4.12. Measured and calculated frequencies of the domain resonances DR$^+$ and DR$^-$, of the domain wall resonance DWR and of the ferrimagnetic resonance FMR vs in–plane induction B$_{||}$ for film no. 5 ([110]) and three different orientations of the stripe domains.

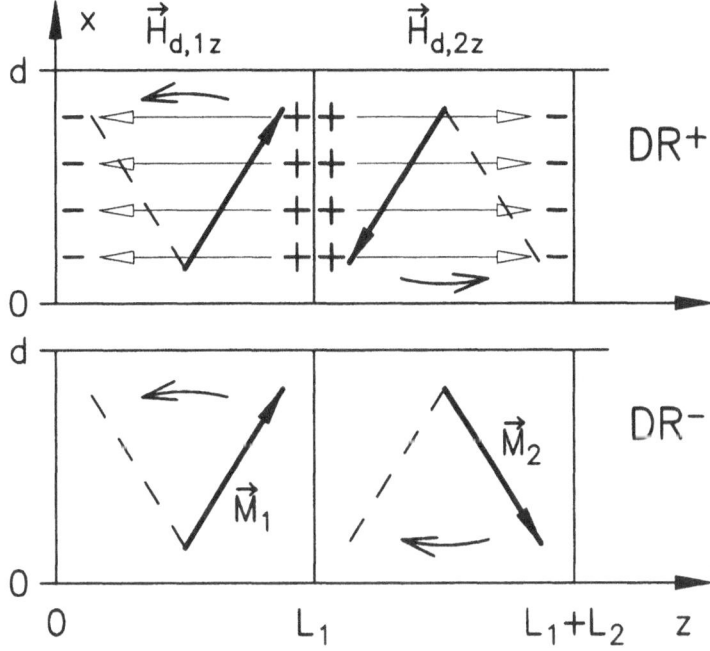

Figure 4.13. Motion of the magnetization of neighbouring domains for the domain resonances DR$^+$ and DR$^-$ at B$_{||}$ = 0.

$$\omega_{DR^+} \approx \gamma \sqrt{\frac{2\,K_u}{M_s}\left(\frac{2\,K_u}{M_s} + \mu_0\,M_s\right)}.$$

Similar approximations can be made for [110] oriented films. For the domain wall resonance DWR, high frequencies of nearly 2 GHz are measured. In this case the high frequencies are mainly due to the large saturation magnetization M$_s$ [52].

Hybridization between the DWR and the two branches of DR is always observed. However, hybridization between the two DR branches depends on the orientation of the stripe domains in the film plane. Energy terms which induce coupling between DR$^+$ and DR$^-$ are the cubic and the orthorhombic anisotropy.

For a [111] oriented film (film no. 4) a contour plot of the cubic anisotropy energy is shown in Fig. 4.14a using stereographic projection onto the (111) plane. All crystallographic equivalent [100] directions are hard directions for the magnetization; they are marked by a solid square symbol. The easy [111] directions are marked by a solid triangle. If the domains are parallel to the straight lines in Fig. 4.14a no hybridization occurs. Along

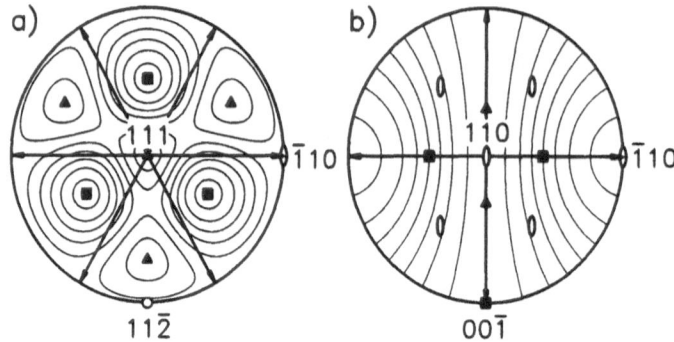

Figure 4.14.   Contour plot of the cubic (a) and the orthorhombic anisotropy energy (b) using stereographic projection on the planes (111) and (110) for films no. 4 and 5, respectively.

these lines Φ_c equals an odd integer multiple of 30° so that in the cubic anisotropy energy (equ.(4.6)) the azimuth dependent term

$$K_1 \frac{\sqrt{2}}{3} \sin^3 \Theta_i \cos \Theta_i \cos 3(\Phi_i - \Phi_c),$$

vanishes for $\Phi_{i0} = 0$. Along all other directions of the stripe domains DR^+ and DR^- are coupled and thus hybridization occurs.

Similarly for a [110] oriented film (film no. 5) a contour plot of the orthorhombic anisotropy energy is presented in Fig. 4.14b. No hybridization occurs along the two straight lines when the domains are parallel to the $[00\bar{1}]$ or $[\bar{1}10]$ directions. These lines correspond to $\Phi_c = 0$ and $\Phi_c = 90°$, where the orthorhombic anisotropy energy is at minimum or maximum, respectively. For $\Phi_{i0} = 0$ the orthorhombic anisotropy energy does not depend on the direction of the magnetization in this case and no coupling between DR^- and DR^+ occurs.

The cubic anisotropy also contains terms which induce coupling between DR^+ and DR^- for [110] oriented films. However, in the present case this coupling is much weaker than that induced by the orthorhombic anisotropy and can be neglected.

For optical modulation applications, the phase relation between the precessing magnetizations of neighbouring domains is essential. It can be derived from the eigenvectors, determined by numerical calculation of the coupled equations of motion (4.4). In Fig. 4.15 this phase relation is presented for samples no. 4 and 5, respectively. The precessing magnetization is projected onto the plane perpendicular to the static equilibrium directions determined by Θ_{i0} and Φ_{i0}. In this plane the trace of the moving magne-

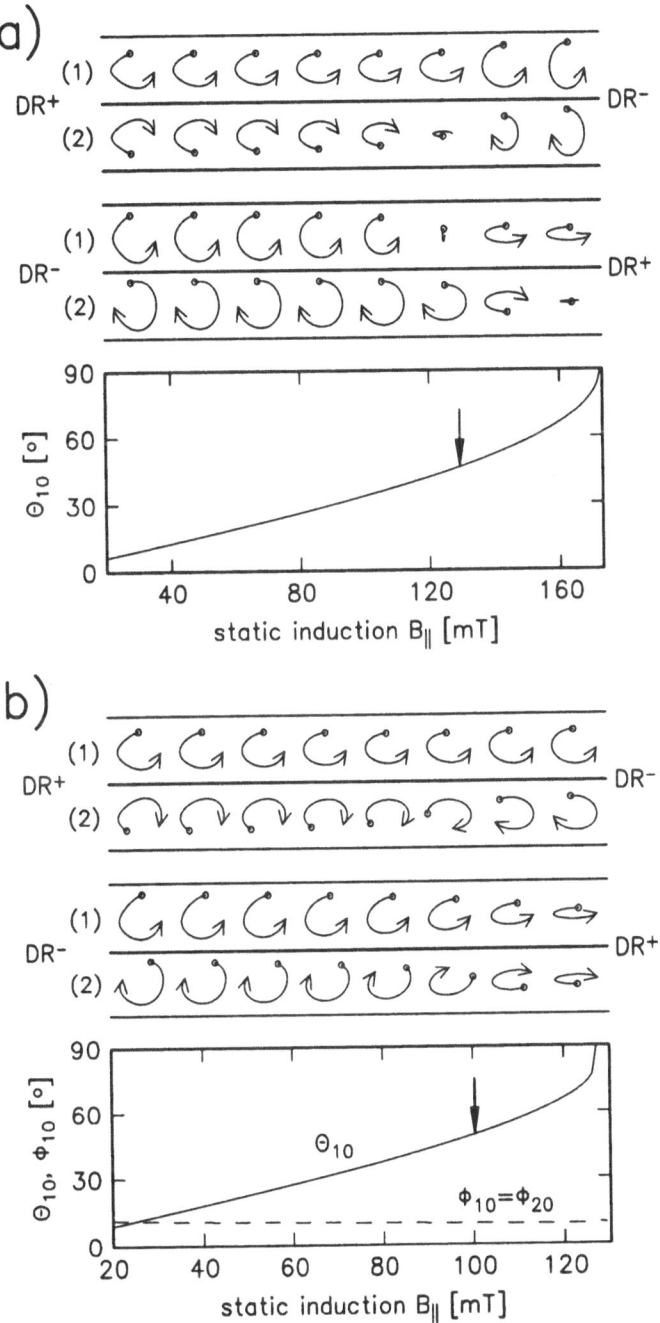

Figure 4.15. Calculated traces of the component of the precessing magnetization in the plane perpendicular to its equilibrium direction; a) film no. 4, b) film no. 5. The lower curves in a) and b) show the variation of Θ_{10} and $\Phi_{10} = \Phi_{20}$ ([110] film no. 5) with the static induction $B_{\parallel}$.

tization is shown for a number of fixed static inductions $B_{\parallel}$. The direction of viewing is antiparallel to the equilibrium direction of the magnetization for domains (1), while it is parallel to the static magnetization for domains (2) . The stripe domains are aligned along $\Phi_c = 0$ in Fig. 4.15a and along $\Phi_c = 45°$ in Fig. 4.15b. Hybridization of the domain resonances occurs in both cases. The precession amplitude of the magnetization is normalized, so that the maximum precession angle is 4° for DR^+ and DR^- at each value of $B_{\parallel}$. For the domains (1) always the same starting position is chosen.

The variation of Θ_{10} is shown in the lower part of each figure. For sample no. 5 also the variation of $\Phi_{10} = \Phi_{20}$ is presented, while the azimuths are zero for the [111] oriented film. The region of hybridization is marked by an arrow.

The domain resonances DR^+ and DR^- can be characterized by the phase difference of the precessing magnetization in the domains (1) and (2). On passing through the regions of hybridization in Fig. 4.15, one observes that this phase difference and thus the domain resonances interchange. If DR^+ and DR^- are not coupled, their resonance branches cross preserving their phase character.

4.3.2. *Dynamic mode conversion*

The coupled equations of motion are solved numerically, yielding the eigenfrequencies and eigenvectors of the resonances $DR^{\pm}$ and DWR. Then from the static and dynamic components of the magnetization the ϵ tensor is derived. The TE and TM modes, the phase mismatch $\Delta\beta$, and the dynamic conversion efficiency are determined at every instantaneous position of the precessing magnetization. This approximation is justified, because the coupling length z_{cl} is typically 4 mm in the experiments and the modes need at 1 GHz only about 3% of the rf period to propagate this distance. If Fourier analysis is applied to the calculated data of a complete precession cycle, the contributions of the fundamental frequency and of higher harmonics to the dynamic conversion are obtained. In the following only the diffraction of zero order is taken into account.

Fig. 4.16 shows the geometry used. The optical modes propagate with propagation constant $|\beta|$ at an angle τ with respect to the plane of the domain walls along the z_c or $[\bar{1}10]$ axis. The precession angle is denoted by θ_p. Neglecting cubic anisotropy the polar equilibrium angles of the magnetization in the domains D_1 and D_2 are related by $\Theta_{10} = \pi - \Theta_{20}$; in the following Θ_0 is written instead of Θ_{10}.

As an example, Fig. 4.17 presents the calculated dynamic conversion efficiencies η_{dyn} for the domain resonances $DR^{\pm}$ at the fundamental precession frequency. Material parameters similar to those of film no. 7 are used and a constant precession angle of $\theta_p = 5°$ is assumed. The coupling length

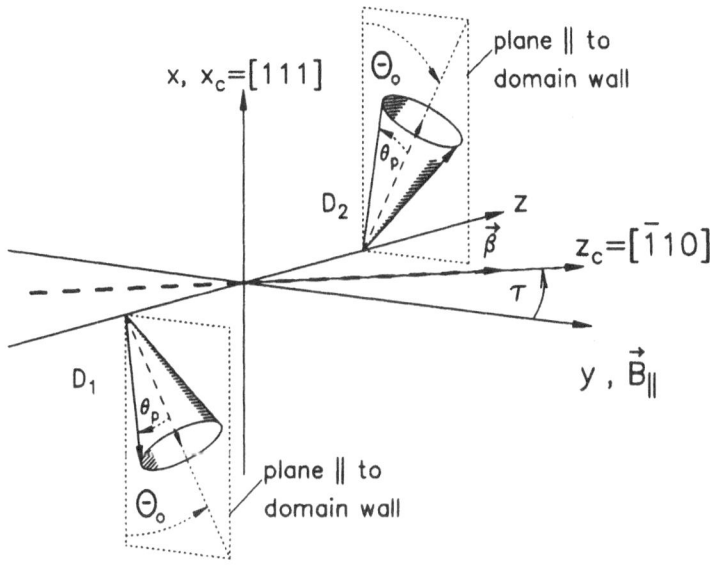

Figure 4.16. Geometry used to calculate the dynamic mode conversion by domain resonances.

is chosen as $z_{cl} = 3.6$ mm, which is twice the maximum coupling length $z_{cl,max}$ induced by Faraday rotation at $\Delta\beta = 0$. The contribution of the Cotton–Mouton effect (ϵ^{LB}) to the dynamic conversion is about 20 times less than that of the Faraday effect (ϵ^{CB}). The maxima occur at propagation angles $\tau = 0°$ and small equilibrium angles Θ_0 for DR$^+$ and at $\tau \approx 75°$ and large angles Θ_0 for DR$^-$.

Fig. 4.18 shows the calculated dynamic conversion efficiencies $\eta^{2\nu}_{dyn}$ at the first harmonic frequency using the same material parameters as in Fig. 4.17. The maximum conversion efficiencies are much less in this case. For DR$^+$ the maximum is obtained for propagation parallel to the domains and at small equilibrium angles Θ_0. For DR$^-$ the modes must propagate at 90° to the domains to achieve maximum conversion.

In the experimental investigation, first the static TE–TM mode conversion is measured as function of coupling length and of bias induction B$_{\|}$, where the magnetization is adjusted parallel to mode propagation. Fitting the experimental results to equ.(4.1) yields the parameters $M_s G_{44}$, $M_s \Delta G$, ϵ_{xx} and $\Delta\epsilon_\perp$. This is demonstrated in Fig. 4.19.

Before measuring dynamic mode conversion, the frequencies of the resonances DR$^\pm$, DWR and FMR are determined as function of B$_{\|}$ using rf techniques. For this purpose planar antennas are used as described above.

To study dynamic mode conversion by DR$^\pm$, light is coupled into and

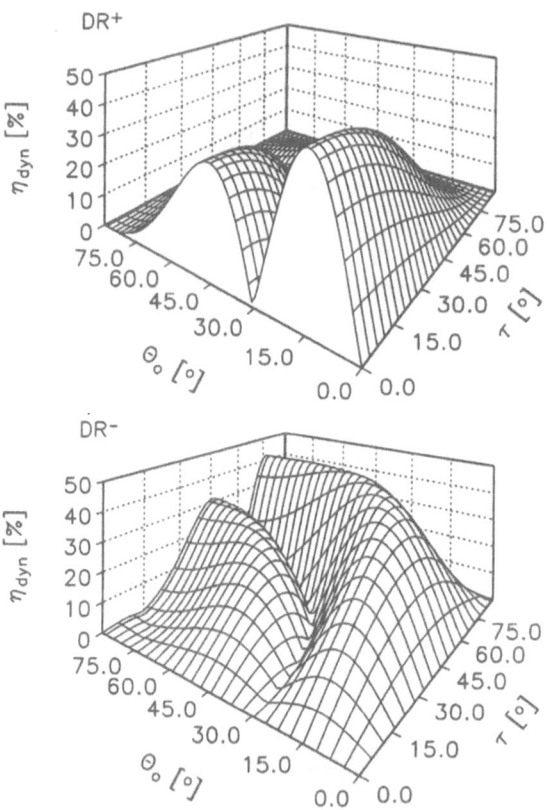

Figure 4.17. Calculated dynamic conversion efficiencies η_{dyn} at the fundamental frequencies of the domain resonances DR^+ and DR^- as function of the equilibrium angle Θ_0 and the propagation angle τ.

out of the waveguiding garnet film by rutile prisms. The optical output is focused into a photodetector having bandwidth of 5 GHz. The signal of the photodetector is amplified and fed to a spectrum analyzer.

In the region of the garnet film between the optical coupling prisms the resonances $DR^\pm$ are excited by an antenna which consists of a metal strip of width 2 mm while its length is about twice the maximum static coupling length $z_{cl,max}$. This metal strip is used as short circuit of a coaxial line so that the rf current flows parallel to its long edge. The strip is pressed on top of the film. The rf magnetic field of the antenna inside the film is approximately linearly polarized. It can be decomposed into two counter rotating circularly polarized rf fields which excite the domain resonances in the domains D_1 and D_2. If the metal strip is parallel to the stripes,

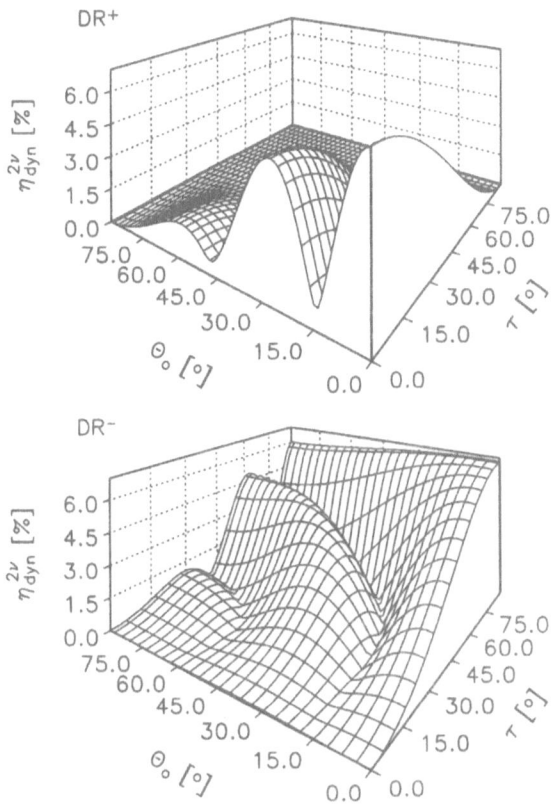

Figure 4.18. Calculated dynamic conversion efficiencies $\eta^{2\nu}_{dyn}$ at the first harmonic frequencies of the domain resonances DR^+ and DR^- as function of the equilibrium angle Θ_0 and the propagation angle τ.

the resonance DR^- is excited, if it is perpendicular to the stripes, DR^+ is excited (see e. g. Fig. 4.25 below). The antenna is connected to an rf source which can be operated between 0.1 and 5 GHz at power levels up to 1 W.

Fig. 4.20 presents the optical measurements of the dynamic TE–TM mode conversion of film no. 6 by the resonances DR^+ and DR^- as function of the static induction $B_{||}$. The mode pair of order m=17 is used because it nearly fulfills phase matching. Mode propagation is parallel to the stripes for DR^+ and at 45° to the stripes for DR^-. Also data taken by rf technique are shown. If $B_{||}$ exceeds the collapse induction $B^{coll}_{||} = 88$ mT, the film is saturated and only the uniform precession (FMR) is excited. The solid curves are calculated where only the material parameters given in Table 2.1 are used. The measurements by rf technique can be taken all along the re-

454

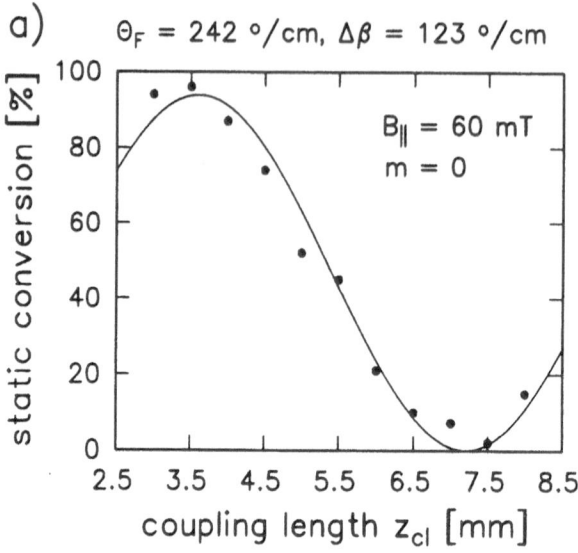

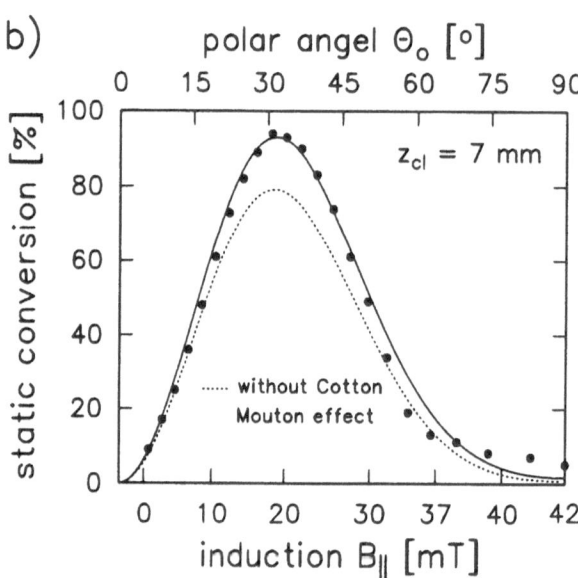

Figure 4.19. Measured static conversion efficiencies of film no. 8 as function of the coupling length (a) and of in-plane induction $B_{\parallel}$ (b) using the TE–TM mode pair of order m = 0. The curves are calculated and fit to the measurements to determine material parameters.

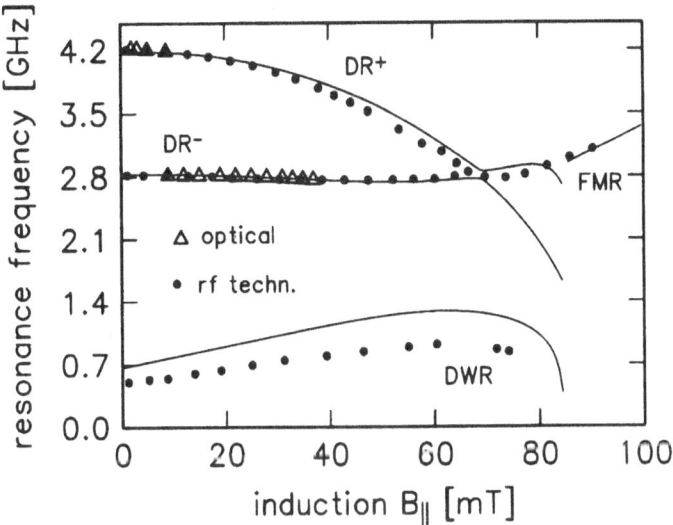

Figure 4.20. Measured resonance frequencies of DR$^{\pm}$ and of DWR of film no. 6 vs in-plane induction B$_{||}$. The solid curves are calculated.

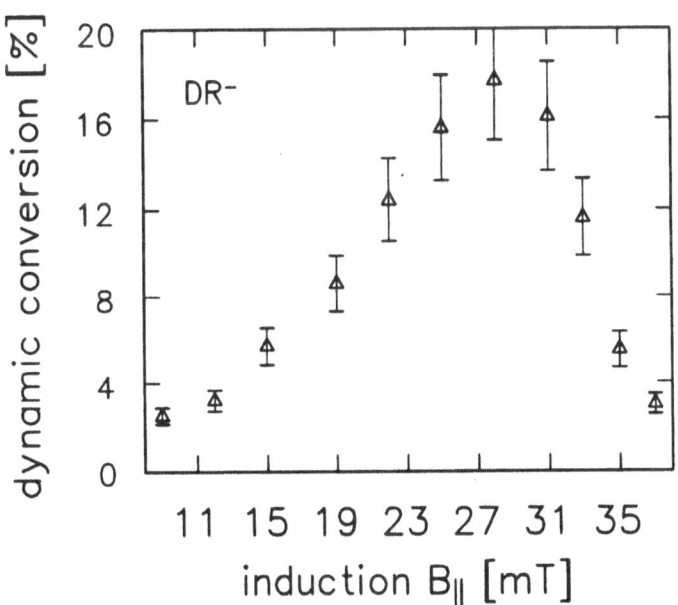

Figure 4.21. Measured dynamic conversion efficiency of film no. 6 vs in-plane induction B$_{||}$ using the DR^{-} branch. Coupling length is 4 mm, propagation angle $\tau = 45°$.

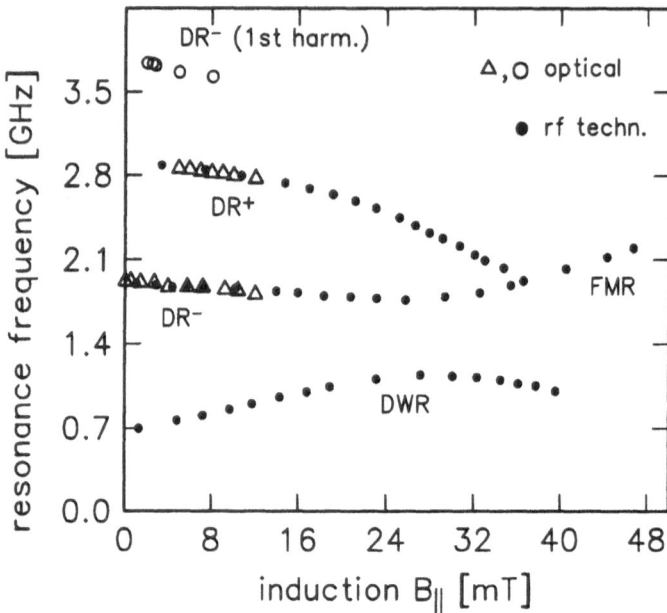

Figure 4.22. Measured resonance frequencies of DR$^\pm$, DWR and FMR of film no. 7 vs in–plane induction B$_{||}$; coupling length z_{cl} = 4 mm.

sonance branches. However, the signal of the optical measurements vanishes at large in–plane inductions B$_{||}$. Fig. 4.21 shows the dynamic conversion efficiencies obtained for the DR$^-$ branch of Fig. 4.20 at a coupling length of z_{cl} = 4 mm. A maximum efficiency of 18% is observed. For the DR$^+$ branch the maximum efficiency is only about 4%.

Similar results are obtained for the other samples. Fig. 4.22 presents the measurements for film no. 7 using the TE–TM mode pair of order 0. For DR$^-$ a maximum efficiency of 8% at B$_{||}$ = 5 mT is obtained. If the modes propagate perpendicular to the stripes, dynamic conversion is observed at the first harmonic of the DR$^-$ resonance in accordance with Fig. 4.18. Calculated resonance curves do not agree very well with the experiments, because this sample has a low Q factor of only 0.45.

The linewidths of the resonances DR$^+$ and DR$^-$ determined optically are quite large. They lie in the range between 50 and 200 MHz for the samples investigated.

The dynamic conversion efficiency can be calculated. However, the precession angle θ_p is not known. Fitting a constant angle θ_p to the measurements yields no satisfactory results. Therefore, it is assumed that θ_p is not constant. Then the precession angle is determined as function of in–plane

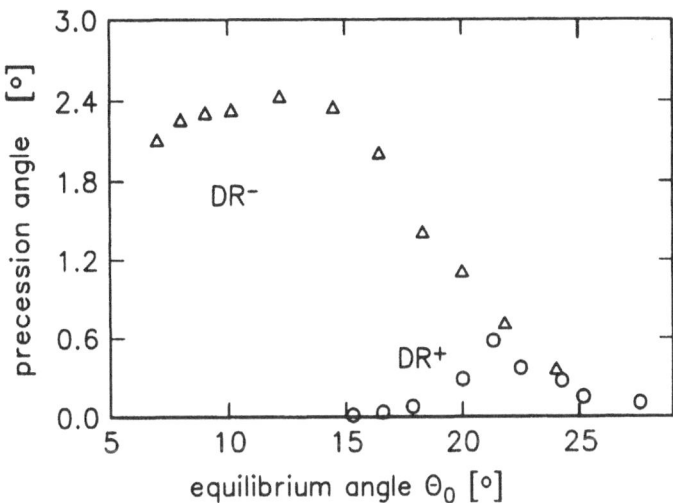

Figure 4.23. The precession angle θ_p derived by fitting the experimental results to calculations for film no. 7.

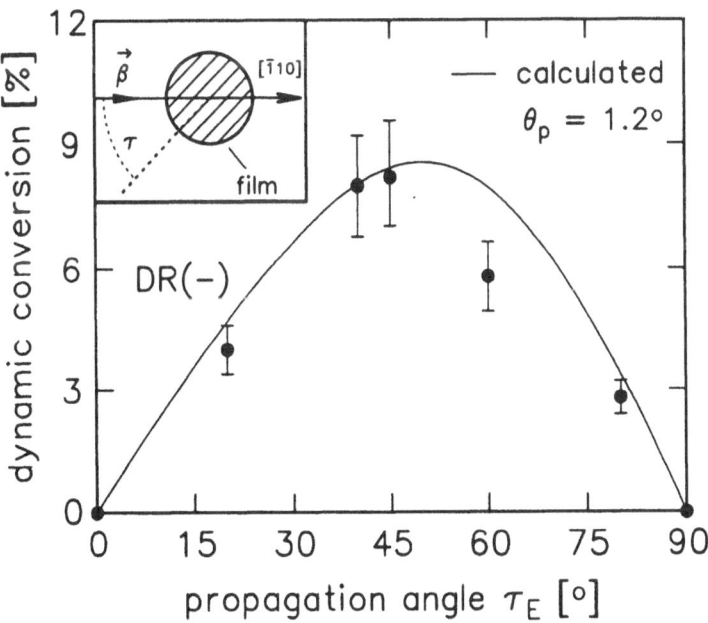

Figure 4.24. Measured and calculated dynamic conversion efficiencies of film no. 6 vs propagation angle τ ($B_{||} = 17$ mT, frequency $= 2.8$ GHz, coupling length $z_{cl} = 4$mm).

induction $B_{||}$ or of the equilibrium angle Θ_0 by comparing theory and experiment. The results are shown in Fig. 4.23 for film no. 7.

The precession angles vanish at large in–plane inductions $B_{||}$ or large equilibrium angle Θ_0. Although rf input power levels of 1 W are used, they are quite small. Furthermore, the precession angles of DR^+ are much less than those of DR^-.

Fig. 4.24 shows the measured and calculated dynamic conversion efficiencies of film no. 6 as function of the angle τ between the directions of mode propagation and stripe domains. The in–plane induction and the excitation frequency are kept constant. Therefore the equilibrium angle Θ_0 and the precession angle θ_p are also constant. The solid curve is calculated using $\theta_p = 1.2°$. The observed dependence of the precession angle θ_p on the equilibrium angle Θ_0 (s. Fig. 4.23) is the reason, why maximum conversion for DR^- is measured at propagation angles τ near 45°, contrary to the calculated results of Fig. 4.17 where θ_p is assumed constant.

4.4. DYNAMIC DEFLECTION BY DOMAIN RESONANCES

The domain resonances induce a spatially periodic lattice which may act as dynmaic coupling structure for propagating optical modes. This structure can be used for mode deflection by Bragg scattering; concurrently light modulation at the resonance frequency or at the first harmonic is obtained. As an example the in–plane components of the precessing magnetization of DR^- are shown in Fig. 4.25 at four different moments during one precession period at $B_{||} = 0$. The dynamic components m_y of the magnetization along the stripe domains describe a rectangular lattice which induces dynamic coupling at twice the precession frequency. Only TE–TM mode coupling by Faraday rotation is considered in the following.

Furthermore, phase matching between coupled TE and TM modes can be automatically fulfilled using anisotropic scattering as sketched in Fig. 4.26. Thus in principle, 100% dynamic conversion efficiency may be achieved.

The conversion efficiency at the end of the interaction length z_{cl} is given by:

$$\eta = \frac{\kappa^2}{\kappa^2 + (\Delta\beta/2)^2} \sin^2\left[z_{cl}\sqrt{\kappa^2 + (\Delta\beta/2)^2}\right] .$$

In this case the coupling constant κ is:

$$|\kappa| = \frac{2(\pi/\lambda)^2}{\epsilon_0\sqrt{\beta_{TE}\cos\Theta_{TE}^B\,\beta_{TM}\cos\Theta_{TM}^B}}\,\vec{e}^{TE}\,\epsilon_{x_c y_c}^{(1)}\,\vec{e}^{TM} ,$$

where $\epsilon_{x_c y_c}^{(1)}$ is the Fourier component of $\epsilon_{x_c y_c}$ of first order and the phase mismatch $\Delta\beta$ is determined by the deviation $\Delta\Theta$ from the Bragg angles

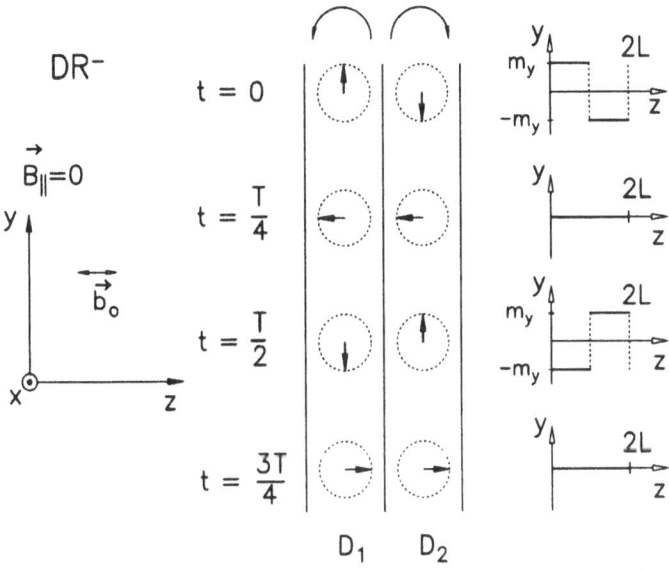

Figure 4.25. Dynamic components of the magnetization in neighbouring domains at four moments during a precession period of DR^-.

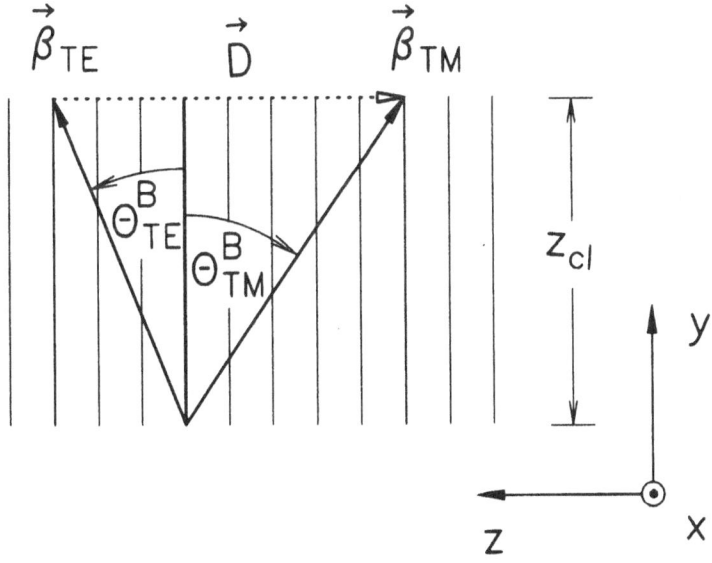

Figure 4.26. Geometry of anisotropic Bragg scattering in a planar waveguide; $|\vec{D}| = \pi/L$.

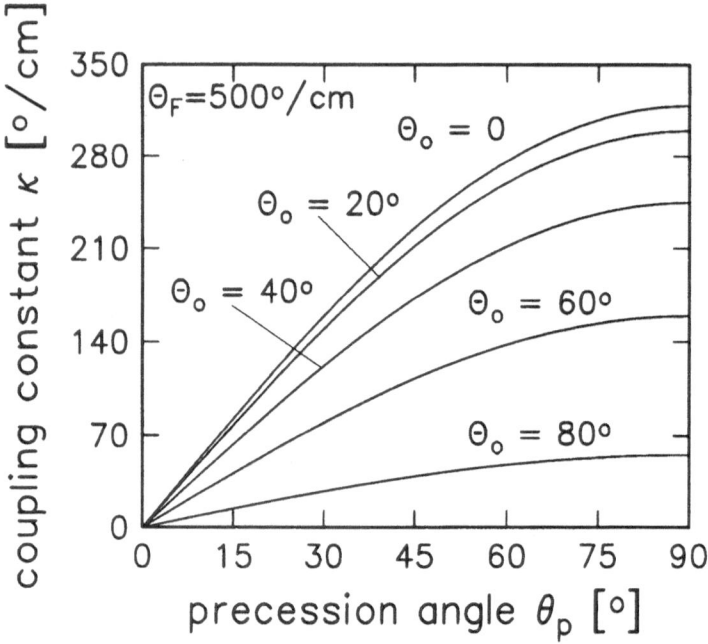

Figure 4.27. Calculated dependence of the coupling strength κ on the precession angle θ_p; parameter is the equilibrium angle Θ_0 of the magnetization in the domains.

Θ_{TE}^B and Θ_{TM}^B:

$$\Delta\beta = |\beta_{TE} \cos(\Theta_{TE}^B + \Delta\Theta) - \beta_{TM} \cos(\Theta_{TM}^B - \Delta\Theta)| .$$

κ determines the coupling length $z_{cl,max}$ for 100% conversion. The coupling strength depends on the equilibrium angle Θ_0 of the magnetization in the domains and strongly on the precession angle θ_p. This behaviour is demonstrated in Fig. 4.27.

The precession angles observed in the experiments described above are quite small, only a few degrees. Thus κ is small and z_{cl} may become such large that the required perfect periodicity of the dynamic structure can be perturbed by some lattice defects. Dynamic deflection by domain resonances is therefore difficult to measure, unless mode coupling has been increased considerably.

On the other hand, very large precession angles of up to 80° are observed in ferrimagnetic resonance. The reason for this discrepancy is not known and still under investigation.

5. Conclusion

The main applications of magnetic garnet films in integrated optics aim at nonreciprocal components. These are the only materials suited for such devices. Although research has been carried out for many years there is no practical solution available at present. The concepts discussed in this article present new possibilities to construct integrated isolators and circulators, but many details of the experimental realization have to be studied further.

It has been shown experimentally that modulation at GHz frequencies can be obtained using dynamic TE-TM mode conversion induced by the resonances of a stripe domain lattice. To increase the modulation efficiency and to achieve dynamic deflection concurrently to the modulation, the mode coupling must be enlarged. Two methods will be studied to achieve this goal.

i) Lattices of parallel stripe domains are generated where already a strong static in-plane component of the magnetization exists with opposite sign in neighbouring domains. In this way a static deflection occurs which can be modulated efficiently by the excitation of the domain resonances. Garnet films with such properties can be grown by liquid phase epitaxy. As such films have Q-factors less than 0.5 the hybridization model must be improved further. Especially the calculation of the wall structure needs refinement in this case.

ii) Magnetostatic waves of the stripe domain lattice will be excited instead of the uniform mode and the interaction with optical modes is studied. Especially when the waves propagate perpendicular to the stripes at the same wavelength as the lattice period, strong interaction is expected.

Finally there remains one problem which is not addressed in this article: the optical absorption. For all magnetooptical applications of garnet films it is essential that it is very low. In the infrared window between 1.2 and 5 μm wavelength this requirement is fulfilled for bulk iron garnet crystals [67]. However, garnet films grown by liquid phase epitaxy using lead-oxide flux show an additional strong and broad absorption in the near infrared region [68, 69, 70, 71, 72]. It is caused by the incorporation of non-three-valent impurity ions like Pb^{2+}, Pb^{4+} or Pt^{4+} into the crystal. Charge compensation among these impurities leads to a low absorption. Also special annealing procedures can reduce the absorption considerably. However, further investigations of this problem are necessary, especially if material properties have to be optimized by varying growth parameters and chemical compositions.

462

Acknowledgements

Deutsche Forschungsgemeinschaft (SFB 225) has financed a substantial part of the research reported in this articel. We also thank Prof. W. Tolksdorf and Dr. S. Sure for the preparation of garnet films.

References

1. Dötsch, H., Lührmann, B., Sure, S., Winkler, H.P., and Tolksdorf, W. (1992) Properties of magnetic garnet films applied for optical communication, *Proc. ICF 6*, 1592–1596, Tokyo and Kyoto.
2. Dötsch, H., Hertel, P., Lührmann, B., Sure, S., and Winkler, H.P. (1992) Magnetic garnet films for applications in optical communication, *Proc. ICF 6*, 1507–1512, Tokyo and Kyoto.
3. Hansen, P., Witter, K., and Tolksdorf, W. (1983) Magnetic and magneto–optic properties of lead and bismuth–substituted yttrium iron garnet films, *Phys. Rev. B* **27**, 6608–6625.
4. Hansen, P. and Krumme, J.P. (1984) Magnetic and magneto–optical properties of garnet films, *Thin Solid Films* **114**, 69–107.
5. Gomi, M., Satoh, K., and Abe, M. (1988) Giant Faraday rotation of Ce–substituted YIG films epitaxially grown by rf sputtering, *Jap. J. Appl. Phys.* **27**, L1536 – L1538.
6. Hansen, P., Witter, K., and Tolksdorf, W. (1984) Magnetic and magneto–optic properties of bismuth and aluminum–substituted iron garnet films, *J. Appl. Phys.* **55**, 1052–1061.
7. Tolksdorf, W., Dammann, H., Pross, E., Strocka, B., Tolle, H.J., and Willich, P. (1987) Growth of yttrium iron garnet multi–layers by liquid phase epitaxy for single mode magneto–optic waveguides, *J. Crystal Growth* **83**, 15–22.
8. Okuda, T., Katayama, T., Kobayashi, H., Kobayashi, N., Satoh, K., and Yamamoto, H. (1990) Magnetic properties of $Bi_3Fe_5O_{12}$ garnet, *J. Appl. Phys.* **67**, 4944–4946.
9. Campos, C., Englich, J., Lütgemeier, H., Marysko, M., Novak, P., and Zinn, W. (1992) Magnetization process and site preference of Bi in YIG:Bi epitaxial films studied by ^{57}Fe NMR, *J. Magnetism & Magn. Mat.* **104–107**, 431–432.
10. Winkler, G. (1981) *Magnetic Garnets*, Vieweg Tracts in Pure and Applied Physics, 109.
11. Hansen, P. and Witter, K. (1985) Growth–induced uniaxial anisotropy of bismuth–substituted iron garnet films, *J. Appl. Phys.* **58**, 454–459.
12. Lenz, H., Hansen, P., and Tolksdorf, W. (1989) Growth–induced magnetic and optic anisotropy in bismuth–substituted iron garnet films, *Appl. Phys. Lett.* **54**, 2484–2486.
13. Ando, K., Takeda, N., Koshizuka, N., and Okuda, T. (1985) Annealing effects on growth–induced optical birefringence in liquid–phase–epitaxial–grown Bi–substituted iron garnet films, *J. Appl. Phys.* **57**, 1277–1281.
14. Vassalo, C. (1991) *Optical Waveguide Concepts*, Elsevier, Amsterdam.
15. Shamonin, M. and Hertel, P. (1994) Analysis of nonreciprocal mode propagation in magneto-optic rib-waveguide structures with the spectral index method, *Appl. Opt.* **33**, 6415-6421.
16. Shintaku, T. (1995) Integrated optical isolator based on nonreciprocal higher-order mode conversion, *Appl. Phys. Lett.* **66**, 2789-2791.
17. Dammann, H., Pross, E., Rabe, G., Tolksdorf, W., and Zinke, M. (1986) Phase matching in symmetrical single–mode magneto–optic waveguides by application of stress, *Appl. Phys. Lett.* **49**, 1755-1757.
18. Monerie, M., Leclert, A., Anizan, P., Moisan, G., and Auvray, P. (1976) Dispositifs magnetooptiques en couches minces a accord de phase: utilisation d'une double heteroepitaxie de grenats ferromagnetiques, *Opt. Commun.* **19**, 143-146.

19. Warner, J. (1975) Nonreciprocal magnetooptic waveguides, *IEEE Trans. Microwave Theory & Techn.* **MTT-23**, 70-78.
20. Tien, P.K., Martin, R.J., Wolfe, R., Le Craw, R.C., and Blank, S.L. (1972) Switching and modulation of light in magneto-optic waveguides of garnet films, *Appl. Phys. Lett.* **21**, 394-396.
21. Dammann, H., Pross, E., Rabe, G., and Tolksdorf, W. (1990) 45° waveguide isolators with phase mismatch, *Appl. Phys. Lett.* **56**, 1302-1304.
22. Hemme, H., Dötsch, H., und Hertel, P. (1990) Integrated optical isolator based on nonreciprocal mode cutoff, *Appl. Opt.* **29**, 2741-2744.
23. Auracher, F. and Witte, H.H. (1975) A new design of an integrated optical isolator, *Opt. Commun.* **13**, 435-438.
24. Okamura, Y., Inuzuka, H., Kikuchi, T., and Yamamoto, A. (1996) Nonreciprocal propagation in magnetooptic YIG rib waveguides, *J. Lightwave Technol.* **4**, 711-714.
25. Mizumoto, T., Mashimo, S., Ida, T., and Naito, Y. (1993) In-plane magnetized rare earth iron garnet for a waveguide optical isolator emplying nonreciprocal phase shift, *IEEE Trans. Magn.* **MAG-29**, 3417-3419.
26. Mizumoto, T., Oochi, K., Harada, T., and Naito, Y. (1986) Measurement of nonreciprocal phase shift in a Bi-substituted $Gd_3Fe_5O_{12}$ film and application to waveguide-type optical circulator, *J. Lightwave Technol.* **4**, 347-352.
27. Erdmann, A., Shamonin, M., Hertel, P., and Dötsch, H. (1994) Design of nonreciprocal couplers for integrated optics, *Proc. SPIE* **2150**, 183-192.
28. Nicol, A. and Davis, L.E. (1985) Nonreciprocal coupling in dielectric image lines, *Proc. IEE* **128**, 269-270.
29. Shamonin, M., Lohmeyer, M., Hertel, P., and Dötsch, H. (1996) Radiatively coupled magneto-optic waveguides, *Proc. SPIE* **2695**, 355-361.
30. Okamura, Y., Negami, T., and Yamamoto, S. (1985) A design for a nonreciprocal phase shifter, *Opt. Quant. Electron.* **17**, 195-199.
31. Inuzuka, H., Okamura, Y., and Yamamoto, S. (1989) Improvements in the differential phase shift of magneto-optic waveguides with high refractive index overlayers, *Appl. Phys. Lett.* **54**, 406-408.
32. Gerhardt, R., Sure, S., Dötsch, S., Linkewitz, T., and Tolksdorf, W. (1993) Optical properties of bismuth and gallium substituted thulium iron garnet films, *Opt. Commun.* **102**, 31-35.
33. Wallenhorst, M., Niemöller, M., Dötsch, H., Hertel, P., Gerhardt, R., and Gather, B. (1995) Enhancement of the nonreciprocal magneto-optic effect of TM modes using iron garnet double layers with opposite Faraday rotation, *J. Appl. Phys.* **77**, 2902-2905.
34. Shamonin, M., Lohmeyer, M., Hertel, P., and Dötsch, H. (1996) Magneto-optic waveguides: modeling and applications, *Proc. SPIE* **2695**, 344-354.
35. Shamonin, M. (1995) Nonreciprocal mode propagation in magnetooptic waveguides, Ph.D. thesis, University of Osnabrück.
36. Shamonin, M., Erdmann, A., Hertel, P., and Dötsch, H. (1993) A note on the analysis of nonreciprocal phase shifters by the spectral index method, *Microwave Opt. Technol. Lett.* **6**, 790-792.
37. Erdmann, A., Shamonin, M., Hertel, P., and Dötsch, H. (1993) Finite difference analysis of gyrotropic waveguides, *Opt. Commun.* **102**, 25-30.
38. Koshiba, M. and Zhuang, X.P. (1993) An efficient finite-element analysis of magnetooptic channel waveguides, *J. Lightwave Technol.* **LT-11**, 1453-1458.
39. Shamonin, M. and Hertel, P. (1995) Analysis of nonreciprocal phase shifters for integrated optics by the Galerkin method, *Opt. Eng.* **34**, 849-852.
40. Erdmann, A. and Hertel, P. (1995) Beam-propagation in magnetooptic waveguides, *IEEE J. Quant. Electron.* **QE-31**, 1510-1516.

41. Yamamoto, S., Koyamada, Y., and Makimoto, T. (1972) Normal–mode analysis of anisotropic and gyrotropic thin–film waveguides for integrated optics, *J. Appl. Phys.* **43**, 5090–5097.

42. Yariv, A. (1973) Coupled–mode theory for guided–wave optics, *IEEE J. Quant. Electr.* **QE–9**, 919–933.

43. Fisher, A.D., Lee, J.N., Gaynor E.S., and Tveten, A.B. (1982) Optical guided–wave interaction with magnetostatic waves at microwave frequencies, *Appl. Phys. Lett.* **41**, 779–781.

44. Tamada, H., Kaneko, M., and Okamoto, T. (1987) TM–TE optical–mode conversion induced by a transversely propagating magnetostatic wave in a $(BiLu)_3 Fe_5 O_{12}$ film, *J. Appl. Phys.* **64**, 554–559.

45. Talisa, S.H. (1988) The collinear interaction between forward volume magnetostatic waves and guided light in YIG films, *IEEE Trans. Magn.* **MAG–24**, 2811–2813.

46. Tsai, C.S, and Young, D. (1990) Magnetostatic forward volume wave based guided–wave magneto–optic Bragg cells and applications to communications and signal processing, *IEEE Trans. Microwave Theory & Techn.* **MTT–38**, 560–569.

47. Tsai, C.S. and Young, D. (1989) Wideband scanning of a guided–light beam and spectrum analysis using magnetostatic waves in yttrium iron garnet–gadolinium gallium garnet waveguide, *Appl. Phys. Lett.* **54**, 196–198.

48. Wang, C.L., Pu, Y., and Tsai, C.S. (1992) Permanent magnet–based guided–wave magnetooptic Bragg cell modules, *J. Lightw. Techn.* **10**, 644–648.

49. Tsai, C.S. and Young, D. (1985) Noncollinear coplanar magneto–optic interaction of guided optical wave and magnetostatic surface waves in yttrium iron garnet–gadolinium gallium garnet waveguides, *Appl. Phys. Lett.* **47**, 651–654.

50. Neite, B. and Dötsch, H. (1987) Dynamical conversion of optical modes in garnet films induced by ferrimagnetic resonance, *J. Appl. Phys.* **62**, 648–652.

51. Argyle, B.E. and Malozemoff, A.P. (1972) Experimental study of domain wall response to sinusoidal and pulsed fields, *AIP Conf. Proc.* **10**, 344–348.

52. Blanke, K., Lührmann, B., Wallenhorst, U., Dötsch, H., and Tolksdorf, W. (1991) High–frequency modulation and deflection of light by oscillating domain lattices, *phys. stat. sol. (a)* **124**, 359–369.

53. Polder, D. and Smit, J. (1979) Resonance phenomena in ferrites, *Revs. Modern Phys.* **25**, 89–90.

54. Artman, J.O. and Sharap, S.H. (1979) Domain mode ferromagnetic resonance in materials with K_1 and K_u^a, *J. Appl. Phys.* **50**, 2024–2026.

55. Bi, S.Y., Seagle, D.J., Myers, E.C., Charap, S.H., and Artman, J.O. (1982) Domain mode FMR for H normal to (111) specimens – theory and experiment, *IEEE Trans. Magn.* **MAG–18**, 1337–1339.

56. Lührmann, B., Dötsch, H., and Sure, S. (1993) High–frequency excitations of stripe domain lattices in magnetic garnet films, *Appl. Phys.* **A57**, 553–559.

57. Bosse, A., Lührmann, B., Dötsch, H., Sure, S., and Tolksdorf, W. (1994) Light modulation by resonance excitations of periodic stripe domain lattices in magnetic garnet films, *phys. stat. sol. (a)* **141**, 417–427.

58. Winkler, H.P., Dötsch, H., Lührmann, B., and Sure, S. (1994) Dynamic conversion of optical modes in magnetic garnet films induced by resonances of periodic stripe domains, *J. Appl. Phys.* **76**, 3272–3278.

59. de Leeuw, F.H., van den Doel, R., and Enz, U. (1980) Dynamic properties of magnetic domain walls and magnetic bubbles, *Rep. Progr. Phys.* **43**, 689–783.

60. Lührmann, B., Ye, M., Dötsch, H., and Gerspach, A. (1991) Nonlinearities in the ferrimagnetic resonance in epitaxial garnet films, *J. Magnetism & Magn. Mater.* **96**, 237–244.

61. Ramesh, M., Jedryka, E., Wigen, P.E., and Shone, M. (1985) Coupled oscillations of domain–domain wall system in garnet films, *J. Appl. Phys.* **57**, 3701–3703.

62. Ramesh, M., Pust, L., and Wigen, P.E. (1986) The effect of cubic anisotropy on the coupled domain–domain wall oscillations, *J. Magnetism & Magn. Mater.* **54–57**, 1205–1206.

63. Ramesh, M. and Wigen, P.E. (1988) Ferromagnetodynamics of parallel stripe domains–domain walls system, *J. Magnetism & Magn. Mater.* **74**, 123–133.

64. Malozemoff, A.P. and Slonczewski, J.C. (1979) *Magnetic Domain Walls in Bubble Materials*, Academic Press, Chap. V.

65. Kooy, C. and Enz, U. (1960) Experimental and theoretical study of the domain configuration in thin layers of $BaFe_{12}O_{19}$, *Philips Res. Repts.* **15**, 7–29.

66. Morkowski, J., Dötsch, H., Wigen, P.E., and Yeh, R.J. (1981) Domain wall oscillations in magnetic garnet films, *J. Magnetism & Magn. Mater.* **25**, 39–55.

67. Wood, D.L. and Remeika, J.P. (1967) Effect of impurities on the optical properties of yttrium iron garnet, *J. Appl. Phys.* **38**, 1038–1045.

68. Jovanovic, C., Sure, S., Clausing, E., Scharfschwerdt, C., Neumann, M., Alwes, H., Lorenz, K., Dötsch, H., Tolksdorf, W., and Willich, P. (1992) Influence of growth conditions and annealing parameters on the near–infrared optical absorption of epitaxial magnetic garnet films, *J. Appl. Phys.* **71**, 436–440.

69. Balbashov, A.M., Baktheuzov, V.E. and Tsvetkova, A.A. (1981) Effect of Impurities on the Absorption Spectra of Pb–Containing Garnet Films, *Zh. Prikl. Spektrosk.* **34**, 537–539.

70. Randoshkin, V.V. and Chervonenkis, A.Ya. (1985) Charge neutralization and the electromagnetic properties of iron garnets, *Sov. Phys.–Tech. Phys.* **30**, 796–799.

71. Scott, G.B. and Page, J.L. (1977) Pb valance in iron garnets, *J. Appl. Phys.* **48**, 1342–1349.

72. Hibiya, T. and Nakajima, J. (1983) Optical absorption of liquid phase epitaxial garnet films at $1.3\mu m$ wavelength for magneto–optic application, *J. Appl. Phys.* **54**, 7110–7113.

INTERACTIONS BETWEEN OPTICAL GUIDED MODES AND NONLINEAR MAGNETOSTATIC WAVES

DANIEL D. STANCIL

Dept. of Electrical and Computer Engineering,
Carnegie Mellon University, Pittsburgh, PA 15213.

AND

ANIL PRABHAKAR

Dept. of Physics,
Carnegie Mellon University, Pittsburgh, PA 15213.

1. Introduction

The interaction between optical guided modes and magnetostatic waves (MSWs) in thin films was demonstrated in the early 1980's [1]. Since then, this technology has made several advances both in terms of materials as well as possible device applications. While the interaction is analogous to that observed in acousto-optic (AO) devices, the MSW - optical devices can operate at microwave frequencies as high as 30 GHz [2]. The use of epitaxial yttrium iron garnet (YIG) and Bismuth-substituted $[BiLu]_3Fe_5O_{12}$ films grown on the commonly used gadolinium gallium garnet (GGG) substrate have made it attractive to look at the MSW - optical scattering using dielectric waveguide modes in the thin film. The earliest reported observations were the collinear interaction of optical guided modes with magnetostatic surface waves [1, 3]. Later, the interaction with magnetostatic forward volume waves was reported [4] as well as the transverse interaction with surface waves [5]. These reports sparked a growing interest in MSW - optical devices with several laboratories becoming involved in the different aspects the research.

The MSW - optical interaction can serve as a frequency shifter as has been previously demonstrated [6, 7, 8]. By tuning the bias magnetic field and the microwave frequency simultaneously, it is possible to obtain an optical frequency shifter that has advantages over its AO counterpart for certain applications. The MSW - optic frequency shifter is inherently single-

R. Marcelli and S.A. Nikitov (eds.), Nonlinear Microwave Signal Processing: Towards a New Range of Devices, 467–485.
© *1996 Kluwer Academic Publishers.*

sideband like the AO device, but also provides a much wider tunable frequency shift in the 2-12 GHz range [7]. Using the dual tuning mechanism, we are able to maintain a constant angular deflection of the output beam [8], a property that will play a crucial role in any application involving the use of optical fibers. With gigahertz bandwidths and high tuning speeds, the MO frequency shifter has possible applications in coherent optical communication systems and optical heterodyne spectroscopy.

The interaction between the magnetostatic wave and the optical guided modes depends on the input microwave power fed to the device. A device with high optical conversion efficiency would operate at a high microwave power. Unfortunately, at high power levels the linear theory that predicts a single resonance frequency for the magnetostatic waves is no longer applicable. The lowest order nonlinear effect is a shift in the frequency of the microwave passband. Higher order effects involve the excitation of parametric spin waves that manifest themselves as a modulation on the output microwave signal amplitude. Investigations are in progress to determine the impact of these parametric spin waves on the MSW - optical interaction. In this chapter, we concentrate on the the lowest order nonlinear effects in the MSW - optical interaction that result from the frequency shift in the microwave passband. We do however determine the thresholds for higher order effects and show that they lie within the operating region of a typical MSW - optical device.

2. The Linear Device

The presence of magnetostatic waves in a magnetic garnet film causes small time-varying perturbations in the directions of the magnetization. These perturbations affect the optical permittivity of the medium. The largest variations in the permittivity are linearly dependant on the magnetization and give rise to the Faraday effect. We refer the reader to other sources in the literature for a thorough treatment of the Faraday effect and its influence on the optical properties of the medium [9, 10, 11, 12]. In optical thin-film waveguides, the Faraday rotation manifests itself as mode conversion between TE and TM optical modes. This conversion must also satisfy the laws of energy and momentum conservation described by the following equations,

$$\omega_{TM} = \omega_{TE} \pm \omega_{MSW} \tag{1}$$

$$\vec{\beta}_{TM} = \vec{\beta}_{TE} \pm \vec{\beta}_{MSW} \tag{2}$$

where ω denotes the frequency and $\vec{\beta}$ the wave-vector of the respective wave. The conversion itself can be described mathematically by a pair of coupled mode equations[12],

$$\frac{\partial A}{\partial z} = \kappa B(z)e^{-i\Delta z} \tag{3}$$

$$\frac{\partial B}{\partial z} = \kappa A(z)e^{+i\Delta z} , \tag{4}$$

where A and B are the amplitudes of the two optical modes, the mismatch in their wave-vectors is $\Delta = \left[\vec{\beta}_A - \vec{\beta}_B \mp \vec{\beta}_{MSW}\right] \cdot \hat{z}$ and κ is the coupling coefficient given by

$$\kappa = -i\frac{\Phi_F}{M_s d}\sqrt{\frac{2P}{\omega_q \mu_0}}. \tag{5}$$

Φ_F is the Faraday rotation (rotation angle per unit length of propagation in the medium), M_s the saturation magnetization, P the power coupled into the MSWs (in mW/mm), ω_q the interaction frequency, d the thickness of the film and μ_0 the permeability of free space. Since the coupling coefficient κ largely determines the strength of the MSW - optical interaction, we find that for a given medium, the efficiency of the device can be improved by using the $\kappa \propto \sqrt{P}$ dependence on microwave power. Unfortunately, as we increase P, the interaction frequency ω_q also changed. The effect of a changing resonance frequency on the MSW - optical interaction will be the primary subject of our discussion for the rest of this chapter.

3. Nonlinear Effects

Lowest order nonlinear effects in the magnetostatic wave - optical interaction are attributed to a shift in the resonance frequency of the propagating MSWs. For a film magnetized perpendicular to the plane, the internal field is reduced by a demagnetizing field numerically equal to the saturation magnetization, M_s. Magnetostatic waves with $k \to 0$ are associated with a nearly-circular precession of the magnetization about the film normal. If the cone angle of the precession is θ, as the amplitude of the oscillation increases, the perpendicular component of the magnetization is reduced by the factor $\cos\theta$. This reduces the demagnetizing field and causes an increase in the frequency of the $k \to 0$ band edge by the amount

$$\Delta\omega = \omega_M(1 - \cos\theta) \approx \omega_M\frac{\theta^2}{2}. \tag{6}$$

where $\omega_M = -\gamma\mu_0 M_s$ and $\gamma = -2\pi(28 \text{ GHz/T})$ is the gyromagnetic ratio. For nearly circular precession, the cone angle θ can be related to the MSW power using the approximation [12]

$$|\theta| = \frac{|m|}{M_s} \approx \frac{2}{M_s d}\sqrt{\frac{2P}{\omega\mu_0}} , \tag{7}$$

where P is the power per unit width along the MSW beam. Equations (6) and (7) are used to determine the frequency shift in the MSW passband as we increase the input power to the transducer. We note however that the above expressions were derived assuming that $\cos\theta \approx 1 - \frac{\theta^2}{2}$, an approximation that is valid as long as $\frac{\theta^2}{2} >> \frac{\theta^4}{24}$. Using (7) with $d = 7\mu m$, $\omega = 2\pi(5 \text{ GHz}), M_s = 140 \text{ kA/m}$, a bidirectional transducer of length 1cm and an input power of 35 dBm, we obtain $\theta = 0.2$ radians. Hence, the assumption of a small cone angle is valid for all the power levels used in our experiments. In Sec. 3.1 we shall see that the shift $\Delta\omega$ is easily observed by monitoring the MSW passband as we increase the input power to the transducer. This shift has also been observed in experiments conducted on MSW delay lines [13].

The interaction condition between the light beams and the MSWs is determined by the momentum conservation equation (2). Again, as we increase the microwave power, the component of M_0 along the direction of the applied magnetic field (H_{dc}) is reduced to $M_s\cos\theta$. For small θ, the dispersion relation for MSWs in the long wavelength limit is approximated to be [14]

$$\beta_q = \frac{4}{\omega_M d}(\omega_q - \omega_{dc} + \omega_M) \tag{8}$$

where $\omega_{dc} = -\gamma\mu_0 H_{dc}$ and ω_q is the interaction frequency. The resulting shift in MSW wavenumber can now be obtained as

$$\Delta\beta_q = \frac{\partial\beta_q}{\partial M_s}\Delta M = \frac{4(\omega_{dc} - \omega_q)}{\omega_M d}(\cos\theta - 1) . \tag{9}$$

The original wavenumber is restored by increasing the frequency by the amount

$$\Delta\omega_q = -(\frac{\omega_M d}{4}\cos\theta)\Delta\beta_q . \tag{10}$$

For small precession angles, using (7), (9) and (10), we find that restoring the original momentum conservation condition can be accomplished by shifting the interaction to a higher MSW frequency given by

$$\omega'_q = \omega_q + \frac{\omega_{dc} - \omega_q}{\omega_q}\frac{4P}{\mu_0 M_s^2 d^2} . \tag{11}$$

3.1. VARIATIONS IN THE MICROWAVE PASSBAND

The following experiment was conducted on a YIG film with $d=7.4\mu m$ and $M_s=140$ kA/m. The film was placed in contact with an input microstrip transducer in the presence of an external magnetic field of 3.50 kG that

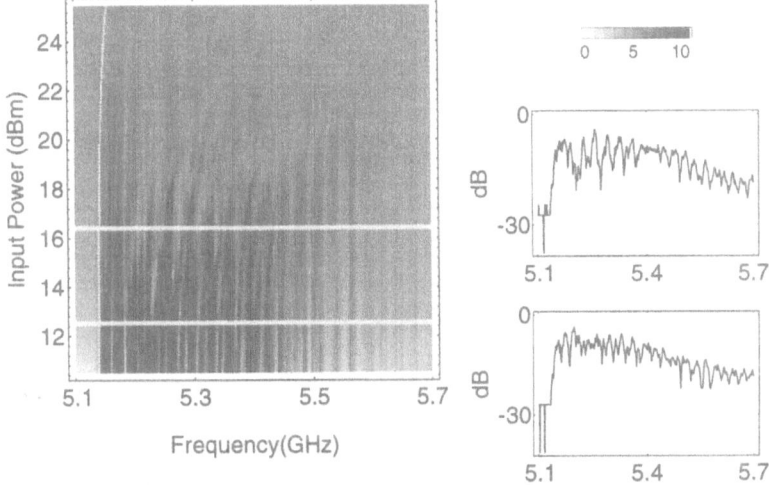

Figure 1. A density plot of the transmission characteristics (S_{12})over the fre-quency-power parameter space. The curvature of the lower edge of the passband was determined theoretically and appears as a strong white curve. In the top right corner is the grey level scheme used to generate the density plot, with minimum transmission cor-responding to white and maximum to black. The horizontal lines mark the power levels at which the two passbands at the right were measured. While the data in the density plot is on a linear grey scale, the passbands have been shown on a logarithmic scale.

was perpendicular to the plane of the film. A second transducer, 3mm from the first, monitored the output signal from the device and fed it to a digital oscilloscope via a diode detector. By sweeping the input frequency between 5.1 and 5.7 GHz, we obtain the the transmission characteristics of the film in the form of a passband. Typical passbands are shown in the right half of Fig. 1. By measuring a number of such passbands we can capture the transmission characteristics over frequency-power parameter space. In the left half of Fig. 1, we see a density plot of the transmission characteristics and notice that the density plot captures the features of the two passbands on the right. (The horizontal white stripes mark the locations of the pass-bands in the density plot.) A density plot is merely a top-down view of a 3-D plot with grey scales being used to denote the height of the contour. The data in Fig. 1 was calculated by taking the ratio of the output and input microwave power levels. This ratio is similar to the S_{12} parameter used to characterize typical microwave circuits. Bright regions in the den-sity plot correspond to areas of weak transmission. The variations in the passbands appear as a global phenomenon with dark regions being inter-spersed with brighter regions. At low power, the almost periodic pattern of bright and dark is attributed to the existence of either multi-path interfer-

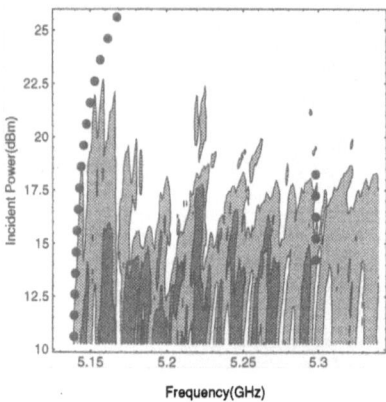

Figure 2. Transmission characteristics over a small region of the parameter space. The data is shown on a 3 level contour plot. The edge of the passband is located at 5.14 GHz. The dark spots that follow a curve are the calculated locations for the edge of the passband as we increase the input microwave power. The vertical dark spots at 5.3 GHz are the locations in parameter space where samples of the output microwave frequency spectrum were obtained. These spectra are shown in Fig. 3 and demonstrate the correlation between the finger-like region of weak transmission and the onset of auto-oscillations.

ence or dipole gaps [15] in the passband. As we increase the input power level, the periodicity is broken and instead long "finger-like" stretches of poor transmission make their appearance. We believe this phenomenon to be a consequence of the parametric excitation of spin waves, a high-order effect. For our present purpose, we concentrate on the variations in the edge of the passband, clearly visible in Fig. 1. The bright curve is a predicted shift in the edge, calculated using (6) and (7). There is a good agreement between the experimentally observed shift in the passband and the corresponding theoretical prediction. This comparison between the theoretical and experimental values has been highlighted in Fig. 2.

Fig. 2 is a magnified view of the transmission characteristics over a small frequency range. We use a 3 level contour plot to highlight the formation of finger-like regions of low transmission around 15 dBm. This is also the power level corresponding to the second Suhl instability at the main resonance[16]. At power levels higher than the instability threshold, parametric spin waves are excited and appear in the output frequency spectrum as secondary peaks around the primary resonance frequency. Alternatively, the parametric excitations manifest themselves in a time series signal as a low frequency modulation ($\sim$100 kHz) and are referred to as auto-oscillations. We have observed that the finger-like regions of low transmission can also be associated with the auto-oscillations. For example, the dark spots in Fig. 2 at 5.3 GHz are the locations in parameter space where the out-

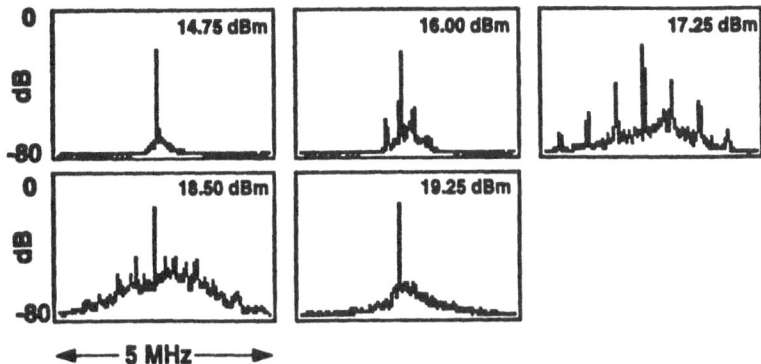

Figure 3. Microwave frequency spectrum measured on a HP8529D Spectrum Analyzer. The input frequency was 5.3 GHz while the the input power was stepped between 14.75 dBm and 19.25 dBm.

put frequency spectrum was monitored. These spectra are shown in Fig. 3 and we observe the onset of auto-oscillations as we traverse the finger of low transmitted power. Zhang and Suhl attribute the auto-oscillations to a Hopf bifurcation of the entire spin-wave manifold at the main resonance [17]. We have observed that the onset of auto-oscillations occurs at the tip of a finger-like region[18]. Our results indicate that the two phenomena, the onset of auto-oscillations and the global variations in transmitted characteristics, are manifestations of the same underlying physical principles. It is worthwhile to note that fingers of auto-oscillation have also been observed in experiments on circular YIG discs [19]. A detailed investigation of these phenomena is currently in progress. Note that the auto-oscillations seen in Fig. 3 are approximately 20 dB weaker than the primary signal. While we expect that the contributions of the parametrically excited spin waves in a MSW - optical device will be small [20], they may still be significant in some practical applications.

3.2. THE TRANSVERSE MSW - OPTICAL INTERACTION

The experimental configuration shown in Fig. 4 was used to characterize the nonlinear effects in the transverse MSW - optical interaction [21]. A $Bi_{0.8}Lu_{2.2}Fe_5O_{12}$ film of thickness $6.3\mu m$ was used in the experiment. The film had a FMR linewidth of about 3 Oe, measured at 9.1 GHz. The sample dimensions were 5mm x 15 mm, and the two 5mm edges were polished to allow coupling to the optical modes. Forward volume waves were excited by the transducer in the presence of an external bias field of 0.457 T (4.57 kG) and an optical beam with a wavelength of $1.3\mu m$ was edge-coupled into the film.

We use the term interaction spectrum to refer to the intensity of the

474

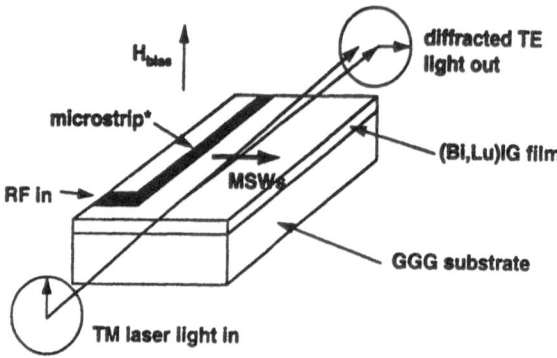

Figure 4. Geometry for the transverse MSW - optical interaction. The microstrip is actually defined on a 254 mm thick alumina substrate backed by a ground plane. The ground plane is above the device in the orientation shown [21].

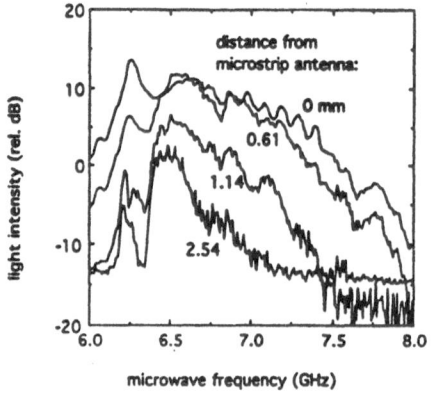

Figure 5. Interaction spectra measured at different distances from the transducer. The input power to the microstrip was 25.5 dBm (355mW) corresponding to 71 mW/mm for each propagation direction [21].

converted output beam as a function of microwave frequency. Several such spectra measured at different distances from the transducer are shown in Fig 5. By tracking a specific feature on the spectra, in this case a peak at the low frequency edge, we monitor variations in the interaction spectrum as we change the experimental parameters.

Fig. 6 shows the interaction spectrum as the optical beam passes directly beneath the transducer for different incident microwave power levels. The shift of the peak identified by the arrow is readily apparent as power is increased. The frequency of this peak as a function of MSW power is shown in Fig 7. The squares and diamonds represent two separate measurements differing from each other only by an error introduced while resetting the magnetic bias field. An experimental fit to the data yields a slope of 0.384

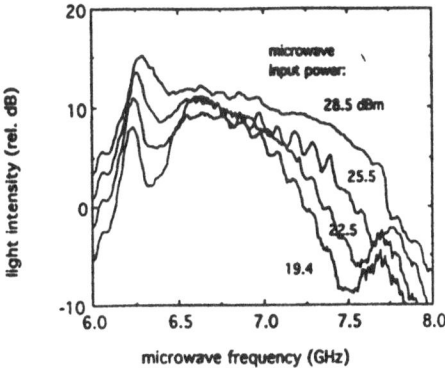

Figure 6. Interaction spectra taken with the optical beam directly beneath the transducer for various values of total microwave input power [21].

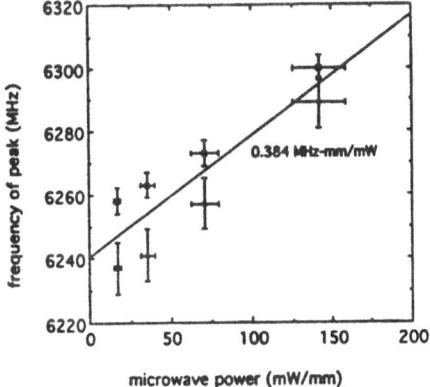

Figure 7. Frequency shift of the peak identified in Fig 6 as a function of microwave power. The power is normalized to the length of the transducer (5mm) [21].

MHz-mm/mW. This slope can be compared with the theoretical value of $\frac{\Delta f}{P} = 0.513$ MHz/mm calculated using (6) and (7). Though there remains a sizeable discrepancy in the experimental and theoretical values, there is a reasonable agreement in their orders of magnitude. The discrepancy could be a result of thermal effects [21].

The transverse interaction geometry facilitates yet another experiment. By tracking the intensity of the peak in Fig 5 as a function of distance from the transducer, we can obtain information about the MSW wave profile and its decay characteristics. Though not directly related to the frequency shift measurements in the transverse interaction, this experiment proved invaluable in understanding the frequency shift in the collinear interaction geometry. The results of this experiment and its utility will be discussed in

476

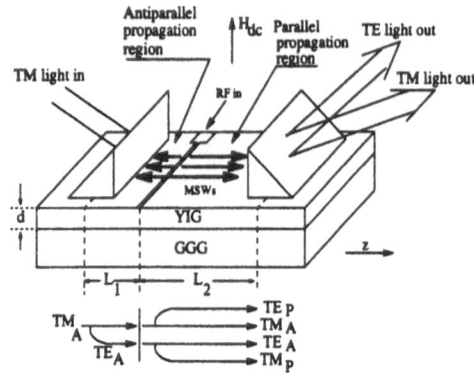

Figure 8. Device geometry used in the MSW-optical interaction with a schematic sketch showing the different optical modes and their regions of interaction [23].

the following section.

3.3. THE COLLINEAR MSW - OPTICAL INTERACTION

The effects of a shift in MSW frequency have also been observed in the collinear MSW - optical interaction[22]. However, the geometry of the device made a characterization of the nonlinear effects difficult. With the help of data obtained in the MSW passband measurements and in the transverse MSW - optical interaction, it became possible to understand and predict the nature of the collinear interaction at high microwave power levels [23].

The device configuration used to model the collinear magnetostatic - optical wave interaction is shown in Fig. 8. The magnetostatic waves (MSWs) interact with incident TM polarized light (identified as TM_A). The guided light initially propagates antiparallel to the MSWs. Once the light crosses the transducer, it begins to propagate parallel to the MSWs. The subscripts A and P describe modes that originate in the antiparallel and parallel regions shown in Fig. 8. The MO interaction lengths are controlled by the distances of the input and output prisms from the transducer. In our model, the reflection of MSWs from the sample ends is assumed to be negligible. For the experiments to be considered, this is a reasonable approximation owing to MSW attenuation and the much greater pathlength for reflected waves. By satisfying energy conservation, the input light is coupled into three other optical modes with different frequencies, as shown schematically in Fig. 8:

$$\omega_{TE_A} = \omega_{TM_A} - \omega_{MSW}$$
$$\omega_{TE_P} = \omega_{TM_A} + \omega_{MSW}$$
$$\omega_{TM_P} = \omega_{TE_A} - \omega_{MSW} = \omega_{TM_A} - 2\omega_{MSW} \ . \tag{12}$$

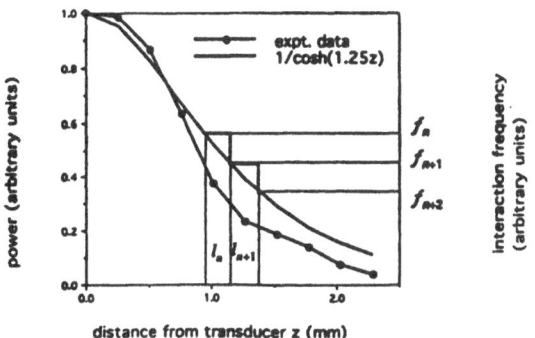

Figure 9. MSW attenuation close to the transducer. The experimental data is fit to a curve of the form $1/\cosh(1.4\,z)$. Since the microwave interaction frequency is proportional to power (Eq. 11), this curve is used to find the distance from the transducer z_n at which maximum conversion occurs for frequency ω_n. The lengths of the uniform coupled lines in the discrete approximation are determined by $l_n = z_{n+1} - z_n$ [23].

The final output is measured using an analyzer and detector configured to detect only TE polarized light.

One expects the microwave power to be a maximum close to the transducer and to fall off rapidly at large distances. A continuous first derivative at z=0 (i.e. at the center of the transducer) and an exponential decay for large z are two features that one expects for the power in a real device. A simple analytic function with this behavior is $1/\cosh(\xi z)$. Fig. 9 shows a comparison between this analytic function and experimental data obtained from the interaction spectra of Fig 5 for the $(BiLu)_3Fe_5O_{21}$ film in the transverse interaction geometry. FMR studies have shown that the YIG film being used for the simulations in the present paper has a microwave attenuation, ξ, approximately one-third that of the $(BiLu)_3Fe_5O_{21}$ film. For computational efficiency, we use the lowest order approximation to $1/\cosh(\xi z)$ given by $P(z) = P_0/(1+\xi^2 z^2)$. While this approximation has a small effect on the quantitative results, the qualitative features of the interaction frequency response remain unchanged. MSW loss measurements reported in the literature are usually in terms of exponential decay rates and it is necessary to appropriately scale ξ so as to correctly describe MSW attenuation in the film. For a film 10 mm long, a value of $\xi=0.35$ mm^{-1} corresponds approximately to an exponential decay rate of 0.25 mm^{-1}.

The TE and TM polarized beams interacting with magnetostatic forward volume waves exchange energy in a manner described by the coupled mode theory outlined in Sec. 2. For the collinear interaction, the film is treated as a large number of cascaded segments, with the length of each segment determined by the discrete frequency steps used in the simulations (Fig. 9). For example, if the maximum interaction at frequency ω_n occurs at position z_n, then the contribution to the total mode conversion from this region is modeled using coupled uniform modes over a length

478

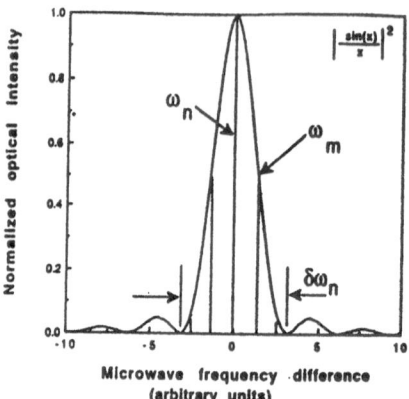

Normalized optical intensity

Microwave frequency difference
(arbitrary units)

Figure 10. Discretization of the bandwidth of interaction around the primary interaction frequency ω_n [23].

$l_n = z_{n+1} - z_n$ with power $P(z_n)$. The strongest interaction in the nth segment occurs within a full bandwidth $\delta\omega_n = \pi\omega_M d/l_n$ around ω_n [24], as shown in Fig. 10. The frequency step $\omega_{n+1} - \omega_n$ is chosen to obtain satisfactory numerical convergence. In each segment we solve the coupled mode equations at every frequency ω_m such that $|\omega_m - \omega_n| \leq \delta\omega_n/2$. For computational reasons, we only consider the mode conversion that occurs within this bandwidth. We believe this gives the dominant behavior of the response, although including a wider bandwidth may affect the fine structure of the interaction passband. The output amplitudes at the mth frequency for the nth segment are calculated using the following solutions to the coupled mode equations (4) [12].

$$A_{mn}(l_n) = [\frac{\kappa_n B_{mn}(0) + (i\Delta_{mn}/2)A_{mn}(0)}{\beta_{mn}}\sin(\beta_{mn}l_n)$$
$$+ A_{mn}(0)\cos(\beta_n l_n)] \cdot e^{-i\Delta_{mn}l_n/2} \qquad (13)$$

$$B_{mn}(l_n) = [\frac{\kappa_n A_{mn}(0) - (i\Delta_{mn}/2)B_{mn}(0)}{\beta_{mn}}\sin(\beta_{mn}l_n)$$
$$+ B_{mn}(0)\cos(\beta_{mn}l_n)] \cdot e^{i\Delta_{mn}l_n/2} , \qquad (14)$$

where $A_{mn}(0) = A_{m,n-1}(l_{n-1})$ and $B_{mn}(0) = B_{m,n-1}(l_{n-1})$ are the inputs to the nth segment. The phase mismatch Δ_{mn} and phase parameter β_{mn} are estimated as [24]

$$\Delta_{mn} = 4 \cdot (\omega_m - \omega_n)/(\omega_M d) \qquad (15)$$

$$\beta_{mn} = \sqrt{\kappa_n^2 + \Delta_{mn}^2/4} . \qquad (16)$$

A discrete form of (5) is used to calculate the coupling constant κ_n.

$$\kappa_n = \frac{\Phi_F}{M_s d}\sqrt{\frac{2P(z_n)}{\omega\mu_0}} , \qquad (17)$$

479

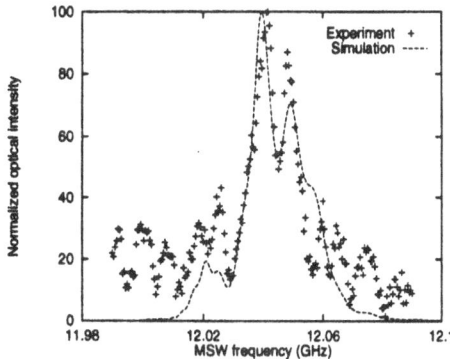

Figure 11. Comparison between experimental data and the results of a simulation using L₁=4mm, L₂=7mm, P=300mW/mm, H_dc=487kA/m (6.12kOe), ξ=0.35/mm and ω_q−2π(12 GHz). Each curve is normalized to its maximum value. The experiment was conducted with an input power of 35dBm (3.16W) being fed to a transducer 5mm long (P=316mW/mm assuming uniform radiation and bidirectional transducers) and with an external magnetic field of 487kA/m [23,26].

where $\Phi_F = 210°/mm$ is the Faraday rotation in YIG for an optical beam with wavelength 1.3μm [25]. A more accurate calculation of κ_n would involve evaluating (17) at each frequency ω_m within the bandwidth $\delta\omega_n$ and repeating the calculation for every segment l_n. However, we notice that the final bandwidth of the frequency response is less than 100 MHz. We choose to calculate κ at the interaction frequency near the transducer and merely scale it thereafter by $\sqrt{P(z_n)/P(z_0)}$. The errors in κ_n introduced by such an approximation are less than 10% and are outweighed by the significant improvement in computational performance.

3.3.1. *Simulation Results*

The output of the MO device as measured by a polarized detector would be the sum of the intensities of different optical modes with the same polarization. (Here we assume that the detector is not fast enough to respond to the microwave beat frequencies.) The experiments were conducted by Nykolai Bilanuik on a YIG sample. Simulations were run for a device having a total interaction length of 11 mm, a film with d=7.4 μm, M_s=140 kA/m ($4\pi M_s$=1.76 kG), $|\gamma|$=2π(28 GHz/T), H_{dc}=300 kA/m (3.77 kOe) and ω_q=2π(6.2 GHz).

A seemingly peculiar formation of multiple peaks in the interaction frequency response has been observed experimentally [26]. Fig. 11 allows us to compare the results of our simulations against experimental data. The experimental curve was obtained by measuring the decrease in intensity of the TM output beam, whereas the theoretical curve represents the increase in the intensity of the TE output beam. Assuming the light converted from

480

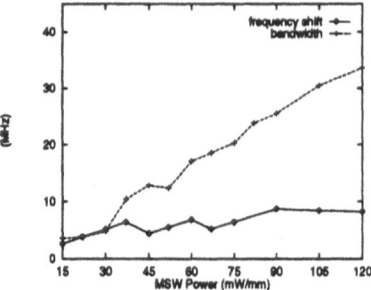

Figure 12. Effects of high microwave power on MO interaction frequency and interaction bandwidth. The power fed into the transducer is varied while $\xi=0.35$/mm, $L_1=0$mm and $L_2=11$mm are fixed [23].

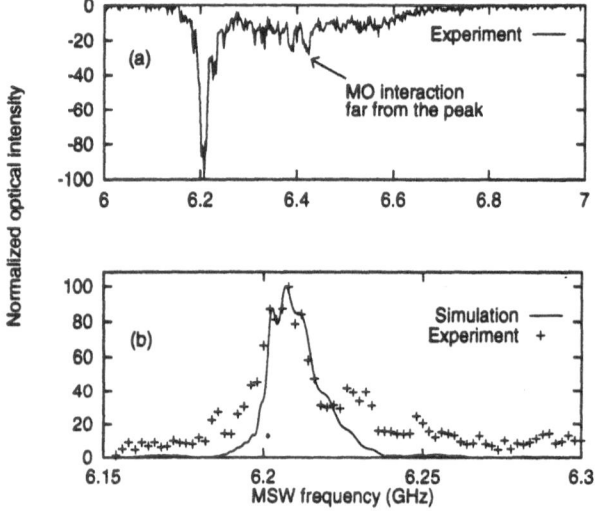

Figure 13. (a) Experimental data showing a broad bandwidth of MO interaction. The experiment was conducted with an input power of 27dBm (P=50mW/mm assuming uniform radiation and bidirectional transducers of length 5mm), $H_{dc}=294$kA/m (3.70kOe) and d=7.4μm. Data was collected by sweeping the microwave frequency in steps of 2MHz. (b) Experimental data near the main peak and its corresponding simulation with parameters $L_1=2$mm, $L_2=8.5$mm, P=50mW/mm and $H_{dc}=300$kA/m (3.77kOe). Both sets of data have been normalized to their maximum value [22,23].

the TM polarization appears in the TE polarization, these curves should be vertically inverted images of each other. In the figure the experimental curve has been inverted to facilitate comparison with theory. The simulations were undertaken with parameter values that attempt to match those used in the experiment. The splitting of the interaction peak is clearly evident in both the experimental and theoretical curves.

Fig. 12 shows the change in the full width at half maximum of the interaction spectrum and the corresponding shift in the frequency of the primary peak as we increase the input microwave power levels. In addition to the spreading of the main peak, some experiments have shown the appearance of a weaker interaction band extending upto 100 MHz from the main peak [22, 27]. Fig. 13 is one example of such an interaction [22]. As in Fig. 11, the experimental curve was obtained by measuring the decrease in intensity of the TM output beam. The present theory gives us a reasonable description of the width of the main peak but cannot explain the origin of the broader features in the experimental curve.

4. Designing an efficient MSW - optical device

The demonstration of an electronically tunable MSW - optical frequency shifter was originally conducted on a YIG sample [7]. The low Faraday rotation of YIG placed restrictions on the efficiency of any reasonably sized device. The search for a higher interaction efficiency has led to research on new ferrimagnetic materials with an emphasis on Bismuth-substituted garnet films. TM → TE optical mode conversion with an efficiency as high as 40% was demonstrated using $[BiLu]_3Fe_5O_{12}$ films [28]. Recently, a frequency shifter with an efficiency of 8% was demonstrated on a Bi-YIG film[8]. Researchers have also fabricated high performance magneto-optic waveguide isolators using Bismuth substituted films[29, 30, 31]. An early overview of magneto-optic isolator materials and devices is given in [32].

Our studies on YIG and Bismuth substituted films at high microwave power levels have shown that the higher microwave attenuation in the latter films make them less susceptible to the parametric excitation of spin waves. The threshold for the Suhl instability at the main resonance has been observed to be about 10 dB higher than that in the YIG films. This resistance to chaotic propagation is a feature that is highly desirable in efficient single sideband frequency-shift applications. Bismuth doped films also commonly exhibit strong spin-wave resonance absorption lines in the microwave spectrum. These absorption lines are undesirable for many device applications, but can be minimized by properly orienting the film in the magnetic bias field [33].

The MSW - optical frequency shifter is analogous to acousto - optic devices that employ acoustic waves to diffract an optical beam. As a case study, we calculate the effects of high microwave power levels on an MSW - optical frequency shifter. As previously mentioned, one of the advantages of the MSW - optical device is that the deflection angle can be held fixed as the frequency is varied. This is accomplished by synchronizing the frequency variation with a compensating change in the magnetic bias field. At high power levels, an additional bias field correction is necessary to compensate for the nonlinear frequency shift. As indicated in equation (2), the MSW wave-vector $\vec{\beta}_q$ couples the TE and TM optical guided modes. Assuming an output optical beam polarized in the TM direction, the deflection angle is given by $\phi = \tan^{-}1\frac{\beta_q}{\beta_{TM}}$. At a constant input microwave power level, we can keep the deflection angle fixed for different frequencies by ensuring that we always operate at a the point in the passband with the same wavenumber β_q. Using the dispersion relation in the long wavelength limit given by (9), we maintain a constant deflection angle at the new interaction frequency $\tilde{\omega}_q$ by increasing the external bias field to

$$\tilde{H}_{dc} = H_{dc} + \frac{\tilde{\omega}_q - \omega_q}{|\gamma|\mu_0} . \tag{18}$$

If we now increase the input microwave power by ΔP, the passband shifts by $\Delta \omega$ as calculated using (6) and (7). The corresponding shift $\Delta \beta_q$ again affects the deflection angle. We correct for this nonlinear shift by reducing the external bias field by

$$\Delta H_{dc} = \frac{\Delta \omega}{|\gamma|\mu_0} = \frac{4\omega_M \Delta P}{|\gamma|\omega_q \mu_0^2 M_s^2 d^2} . \tag{19}$$

Finally, the external bias field required to maintain a constant deflection angle ϕ is calculated as $\tilde{H}_{dc} - \Delta H_{dc}$. The dual tuning mechanism comprising of the microwave frequency and the external bias magnetic field can be used in the above manner to ensure an optimal magnetostatic wave - optical interaction and a constant angular deflection of the optical beam over a gigahertz frequency range. It follows that since both the power and bias field influence the interaction, these two degrees of freedom can be used to keep both the amplitude and the angle of the diffracted beam constant as the frequency is varied.

5. Conclusion

In the preceding sections, we described the excitation of magnetostatic waves at high microwave power levels. The lowest order nonlinear effect is a shift in the resonance frequency of the magnetostatic waves. The frequency shift is clearly observed as we monitor the microwave transmission characteristics of the device. By retaining the second order terms for the demagnetizing field, we were able to predict the shift in the passband and found that the experimental observations closely followed our theoretical predictions. We introduced a new technique of monitoring variations in the passband using density plots over the frequency-power parameter space. While the frequency shift is clearly visible in the density plots, we also observed that the plots capture higher order nonlinear effects in the microwave spectrum. In particular, the threshold for the Suhl instability at the main resonance and the excitation of parametric spin waves beyond the threshold manifest themselves as sudden global variations in the density plot. A detailed analysis of these higher order effects is under investigation.

In the case of the MSW - optical interaction, the frequency shift in the microwave passband alters the phase matching condition between the optical guided modes and the MSWs. The phase matching is restored by shifting the interaction frequency to compensate for the nonlinear frequency shift. In the transverse interaction, this phenomenon appears as a shift in the interaction spectrum. By tracking a specific feature in the interaction spectrum, we calculated the frequency shift and found that it was in good agreement with the theoretical values that we predicted. The interaction spectrum in the transverse geometry was also used to study how the MSW power decayed away from the transducer. This data was then used to understand the nonlinear effects in the collinear geometry.

MSW-optical experiments conducted in the collinear geometry revealed that the interaction spectrum consisted of multiple lobes. The interaction did not occur at a unique interaction frequency, but was spread over a bandwidth of approximately 10 MHz. This spread could not be explained by a spatially uniform frequency shift in the microwave passband. Using data obtained in the transverse interaction, we modeled the MSW power as it decayed away from the transducer and introduced a spatially non-uniform frequency shift. We simulated the MSW - optical interaction with parameter values that were comparable to those used during the experiments. The formation of multiple peaks in the interaction spectrum and an increase in the bandwidth were observed. The results of our simulations were compared with those obtained experimentally and a strong correlation was observed between the two interaction spectra.

On the basis of our theoretical calculations and our experimental observations, we conclude that the lowest order nonlinear effects in the microwave transmission characteristics have a significant effect on the MSW - optical interaction. Since we believe that the use of high microwave power levels is necessary for an improvement in the efficiency of MSW - optical devices, it also becomes imperative for us to account for the nonlinear effects while designing any practical device. The theoretical expressions given in Sec. 3 appear to describe the frequency shift accurately and can be used to appropriately compensate for any nonlinear behavior that is observed in the MSW - optical interaction.

6. Acknowledgments

We would like to thank J. Peruyero and D. Adam for providing us with the samples used in our experiments. We gratefully acknowledge A. Cash and

N. Bilaniuk for many helpful discussions and for permission to use some of their experimental data on the magneto - optic interaction. This work was supported by the National Science Foundation, U.S.A., under Grant No. 9206817.

References

1. Fisher, A.D., Lee, J.N., Gaynor, E.S. and Tveten, A.B. (1982) Optical guided-wave interactions with magnetostatic waves at microwave frequencies, *Appl. Phys Lett.*, **41**, 779.
2. Owens, J.M., Ataiiyan, Y.J., Avitabile, G. and Carter, R.L. (1986) Ka band magnetostatic wave delay lines, *Proc. IEEE Ultrasonics Symp.*, 183.
3. Tsai, C.S. (1983) Hybrid integrated optic modules for real-time signal processing, *Proc. of NASA Optical Information Processing Conf. II*, NASA Conf. Pub. No. **2302**, 149.
4. Fisher, A.D., Gaynor, E.S. and Lee, J.N. (1983) Magnetostatic wave devices for integrated-optical signal processing, *IEEE Ultrasonics Symp.*, **1**, 226.
5. Tsai, C.S., Young, D., Chen, W., Adkins, L., Lee, C.C. and Glass H. (1985) Noncollinear coplanar magneto-optic interaction of guided optical wave and magnetostatic surface waves in yttrium iron garnet - gadolinium gallium garnet waveguides, *Appl. Phys. Lett.*, **47**, 651.
6. Stancil, D.D. (1989) Optical frequency shifter using magnetostatic waves, U.S. Patent no. **4,796,983**.
7. Bilaniuk, N. and Stancil D.D. (1990) An optical frequency shifter using magnetostatic waves, J. Appl. Phys. **67**, 508.
8. Pu, Y. and Tsai, C.S. (1993) Wideband electronically tunable integrated magneto-optic frequency shifter at X band, *Appl. Phys. Lett.* **62**, 3420.
9. Prokhorov, A.M., Smolenskii, G.A. and Ageev, A.N. (1984) Optical phenomena in thin-film magnetic waveguides and their technical application, *Sov. Phys. Usp.*, **27**, 339.
10. Stashkevich, A.A. (1989) Waveguide interaction of light with spin waves in a ferromagnetic film, *Sov. Phys. Jour.*, **32**, 241.
11. Yariv, A. (1973) Coupled-mode theory for guided-wave optics, *IEEE J. Quant. Elect.*, **QE-9**, 919.
12. Stancil, D.D. (1993) *Theory of Magnetostatic Waves*, Springer Verlag, New York.
13. Tsankov, M.A., Chen, M. and Patton, C.E. (1994) Magnetostatic wave dynamic magnetization amplitude response in yttrium iron garnet films at high microwave power levels, **HE-10**, Sixth Joint MMM-Intermag Conference, Alberquerque, NM.
14. Stancil, D.D. and Bilaniuk, N. (1995) Collinear interaction of optical guided modes with microwave spin waves in magnetic films, in *High Frequency Processes in Magnetic Materials*, G. Srinivasan and A. Slavin, Ed., World Scientific Publishing Co., pp 357-393.
15. Wolfram, T. and DeWames, R.E. (1971) Magnetoexchange branches and spin-wave resonance in conducting and insulating films: perpendicular resonance, *Phys. Rev. B*, vol.4, no. 9, 3125.
16. Suhl H. (1957) The theory of ferromagnetic resonance at high signal powers, *J. Phys. Chem. Solids*, **1**, 209.
17. Zhang, X.Y. and Suhl, H. (1988), Theory of auto-oscillations in high power ferromagnetic resonance, Phys. Rev. B, **38**, 4893.
18. Prabhakar, A. and Stancil, D.D. (1996, in press) Variations in auto-oscillation frequency at the main resonance in rectangular YIG films, J. Appl. Phys, **79**(8).
19. McMichael, R.D. and Wigen, P.E. (1990), High power ferromagnetic resonance without a degenerate spin-wave manifold, Phys. Rev. Lett., **64**, 64.

20. Stashkevich, A.A. (1995), Interaction of light with a nonlinear spin-wave in a normally magnetized ferromagnetic film, paper **DD-06** presented at MMM'95 Conf., Philadelphia.

21. Cash, A.F. and Stancil, D.D. (in press) Measurement of magnetostatic wave profiles using the interaction with transverse optical guided waves, *IEEE Trans. Mag*, based on paper **HE-11**, Sixth Joint MMM-Intermag Conference 1994, Alberquerque, NM.

22. Bilaniuk, N. (1992) A study of the collinear interaction between magnetostatic waves and optical guided modes in garnet thin films, *PhD Thesis*, Carnegie Mellon University.

23. Prabhakar, A. and Stancil, D.D. (in press) Effects of High Microwave Power on Collinear Magnetostatic - Optical Wave Interactions, *IEEE Trans. Magn.*

24. Bilaniuk, N. and Stancil, D.D. (1989) Effective interaction lengths in the collinear magnetostatic wave - optical interaction, *SPIE Integrated Optics and Optoelectronics*, **1177**, 365.

25. Stancil, D.D. (1991) Optical-magnetostatic wave coupled-mode interactions in garnet heterostructures, *IEEE J. Quant.Elect.*. **QE-27**, 61.

26. N. Bilaniuk, unpublished.

27. Talisa, S.H. (1988) The collinear interaction between forward volume magnetostatic waves and guided light in YIG films, *IEEE Trans. Magn.,*, **24**, 2811.

28. Tamada, H., Kaneko, M. and Okamoto, T. (1988) TM-TE optical-mode conversion induced by a transversely propagating magnetostatic wave in $(BiLu)_3F_5O_12$ film, *J.Appl. Phys.*, **64**, 554.

29. Ando, K., Okoshi, T. and Koshizuka, N. (1988) Appl. Phys. Lett., vol. **53**, no. 1, 4.

30. Wolfe, R., Lieberman, R.A., Fratello, V.J., Scotti, R.E. and Kopylov, N. (1990) Appl. Phys. Lett., vol. **56**, no. 5, 426.

31. Levy, M., Illic, I., Scarmozzino, R., Osggod, R.M., Wolfe, R., Gutierrez, C.J. and Prinz, G.A. (1993) Thin-film-magnet magnetooptic waveguide isolator, IEEE Photonics Tech. Lett., vol. **5**, no. 2, 198.

32. Tsushima, K. (1988) Overview of magneto - optic isolator materials and devices, J. Appl. Phys., vol. **63**, no. 8, 3118.

33. Chernakova, A.K., Cash, A., Peruyero, J. and Stancil, D.D. (1994) Orientation dependence of dipole gaps in the magnetostatic wave spectrum of Bi - substituted iron garnets, *J. Appl. Phys.* **64**, 554.

INTEGRATED MAGNETOOPTIC DEVICES WITH APPLICATIONS TO RF SIGNAL PROCESSING AND COMMUNICATIONS[*]

CHEN S. TSAI

Department of Electrical and Computer Engineering and
Institute for Surface and Interface Sciences
University of California, Irvine, CA 92717, U.S.A.

Abstract

A number of significant advances have been made recently in MSW-based integrated magnetooptics. A new class of optical modulator called integrated MO Bragg cells which are potentially capable of providing desirable features similar to that of the now prevalent AO Bragg cells, but with potentially superior performance characteristics, have been realized. These advances have paved the way for construction of a variety of MO Bragg cell-based integrated optic devices, such as high-speed optical scanners and space switches, tunable carrier frequency RF spectrum analyzers and correlators, and wideband optical frequency shifters and modulators. In this paper a review on the integrated MO Bragg cells and their potential applications to RF signal processing and communication is presented.

1. Introduction - Why Integrated Magnetooptic Devices?

The increasing needs for analog integrated devices that function at ultrahigh frequency ranges of electromagnetic radiation, from 10 GHz and up, for both military and civilian uses are by now very well documented. More recently, there

---

[*] This work was supported by the ONR.

R. Marcelli and S.A. Nikitov (eds.), Nonlinear Microwave Signal Processing: Towards a New Range of Devices, 487–507.
© 1996 *Kluwer Academic Publishers.*

also has been increasing R and D interests and activities in photonics-based analog radio frequency (RF) technology. For example, photonics-based analog devices required for the following tasks are being sought: 1). Multiband RF signal distribution operating from 1 to 100 GHz, demonstrating RF frequency independence, 2). Tunable Optical Heterodyne Transmitter/Receiver operating from 1-100 GHz with incremental tuning capability of 1 GHz, and Channelizers capable of operation in the 1-100 GHz range with 1 GHz resolution, and 3). Optical Null Steering Antenna Processor with wideband null steering implemented at the RF frequency. Thus, the relevancy of the analog integrated magnetooptic (MO) devices to be presented in this paper is clearly shown as follows.

Guided-Wave Magnetooptics (GWMO) upon which the integrated MO devices is based is a field similar to, but considerably less developed than Guided-Wave Acoustooptics (GWAO) [1]. Fig. 1 shows the basic configuration for guided-wave acoustooptic (AO) Bragg diffraction upon which a whole array of integrated AO devices [2] are based. Since the early 1970's a large number of studies have been focused on GWAO in which both the light and the surface acoustic waves (SAW) are confined to a small depth in suitable solid substrates. This focus on GWAO was a natural outgrowth of the guided-wave optics science and technology and the SAW device technology developed earlier. These studies on GWAO have already generated much fruitful results. For example, the resulting wide-band planar AO Bragg modulators (cells) and deflectors [1] were widely used in earlier years in the development and realization of microoptic modules for real-time processing of radar signals, e.g., the integrated-optic RF spectrum analyzers [1]. However, it should be noted that like the SAW devices the operating carrier frequencies of such guided-wave AO devices are limited to the VHF and UHF regions, namely a couple of GHz. Consequently, for such AO devices to be used at higher carrier frequencies, down-conversion techniques must be employed which in turn require much more hardware, and thus result in greater complexity and higher cost.

GWMO is concerned with MO interactions between the guided-optical waves and the magnetostatic waves (MSWs) in suitable magnetic substrates such as the Yttrium Iron Garnet-Gadolinium Gallium Garnet (YIG-GGG) waveguides [3]. The science and technology of MSW-based GWMO has become an international effort [3]. MSWs can be readily and efficiently generated by applying a microwave signal to a microstrip transducer deposited directly on the YIG layer or brought over it. For the YIG-GGG substrate, the carrier frequency of the MSW can be tuned,

Table1. <u>Applications of Integrated Acoustooptic Devices</u>

I. **Communications**
    Light-beam modulation and deflection
    Multiport switching
    Space-, Time- and wavelength-division multiplexing/demultiplexing
    Tunable optical wavelength filtering
    Optical-frequency shifting and heterodyne detection
    Optical interconnect

II. **Radio-Frequency Signal Processing**
    Spectral analysis or Fourier transform
    Pulse compression
    convolution
    Time- and space-integrating correlations
    Adaptive filtering
    Ambiguity function

III. **Computing**
    Matrix-vector multiplication
    Matrix-matrix multiplication
    Programmable analog and digital correlation/filtering
    Optical bistability

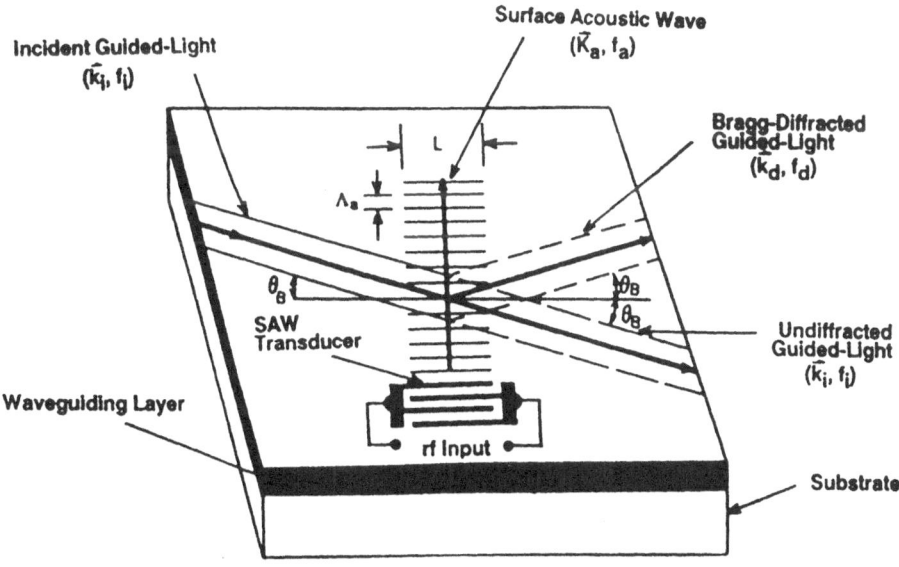

Fig. 1 Configuration of A Basic Planar Waveguide Acoustooptic Bragg Cell Modulator

490

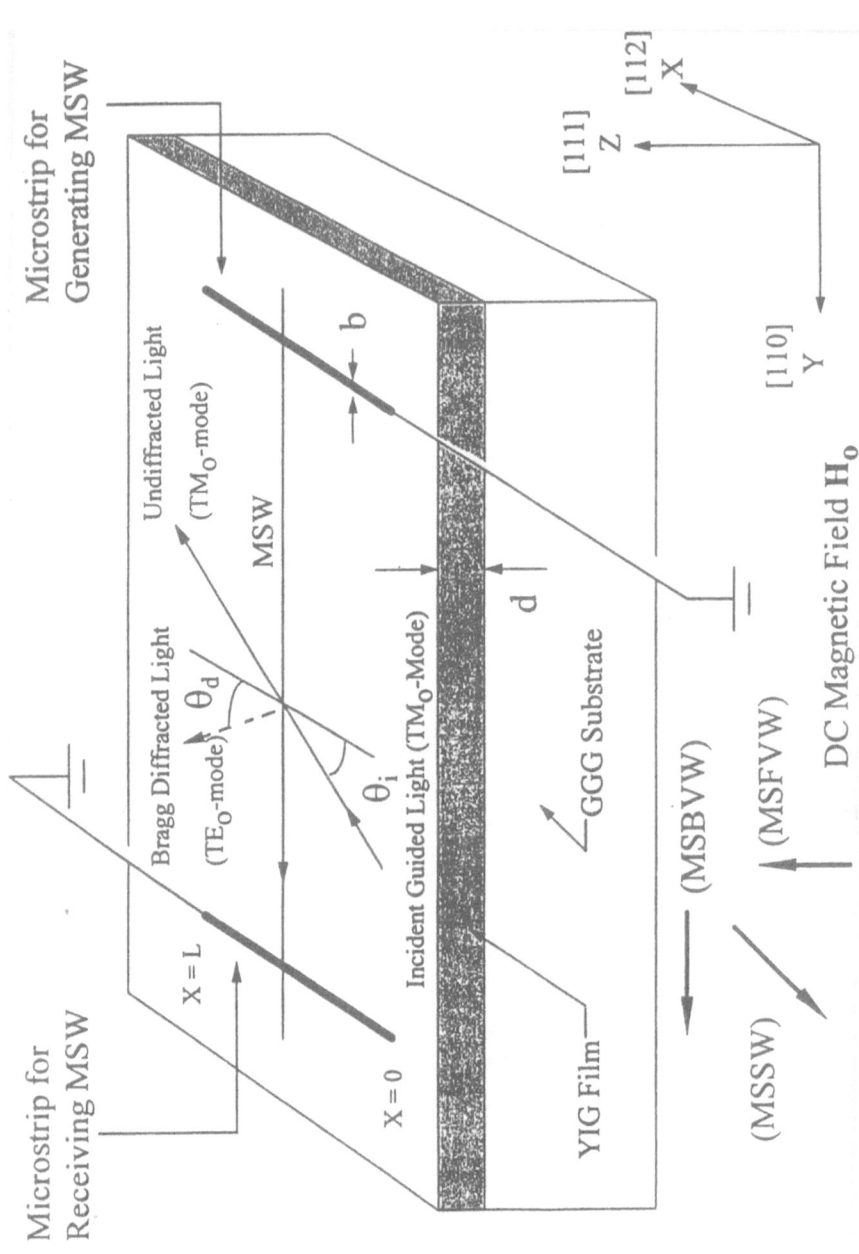

Fig. 2 Noncollinear Coplanar Guided-Wave Magneto-Optic Bragg Diffraction From Magnetostatic Waves in YIG-GGG Substrate

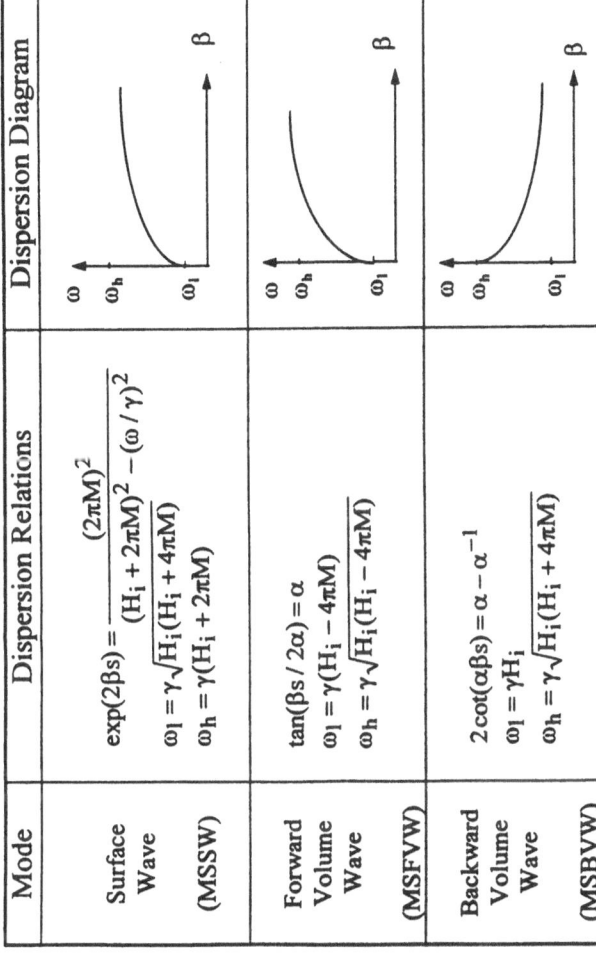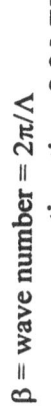

Mode	Dispersion Relations	Dispersion Diagram
Surface Wave (MSSW)	$\exp(2\beta s) = \dfrac{(2\pi M)^2}{(H_i+2\pi M)^2 - (\omega/\gamma)^2}$ $\omega_l = \gamma\sqrt{H_i(H_i+4\pi M)}$ $\omega_h = \gamma(H_i+2\pi M)$	
Forward Volume Wave (MSFVW)	$\tan(\beta s/2\alpha) = \alpha$ $\omega_l = \gamma(H_i-4\pi M)$ $\omega_h = \gamma\sqrt{H_i(H_i-4\pi M)}$	
Backward Volume Wave (MSBVW)	$2\cot(\alpha\beta s) = \alpha - \alpha^{-1}$ $\omega_l = \gamma H_i$ $\omega_h = \gamma\sqrt{H_i(H_i+4\pi M)}$	

$\alpha = [(\omega/\gamma)^2 - H_i^2]\cdot[H_i(H_i+4\pi M)-(\omega/\gamma)^2]^{-1}$,

$4\pi M$ = YIG saturation magnetization,

H_i = internal field = $H_{external} + H_{anisotropic}$,

β = wave number = $2\pi/\Lambda$

γ = gyromagnetic ratio = 2.8 MHz/Oe

s = YIG film thickness

Fig. 3 Excitation and Propagation of Magnetostatic Waves in YIG Films

typically from 0.5 to greater than 20 GHz, by simply varying an external bias magnetic field. MO interactions result from the moving optical gratings induced by the MSW through the Faraday and Cotton-Mouton effects [4] in a manner similar to the guided-wave AO interactions in which the SAW induces moving optical gratings through the photoelastic effect [1]. The much higher carrier frequencies and the much larger degree of their tunability that associate with the integrated MO devices in comparison to the integrated AO devices clearly show the relevancy of the former.

2. Noncollinear coplanar MSW-Based MO Bragg Diffraction And Resulting MO Bragg Cell Modulator

Figure 2 shows the interaction geometries for noncollinear coplanar guided-wave MO Bragg diffraction, depicting the distinct directions of the bias magnetic field (H_o) for the three cases that involve, respectively, the magnetostatic surface wave (MSSW), the magnetostatic forward volume wave (MSFVW), and the magnetostatic backward volume wave (MSBVW). Microstrip transducers are commonly used for excitation of the MSWs. The required orientations for the microstrip transducer and the bias magnetic field for excitation of each of the MSW modes in a YIG film are also depicted in Fig. 3. The corresponding dispersion relations and diagrams are also provided in the figure. It is to be noted that both the lower and the upper frequency bounds (ω_ℓ and ω_h) increase with the bias magnetic field and that ω_ℓ for the MSBVW is higher than that of the MSSW and the MSFVW by about $\gamma 4\pi M$ in which γ is the gyromagnetic ratio (2.8 MHz/Oe) and M is the saturation magnetization of the YIG film. For the commonly used YIG samples, this frequency differential is about five GHz. Also, while the direction of group velocity is the same as the phase velocity for both the MSSW and the MSFVW, the group velocity of the MSBVW is in a direction opposite to that of the phase velocity.

It is relatively simple to facilitate efficient and wideband excitation of the MSWs using the microstrip transducers. Figure 4 shows an example that provides an insertion loss (conversion of RF power to MSW power) of about -10 dB and a -3 dB bandwidth greater than 1.0 GHz. Figure 5 shows the corresponding dispersion diagram for the same range of carrier frequency, indicating a wide range of

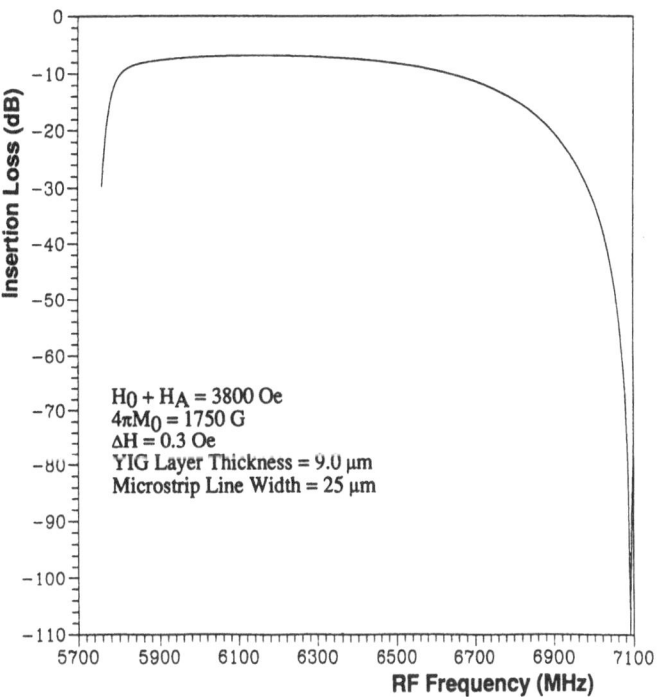

Fig. 4 **Calculated Frequency Response Of Microstrip Transducer Insertion Loss**

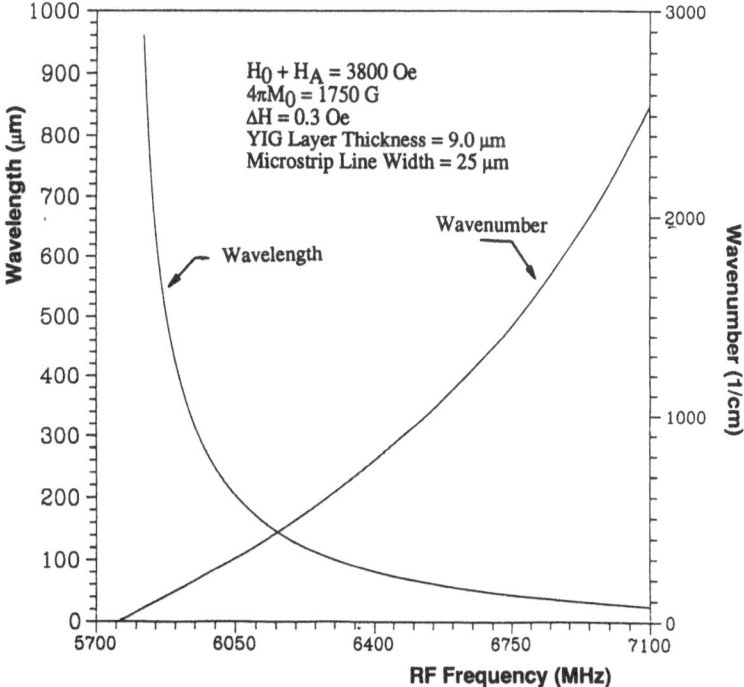

Fig. 5 **Computer Generated Wavenumbers And Wavelengths Of MSFVW In Pure YIG Waveguide**

variations in the wavenumber and the corresponding wavelength for the MSFVW. The dispersion diagram also shows that a very large range of phase velocity is associated with the MSFVW, e.g., one to three orders of magnitude higher than that of the SAW. Consequently, the transit time of the MSW and, thus, the speed of the resulting MO space switches and scanners will be one to three orders of magnitude faster than their AO counterparts.

As stated in the Introduction, propagation of the MSW generated by the input microstrip transducer induces moving optical diffraction gratings in the YIG-GGG waveguide through the Faraday and Cotton-Mouton effects. Referring to Fig. 6, portions of the input guided-light wave, incident upon the gratings at Bragg angle, is diffracted and frequency-shifted, and propagates at Bragg angle with respect to the gratings in the plane of the waveguide. The intensity of the Bragg-diffracted light is directly proportional to the power of the MSW before some nonlinear or saturation effect sets in [5]. The diffracted light is scanned on the plane of the waveguide as the carrier frequency of the MSW is varied. It is important to note that in contrast to guided-wave isotropic AO Bragg diffraction, the output angle of the diffracted light does not vary linearly with the carrier frequency of the MSW as a result of the dispersive property of the MSW. Also, for the same reason the diffracted light is scanned on the plane of the waveguide as the bias magnetic field is varied, while the carrier frequency of the MSW remains fixed. In analogy with the AO Bragg cell modulators, the resulting devices are called the MO Bragg cell modulators [1, 6].

3. Recent Advances

A number of advances toward realization of high-performance integrated MO Bragg cell-based devices have been made recently.

3.1 ANALYSIS FOR DESIGN OF HIGH-PERFORMANCE MO BRAGG CELL MODULATORS

A detailed coupled-mode analysis on guided-wave MO Bragg interaction in a practical four-layer YIG/GGG structure as shown in Figure 7 was carried out recently [7]. This analysis has identified the key device parameters and the

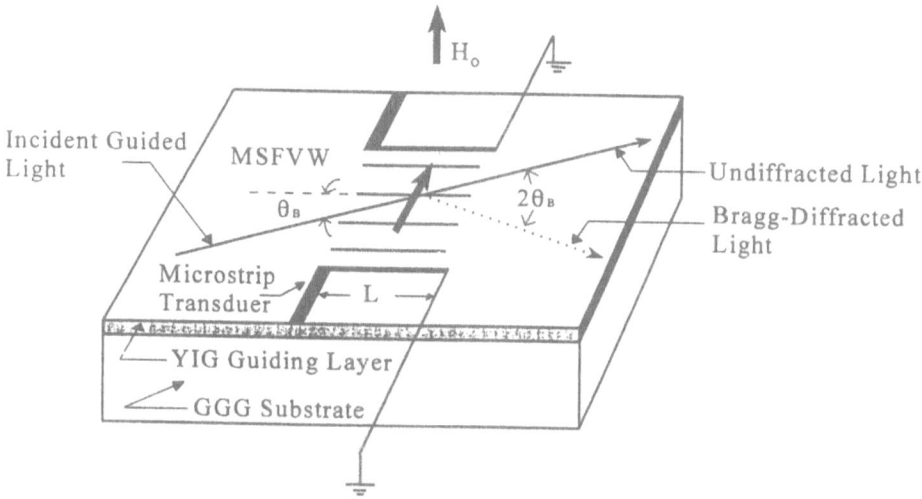

Fig. 6 **A Basic Planar Waveguide Magnetooptic Bragg Cell Modulator**
(For the Case Involving Magnetostatic Forward Volume Waves)

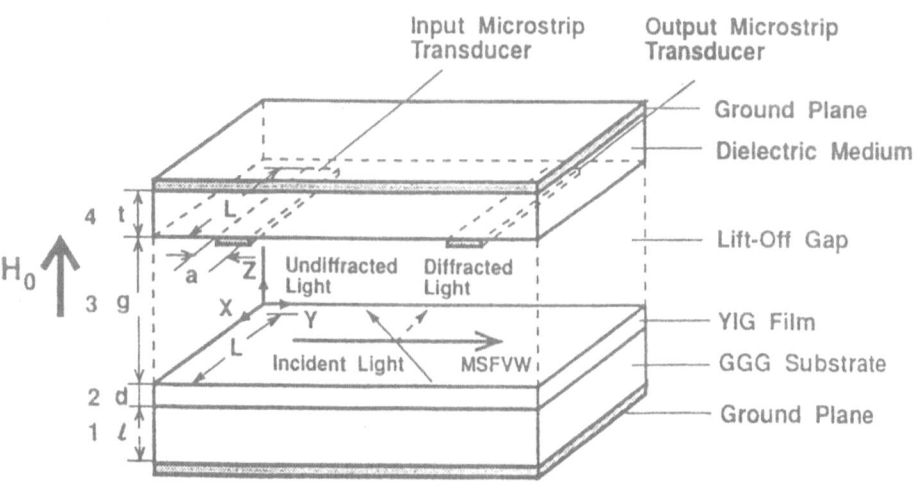

Fig. 7 The YIG-GGG Four-Layer Structure for MSFVW Based Guided-Wave
Magnetooptic Bragg Diffraction

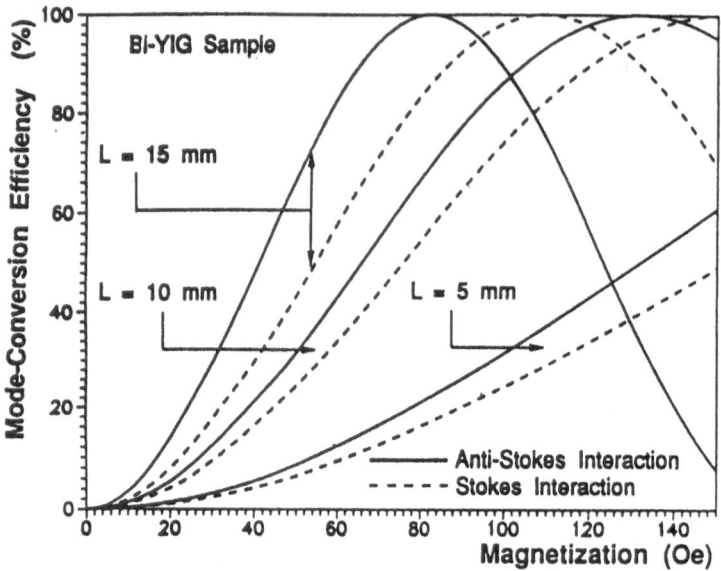

Fig. 8 Bragg Diffraction Efficiency of Guided-Wave Magnetooptic
Bragg Diffraction Versus RF Magnetization of the MSFVW

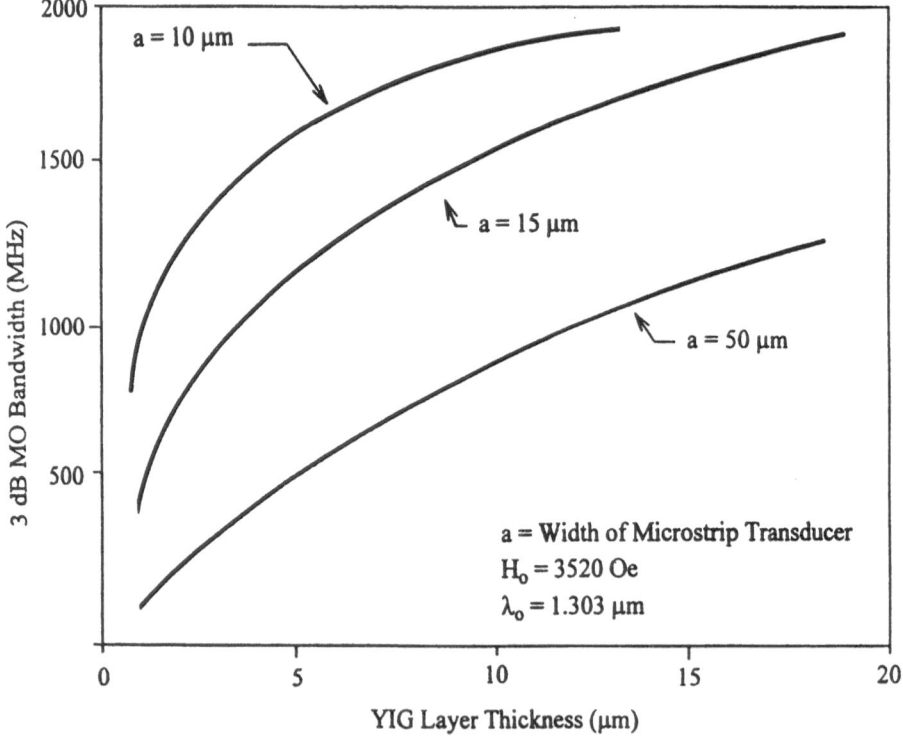

Fig. 9 Calculated Magnetooptic Bragg Bandwidth Versus YIG Layer Thickness
with the Width of Microstrip Transducer as a Parameter

guidelines for design and construction of high-performance guided-wave MO Bragg cell modulators. Only the results on two most important device performance parameters, the diffraction efficiency and the interaction bandwidth, are given here. Figures 8 and 9 show sample calculations based on this analysis. First, we note that a maximum MO Bragg diffraction efficiency of 12% was previously measured in a device that utilized a 0.5 cm interaction length in a Bi-doped YIG/GGG waveguide [6]. The required RF drive power was about 2.0 watt. Thus, Figure 8 suggests that at the same RF drive power density the corresponding diffraction efficiency will be increased to about 44 and 85%, respectively, when the interaction length is enlarged by a factor of two and three. Figure 9 shows that for a commonly used YIG layer thickness of about 3.25 μm, the MO Bragg bandwidths of 300, 850, and 1,500 MHz, respectively, will be obtainable with the width of microstrip transducer equal to 50, 15, and 10 μm. Thus, a device bandwidth greater than one gigahertz is achievable as previously demonstrated [8].

3.2 MINIATURIZED MO BRAGG CELL MODULATORS

It is clear that in order to facilitate real-world applications of the guided-wave MO Bragg cell modulators technologies for their miniaturization, integration and packaging must be devised and developed. Toward this goal the first miniaturized MO device module using a pair of small permanent magnets together with a pair of current-carrying coils in a compact magnetic circuit were realized recently [9]. Figure 10 is a photograph of the miniaturized MO device module. Fig. 11 shows a sketch of the compact magnetic circuit. The pair of permanent magnets, each 1" x 1" x 0.25" in size, were made of samarium-cobalt (Sm-Co). The pole pieces of smaller cross-sectional area were used to concentrate the magnetic flux and provide a uniform magnetic field in the air gap where the MO Bragg cell sample was inserted. The current-carrying coils were wound on nylon robins and placed around the pole pieces in order to facilitate electronic tuning of the magnetic field in the air gap. Using this simple magnetic circuit, a uniform bias magnetic field (H_o) at the gap was readily varied from 1600 to 4126 Oe. The corresponding tuning range for the carrier frequency of the MSFVWs was measured to be from 2.0 to 12.0 GHz. For example, as shown in Figure 12 the measured changes in the bias magnetic field (ΔH_{oe}) induced by a current of 0.85 amp was as high as 2446 Oe at an air gap of 6.16 mm. These current-controlled changes in the bias magnetic field provided a

Fig. 10 Photograph of Miniaturized Magnetooptic
Bragg Cell Modulator

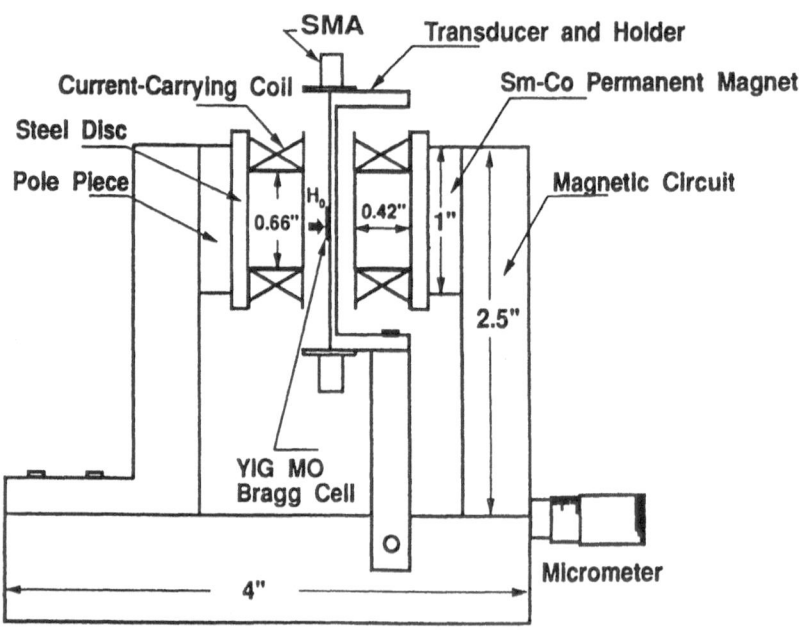

Fig. 11 Compact Magnetic Circuit Used in Realization of
Miniaturized Magnetooptic Bragg Cell Modulator

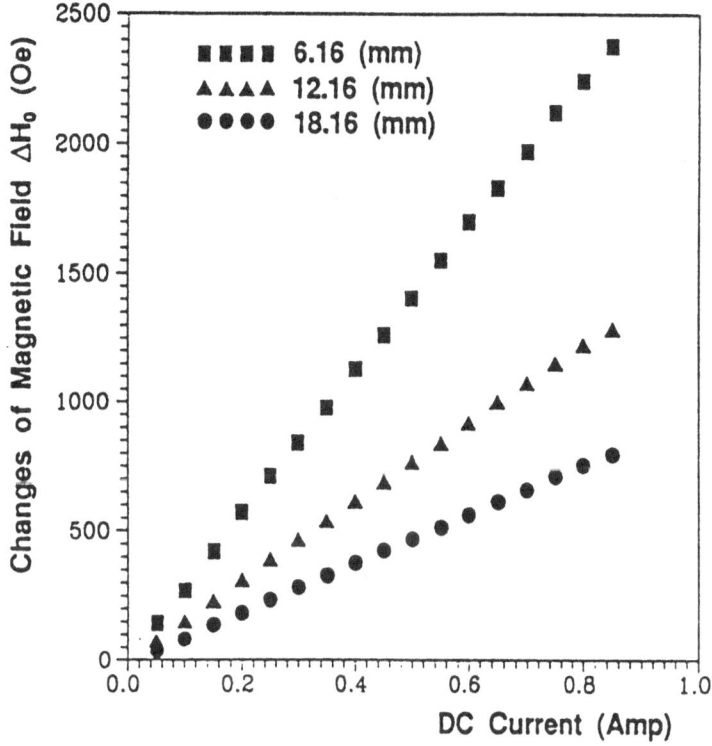

Fig. 12 Measured Changes of Magnetic Field Induced by Current-Carrying Coils vs. DC Current at Three Different Air Gaps

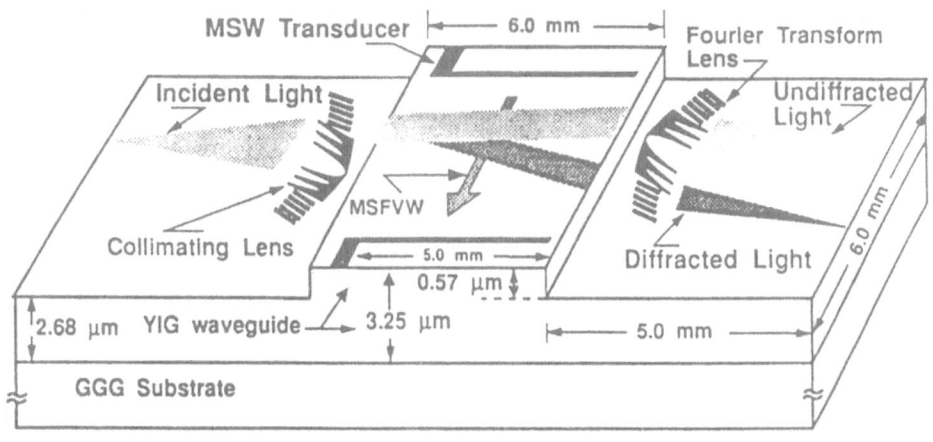

Fig. 13 Integrated Magnetooptic Bragg Cell Device Module in Bi-Doped YIG-GGG Tapered Waveguide Structure

carrier frequency tuning bandwidth as high as 6.85 GHz. Clearly, even larger changes in the bias magnetic field, and thus larger tuning bandwidths for the carrier frequency could be obtained at a smaller air gap.

Using the compact magnetic circuit just described miniaturized MO Bragg cell modulators using the MSFVW and the MSBWV have been realized. A brief description on each now follows:

3.2.1 Compact MSFVM-based MO Bragg Cell Modulator

Both pure and Bi-doped YIG-GGG waveguide samples having, respectively, 4.2 and 3.5 μm YIG layer thickness were used in the construction of compact MSFVW-based MO Bragg cell modulators. Measurements of performance characteristics were carried out at an optical wavelength of 1.303 μm using the setup and procedure previously reported [9]. The measured performances were comparable to those obtained using a bulky electromagnet. For example, the measured diffraction efficiency at a carrier frequency of 7.6 GHz for a Bi-doped YIG Bragg cell was 7.14% at an RF drive power of 28.2 dBm (660 mW) or a corresponding calculated MSFVW power of 20 mW [9]. A dynamic range of 40 dB was also reproduced. However, for RF drive powers higher than 28.2 dBm, some deviation from linearity between the diffraction efficiency and the RF drive power was observed [6]. Based on the theoretical predictions [7], compact MO Bragg cell modulators with even higher carrier frequency, larger range of tunable carrier frequency, higher diffraction efficiency, and smaller physical size can be constructed. The compact MSFVW-based MO Bragg cell modulators have been used to demonstrate applications in light beam scanning and RF spectral analysis.

3.2.2 Compact MSBVW-based MO Bragg Cell Modulator

The compact magnetic circuit described previously was also utilized to construct miniaturized MSBVW-based MO Bragg cell modulator [10]. First, it is to be noted that the direction of the bias magnetic field required for excitation of the MSBVW as depicted in Figure 5 necessitated a considerably greater gap between the two pole pieces. However, as stated in Section 2, the lower bound frequency of the MSBVW is higher than that of the MSFVW by about $\gamma 4\pi M$. For pure YIG and Bi-YIG samples, we have $\gamma 4\pi M$ = 1750 Oe and 1800 Oe, respectively, and the corresponding $\gamma 4\pi M$ are 4.90 GHz and 5.04 GHz.

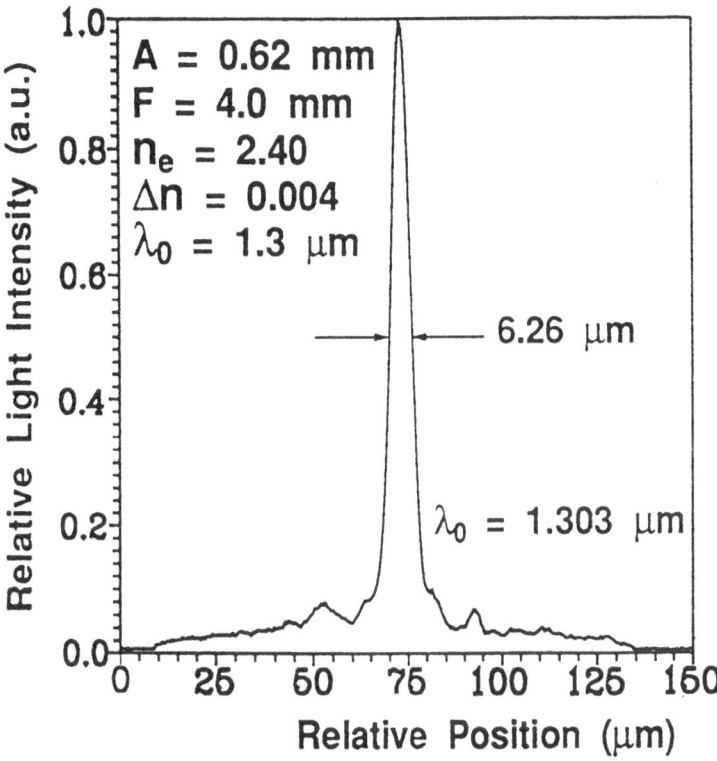

Fig. 14 **Measured Profile of Focused Light Spot for Curved**
Hybrid Lens Pair

For example, the center carrier frequencies of the MSBVWs were measured to be 4.78 and 6.10 GHz, respectively, at the bias magnetic fields of only 1,000 and 1,900 Oe [10]. Such relatively low magnetic fields were readily provided by the compact magnetic circuit at a gap greater than 8.0 mm. Thus, it was possible to insert the MO Bragg cell sample of adequate dimensions into the air gap of the same magnetic circuit to realize miniaturized MSBVW-based MO Bragg cell modulators.

The resulting MSBVW-based MO Bragg cell modulators in both pure and Bi-doped YIG-GGG samples have demonstrated light beam scanning from the UHF band (< 2 GHz) to the X-band (> 10 GHz). The measured performances of the MSBVW-based MO Bragg cell modulators, including the center frequency and its tuning range and the corresponding tuning range of the bias magnetic field, the location of the first passband, the diffraction efficiency, the microwave drive power, and the light beam scan angle were found to be superior to that of the MSFVW-based MO Bragg cell modulators.

3.3 INTEGRATED MO BRAGG CELL MODULATOR MODULES

Aside from the miniaturization of the magnetic circuit as described in Section 3.2, it is also essential to incorporate a collimation-focusing waveguide lens pair in a common YIG-GGG substrate. As in integrated AO device modules [2], such lens pair is required in a variety of integrated MO device modules such as RF signal processors [1], optical space switches and scanners [11], and optical frequency shifters and modulators [12]. Recently, a curved hybrid ion-milled collimation-focusing lens pair [13] of parabolic contour was integrated with a MO Bragg cell modulator in a common YIG-GGG taper waveguide substrate 6.0 x 16.0 mm^2 in size as shown in Fig. 13 [14].

Flat hybrid waveguide lenses of straight line contour which combine analog Fresnel and chirp gratings were devised and fabricated earlier in GaAs, LiNbO$_3$ and YIG-GGG waveguide substrates using ion-milling technique [15-17]. Such ion-milled lenses had provided both high throughput efficiency and near diffraction-limited focal spot size for a light beam propagating within a small angle from the lens axis. However, when the light beam propagated at a larger angle from the lens axis the overall performance was significantly degraded due to the high degree of coma incurred. The curved hybrid lenses have demonstrated capabilities for larger angular field of view and lower level of coma [18]. Furthermore, such curved hybrid lenses were recently formed in a tapered YIG-GGG waveguide structure to accommodate simultaneously the requirements for large MO Bragg bandwidth and high lens throughput, and thus facilitated the realization of integrated MO Bragg cell modulator modules [14] as depicted in Fig. 13. The initial layer thickness of the Bi-doped YIG waveguide sample 6.0 x 16.0 mm^2 in size was 3.25 μm. The two end regions of the tapered waveguide were ion-milled down to 2.68 μm in several steps in order to produce a gradual transition, and thus ensure a high transmission for the light beam. The curved hybrid lenses with 4.0 mm focal length and 0.8 mm aperture were then fabricated onto the two end regions (each 5.0 mm in length) using the ion-milling technique. As seen from Fig. 14, the measured focal spot profiles obtained at 1.3 μm wavelength from the curved hybrid lenses show practically no coma with sidelobe levels lower than 12.3 dB from the main lobe for the incident light angle up to ± 3.5 degrees from the lens axis. An MO Bragg cell modulator was subsequently constructed by incorporating a microstrip transducer in the central region of the tapered waveguide. The compact magnetic circuit described previously in Section 3.2 was used to provide the required bias magnetic field for saturation of the YIG layer as well as excitation and tuning of the carrier frequency of the MSFVWs ranging from 2 to 12 GHz. Typically performance figures such as a bandwidth of 260 MHz at the center carrier frequency of 10.0 GHz, a diffraction efficiency of 2.0% at one watt RF drive power, and a dynamic range of 30 dB were measured with the resulting integrated MO Bragg cell modulator module. Again, the device module was used to demonstrate light beam scanning and switching, and RF spectral analysis.

4. Novel Applications to RF Signal Processing and Communication

All of the integrated AO devices that are based on the AO Bragg cell modulator [1,2] can also be realized using the MO Bragg cell modulator. Accordingly, all applications listed in Table 1 that have been demonstrated using the integrated AO devices may also be accomplished using the integrated MO devices. However, in comparison to their AO counterparts, the unique advantages associated with the MO devices are: 1) A much larger range of tunable carrier frequencies (0.5 to 20 GHz and higher) may be obtained by varying the bias magnetic field. Such high and tunable carrier frequencies with the MO devices allow <u>direct</u> processing at the carrier frequency of wideband RF signals and, thus, eliminate the need for <u>indirect</u> processing via frequency down-conversion as is required with the AO devices [1, 19, 20], 2) A large MO bandwidth may be realized by means of a simpler transducer, 3) Much higher and electronically tunable modulation/switching and scanning speeds are achievable because as mentioned in Sect. 2 the velocity of propagation for the MSWs can be higher than that of the SAWs by one-to three-orders of magnitude, and 4) The dispersive nature of the MSWs provides potential for implementation of unique signal processing functions. Consequently, a variety of unique applications in the areas of real-time RF signal processing and optical communications such as the following are anticipated: 1. Electronically tunable RF spectral analysis at X-band and beyond [21], 2. Multicarrier frequency demultiplexing [22], 3. Nanosecond optical space switching [3, 21], and 4. High-speed wideband optical frequency shifting and modulation [23].

5. Conclusion and Future Directions

A number of significant advances have been made recently in MSW-based guided-wave magnetooptics. These advances include growth of high-quality pure YIG-GGG waveguides and quality improvement in Bi-doped YIG-GGG waveguides, design and fabrication of efficient and wideband transducers for MSWs, theoretical analysis on noncollinear coplanar guided-wave MO Bragg diffraction, realization of compact MO Bragg cell modulators, realization of ion-milled waveguide lenses and integration with the MO Bragg cell modulator, and demonstration of their applications in light beam modulation and scanning/switching, RF spectral analysis, and optical frequency shifting and modulation at the X-band. Thus, a new class of optical devices called MO Bragg cells which are potentially capable of providing desirable features similar to that of the now prevalent AO Bragg cells, but with potentially superior performance characteristics, have been realized. These advances have paved the way for realization of a variety of MSW-based integrated MO device modules, such as high-speed optical scanners and space switches, tunable carrier frequency band RF spectrum analyzers and correlators, and wideband optical frequency shifters and modulators. It may also be possible to realize tunable optical filters similar to AO tunable filters [24].

Finally, to expedite realization of the aforementioned integrated MO device modules and their applications, further advances such as those listed in the following are needed: 1. schemes for enhancement of MO Bragg diffraction efficiency, and thus reduction of the RF drive power requirement for the resulting integrated optic device modules, 2. development of robust coupling techniques between the YIG-GGG waveguide and the laser and/or the photodetector, 3. development of relevant packaging technology, and 4. realization of laser sources in Er-doped YIG-GGG substrates.

9. References

1. See, for example, Tsai, C.S. (1979) Guided-wave acoustooptic Bragg modulators for wide-band integrated optic communications and signal processing, *IEEE Trans. on Circuits and Systems.* CAS-**26**, 1072-1098; Tsai, C.S. (1990) *Guided-Wave Acousto-Optics*, in *Springer Series in Electronics and Photonics*, Vol. 23, C.S. Tsai (Ed.), Springer-Verlag.

2. Tsai, C.S. (1992) Integrated acoustooptic circuits and applications, *IEEE Trans. Ultrasonics, Ferroelectrics and Frequency Control.* **39**, 529-554.

3. See, for example, the many references cited in Tsai, C.S. and Young, D. (1990) Magnetostatic-forward-volume wave-based guided-wave magneto-optic Bragg cells and applications to communications and signal processing, *IEEE Trans. on Microwave Theory and Techniques.* **38**, 560-573.

4. (a) Tsai, C.S (1983) Hybrid integrated optic modules for real-time signal processing, in *Proc. of NASA Optical Information Processing Conference II*, NASA Conference Publication, **2302**, pp. 149-164;

    (b) Tsai, C.S., Young, D., Chen, W., Adkins, L., Lee, C.C., and Glass, H. (1985) Noncollinear magnetooptic interaction of guided-optical wave and magnetostatic surfaces waves in YIG/GGG waveguides, *Appl. Phys. Lett.* **47**, 651-654.

5. See, for example,

    (a) Tsai, C.S., Young, D. (1993) Some experimental observations on nonlinear propagation of magnetostatic waves and nonlinear magnetooptic interactions in YIG-GGG waveguides, Presented at the Second International Workshop On Nonlinear Interactions In Magnatic And Magnetooptic Materials, Dec. 12-14, 1993, Costa Mesa, CA.

506

(b) Su, J., Marcelli, R., De Gasperis, P. (1995) Guided optical wave interaction with nonlinear magnetostatic forward volume waves in YIG films, Presented at 1995 INTERMAG, April 18-21, San Antonio, Texas.

(c) Stancil, D.D. and Prabhakar, A., (1995) Interactions between optical guided modes and nonlinear magnetostatic waves, Presented at the NATO Workshop on Nonlinear Microwave Signal Processing, Oct. 3-6, 1995, Rome, Italy.

6. Tsai, C.S., Young, D. (1989) X-band magnetooptic Bragg cells using bismuth-doped yttrium iron garnet waveguides, *Appl. Phys. Lett.* **55**, 2242-2244.

7. Pu, Y. and Tsai, C.S. (1991) RF magnetization of magnetostatic forward volume waves in a YIG-GGG layered structure with application to design of high-performance guided-wave magnetooptic Bragg cells, *International Journal of High-Speed Electronics.* **2**, 185-208.

8. Tsai, C.S., Young, D. (1988) GHz bandwidth magnetooptic interaction in YIG-GGG waveguide using magnetostatic forward volume waves, *Appl. Phys. Lett.* **53**, 1696-1698.

9. Wang, C.L., Pu, Y., and Tsai, C.S. (1992) Permanent magnet-based guided-wave magnetooptic Bragg cell modules, *IEEE J. Lightwave Tech.* **10**, 644-648.

10. Pu, Y., Wang, C.L., and Tsai, C.S. (1991) Magnetostatic backward volume wave-based guided-wave magnetooptic Bragg cells and application to wide-band lightbeam scanning, *IEEE Photonics Technology Letters.* **5**, 462-465.

11. Tsai, C.S. and Le, P. (1992) A 4 X 4 nonblocking integrated acoustooptic space switch," *Appl. Phys. Lett.* **60**, 331-333; Also, Roy, A.K. and Tsai, C.S. (1992) A 8 X 8 symmetric nonblocking integrated acoustooptic space switch module in LiNbO$_3$, *IEEE Photonics Tech. Lett.* **4**, 731-734.

12. Tsai, C.S. and Cheng, Z.Y. (1989) A novel integrated optic frequency shifter using guided-wave acoustooptic Bragg diffraction in cascade, *Appl. Phys. Lett.* **54**, 1616-1618; Cheng, Z.Y. and Tsai, C.S. (1992) A Baseband Integrated Acoustooptic Frequency Shifter, *Appl. Phys. Lett.*, **60**, 12-14.

13. Wang, C.L., Pu, Y., and Tsai, C.S. High-performance curved hybrid lenses in YIG-GGG waveguide, (to be published).

14. Wang C.L. and Tsai, C.S., Integrated magnetooptic Bragg cell modules in YIG-GGG taper waveguide, (to be published *IEEE Photonics Technology Lett.*).

15. Vu, T.Q., Norris, J.A., and Tsai, C.S. (1989) Formation of negative index-change waveguide lenses in LiNbO$_3$ using ion milling, *Opt. Lett.* **13**, 1141-1143.

16. Vu, T.Q., Norris, J.A., and Tsai, C.S. (1989) Planar waveguide lenses in GaAs using ion milling, *Appl. Phys. Lett.* **54**, 1098-1100.

17. Vu, T.Q., Tsai, C.S., Young, D., and Wang, C.L. (1989) Ion-milled lenses and lens arrays in yttrium iron garnet-gadolinium gallium garnet waveguides, *Appl. Phys. Lett.* **55**, 2271-2273.

18. Vu, T.Q., Tsai, C.S., and Kao,Y.C. (1992) Integration of curved hybrid waveguide lens and photodetector array in a GaAs waveguide, *Appl. Opts.* **31**, 5246-5254.

19. Xu, G.D. and Tsai, C.S. (1991) Integrated acoustooptic modules for interferrometric RF spectrum analyzers, *IEEE Photonics Technology Lett.* **3**, 153-155.

20. Abdelrazek, Y., Tsai, C.S., and Vu, T.Q. (1990) An integrated optic RF spectrum analyzer in a ZnO-GaAs-AlGaAs waveguide, *J. Lightwave Tech.* **8**, 1833-1837.

21. Tsai, C.S. and Young, D. (1989) Wideband scanning of guided-lightbeam and RF spectral analysis using magnetostatic forward volume waves in a YIG-GGG waveguide," *Appl. Phys. Lett.* **54**, 196-198.

22. Tsai, C.S. (1994) Integrated acoustooptic and magnetooptic device modules for on-board processing and switching of microwave signals, in *Proc. of the 15th American Institute of Aeronautics and Astronautics Conf. on Satellite Communication Systems*, San Diego, CA., February 28 - March 5, 1994, pp. 1315-1322.

23. Pu, Y. and Tsai, C.S. (1993) Wideband integrated magnetooptic frequency shifter at X-band, *Appl. Phys. Lett.* **62**, 3420-3422.

24. See, for example, the many references cited in Kar-Roy, A. and Tsai, C.S. (1994) Integrated acoustooptic tunable filters using weighted coupling, *IEEE J. Quantum Electron.* **30**, 1574-1586.

Subject Index

Magnetic Storage 305
Autooscillations 355, 381, 467
Chaos 355, 381
Demagnetization 45
Domain Walls 325, 411
Faraday Effect 411, 467, 487
Hopf Bifurcation 381
Instantons 325
Integrated Optics 411
Kinetic Instability 139
Magnetic Garnets 13, 45, 71, 277, 355, 411, 467, 487
Magnetic Resonance 355, 381, 411
Magnetic Storage 3
Magneto-optical Devices 411, 467, 487
Magnetostatic Waves 13, 45, 71, 277, 305, 467, 487
Magnons 139
Magnetization Reversal 3
Microwaves 355
Microwave Devices 13, 45, 71, 165, 277
Microwave Solitons 13, 71, 165, 277, 305, 325
Modulational Instability 277, 467
Nonlinear Magneto-optical Devices 411, 467, 487
Non-linear Transmission Lines 13, 71
Optical Fibers 101
Optical Solitons 101, 274
Parametric Coupling 305
Parametric Instability 130, 165, 213, 253
Perturbation Methods 101
Semiconductor Devices 71
Spin Waves (Linear and Non-linear) 165, 213, 381, 467
Strange Attractors 355, 381
Suhl Instabilities 381
Surface Acoustic Wave Devices 13, 487
Superconductor Devices 13
Spin Wave Interactions 121
Telecommunications 13, 71, 101, 487
Tunneling Phenomena 325
Vortex 325